Genetically Modified and other Innovative Vector Control Technologies

Brij Kishore Tyagi
Editor

Genetically Modified and other Innovative Vector Control Technologies

Eco-bio-social Considerations for Safe Application

Springer

Editor
Brij Kishore Tyagi
VIT University
Vellore, Tamil Nadu, India

ISBN 978-981-16-2963-1 ISBN 978-981-16-2964-8 (eBook)
https://doi.org/10.1007/978-981-16-2964-8

This Springer imprint is published by the registered company Springer Nature Singapore Pte Ltd.
The registered company address is: 152 Beach Road, #21-01/04 Gateway East, Singapore 189721, Singapore

Dedicated to
All the crusaders in our war, in past and in present,
against the deadly and debilitating vectors of human diseases.

Foreword

It gives me great pleasure to pen this foreword applauding the collective efforts of a worldwide consortium of scientific experts towards the publication of this book entitled *Genetically Modified and Other Innovative Vector Control Technologies: Eco-bio-social Considerations for Safe Application* (*Edited* by Prof. Dr. B.K. Tyagi). It represents an ambitious effort to highlight our battle against the world's deadliest messenger, the mosquito, that serves as the vector for the transmission of myriad diseases that potentially place 6.5 billion people in the planet at risk for acquiring one or more of these diseases at some point in their lifetime. These diseases also extract a heavy economic toll in terms of the relentless efforts needed to keep these disease-causing vectors at bay, and sadly they affect the poorer and underprivileged segments of the global society disproportionately. As an example, malaria affects pregnant women and young children under the age of 5 in sub-Saharan Africa, causing the preventable death of a precious child every 2 min globally.

The volume aims to showcase current innovations in vector control, focusing on the successes and challenges in this ongoing arms race between this anthropophilic vector and modes of controlling it. Particularly relevant in this context are both Indian and global goals of eliminating mosquito-borne diseases, such as malaria, by 2030 and 2050, respectively. To meet these stretch goals, it seems clear that existing tools, although successful to varying extents, are losing their edge and are insufficient to solve them. So new emerging technologies, such as the use of Wolbachia-infected mosquito strains, improved sterile insect technologies (SITs) and genetic engineering of mosquitoes for population suppression and/or modification are showing promise, but these must bridge efficacy, safety, social, regulatory, ecological and environmental hurdles.

This edited book taps into the expertise of vector-control scientists and practitioners from around the world to address the following topics—(1) basic principles of the transgenic and para-transgenic manipulations of vectors and their potential impact on humans and the environment; (2) ecological, biological, ethical, legal and social implications of vector control through genetic manipulation; (3) identification of potential hazards, assessment and management of risks posed to humans and the environment, risk/benefit analysis; (4) principles and practices for the assessment and management of biosecurity and biosafety in laboratories and in the field; (5) guiding principles for creation and management of institutional or national biosafety review boards and ethics review committees and (6) an introduction to the development and application of a biosafety regulatory framework and its related legal principles at national levels for securing the development and use of vector control methods based on *Wolbachia* and SIT treatment and/or genetic manipulation.

The book holds the promise of a good read on a very timely topic. I thank the editor, Dr. B.K. Tyagi, and the authors for their efforts and wish them success.

Global Director TIGS—Centre at
inStem: National Centre for Biological
Sciences (NCBS), Bangalore, India

Suresh Subramani

Preface

Vector-borne diseases (accounting for more than 17% of all infectious diseases) are among the most mortifying and debilitating human illnesses caused by parasites, viruses and bacteria that are transmitted by arthropod vectors. These infections not only attenuate human health through chronic suffering, lifelong morbidity, disability and occasional stigmatization, but also emaciate intelligentsia and sap off resources of a country's economy. Every year there are more than 700,000 deaths from diseases such as malaria, dengue, schistosomiasis, human African trypanosomiasis, leishmaniasis, Chagas disease, yellow fever, Japanese encephalitis, and onchocerciasis. Of all these infections, two infections can be singled out, i.e. malaria and dengue. While malaria, against which luckily we have some very effective antidotes and a vaccine is almost on our heels, continues to be the severest infection, causing an estimated 219 million cases globally and resulting in more than 400,000 deaths every year, most of which in children below the age of 5 years in Africa, dengue, for which neither any specific drug nor a vaccine is yet available, has in recent decades emerged as the most prevalent viral infection transmitted by *Aedes* mosquitoes; more than 3.9 billion people in over 129 countries being at risk of contracting dengue, with an estimated 96 million symptomatic cases and an estimated 40,000 deaths every year. Moreover, the burden of these diseases is highest in tropical and subtropical regions, and they disproportionately affect the poorest populations the most. Most of these vector-borne diseases are manageable through vector control. Many of these vectors are bloodsucking insects, which ingest disease-producing microorganisms during a blood meal from an infected host (human or animal) and later transmit it into a new host, after the pathogen has replicated. Often, once a vector becomes infectious, they are capable of transmitting the pathogen for the rest of their life during each subsequent bite/blood meal.

Conventionally, for long decades now insecticides have provided an effective control for various different vectors of human diseases, although serious concerns have been raised about their ill effects on human health, non-target organisms and environment, in addition to serious operational constraint of insecticide resistance development in the target vector species. Other modes of vector control within the

Integrated Vector Management (IVM) scheme vary a great deal in their performance in different eco-climatic conditions. This scenario has warranted urgently search for alternative more specific, unhazardous and eco-friendly control means and ways, procedures, processes and tools which might bring about an appreciable reduction either by suppression or replacement in the vector population so as to cause a complete cessation of the disease transmission. Genetic manipulation, together with various other non-biotechnological tools, offers a great hope in changing the vector biology and behaviour in a way so as to alter the vector population structure and reduce the same below a threshold level. Genetic manipulations such as the transgenesis (e.g. using Release of Insect population having Dominant Lethal, RIDL, gene system or gene drive-based genome editing, e.g. CRISPER/Cas9, to suppress or replace the vector population) or paratransgenesis (e.g. deploying endosymbiont *Wolbachia*-induced cytoplasmic incompatibility for replacing natural vector population) have reinforced hope in finding out an environment-friendly solution to the vector-borne diseases like malaria, dengue, yellow fever, chikungunya and Zika virus. In particular, the advent of CRISPR technology has excited the potential to engineer new game-changing technologies and innovative systems that can be used to control wild populations of mosquitoes. Two developments of particular interest are a self-limiting system termed-precision-guided sterile insect technique (pgSIT) and a homing-based gene drive (HGD). The unique features of these systems can make them valuable tools to control vector mosquitoes in future. All these biotechnological advancements, with social and regulatory acceptance, can join hands with other innovative tools such as nanotechnology and polymer science, Remote Sensing, ovitraps, diagnostics, repellents, long-lasting insecticide bed nets (LLIN), and even insecticides in vogue in effecting vector control in specific ecological settings under the multi-methodical Integrated Vector Management (IVM) scheme.

This book, *Genetically Modified and Other Innovative Vector Control Technologies: Eco-bio-social Considerations for Safe Application* is a unique treatment embodying within its fold all the latest biotechnological advancements in genetic manipulations as well as other innovations towards effective and potential vector control, particularly malaria, dengue and leishmaniasis. Bejewelled with 25 chapters by 78 authors originating from 15 countries (Australia, Austria, Bangladesh, Burkina Faso, Czech Republic, India, Malaysia, Nigeria, Papua New Guinea, Spain, Sri Lanka, Sudan, The Philippines, Uganda, the United States), the book brings forth updated information on all the technologies dealing with genetically modified vectors (GMV) and other innovative, both current and futuristic, vector control technologies, with emphasis on their eco-bio-social (EBS) and regulatory considerations.

The book, one of its kind ever, is written in a simple and comprehensible style, and has been designed to suit all—academicians and researchers, alike.

Vellore, Tamil Nadu, India
March 31, 2021

Brij Kishore Tyagi

Acknowledgement

Writing a multi-authored book is both a personal delight and a great challenge at the same time, and invariably requires a great teamwork. In this regard, I am highly fortunate to have on my side a formidable team of committed authors with a humongous sense of camaraderie. Therefore, foremost, I wish to offer my heartfelt gratitude to all the authors (# 78) from as many as 15 countries whose great sense of sincerity, cooperation and friendliness have been the inexhaustible source of my inspiration throughout to accomplish a rare task like this book, *Genetically Modified and Other Innovative Vector Control Technologies: Eco-bio-social Considerations for Safe Application*.

Above all, my cordial thanks are due to Professor Dr. Suresh Subramani, Global Director, TIGS, UC San Diego, USA/Collaborative Science Chair, inStem, Bangalore for his generous encouragement and candid remarks on the book in the form of a "Foreword".

As always in past, here too, I wish to thank my wife, Ajita, who gave me full freedom of time and discussion so that I could put my whole attention to the accomplishment of this long and meandering journey of bringing together meaningfully a large number of chapters in a cohesive and comprehensible manner for the benefit of all those men and women who have been ever crusading a lifelong war against the mankind's deadliest foe—mosquito, in particular! Other vectors lurk, not far away, and are equally important.

Last but not least, I duly thank the publisher Springer-Nature, Singapore, and their highly competent and able editorial team for their patient guidance throughout the course of processing this book for publication.

About the Book

Vector-borne diseases (accounting for more than 17% of all infectious diseases) are among the most mortifying and debilitating human illnesses (e.g. malaria, dengue, schistosomiasis, human African trypanosomiasis, leishmaniasis, Chagas disease, yellow fever, Japanese encephalitis, onchocerciasis, and plague) caused by parasites, viruses and bacteria that are transmitted by arthropod vectors. These infections have failed the conventional methodologies, and there is an urgent need for alternative more effective, economical and environment-friendly technologies to appreciably reduce the vector populations. Genetics-based technologies such as transgenesis and paratransgenesis have offered a great promise, together with other innovative methodologies (e.g. nanotechnology and polymer science, Remote Sensing, ovitraps, diagnostics, repellents, long-lasting insecticide bed nets (LLIN) and even insecticides in vogue), to either suppress or replace the vector populations below threshold levels.

This book, *Genetically Modified and Other Innovative Vector Control Technologies: Eco-bio-social Considerations for Safe Application*" is a unique treatment embodying within its fold all the latest biotechnological advancements in genetic manipulations as well as other innovations towards effective and potential vector control, particularly malaria, dengue and leishmaniasis. Bejewelled with 25 chapters contributed by 78 authors originating from 15 countries (Australia, Austria, Bangladesh, Burkina Faso, Czech Republic, India, Malaysia, Nigeria, Papua New Guinea, Spain, Sri Lanka, Sudan, The Philippines, Uganda, the United States), the book brings forth updated information on all the technologies dealing with genetically modified vectors (GMV) and other innovative, both current and futuristic, vector control technologies, with emphasis on their eco-bio-social (EBS) and regulatory considerations. The treatment of the book is truly global and is an essential keep on the study table of all serious academicians and vector-borne disease control researchers.

Contents

Editor and Contributors

About the Editor

Brij Kishore Tyagi Born 23rd July 1951, Dr. Brij Kishore Tyagi (PhD 1979) has become emeritus in 2013, with two years' extension till 2015, as Scientist 'G' & Director-in-Charge from the ICMR-Centre for Research in Medical Entomology, Madurai, after distinguished service in medical entomology and vector-borne diseases in most of States and UTs in the country. He has since continued as the Visiting Professor in various universities (2016–2019) and currently is the Advisor, VIT Univ., Vellore (TN) as well as the International Consultant for a BCIL-FAO Sri Lanka Project on GMO Biosafety (2019–2021). He has to his credit >640 scientific titles, besides 37 books, incl. a WHO Training Manual on GM vectors (2015). Recipient of the prestigious WHO-TDR First Prize in worldwide competition (1995) and the ICMR's Dr. MOT Iyengar Memorial Award (2008), he is the Founding President of the Society of Medical Arthropodology (www.soma16.org). He has participated in preparing OECD's Consensus Document on *Aedes aegypti* (2018), is Chairman, RCGM-DBT's Subcommittee for the formulation of biosafety guidelines to conduct and monitor research trials (CRTs) on genetically engineered mosquitoes, and an Expert Member of the DBT-RCGM Committee on GM insects.

Contributors

Achim Nurfarahin Hanini Medical Entomology Unit, Institute for Medical Research, Kuala Lumpur, Malaysia

Adeogun Adedapo Public Health and Epidemiology Department, Nigerian Institute of Medical Research, Yaba, Lagos, Nigeria
Department of Biological Sciences, Faculty of Basic Medical and Applied Sciences, Lead City University, Ibadan, Oyo State, Nigeria

Mohd Adnan Nuradila Medical Entomology Unit, Institute for Medical Research, Kuala Lumpur, Malaysia

Wasi Ahmad Nazni Medical Entomology Unit, Institute for Medical Research, Kuala Lumpur, Malaysia

Ahmad Norazah Infectious Disease Research Centre, Institute for Medical Research, Kuala Lumpur, Malaysia

Irfan Ahmad Shazia Medical Entomology Unit, Institute for Medical Research, Kuala Lumpur, Malaysia

Vibha Ahuja Biotech Consortium India Limited, New Delhi, India

Olalekan Akinbo African Union Development Agency-New Partnership for Africa's Development, Ouagadougou, Burkina Faso

Asim Hasnor Hadi Health Education Unit, Melaka State Health Department, Melaka, Malaysia

Mohd Azam Muhammad Arif Medical Entomology Unit, Institute for Medical Research, Kuala Lumpur, Malaysia

Agapitus Kato Babumba Uganda Virus Research Institute, Entebbe, Uganda

J. Fowmitha Banu Centre for Climate Change Research, Department of Environmental Biotechnology, School of Environmental Sciences, Bharathidasan University, Tiruchirappalli, Tamil Nadu, India

I. Baruah Defense Research Laboratory (DRDO), Tezpur, Assam, India

Shirish R. Barwale Gangabishan Bhikulal Investment and Trading Limited (GBIT), Jalna, Maharashtra, India
Mahyco Private Limited, Jalna, Maharashtra, India

Probal Basu Centre of Polymer Systems, University Institute, Tomas Bata University in Zlin, Zlín, Czech Republic

Sajal Bhattacharya Post Graduate Department of Zoology, Asutosh College, University of Calcutta, Kolkata, West Bengal, India

K. Chandel Defense Research & Development Establishment, Gwalior, Madhya Pradesh, India

Natarajan Chandrasekaran Centre for Nanobiotechnology, Vellore Institute of Technology, Vellore, Tamil Nadu, India

Bharat Char Mahyco Private Limited, Jalna, Maharashtra, India

Jhansi Charles Department of Microbiology, Velammal Medical College Hospital and Research Institute, Madurai, Tamil Nadu, India

Ritwika Chatterjee Post-Graduate Department of Zoology, Asutosh College, University of Calcutta, Kolkata, West Bengal, India

Shaibal K. Dasgupta Gangabishan Bhikulal Investment and Trading Limited (GBIT), Jalna, Maharashtra, India
Tata Institute for Genetics and Society, Bengaluru, Karnataka, India

R. S. Dassanayake Department of Chemistry, Faculty of Science, University of Colombo, Colombo, Sri Lanka

S. D. Denipitiyage Department of Chemistry, Faculty of Science, University of Colombo, Colombo, Sri Lanka

Vas Dev ICMR—National Institute of Malaria Research, New Delhi, India

S. K. Dhiman Defense Research Laboratory (DRDO), Tezpur, Assam, India

P. V. D. Dilani Department of Chemistry, Faculty of Science, University of Colombo, Colombo, Sri Lanka

Diwakar Singh Dinesh Division of Vector Biology and Control, ICMR-Rajendra Memorial Research Institute of Medical Sciences, Patna, Bihar, India

Mohamad Dzomir Ahmad Zainuri Division of Agrotechnology and Biosciences, Malaysian Nuclear Agency, Selangor, Malaysia

Arwa Elaagip Department of Parasitology and Medical Entomology, Faculty of Medical Laboratory Sciences, University of Khartoum, Khartoum, Sudan

Badria Babiker El-Sayed Tropical Medicine Research Institute, National Centre for Research, Khartoum, Sudan

Zurita Federico Department of Genetics, University of Granada, Granada, Spain

Diganta Goswami Defense Research Laboratory (DRDO), Tezpur, Assam, India

M. Govindaraju Centre for Climate Change Research, Department of Environmental Biotechnology, School of Environmental Sciences, Bharathidasan University, Tiruchirappalli, Tamil Nadu, India

Y. I. N. Silva Gunawardene Molecular Medicine Unit, Faculty of Medicine, University of Kelaniya, Ragama, Sri Lanka

Mary Sylvette T. Gunigundo Commission on Higher Education, Manila, Philippines
University of the Philippines, Manila, Philippines

Hamzah Rabizah Entomology and Pest Unit, Melaka State Health Department, Melaka, Malaysia

Rupenangshu Kumar Hazra ICMR-Regional Medical Research Centre, Bhubaneswar, Odisha, India

Mohammad Nazrul Islam Hamdard University Bangladesh, Dhaka, Bangladesh

Shaikh Ismail Shaikh Norman Hakimi Medical Entomology Unit, Institute for Medical Research, Kuala Lumpur, Malaysia

C. Kandasamy Agrochemicals, Chennai, Tamil Nadu, India

Stephan Karl Australian Institute of Tropical Health and Medicine, James Cook University, Smithfield, QLD, Australia
Vector-Borne Diseases Unit, PNG Institute of Medical Research, Madang, Madang Province, Papua New Guinea

Lilian A. De Las Llagas Commission on Higher Education, Manila, Philippines
University of the Philippines, Manila, Philippines

Rosemary Susan Lees LITE (Liverpool Insect Testing Establishment), Vector Biology Department, Liverpool School of Tropical Medicine, Liverpool, UK

Lee Han-Lim Medical Entomology Unit, Institute for Medical Research, Kuala Lumpur, Malaysia

Yoon-Ling Cheong Biomedical Museum Unit, Special Resource Centre, Institute for Medical Research, Kuala Lumpur, Malaysia

Maheswaran Tanusshni Medical Entomology Unit, Institute for Medical Research, Kuala Lumpur, Malaysia

Hamidou Maiga Insect Pest Control Laboratory, Joint FAO/IAEA Division of Nuclear Techniques in Food and Agriculture, International Atomic Energy Agency, Vienna International Centre, Vienna, Austria

Karim Maredia Michigan State University, East Lansing, MI, USA

John M. Marshall Division of Epidemiology & Biostatistics, School of Public Health, University of California, Berkeley, CA, USA
Innovative Genomics Institute, Berkeley, CA, USA

Ruth Mbabazi Michigan State University, East Lansing, MI, USA

Aik-Meng Tan Medical Entomology Unit, Institute for Medical Research, Kuala Lumpur, Malaysia

Prabhakar Mishra Department of Biotechnology, School of Applied Sciences, REVA University, Bengaluru, Karnataka, India

Ipsita Mohanty ICMR-Regional Medical Research Centre, Bhubaneswar, Odisha, India

Amitav Mukherjee Centre for Nanobiotechnology, Vellore Institute of Technology, Vellore, Tamil Nadu, India

Guat-Ney Teoh Medical Entomology Unit, Institute for Medical Research, Kuala Lumpur, Malaysia

Krishna Pandey ICMR-Rajendra Memorial Research Institute of Medical Sciences, Patna, Bihar, India

Prasad Paradkar CSIRO, Australian Centre for Disease Preparedness, Geelong, VIC, Australia

Prabhakargouda B. Patil Gangabishan Bhikulal Investment and Trading Limited (GBIT), Jalna, Maharashtra, India

Bipul Rabha Defense Research Laboratory (DRDO), Tezpur, Assam, India

K. Rangarajan Centre for Climate Change Research, Department of Environmental Biotechnology, School of Environmental Sciences, Bharathidasan University, Tiruchirappalli, Tamil Nadu, India

Animesha Rath ICMR-Regional Medical Research Centre, Bhubaneswar, Odisha, India
KIIT School of Biotechnology, Kalinga Institute of Information Technology, Bhubaneswar, Odisha, India

Moussa Savadogo African Union Development Agency-New Partnership for Africa's Development, Ouagadougou, Burkina Faso

G. N. Silva Department of Chemistry, University of Colombo, Colombo, Sri Lanka

K. Sridhar Centre for Climate Change Research, Department of Environmental Biotechnology, School of Environmental Sciences, Bharathidasan University, Tiruchirappalli, Tamil Nadu, India

I. P. Sunish ICMR-Regional Medical Research Centre, Port Blair, Andaman & Nicobar Islands, India

P. K. Talukdar Defense Research Laboratory (DRDO), Tezpur, Assam, India

S. N. Tikkar Defence Research & Development Establishment, Gwalior, Madhya Pradesh, India

Roshan Kamal Topno Division of Epidemiology, ICMR-Rajendra Memorial Research Institute of Medical Sciences, Patna, Bihar, India

B. K. Tyagi SpoRIC, VIT University, Vellore, Tamil Nadu, India

Váleri N. Vásquez Energy & Resources Group, University of California, Berkeley, CA, USA
Berkeley Institute for Data Science, University of California, Berkeley, CA, USA

Vijay Veer Defense Research Laboratory (DRDO), Tezpur, Assam, India

P. Vignesh Centre for Climate Change Research, Department of Environmental Biotechnology, School of Environmental Sciences, Bharathidasan University, Tiruchirappalli, Tamil Nadu, India

P. D. S. U. Wickramasinghe Department of Chemistry, University of Colombo, Colombo, Sri Lanka

David T. Williams CSIRO, Australian Centre for Disease Preparedness, Geelong, VIC, Australia

Kamlesh K. Yadav Gangabishan Bhikulal Investment and Trading Limited (GBIT), Jalna, Maharashtra, India

R. Yadav Defense Research & Development Establishment, Gwalior, Madhya Pradesh, India

Usha B. Zehr Gangabishan Bhikulal Investment and Trading Limited (GBIT), Jalna, Maharashtra, India
Mahyco Private Limited, Jalna, Maharashtra, India

Chapter 1
Arthropods of Medical Importance: Need for Genetic and Other Innovative Vector Control Technologies, with Emphasis on Eco-biosocial and Environmental Considerations

B. K. Tyagi

Abstract Among the world's known vector groups, viz. arthropods, snails and rodents, the most important vectors originate from arthropods, the jointed legs. Arthropods are doubtlessly regarded as the most dominant creatures on the Earth due largely to their remarkable structural and behavioural diversity, besides humongous species preponderance. Of course, some of these arthropods are serious pests and/or vectors of human and animal diseases—deadly, debilitating and economy destructing. According to an estimate, arthropod species make approximately 80% of the global biological diversity. Born some 350–400 million years ago, they have of course achieved, to the utter envy of all other animal forms, a formidable genetic diversity and robustness so much so that they have virtually captivated pivotal human attention for centuries. They serve as a spectacular model of bioprospecting or laboratory experiments mostly because they are found in abundance, breed prodigiously and are exceptionally easier to culture or cultivate. For the aforesaid reasons, arthropods are also the easy target for genetic manipulations such as the transgenesis (using the release of insect carrying dominant lethal (RIDL) gene system or gene drive-based genome editing, e.g. CRISPR/Cas9, to suppress or replace the vector population) or paratransgenesis (e.g. deploying endosymbiont *Wolbachia*-induced cytoplasmic incompatibility for replacing natural vector population). In particular, the advent of CRISPR technology has excited the potential to engineer new game-changing technologies and innovative systems that can be used to control wild populations of mosquitoes. Two developments of particular interest are a self limiting system termed precision-guided sterile insect technique (pgSIT) and a homing-based gene drive (HGD). The unique features of these systems can make them valuable tools to control vector mosquitoes in the future. All these biotechnological advancements in vector control are designed to fit well in the multi-methodical integrated vector management (IVM) strategy.

B. K. Tyagi (✉)
SpoRIC, VIT University, Vellore, Tamil Nadu, India

B. K. Tyagi (ed.), *Genetically Modified and other Innovative Vector Control Technologies*, https://doi.org/10.1007/978-981-16-2964-8_1

Keywords Arthropods · Genetics · Innovative technologies · Eco-biosocial considerations

1.1 Introduction

Current estimates for the number of species on Earth range between 5.3 million and 1 trillion, although there is a massive degree of uncertainty about an exact number due largely to the lack of a global register. The whole animal kingdom is organized into invertebrates (≥30 million species) and vertebrates (47,000 species), the former constituting over 95% of all animals on the earth. Invertebrates comprise eight phyla: Porifera, Cnidaria, Platyhelminthes, Nematoda, Mollusca, Annelida, Arthropoda and Echinodermata. Over 80% of all invertebrates are arthropods (phylum Arthropoda, the jointed legs) that include insects, mites, ticks, spiders, crustaceans, centipedes and millipedes. Arthropods' enviable dominance, transcending diverse and essential natural processes in Earth's terrestrial, aerial and freshwater ecosystems that is so essential to contribute to the function of the natural world as a self-sustaining biological system, is a fundamental scientific insight among the world's animals. A great deal of scientific information on life comes only from this special group—Arthropoda—which are, in fact, an integral and complex part of the terrestrial and freshwater ecosystems with which the future of humans is inextricably linked. Evolved about 350–400 million years ago, they are considered to possess one of the most dynamic yet significant animal genetic diversities. Accordingly, they serve as a spectacular model of laboratory experiments mostly due to following reasons:

1. They are found in abundance.
2. They breed prodigiously in captivity.
3. They are exceptionally easier to culture.
4. A large number of species of arthropods are serious pests and vectors of plant, human and animal diseases.

For these reasons, arthropods are also the easy target for genetic manipulations such as the transgenesis (e.g. using the release of insect population having dominant lethal (RIDL), gene system, or gene drive-based genome editing, e.g. CRISPR/Cas9, to suppress or replace the vector population) and paratransgenesis (e.g. deploying endosymbiont *Wolbachia*-induced cytoplasmic incompatibility for replacing natural vector population). In particular, the advent of CRISPR technology has excited the potential to engineer new game-changing technologies and innovative systems that can be used to control wild populations of mosquitoes. Two developments of particular interest are a self-limiting system termed precision-guided sterile insect technique (pgSIT) and a homing-based gene drive (HGD). The unique features of these systems can make them valuable tools to control vector mosquitoes in the future. All these biotechnological advancements in vector control are designed to fit well in the multi-methodical integrated vector management (IVM) system.

Within Arthropoda, a wide range of organisms, such as mosquitoes, flies, biting midges, ticks, mites, fleas, bugs, lice and other arthropods, carry and transmit disease-causing organisms, or pathogens, from one host to another, yet mosquitoes are considered the deadliest of all vectors of diseases of public health and veterinary importance. Insects, generally regarded as the more serious carriers of a large number deadly and/or debilitating diseases (e.g. malaria, dengue, yellow fever, chikungunya, Zika virus, Japanese encephalitis, Chagas' disease, onchocerciasis, leishmaniasis, etc.), rule the roost notwithstanding a widely acknowledged public health significance of several arachnids such as the mites and the ticks. These diseases cause significant economic damage and harm with extensive trade loss globally, in addition to putting more than half of the world's human population to the risk of infection (Tyagi 1994, 2015; Pérez-Guerra et al. 2010; Shepard et al. 2014) (Table 1.1).

1.2 Vector Control

Vector mosquito control is an integral part of the disease management. There are several different methods available at both individual and community levels. Mosquito control can be divided into two areas of responsibility: individual and public. Most often it is performed following the integrated mosquito management (IMM) concept. IMM is based on ecological, economic and social criteria and integrates multidisciplinary methodologies into pest management strategies that are practical and effective to protect public health and the environment and improve the quality of life. IMM strategies are employed in concert with insecticide. These include source reduction, which incorporates physical control (digging ditches and ponds in the target marsh) and biological control placing live mosquito fish (*Gambusia* spp.) in the ditches and ponds to eat mosquito larvae. Other non-chemical control methods include invertebrate predators, parasites and pathogens to control mosquito larvae. Adult mosquito biological control by means of birds, bats, copepods, dragonflies and frogs has been applied by various agencies. However, supportive data are anecdotal, and there is no documented study to show that either of these predators consume enough adult mosquitoes to be effective control agents.

Pesticides continue to hold the main plank against vector-borne diseases especially during epidemics. The use of insecticides is one of the main tools employed against pathogen vectors by governmental campaigns, private companies and households. The intense use of these chemicals, however, has been selecting resistant populations of several arthropod species all around the world. Insecticide resistance is now claimed to be one of the main threats to the control of arthropod vectors of pathogens, especially mosquitoes. Although new strategies of vector control are currently being tested under bad conditions, the use of insecticides will retain an important role in the near future at least. In this sense, the knowledge of processes leading to resistance is crucial to try to revert the resistant status of natural populations, avoiding resistance to new compounds and maintaining this strategy

Table 1.1 Some of the most deadly and/or debilitating arthropod vectors of human diseases

Arthropod	Vector	Virus	Disease	Geographical distribution
Mosquito	*Aedes aegypti* and *Ae. albopictus*	Dengue virus 1–4	Dengue	Tropical region
	Aedes aegypti and *Ae. albopictus*	Yellow fever virus	Yellow fever	Africa and South America
	Culex tritaeniorhynchus	Japanese enceph-alitis virus	Japanese encephalitis	Asia and Pacific
	Culex annulirostris	Murray Valley encephalitis virus	Murray Valley encephalitis	Australia and Papua New Guinea
	Culex spp.	Rocio virus	Rocio	America and Brazil
	Culex pipiens and *Cx. quinquefasciatus*	St. Louis encephalitis virus	St. Louis encephalitis	North and South America
	Culex spp.	West Nile virus	West Nile	Africa, Asia, Europe and America
	Culex spp.	Rift Valley virus	Rift Valley fever	Africa and Mid-dle East
	Aedes triseriatus	La Crosse virus	La Crosse encephalitis	North America
	Aedes triseriatus	California encephalitis virus	California encephalitis	North America, Europe and Asia
	Aedes aegypti and *Aedes albopictus*	Chikungunya virus	Chikungunya	Africa and Asia region
	Aedes aegypti	Zika virus	Zika	Most of tropical and subtropical countries, including India
	Culex annulirostris, *Ae. vigilax* and *Ae. camptorhynchus*	Ross River virus	Ross River	Australia and South Pacific
	Haemagogus spp.	Mayaro virus	Mayaro	South America
	Anopheles funestus and *Anopheles gambiae*	O'nyong-nyong virus	O'nyong-nyong fever	Africa
	Culex spp.	Sindbis virus	Sindbis	Africa, Egypt, Israel, Australia and the Philippines
	Aedes notoscriptus	Barmah Forest virus	Barmah Forest	Australia
	Culiseta melanura and *Cs. morsitans*	Eastern equine encephalitis virus	Eastern equine encephalitis	North, Central and South Amer-ica and the Caribbean

(continued)

Table 1.1 (continued)

Arthropod	Vector	Virus	Disease	Geographical distribution
	Culex tarsalis	Western equine encephalitis virus	Western equine encephalitis	The Americas
	Aedes taeniorhynchus	Venezuelan equine encephalitis virus	Venezuelan equine encephalitis	The Americas
	Anopheles spp.	*Plasmodium* spp.	Malaria	Tropical and subtropical region
	Anopheles, *Culex* spp. and *Mansonia* spp.	*Wuchereria bancrofti*, *Brugia malayi* and *Brugia timori*	Human lymphatic filariasis	South East Asia and Africa
Ticks	*Haemaphysalis spinigera*	Kyasanur Forest disease virus	Kyasanur Forest diseases	South Asia (India) and Saudi Arabia
	Dermacentor reticulatus, *D. marginatus* and *Ixodes persulcatus*	Omsk haemorrhagic fever virus	Omsk haemorrhagic fever	Western Siberia regions of Omsk, Novosibirsk, Kurgan and Tyumen
	Ixodes scapularis, *I. ricinus* and *I. persulcatus*	Tick-borne encephalitis virus (Russian spring-summer encephalitis virus)	Tick-borne disease	Europe and Asia
	Hyalomma spp.	Crimean-Congo haemorrhagic virus	Crimean-Congo haemorrhagic fever	East and West Africa, the Balkans, the Middle East and Asia (including India)
	Ornithodoros spp.	Borrelia bacteria	Tick-borne relapsing fever	Africa, Spain, Saudi Arabia, Asia, Canada and the western United States
	Deer tick	Spirochetal bacteria	Lyme disease	North America and Europe
	Ixodes holocyclus and *I. tasmani*	Rickettsia bacteria	Queensland tick typhus	Australia
	Dermacentor ticks	*Rickettsia rickettsii*	Rocky Mountain spotted fever	America
	Ixodes ricinus	*Rickettsia helvetica*	Helvetica spotted fever	Sweden, Switzerland and France
	Lone star tick	*Anaplasma phagocytophilum*	Granulocytic anaplasmosis	East, Southeast and Midwest United States

(continued)

Table 1.1 (continued)

Arthropod	Vector	Virus	Disease	Geographical distribution
	Dermacentor andersoni and *D. variabilis*	*Francisella tularensis*	Tularaemia	North America, Europe and Asia
	Dermacentor andersoni	Colorado tick fever virus	Colorado tick fever	America
	Ixodes scapularis and *I. pacificus*	*Babesia microti* and *B. equi*	Babesiosis	America
	Brown dog tick (*Rhipicephalus sanguineus*), Rocky Mountain wood tick (*Dermacentor andersoni*) and lone star tick (*Amblyomma americanum*)	*Coxiella burnetii*	Q fever	Eastern United States
Midges	*Culicoides paraensis*	Oropouche virus	Oropouche	Central and South America
Sandfly	Phlebotomine sandflies	Protozoan parasites	Leishmaniasis	East and North Africa and Europe
Black fly	Black fly	*Onchocerca volvulus*	Onchocerciasis (river blindness)	Africa, America and Yemen
Tsetse fly	*Glossina*/tsetse fly	*Trypanosoma brucei gambiense* and *Trypanosoma brucei rhodesiense*	African trypanosomiasis/ sleeping sickness	Africa
Triatomine bugs	Triatomine/kissing bugs	*Trypanosoma cruzi*	American trypanosomiasis (Chagas' disease)	America, Canada, European and Western Pacific countries
Fleas	Fleas that infest rats	*Rickettsia typhi*	Murine typhus	America
	Rat via fleas	*Yersinia pestis*	Plague	Asia, Africa and South America
Lice	*Pediculus humanus corporis*	*Rickettsia prowazekii*	Epidemic typhus	Africa, America and Asia
	Lice	*Rickettsia* and *Borrelia*	Epidemic relapsing fever	Ethiopia and Sudan
Mites	*Leptotrombidium* spp. (red mites)	*Orientia tsutsugamushi*	Scrub typhus	Asia-Pacific region

as an effective alternative for arthropod control. Many mosquito vector species have lately developed resistance against most of conventional insecticides (e.g. DDT, BHC, malathion and certain pyrethroids), and likelihood of availability of newer insecticides is highly precarious, cost inhibitive, tedious and long exercise. They may be applied to control larvae (larvicides) or adults (adulticides). Applications of adulticides or larvicides are made after the presence of mosquitoes has been demonstrated by surveillance procedures. Application is made by prescribed standards. All insecticides must have the name and amount of active ingredient (AI) appearing on the label; examples are DEET, DEPA and pyrethroids. No pesticide is 100% safe and care must be exercised in the use of any pesticide. Material Safety Data Sheets (MSDS) contain basic information about a product intended to help you work safely with the material.

Mosquitoes mediate transmission of some of the world's deadliest and most debilitating vector-borne diseases that are prevalent in entire tropics and subtropics but are also gradually extending their distribution to temperate regions. India occupies a very special place in the history of malaria research since it was here that the most important discovery of the inextricable relationship of mosquito and malaria parasite was first demonstrated, least to say about the plight of people due to scourge of malaria (Tyagi 2020; Tyagi et al. 2020). While in the pre-DDT era, malaria alone was responsible for about 1 lakh deaths and over 100 million cases in India each year, resulting into unlimited human misery including economy as well as physical and intellectual deprivation (Sinton 1935; Sharma and Mehrotra 1986). With the discovery of insecticidal property of DDT in 1939 and its availability for health programmes by the late 1940s, it became the mainstay for controlling vector mosquitoes and reducing the disease burden. However, soon it was discovered that the main malaria vector, *An. culicifacies*, which effected almost two-thirds of the total malaria cases in the country, had developed resistance against DDT, rendering it ineffective even when deployed at the maximum permissible limit of 2 g (ai)/sq. m. This situation warranted immediate deployment of alternative insecticides, notwithstanding some of them being highly toxic to humans, birds and fishes (e.g. dieldrin), on one hand, and others less effective and long-lasting but highly expensive, on the other. The control of vector mosquitoes appeared a near possibility when less than 0.1 million cases without single death in the early 1960s were reported under the command of National Malaria Eradication Programme (NMEP). The euphoria was however short-lived, and malaria resurged with a vengeance within next few years, by the mid-1970s. In the ensuing years, it became extremely difficult to control vector mosquito populations with insecticides like DDT, BHC, malathion (as adulticides) and Abate and Baytex (as larvicides) due to several technical and administrative bottlenecks, although most of these insecticides continue to be used selectively even now. Lately the use of these and chemicals in public health has been continuously opposed due to their multifarious impacts on human health and environment.

1.3 Necessity for Alternative Vector Control Strategies

Insecticides continued to be the main plank in controlling vectors and vector-borne diseases and were particularly resorted to during the emergency of epidemics or vector nuisance by the national agencies, despite the fact that the extensive usage of insecticides was no longer effective in bringing about the desired vector population abatement, on one hand, and their deleterious effects on the human health and the environment safety, on the other. Due to injudicious use for decades in both agriculture and health programmes, the insecticides have of late been discouraged mainly for three reasons:

1. First, throughout the globe, 392 arthropod species (of which 136 or 35% of the species are vectors of diseases of public health importance) have developed resistance of some kind against the insecticides currently deployed in vector control (Table 1.2).

   Indian scenario of extensive insecticide usage in malaria control soon after independence in 1947 under the National Malaria Control Programme (established 1953) and subsequently under the rechristened National Malaria Eradication Programme (established 1958), economics of malaria involved, offers a glaring example for opposing insecticides. *Anopheles culicifacies*, the most important vector contributing more than 65% of malaria cases in India, has become resistant to DDT, BHC and malathion in several parts of the country endemic to malaria.
2. Secondly, astronomical costs associated with the insecticides are a serious concern globally. During the 1990s, an investment worth US $20 million was slated to produce one insecticide, notwithstanding the chance for discovering a potential insecticide stood at only 1 out of 1500 compounds screened (0.06%). Further, nearly US $10 billion worth insecticide usage was estimated to have incurred on environmental and societal damages. Obviously, therefore, the production of insecticides and their subsequent deployment in vector control operations are highly challenging and discouraging for any government, particularly in developing nations like India. Major inputs of insecticides are in agriculture; nevertheless a total of 77.75 mt DDT, 123 mt BHC and 68.7 mt malathion had been

Table 1.2 Some important mosquito vectors having developed resistance against insecticides

S. no.	Vector mosquito group	Quantum of species	Insecticide type
1	Anopheline mosquitoes	47	Dieldrin (and BHC)
		34	DDT
		4	Carbamates
		10	Organophosphate compounds
		7	Synthetic pyrethroids
2	Culicine mosquitoes	42	Dieldrin (and BHC), DDT, carbamates, organophosphate compounds and synthetic pyrethroids

applied under various health programmes in India between 1960 and 1980 only. The Government of India was at that time keeping a layout of approximately Rs 2000–3000 million (>50% of the total budget earmarked for health) annually on control of malaria alone and ever since there has been persistent demand for higher allocations each subsequent year.

3. Thirdly, incessant, injudicious and overuse of insecticides like DDT, the cheapest among all available organic pesticides (110 mt DDT had been deployed for vector control against only 38.23 mt DDT in agriculture in India till 1978), has caused an irreversible catastrophic effect on both the human health and the environment. In spite of a ban on the use of DDT in many advanced countries on account of its deleterious effects on the human health and the environment, the insecticide was preferred in countries such as India and others in Asia as well as Africa, citing its low mammalian toxicity, low costs and high degree of residuality on sprayed surfaces. This, however, proved dangerous since traces of the insecticide were often detected in the human blood, milk and tissue. According to one estimate in the mid-1980s, DDT occurred 20 times the permissible limit in the vegetables, while the bottled milk contained 4.8–6.3 ppm of DDT (7.2–9.5 times the permissible limit) in Delhi and Maharashtra State, respectively. Other insecticides were rather more injurious since a popular drink like tea samples from Assam, for example, had 4.280 ppm fenpropathrin against the UN food standards, CODEX, fixing its MRL at 2 ppm. All insecticides have close interaction with the environment, and their use in any way is bound to affect man and his associates severely.

These arguments prompted scientists to innovatively discover alternative tools, processes and approaches which are arguably more cost-effective, less hazardous and community friendly. Moreover, the manner in which these products are developed using modern biotechnological or genetics-based technologies, some of them do not obviously pose direct risks of development of resistance in the target vector species or any visible hazard to the environment. In the following lines, a summary of more important methodologies within the precinct of integrated vector management (IVM) is offered to understand the future scope of development of effective vector control strategies.

1.3.1 Chemical-Based Measures

Some of the ongoing, under-trial as well as promising-looking methodologies are listed below.

1.3.1.1 Insect Growth Regulators (IGRs)

The third-generation insecticides comprise a very special group of compounds, referred to as insect growth regulators (IGRs), that function by interfering with the hormonal mechanisms of target organisms and result into various different kinds of morphological, anatomical and physiological abnormalities so that the target species does not reach the final, viable adult stage of development. Essentially insect growth regulators tend to break the life cycle of a vector species and play an important role in integrated vector management (IVM) or insect resistance management (IRM) programmes. Since these compounds function in an altogether different manner, the development of resistance in vector species is very unlikely, although a few exceptions, such as diflubenzuron, do exist. Even in cases where the possibility of resistance development looks bright, the process is slow, and the phenomenon can be procrastinated by resorting to certain synergists like piperonyl butoxide (PPB) and *S*,*S*,*S*-tributyl phosphorothioate (TP). The IGR compounds have diverse range of actions; they can intervene with the growth or moulting (ecdysones) and the chitin synthesis in arthropods. A large number of synthetic as well as plant-based natural IGRs, both juvenoids and chitin synthesis inhibitors (e.g. diflubenzuron, methoprene, fenoxycarb), have been evaluated for their vector control activities, but only very few of these were found effective and commercially viable in varied breeding habitats (Tyagi et al. 1985, 1987; Amalraj et al. 1988a, b).

1.3.1.2 Surface-Active Agents

These are the non-ionic products characterized by specific traits of high dispersive capacity, biodegradability and low toxicity. Several products like Arosurf ISA 20E [polyoxyethylene (2) isostearyl ether (2) mole ethoxylate of isostearyl alcohol] are commercially available and widely deployed in difficult and more inaccessible sites like large lakes and marshlands prone to heavy mosquito breeding throughout (Das et al. 1986a). These products form a monomolecular film on the water surface and bring about mortality in the immatures by a "physical" (anoxia) rather than a chemical factor. This characteristic of surface-active agents averts the phenomenon of resistance development *hook, line and sinker!* observed with the chemical larvicides. The non-ionic monomolecular films also characteristically lower the surface tension of the treated waterbody (25–29 cm/dynes) resulting in contraction or suffocation in the tracheal system of the larvae and pupae, on one hand, and disallowance of inflation of the wings of the emerging adults, on the other. When used at a rate of 11.2 L/h, Arosurf, either alone or in combination with a fast-acting and residual larvicide like fenthion, effectively renders control of deadly vectors such as *Culex quinquefasciatus*, *Anopheles stephensi* and *Aedes aegypti* in varied types of breeding habitats such as cesspits, cesspools, drains and wells.

1.3.1.3 Insecticide-Impregnated Paint, Fabric and Bed Nets

The impregnated paint technology was innovated in the mid-twentieth century primarily to disinfect places like hospitals, aerodromes, movie halls, convention theatres, hotels, cinema theatres, ships, etc. and make them pest- and/vector-free on a long-term basis. This technology, involving embedding of long-lasting microencapsulated insecticides, or active ingredients (AIs), in the paint matrix with minute openings channelling a gradual release on the surface of the dried paint, which help to prevent its fallout from the surface, became a sure way to safeguard such infrastructures as stated above in Europe and the United States where they were promoted against nuisance pests that dwell on walls and ceilings. In India, however, the technology has never been deployed on large scale, and the first ever evaluation of a commercially available insecticidal pint was carried out by Das et al. (1986b) who have demonstrated utility of Vernacide against both the filariasis vector mosquito, *Culex quinquefasciatus*, and the household pest, *Periplaneta americana*, in urban conditions with a 100% effect for over 1 year. A major advantage of this technology is that the paint retained its effectiveness after a cold-water wash to the painted surface, although the insecticidal paints are available in both water-base and oil-base formulations.

Earlier to insecticidal paints, insecticide-impregnated fabrics have also been developed in various different shapes and forms, for example, hood, face netting, gowns, long sleeves, trousers, gloves and socks primarily to protect from the bites of mosquitoes and other flies. In malaria-endemic tribal villages of Koraput district of Orissa (now Odisha), an effect of bed nets impregnated with permethrin (at 0.5 g (ai)/sq. m.) on the reduction of vector density of *An. culicifacies* was evaluated for more than 3 months. Similar investigations to enhance usefulness of fabrics, ropes and/or bed nets impregnated with cyfluthrin, deltamethrin and lambda-cyhalothin were conducted in various parts of the country by Sharma and Yadav (1995), Sahu et al. (2003) and Jambulingam et al. (1989).

1.3.1.4 Controlled Slow-Release Formulations

In vast stretches of polluted lakes and marshy breeding habitats, spraying of insecticides (e.g. fenthion or Baytex) is generally non-productive and inhibitive to achieve desired suppression of vector population. To overcome this serious operational constraint, slow-release formulations of varied types have been developed and made commercially available in different forms. A slow-release formulation has an active ingredient material and a carrier material which is neutral and biodegradable. As a result, a larvicide with a relatively high concentration dose that is imbibed in the base biodegrades carrier material which by a process of larvicide dilution spontaneously releases a desired concentration on a long-term basis in the breeding habitat for an eventual abatement of the targeted vector species. As a rule, such slow-release formulations as briquettes are sprayed on weekly basis to achieve a substantial vector control. Such methodologies have been found quite cost-effective and time saving in the long run in controlling serious lymphatic filariasis and Japanese encephalitis transmitting vector species as *Culex quinquefasciatus* and *Cx. tritaeniorhynchus*.

1.3.2 Non-chemical Measures

In certain specific breeding habitats, for example, unused wells, cess pools and cesspits, quite a few non-chemical or physical measures or products like expanded polystyrene (EPS) beads—synthetic products of a very light thermocol-like material that are inert, non-toxic, non-wettable and resistant (do not interact) to seawater, salt solutions, soap and wetting agents and of spherical shape or form measuring approximately 0.2–0.8 mm in diameter—have been found highly effective to cease the breeding mosquitoes like *Cx. quinquefasciatus* and *An. stephensi*. The EPS is widely available commercially at a price of Rs 80/kg in major metropolises. The major advantage using the EPS is that they are non-biodegradable and can remain effective for months in the treated waterbodies such as the unused or discarded wells and latrine pits. On application, the EPS beads being superlatively light keep buoyant irrespective of the level of the breeding habitat's water at any time. However, they need to be protected in field from direct exposure to the sunlight since they can turn yellowish and brittle. The EPS beads, when applied at 500 g to 1 kg/m^2 in different habitats, control a serious nuisance and vector mosquitoes like the primary filariasis vector *Cx. quinquefasciatus*, a great deal in unused wells with turbid and/or polluted water and latrine pits, and the urban malaria vector, *Anopheles stephensi* in live wells with good drinking water (Curtis and Minjas 1985; Dua et al. 1989; Sivagnaname et al. 2005; Sunish et al. 2018). In addition to expanded polystyrene (EPS) beads, Soltani et al. (2008) applied shredded waste polystyrene (SWAP) beads for control of mosquitoes.

1.3.3 Biological Control Agents

Having due cognizance of the fact that insecticide usage is non-effective against the target vector species in the long run, on one hand, and detrimental to the human health and the environment, on the other hand, safer strategies like the use of biological means were advocated as a strong alternative control measure, albeit on cost of time and huge investments.

1.3.3.1 Bacterial Agents

Although several microbes have been deeply explored for their vector control potential, finally two of these, i.e. *Bacillus thuringiensis* H-14 (var. *israelensis*) and *B. sphaericus*, commonly called *Bti* an *Bs*, respectively, have been able to be commercially developed and available throughout the world. Many countries, especially Germany, exterminated the nuisance of pestilent and vector mosquitoes such as *Aedes vexans* and *Culex pipiens pipiens*, breeding in extensive marshes and polluted waterbodies, in particular, with the help of *Bacillus thuringiensis* subsp. *israelensis* or, in short, *Bti* (Becker and Ludwig 1983). Commercially *Bti* is sold as Bactimos (at 1 g/sq. m.) or Tekner (at 3 mL/sq. m.) and is available in different formulations (e.g. liquid concentrate and water-dispersible powder). Although the *Bs* is considered highly effective against all stages of anophelines in rather fresh waters, the Bti is particularly able to control culicines breeding in highly polluted breeding habitats such as underground drains, etc.

1.3.3.2 Fungal Agents

For a long time, entomogenous fungi have been appreciably known for killing the host insect, both immature and/or adult populations, under natural conditions, and thus exhibiting excellent biocontrol potential against the vector mosquitoes. Many groups' fungi, such as *Culicinomyces clavisporus* and *Tolypocladium cylindrosporum* (fungi imperfecti), *Lagenidium giganteum* (*Oomycetes*) and several species of *Coelomomyces* (*Chytridiomycetes*) had been field-tested. As far as India is concerned, species belonging to *Culicinomyces*, *Lagenidium* and *Coelomomyces* and derived from natural sources have been extensively worked upon. For instance, *Coelomomyces* spp. and *Lagenidium* spp. conidia in water-based suspension were applied repeatedly to the breeding season with manually operated pressure sprayers or motor-driven knapsack sprayers at the application rate of 10^{13} conidia/ha and 1010 zoospore/ha during morning hours. Another entomopathogenic fungus, viz. *Coelomomyces*, was introduced by placing in medium sporangia or moribund larvae that contained resistant sporangia at a rate of 1000 infected moribund larvae/ha which was good enough to check the larval breeding effectively in the early stage. In this case as well, repeated applications were required. Currently certain fungi are being investigated for their paratransgenic possibilities and cause suppression in vector populations.

1.3.3.3 Nematodes Against Mosquitoes

Certain helminths belonging to the family Mermithidae have been documented for attacking and destroying vector larvae especially in paddy field-breeding vector mosquitoes of Japanese encephalitis. Two species such as *Romanomermis culicivorax* and *R. iyengari* whose infective preparasite stage (nema) actually penetrates the mosquito larvae have been bioprospected in Indian laboratories and extensively explored for their antilarval kill *Cx. vishnui* subgroup species larval populations breeding in paddy fields near Bangalore (now Bengaluru), southern India. After 6–8 days when the parasite emerges out as a postparasite, this invariably results into the death of the immature mosquito form. Operational potential of these mermithids faces a unique challenge as the infective preparasite stage, nema, can tolerate but very low levels of salinity, alkalinity and pollution, restricted safely for application only in fairly freshwater bodies such as the stagnated fresh water in rice fields.

1.3.3.4 Larvivorous Fishes and Predaceous Arthropods

A large number of larvivorous fishes, for example, *Cyprinus carpio*, *Ctenopharyngodon idella*, *Catla catla* and *Gambusia affinis*, prevalent over a wide range of ecological habitats, are effective predators and found to have a special penchant for destroying the *Anopheles* mosquitoes and to some extent also *Culex*, *Aedes* and *Armigeres* populations. The greatest advantage in applying these fishes in vector control is their miniscule operational costs involved in the integrated control programmes in the long run.

Besides, among arthropods many insects which are predaceous by nature, for example, *Anisops bouvieri* (Notonectidae), *Toxorhynchites splendens* (a non-biting mosquito) and dragonfly larvae of *Crocothemis servilia* and *Bradinopyga geminata* have been extensively experimented for their potential to control vector larval population under specific natural conditions (Tyagi 2020; Corbet 1986). Single *Tx. splendens* fourth instar can consume 114 larvae of the vector species in laboratory conditions. Sebastian et al. (1980) deployed inundatively dragonfly *Crocothemis servilia* to successfully control the yellow fever/dengue vector, breeding in bamboo stumps, ornamental vases and other water-holding pots in Myanmar. A dragonfly 12th instar larva can consume approximately 133 ± 2 all stages of larvae of *Aedes aegypti* per 24 h. In case of dragonfly *Bradinopyga geminata* controlling dengue/ chikungunya vector *Ae. aegypti*, in the laboratory, the feeding rate of eighth instar manifested maximum predation on first instar larvae (86%), followed by second, third and fourth instars (72%, 66% and 48%), respectively (Venkatesh and Tyagi 2015).

1.3.3.5 Cyclops as Predator of *Aedes aegypti*

Some forms of copepod arthropods, e.g. the cyclops, which are otherwise infamously known for harbouring the larvae of *Dracunculus medinensis* - the helminth parasite causing Guinea Worm infection, can be still exploited to control dengue fever (the most widely spread mosquito-borne viral disease in the world, transmitted by *Aedes* mosquitoes), have been profitably resorted to modulate as potential predators on *Aedes aegypti* and *Ae. albopictus*. Realizing that these vectors breed mainly in potable waters in which case larviciding the habitat is generally not recommended, save for the fortnightly/weekly treatment in certain sites by temephos at 0.5–1 ppm, Vu et al. (2005) and Tuno et al. (2020) deployed *Mesocyclops* spp. and others to predate and control mosquito larvae in several villages in Vietnam.

1.4 Repellents

One of the oldest known methods to prevent contact with the nuisance-causing and disease-transmitting mosquitoes is the topical application of repellents, both natural and synthetic (grouped into different categories based on the functional groups present), the latter being based on chemical products with an offensive smell or taste to mosquitoes. Natural repellents or plant-derived compounds are the phytochemicals (e.g. nitrogen-containing compounds, terpenoids, phenolics, proteinase inhibitors and growth regulators) originated by the plant in support of its own safeguard. Plants such as grasses belonging to *Cymbopogon* spp. under Poaceae family are among the most favoured to yield repellents, although other families such as Lamiaceae, Fabaceae, Rutaceae, Myrtaceae and Asteraceae also possess species that are repellent rich in properties. Before the discovery of synthetic repellent products, there used to be a flourishing entrepreneurship, especially with military personnel posted in the frontier areas being the major consumer, producing and selling natural aromatic/essential oils and creams containing a melange of citronella, camphor and paraffin, notwithstanding a relatively shorter period of effect being mostly volatile. In recent decades, the most popular synthetic repellents being used worldwide include indalone, dimethyl phthalate, 2-ethyl-1,3-hexanediol (Rutgers 612) and *N*,*N*-diethyl-*m*-toluamide (DEET). The Indian scientists have come up with their own commercially viable alternative, *N*,*N*-diethyl phenyl acetamide (DEPA), innovatively developed in the 1980s. Besides low costs and availability in a range of formulations, DEPA has an added advantage in providing protection against a wide spectrum of haematophagous arthropods including tick, mites, sandflies, black flies and leeches as well.

1.5 Insect Traps

Effective trap mechanisms, originally designed to measure density but later the scope advanced to render surveillance of *Aedes* vectors and lethality to adults or larvae of *Ae. aegypti* if the ovistrip was treated with an insecticide such as fast-acting deltamethrin, have been developed against all flying pests and vectors, and mosquitoes are no exception. During field trials in Brazil, more than 89% mortality in adults and 99% mortality in larvae of *Ae. aegypti* were achieved during 1 month only. There are traps (autocidal ovitraps) which allow oviposition but prevent adult emergence and sticky ovitraps which trap mosquitoes on landing. Such traps have been extensively used in multistoried buildings in dengue-epidemic-prone Singapore where usual surveillance and/or control of *Ae. aegypti* by conventional fogging and/or other methods was not feasible. Since, despite best service provided by these traps, both the dengue outbreaks and the preponderance of the vector *Ae. aegypti* were unstoppable, newer approaches to killing mosquitoes in a non-toxic way were evolved by developing a device that burns propane, thus generating carbon dioxide, warmth and water vapour which draws the mosquitoes towards the propane flame, where they are then sucked into a net or holder where they are trapped. More recently attractant-based mosquito traps have been developed which work on the principle of disabling mosquito's odour receptor as a result of which the host-seeking female mosquitoes are guided by attractant odours released by their target. Different odour-disabling or deception traps like black hole mosquito and midge trap and Jakmax insect trap use either heat or CO_2, ammonia, lactic acid and other carboxylic acids naturally present in the warm-blooded animals' body odour and sweat and trapping mosquitoes, sandflies, midges and other flies as well as termites.

1.6 Strategies Based on Genetic Manipulation

All conventional methodologies have their own advantages and/or disadvantages with respect to a plethora of factors such as targeted vector species, time, place and ecological settings, human temperament, operational constraints including quality control, adequate personnel and tools to spray insecticides or execute inundative releases of biocontrol organisms, etc. To overcome most of these avoidable reasons, genetic modification (GM) of vector mosquitoes, working on the principles of causing either population suppression or population replacement, is considered one of such technologies which may be mainly used to either suppress or replace the wild populations of a vector so as to decrease vector populations or reduce the vector's ability to transmit (Tyagi 2015, 2020). Creating chromosomal aberrations in male mosquitoes rendering them nonviable through irradiation was among the earliest such technologies known as sterilized insect technique (SIT). However, since the earlier SIT technology failed to maintain the reproductive fitness of the sterilized males in the field in competence to wild populations of males, it had to be advanced genetically using modern biotechnological techniques. The modern GM technologies already under laboratory (Patil et al. 2014) or field-testing in countries like Brazil Cayman islands, Florida, Mexico, Malaysia, etc. encompass multiple approaches, viz. transgenesis implying the Release of Insect carrying Dominant Lethal (RIDL) gene system and gene editing using CRISPR/Cas9, *Wolbachia*-induced cytoplasmic incompatibility (CI) and classical radiation-induced male sterility, while population-manipulative technologies include Medea-based gene drive, under-dominance gene drive, homing endonuclease, *Wolbachia*-mediated heritable biocontrol or genetically modified midgut bacteria and transposable element like PiggyBac. Most of the technologies require essentially, firstly, the release of large volume of mosquitoes into the environment at different period of interval to either suppress or replace mosquito population and, secondly, different region-based field-testing under independent monitoring to gain the acceptability of society. These GM technologies are generally proven effective, albeit certain biological, ecological and sociological concerns. Three statements from social communities that often confront the GM technologies are (1) there is post-release increase of vector population; (2) ecological effects were also reported, such as horizontal transfer events; and

(3) worsening pathogenesis is induced by natural *Wolbachia* (Nazareth et al. 2020), about which the technologies have to be prepared to answer and convince through scientific data.

1.7 Integrated Vector Control Strategy

Since no individual vector control technology, however, modern and sophisticated in approach, is cent per cent effective, including also the acceptability of factors such as availability, operationality and cost-effectiveness, a cocktail sort of multicompartmental strategy, which adjusts within its fold all the available complementing technologies, allows each methodology to function up to its potential in specific eco-biosocial settings. Thus, integrated vector management or IVM, based on sound ecological principles and integrated multidisciplinary methods in developing ecosystem management strategies that are practical, effective, economical and protective to both public health and environment, is precisely defined as "the utilization of all technological and management advancements to bring about an effective degree of vector population's suppression in a cost-effective manner". An integrated approach involving all the major and effective mosquito control strategies which are also sustainable and target oriented for a vast range of different mosquito species is shown in Fig. 1.1.

Two major examples of IVM could be cited from (1) a successfully demonstrated strategy based on "Gandhian philosophy" advocating environmental means to control malaria in Kheda District, Gujarat State, by source reduction of breeding habitats of the main vector *Anopheles culicifacies*, during the 1980s, and (2) the control of brugian filariasis in Shertallai District, Kerala State (Sharma et al. 1986; Rajagopalan et al. 1989). Yet, another fine example is quotable from a study on control of bancroftian filariasis in an urban agglomeration in Puducherry, India (Rajagopalan and Das 1988). Keeping in view a generally deteriorating infrastructure of public healthcare in metropolises worldwide, especially in developing nations, master plan for mosquito control needs to be invariably developed (Rajagopalan et al. 1987).

The IVM, offering a desired degree of control at a lower cost or with a greater long-term benefit, invariably emphasizes on local needs and problems, and, therefore, the operational strategy changes from area to area accordingly. The success of an IVM strategy rests largely on the extent of health education, community participation and supportive legislation or regulations. In essence, an IVM strategy might have scores of effective and specific tools, but they are all integrative in approach, implying that each tool or measure or technology is complementarily integrated to the rest of methodology.

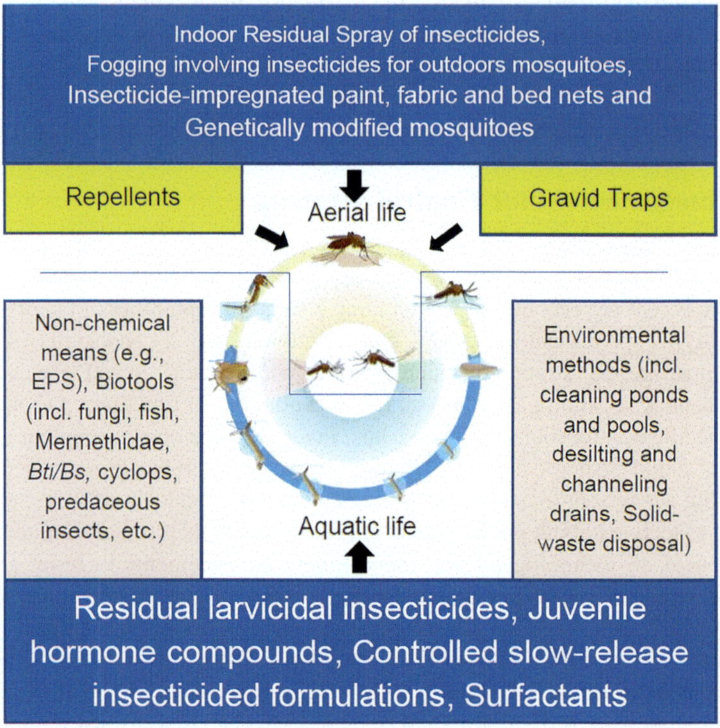

Fig. 1.1 Sustainable and target-oriented integrated approach for effective mosquito control using different technologies under integrated vector management (IVM). (*Source*: This work)

References

Amalraj D, Vasuki V, Sadanandane C, Kalyanasundaram M, Tyagi BK, Das PK (1988a) Evaluation of two new juvenile hormone compounds against mosquito vectors. Indian J Med Res 87:19–23

Amalraj D, Vasuki V, Kalyanasundaram M, Tyagi BK (1988b) Laboratory and field evaluation of three insect growth regulators against mosquito vectors. Indian J Med Res 87:24–31

Becker N, Ludwig M (1983) Mosquito Control in West Germany. Bull Soc Vector Ecol 8:85–93

Corbet PS (1986) Using dragonflies to suppress mosquitoes in domestic water-storage containers. Waterlines 4:10–11

Curtis CF, Minjas J (1985) Expanded polystyrene for mosquito control. Parasitol Today 1(1):36. https://doi.org/10.1016/0169-4758(85)90106-1

Das PK, Tyagi BK, Somachari N, Vasuki V (1986a) Efficacy of Arosurf, a molecular surface film in controlling *Culex quinquefasciatus* Say, *Anopheles stephensi* Liston and *Aedes aegypti* (L.). Indian J Med Res 83:271–276

Das PK, Tyagi BK, Kalyasundaram M (1986b) Vernacide, a new insecticide impregnated paint for controlling mosquito vector, *Culex quinquefasciatus* and cockroach, *Periplaneta americana*. Indian J Med Res 83:268–270

Dua VK, Sharma SK, Sharma VP (1989) Use of expanded polystyrene beads for the control of mosquitoes in an industrial. complex at Hardwar, India. J Am Mosq Contr Assoc 5(4):614–615

Jambulingam PK, Gunasekharan K, Sahoo SS, Hota PK, Tyagi BK, Kalyanasundaram M (1989) Effect of permethrin-impregnated bednets in reducing population of malaria vector, *Anopheles culicifacies* in a tribal village of Orissa State (India). Indian J Med Res 89:48–51

Nazareth T, Craveiro I, Moutinho A, Seixas G, Gonçalves C, Gonçalves L, Teodósio R, Sousa CA (2020) What happens when we modify mosquitoes for disease prevention? A systematic review. Emerg Microb Infect 9(1):348–365. https://doi.org/10.1080/22221751.2020.1722035

Patil PB, Niranjan Reddy BP, Gorman K, Seshu Reddy KV, Barwale SR, Zehr UB, Nimmo D, Naish N, Alphey L (2014) Mating competitiveness and life-table comparisons between transgenic and Indian wild-type Aedes aegypti L. Published online in Wiley Online Library. https://doi.org/10.1002/ps.3873

Pérez-Guerra CL, Halasa YA, Rivera R, Peña M, Ramírez V, Cano MP, Shepard DS. (2010) Economic cost of dengue public prevention activities in Puerto Rico. Dengue Bulletin 34:13–23

Rajagopalan PK, Das PK (1988) Filariasis control Demonstration Project. Vector Control Research Centre, Pondicherry. Indian J Med Res 89:418–425

Rajagopalan PK, Das PK, Kalyanasundaram M, Tyagi BK, Arunachalam N, Somachary N, Reddy CBS, Reddy CMR (1987) Bangalore mosquito control project: master plan. Vector Control Research Centre, Pondicherry, 309 pp

Rajagopalan PK, Panicker KN, Pani SP (1989) Impact of 50 years of vector control on the prevalence of Brugia malayi in Shertallai area of Kerala state

Sahu S, Jambulingam P, Vijayakumar T, Kalyanasundaram M (2003) Impact of alphacypermethrin treated bed nets on malaria in villages of Malkangiri district, Orissa, India. Acta Trop 89 (1):55–66

Sebastian A, Thu MM, Kyaw M and Sein MM, (1980) The use of dragonfly larvae in the control of Aedes aegypti. SEast Asian J trop Med pub Health 11(1):104–107

Sharma VP, Mehrotra KN (1986) Malaria resurgence in India: a critical study. Soc Sci Med 22:835–845

Sharma VP, Yadav RS (1995) Impregnating mosquito nets with cyfluthrin. Study in the mining settlements of Orissa, India to control malaria. Public Health 12:8–17

Sharma VP, Sharma RC, Gautam AS (1986) Bioenvironmental control of malaria in Nadiad, Kheda district, Gujarat. Indian J Malariol 23:95–117

Shepard DS., Halasa YA, Tyagi BK, Adhish SV, Nandan D, Karthiga KS, Chellaswamy V, Gaba M, Arora NK, and the INCLEN Study Group (2014) Economic and disease burden of dengue illness in India. Am J Trop Med & Hyg 91(6):1235–1242

Sinton JA (1935) What malaria costs India, nationally, socially and economically. Thacker's Press & Directories, Calcutta, 77 pp

Soltani A, Vatandoost H, Jabbari H, Mesdaghinia AR, Mahvi AH, Younesian M, Hanafi-Bojd AA, Bozorgzadeh S, Abai MR, Pakari A, Shabkhiz H (2008) Use of expanded polystyrene (EPS) and shredded waste polystyrene (SWAP) beads for control of mosquitoes. Iranian J Arthropod-borne Dis 2(2):12–20

Sivagnaname N, Amalraj DD, Mariappan T (2005) Utility of expanded polystyrene (EPS) beads in the control of vector-borne diseases. Indian J Med Res. 122(4):291–6

Sunish IP, Kalimuthu M, Kumar A, Rajendran R, Munirathinam A, Nagaraj N, Arunachalam N and Tyagi BK (2018) Integrated vector control for the elimination of bancroftian filariasis in the villages of Tirukoilur, South India. (In: Lymphatic Filariasis: Epidemiology, Treatment and Prevention - The Indian Perspective. B.K. Tyagi (Ed.), Springer, Singapore, 314 pp

Tuno N, Phong TV, Takagi M (2020) Climate change may restrict the predation efficiency of *Mesocyclops aspericornis* (Copepoda: Cyclopidae) on *Aedes aegypti* (Diptera: Culicidae) larvae. Insects 11(5):307. https://doi.org/10.3390/insects11050307

Tyagi BK (1994) Distribution of arthropod vector-borne communicable diseases and control of their vectors in India. Indian Rev Life Sci 14:223–243

Tyagi BK (extracted from the text) (2015) Biosafety for human health and the environment in the context of the potential use of genetically modified mosquitoes. Training manual. A tool for biosafety training (CD and internet version) based on courses in Africa, Asia and Latin America, 2008–2011. World Health Organization, 240 pp

Tyagi BK (2020) Mosquito hunters: a history of hostilities against man's deadliest foe—the mosquito—since 1881. Scientific Publishers, Jodhpur, 474 pp

Tyagi BK, Kalyanasundaram M, Das PK and Somachary N (1985) Evaluation of a new compound (VCRC/INS/A.23) with juvenile hormone activity against the mosquito vectors. Indian Journal of Medical Research 82:9–13

Tyagi BK, Somachari N, Vasuki V and Das PK (1987) Evaluation of three formulations of a chitin synthesis inhibitor (fenoxycarb) against mosquito vectors. Indian J Med Res 85:161–167

Tyagi BK, Naik Reddya B, Bhattacharya S (2020) Dr Ronald Ross: mosquito, malaria, India and the Nobel Prize—an untold story of the First Indian Nobel Laureate. Scientific Publishers, Jodhpur, 258 pp

Venkatesh A, Tyagi BK (2015) Predatory potential of *Bradinopyga geminata* and *Ceriagrion coromandelianum* larvae on dengue vector *Aedes aegypti* under controlled conditions (Anisoptera: Libellulidae; Zygoptera: Coenagrionidae; Diptera: Culicidae). Odonatologica 42 (2):139–149

Vu SN, Nguyen TY, Tran VP, Truong UN, Le QM, Le VL, Le TN, Bektas A, Briscombe A, Aaskov JG, Ryan PA, Kay BH (2005) Elimination of dengue by community programs using *Mesocyclops* (Copepoda) against *Aedes aegypti* in central Vietnam. Am J Trop Med Hyg 72 (1):67–73

Chapter 2
Field Trials of Gene Drive Mosquitoes: Lessons from Releases of Genetically Sterile Males and *Wolbachia*-infected Mosquitoes

John M. Marshall and Váleri N. Vásquez

Abstract The discovery of CRISPR-based gene editing and its application to homing-based gene drive has been greeted with excitement, for its potential to control mosquito-borne diseases on a wide scale, and concern, for the invasiveness and potential irreversibility of a release. At the same time, CRISPR-based gene editing has enabled a range of self-limiting gene drive systems to be engineered with much greater ease, including (1) threshold-dependent systems, which tend to spread only when introduced above a certain threshold population frequency, and (2) temporally self-limiting systems, which display transient drive activity before being eliminated by virtue of a fitness cost. As these CRISPR-based gene drive systems are yet to be field-tested, plenty of open questions remain to be addressed, and insights can be gained from precedents set by field trials of other novel genetics-based and biological control systems, such as trials of *Wolbachia*-transfected mosquitoes, intended for either population replacement or suppression, and trials of genetically sterile male mosquitoes, either using the RIDL system (release of insects carrying a dominant lethal gene) or irradiation. We discuss lessons learned from these field trials and implications for a phased exploration of gene drive technology, including homing-based gene drive, chromosomal translocations, and split gene drive as a system potentially suitable for an intermediate release.

Keywords Trial · Gene drive · Mosquitoes · Genetically sterile males · *Wolbachia*

J. M. Marshall (✉)
Divisions of Epidemiology and Biostatistics, School of Public Health, University of California, Berkeley, CA, USA

Innovative Genomics Institute, University of California, Berkeley, CA, USA
e-mail: john.marshall@berkeley.edu

V. N. Vásquez
Energy and Resources Group, University of California, Berkeley, CA, USA

Berkeley Institute for Data Science, University of California, Berkeley, CA, USA
e-mail: vnvasquez@berkeley.edu

B. K. Tyagi (ed.), *Genetically Modified and other Innovative Vector Control Technologies*, https://doi.org/10.1007/978-981-16-2964-8_2

2.1 Introduction

The discovery of CRISPR and its application as a gene editing tool has enabled gene drive systems to be engineered with much greater ease (Doudna and Charpentier 2014; Champer et al. 2016). Recent attention has focused on homing-based drive systems and their potential to control mosquito-borne diseases on a wide scale, either by spreading disease-refractory genes (Gantz et al. 2015) or by spreading genes that confer a fitness load or sex bias thereby suppressing mosquito populations (Hammond et al. 2016; Kyrou et al. 2018). However, there is growing awareness of the invasiveness of homing-based drive systems (Noble et al. 2018) and interest in alternatives that could be confined to partially isolated populations and remediated—properties that are well aligned to the conduct of field trials (Marshall and Akbari 2018).

In addition to homing-based gene drive, the increased ease of gene editing has advanced the entire field of gene drive, including systems that, by design, limit their spread in space and time (Marshall and Akbari 2018). Such systems would ideally be capable of enacting local population control by: (a) effectively spreading into populations to the extent required to achieve the desired epidemiological or ecological effect and (b) being recallable from the environment in the event of unwanted consequences, public disfavor, or the end of a trial period. Two varieties of these systems have been recently engineered: (1) threshold-dependent systems that tend to spread when introduced above a certain population frequency (Akbari et al. 2013; Buchman et al. 2018) and (2) temporally self-limiting systems that display transient drive activity before being eliminated by virtue of a fitness cost (Gould et al. 2008; Li et al. 2020).

In this chapter, we discuss considerations for field trials of gene drive systems, with a specific focus on confinement and reversibility criteria, and lessons learned from other genetics-based and biological control systems (Table 2.1). We pay special attention to reciprocal chromosomal translocations (Buchman et al. 2018), as an example of a threshold-dependent system that is confineable and reversible, and then extend our consideration to CRISPR-based homing gene drive systems and temporally self-limiting systems, such as split gene drive (Li et al. 2020), which could be used as confineable and reversible intermediate systems in a development pathway of homing-based systems. While these gene drive systems are yet to be trialed in the wild, lessons can be learned from trials of several varieties of sterile male mosquitoes, specifically, those sterilized through radiation (sterile insect technique, SIT), transfection with *Wolbachia* (incompatible insect technique, IIT) (Zheng et al. 2019; Crawford et al. 2020), and release of insects carrying a dominant lethal (RIDL) gene (Harris et al. 2011; Carvalho et al. 2015), as well as releases of *Wolbachia*-infected mosquitoes for population replacement (Hoffmann et al. 2011). We begin by discussing trials of these systems and discuss threshold-dependent, self-limiting, and nonlocalized gene drive systems in this context.

Table 2.1 Genetics-based and biological mosquito control strategies and their potential to be confineable and reversible

Strategy	Variant	Mechanism of action	Confineable	Reversible
Sterile insect technique (SIT)	Ionizing radiation or chemosterilization	Offspring of released females and males are unviable	Yes	Yes
Wolbachia	Incompatible insect technique (IIT)	Offspring of released males are unviable	Yes, if no females released	Yes, if no females released
	Population replacement	Spreads through population due to cytoplasmic incompatibility	Yes, for moderate-to-high fitness costs	Possibly, for high fitness costs
Release of insects carrying a dominant lethal (RIDL) gene	Bisex RIDL (bi-RIDL)	Both female and male offspring having the RIDL allele are unviable	Yes	Yes
	Female-specific RIDL (fs-RIDL)	Only female offspring having the RIDL allele are unviable	Yes	Yes
Chromosomal translocations	CRISPR or other endonucleases	Translocation heterozygotes with unbalanced chromosome sets are unviable, leading to bistable dynamics	Yes	Yes
CRISPR-based gene drive	Homing-based drive systems	Bias inheritance by cleaving a target sequence and serving as a template for DNA repair, effectively turning a heterozygote into a homozygote	Potentially, but with difficulty	Potentially, but with difficulty
	Split gene drive systems	Components of drive system are split across two loci, leading to transient drive when they co-occur before being eliminated due to a fitness cost	Yes	Yes

2.2 Lessons from Releases of Genetically Sterile Male Insects

Releases of irradiated sterile male insects as a means of population suppression have been discussed since the early twentieth century (Klassen and Curtis 2005), and a transgenic version of this technology was the first transgenic mosquito product to be trialed in the field (Harris et al. 2011). As the first transgenic mosquito release, this intervention has come under high levels of scrutiny and serves as an important case study for potential releases of gene drive mosquitoes. The traditional SIT approach involves mass-rearing insects and applying a carefully calibrated amount of radiation such that their genetic material is mutated to render them sterile while still being able to compete for female mates in the field (Knipling 1955). Upon release, sterile males (preferably the majority of released insects) seek out wild females, essentially wasting their reproductive potential as the females produce no or significantly less viable offspring. Consecutive releases over a sufficiently wide area result in less productive matings and a progressive reduction in insect population size over subsequent generations (Hendrichs and Robinson 2009).

The most widely celebrated application of SIT involved the use of ionizing radiation to eradicate the screwworm fly, *Cochliomyia hominivorax*, from North America—a program that began in 1957 following successful field trials on the Island of Curacao—and continues to this day to prevent reinvasion of the continent (Wyss 2000; Klassen and Curtis 2005). In this intervention, large-scale releases of sterilized insects led the screwworm fly population in the USA to crash within a decade. Subsequent releases progressively shifted the eradication zone southward, eventually covering all of North and Central America by 2001 (Robinson 2002).

The success of the screwworm SIT project motivated the application of SIT to a range of other insect pest species, including mosquito vectors of disease (Knipling 1968). Both irradiation and chemosterilization were initially explored for applications to mosquitoes, and in the 1960s and 1970s, large SIT field trials were conducted using chemosterilized *Culex quinquefasciatus* in India and *Anopheles albimanus* in El Salvador (Klassen and Curtis 2005). The trial in India was halted in the mid-1970s, following false accusations that the project was being used to collect data to engage in biological warfare (Nature 1975), highlighting the importance of effective community and political engagement for international biocontrol programs. Nevertheless, benefits of chemosterilization were demonstrated for this particular intervention due to reduced fitness costs as compared to irradiation.

A significant advancement in SIT technology was ushered with specific DNA changes introduced by the RIDL construct (Thomas et al. 2000; Alphey 2002). Insects sterilized through mutagens are subject to a myriad of random genetic mutations, which are invariably associated with significant fitness costs. In theory, releases of insects carrying (in homozygous form) a dominant lethal gene (RIDL) have essentially the same population impact as SIT—i.e., offspring of released males are unviable—although in a more controlled way that has potential for smaller associated fitness costs. Benedict and Robinson (2003) argued that a transgenic version of SIT should be the first application of transgenic mosquitoes in the wild (as it was), both for enhanced efficacy and for biosafety features—i.e., lethality-inducing transgenes should be quickly eliminated from the environment, causing the intervention to be reversible within a few generations. Quick elimination of transgenes also leads to confinement, since released mosquitoes can only travel so far in a few generations.

Sterile insect approaches based on genetic engineering present more opportunities than those based on mutagenesis, as genes and their associated traits can be modified in a more precise way. The original RIDL strain in *Aedes aegypti*, OX513A, causes lethality in both female and male offspring (bi-RIDL) (Thomas et al. 2000); however, an alternative construct was engineered soon after that only causes female offspring to be inviable (female-specific RIDL, or fs-RIDL). This allows the population-suppressing trait to persist for a few more generations through the male line, while continuing to suppress the female population, which is effective since only female mosquitoes bite and transmit diseases to humans. Furthermore, the introduced trait is late acting, affecting the development of wing muscles in adult females (Fu et al. 2010). This has the benefit that viable population reduction is not seen until the adult stage, delaying the reduction in larval density and hence maintaining high larval mortality rates for longer due to density-dependent competition of larvae in breeding sites (Black et al. 2011).

The first field trials of *Ae. aegypti* mosquitoes having the RIDL construct were conducted using the OX513A strain in Malaysia and the Cayman Islands. In Malaysia, Oxitec Ltd. and the Institute for Medical Research, Malaysia, worked closely with the Malaysian government in conducting a risk assessment. Releases were carried out in an uninhabited area to assess the mortality and dispersal characteristics of released RIDL mosquitoes; however, negative reactions were encountered from nongovernmental organizations and the media, preventing a trial from being conducted in an inhabited area where the impact on wild *Ae. aegypti* populations could be assessed (Enserink 2011).

In the Cayman Islands, Oxitec Ltd. worked with the local Mosquito Research and Control Unit (MRCU), initially conducting smaller releases over the course of 4 weeks to assess the fitness of genetically modified (GM) sterile males relative to wild males and subsequently conducting a population suppression field trial over the course of several months, again using the OX513A *Ae. aegypti* strain. In a lab cage study, GM sterile males were found to be more or less of equal competitiveness in mating with wild females, and the lethality trait was found to be effective in all crosses between GM sterile males and wild females (Harris et al. 2011). Subsequent field releases over a 4-week period found that GM males successfully mated with wild females in the field and fertilized their eggs resulting in unviable offspring; however, the field competitiveness of the GM males was estimated at ~56% that of wild males, albeit with a very wide 95% confidence interval of 3.2–197% (Harris et al. 2011).

The subsequent suppression field trial in the Cayman Islands was carried out across three contiguous areas on Grand Cayman island (denoted areas A, B, and C) over a period of 23 weeks (Harris et al. 2012). The initial goal had been to achieve a 10:1 GM to wild male ratio by releasing across all three areas (55 ha in total); however, production limitations led the actual achieved ratio to be significantly less (~2:1 GM-to-wild males), and a subsequent release in areas A and B still only achieved a ratio of ~5:1 GM-to-wild males. The third phase of the release was carried out solely in area A, achieving a release ratio of ~25:1 GM to wild males and demonstrating the benefit of a smaller trial area. Another benefit of the area A release was that area C served as a control and area B served as a buffer region. Significant population reduction was seen in this phase, with an 80% reduction in the mean ovitrap index in area A relative to areas B and C over the last 7 weeks of the release period (Harris et al. 2012).

Releases of GM sterile males in the Cayman Islands faced some controversy (Nightingale 2010; Enserink 2010); however, the major criticisms concerned the manner in which information about the trials was disseminated, rather than the conduct of the trials themselves. The releases did abide by national regulations, in particular, a draft biosafety bill that had yet to become law, the MRCU obtained a permit from the Cayman Islands Department of Agriculture, and a risk analysis and environmental impact assessment were carried out. The degree of community engagement was questioned; however, with several groups complaining, they had not been given details of the releases in advance (Enserink 2010).

Subsequent releases in Brazil followed a much more transparent approach. From the outset, a joint project was agreed, the Projeto Aedes Transgênico (PAT), between the University of São Paulo and Oxitec Ltd. to explore the potential use of GM sterile male *Ae. aegypti* as a form of urban mosquito control in terms of its social, technical, and operational dimensions. The project was launched by Moscamed, a Brazilian not-for-profit organization dependent on the Brazilian Ministry of Agriculture. The project enjoyed significant support in its early years as the government and public were aware of dengue outbreaks caused by this mosquito, and governmental support showed that they were being proactive in using the latest technology to control these outbreaks. The PAT worked closely with the Brazilian regulatory system to obtain required permits for field activities and adopted a vigorous community engagement campaign including school presentations, public events, interviews on TV and radio, house visits, and involvement of the community in trap monitoring and surveillance (de Campos et al. 2017).

The most well-documented trial of GM sterile male *Ae. aegypti* in Brazil was carried out in the Itaberaba suburb of the city of Juazeiro in Bahia, Brazil. This site had generally low socioeconomic indicators and relied on stored water to a large extent, providing breeding sites for mosquitoes and leading to relatively high dengue transmission. Similar to the Cayman Islands, the study area was divided into treatment areas A and B and a control area, with treatment eventually being restricted to area A in order to maintain sufficiently high release ratios. A Moscamed mass-rearing facility was built specifically for the project, producing millions of GM sterile males over the course of the study. Releases began with a "range finder" phase lasting a little over a month, which allowed the release requirements to be calibrated and estimates of parameters such as male mating competitiveness to be refined. GM male mating competitiveness was estimated to be ~3.1% that of wild males (95% CI: 2.5–3.6%), suggesting that releases for the "suppression" phase would need to be increased ninefold in order to achieve the target of 50% of mating events involving a GM sterile male (Carvalho et al. 2015).

The GM sterile male field trial in Brazil was successful, achieving a ~95% reduction in mosquito density at the release site, albeit with large release requirements of ~140,000 mosquitoes per week over a 5.5 ha control site for ~3 months (Carvalho et al. 2015). Enthusiasm for the GM sterile male approach was initially raised when the Zika outbreak began in 2015; however, an unexpected complication arose as untrue claims began to circulate in social media linking the Zika outbreak to past releases of the GM mosquitoes (de Campos et al. 2017). This draws attention to the importance of an enduring community engagement effort as well as political engagement and stakeholder messaging.

While this is not an invasive technology, these releases of sterile male mosquitoes do provide lessons from which potential field trials of gene drive mosquitoes may learn. Releases both of chemosterilized *Cx. quinquefasciatus* in India and of GM sterile *Ae. aegypti* in Brazil highlight the crucial importance of an effective and sustained community engagement effort. This especially applies to technologies developed in the Global North and applied in the Global South, which provide much potential for community mistrust. Furthermore, releases of GM sterile *Ae. aegypti* in both the Cayman Islands and Brazil highlight the importance of choosing a study site in which the required release sizes can be achieved and in conducting a range finder release phase to refine release requirements. For threshold-dependent gene drive systems, this will be important to determine release sizes that exceed the threshold, while for nonlocalized gene drive systems, this will be important to determine release sizes that are expected to demonstrate population control within the timeframe of the trial.

2.3 Lessons from the *Wolbachia*-based Incompatible Insect Technique

A promising alternative to SIT and GM sterile male releases is IIT, in which male mosquitoes are released that are infected with a *Wolbachia* strain absent from the wild population, resulting in sterile matings with wild females that lack the *Wolbachia* strain due to a phenomenon referred to as cytoplasmic incompatibility (CI) (LePage et al. 2017) (Fig. 2.1). This strategy has proceeded with much less resistance than GM approaches in recent years and serves as a case study for potential releases of novel biological control technologies, particularly regarding the use of factory rearing facilities (Zheng et al. 2019; Crawford et al. 2020).

The first field trial of IIT was conducted in Burma (now Myanmar) in 1967. The technique was seen as an alternative to insecticide-based strategies given the growing insecticide resistance among target species, *Cx. pipiens fatigans*, a vector of lymphatic filariasis (LF) which had proliferated in Southeast Asia at the time (Laven 1967). Despite successful elimination of the vector species from that trial site, the approach has not been deployed operationally until recently due to concern that accidental releases of *Wolbachia*-infected fertile females could result in the *Wolbachia* strain spreading into the population, preventing further suppression efforts. This is because *Wolbachia* is maternally inherited, and in most cases, the only incompatible crosses are between infected males and uninfected females. In 2009 and 2010, however, subsequent trials were carried out in French Polynesia to suppress populations of *Aedes polynesiensis*, a primary vector of LF in the South Pacific (O'Connor et al. 2012). Results from those field experiments showed that (1) *Wolbachia* transfected *Ae. polynesiensis* males successfully competed for mates following release and (2) the trial did not result in population replacement eventuating.

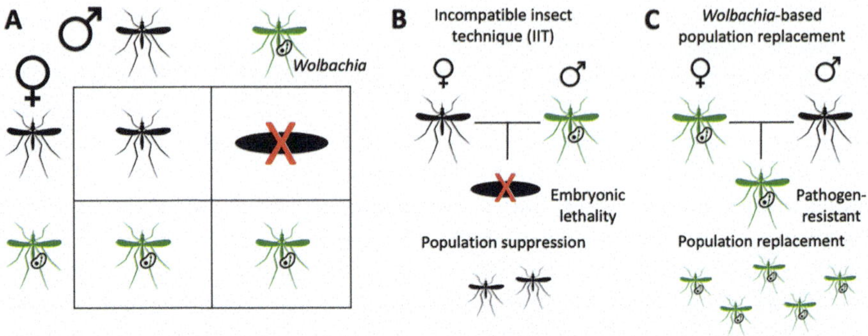

Fig. 2.1 (**a**) The use of *Wolbachia* as a means for both population suppression (incompatible insect technique, IIT) and population replacement hinges on the inheritance pattern in which crosses between *Wolbachia*-infected males and uninfected females produce unviable offspring due to cytoplasmic incompatibility (CI), while crosses involving *Wolbachia*-infected females produce *Wolbachia*-infected offspring due to *Wolbachia* being maternally inherited. (**b**) In IIT, *Wolbachia*-infected males are released into a wild population lacking that strain of *Wolbachia*. This leads to population suppression as mating events involving *Wolbachia*-infected males produce no viable offspring. (**c**) In *Wolbachia*-based population replacement, *Wolbachia*-infected females are included in the release. This leads to population replacement as CI biases inheritance in favor of *Wolbachia* when *Wolbachia*-infected females are present

In the last few years, two factory-scale IIT projects have moved forward to achieve community-scale mosquito population suppression: (1) an IIT program supplemented with sterilizing irradiation (also termed IIT-SIT) in Guangzhou, China (Zheng et al. 2019), and (2) an IIT program supplemented with factory-scale automation of production and sex sorting in Fresno, California (Crawford et al. 2020). The two projects represent different approaches to prevent population replacement: (1) through greatly reducing the fertility of any *Wolbachia*-infected females that may be accidentally released and (2) through using automation and machine learning to reduce the number of accidentally released *Wolbachia*-infected females effectively to zero.

In the IIT-SIT program in Guangzhou, *Aedes albopictus*, the main vector of dengue and other arboviruses in Guangzhou, were generated having an artificial triple *Wolbachia* infection (termed HC), through the addition of the *w*Pip *Wolbachia* strain to the native double infection of the *w*AlbA and *w*AlbB strains of *Wolbachia*. High levels of CI were confirmed such that matings of HC males with wild females produced no viable offspring and maternal transmission of the triple *Wolbachia* infection was confirmed, allowing efficient mass production of HC males. HC males were exposed to low-dose irradiation at the pupal stage to reduce the fecundity of any accidentally released HC females, and semi-field cage studies confirmed that the irradiated HC males effectively competed for mates leading to population suppression, without population replacement occurring due to released HC females. Furthermore, as an additional safety precaution, HC females were shown to be less competent at disease transmission than their wild counterparts (Zheng et al. 2019).

A trial carried out by the Wolbaki Biotech Company in 2016–2017 demonstrated the high degree of population suppression possible when factory rearing of mosquitoes is involved. Irradiated HC males were released on a weekly basis on two riverine islands within the jurisdiction of Guangzhou, with the ratio of released HC to wild males varying between 8.7:1 and 15.8:1 over the 38-week intervention period. Population suppression was highly successful, achieving a >94% reduction in the number of hatched eggs per ovitrap, as compared to control sites, and an 83–94% reduction in the number of wild adult females caught per trap. The success of the program also led to a significant increase in community support, with interviews suggesting 13% of residents were supportive prior to the intervention (notably, with 76% being neutral) and 54% were supportive following the intervention (Zheng et al. 2019).

The IIT program in Fresno, CA, showcased the role that large-scale, automated rearing and sex sorting of mosquitoes can play in increasing the scale of an IIT intervention. In this case, *Ae. aegypti*, the main arboviral vector through much of the Americas, was transfected with the *w*AlbB strain of *Wolbachia*, and sterility of crosses between infected males and wild females was confirmed. An automated larval rearing system was designed that, at maximum capacity, was able to produce almost 3 million pupae per week. A multistep sex-separation process was then designed that removed 95% of females at the pupal stage and the remainder at the adult stage based on a machine learning algorithm informed by photographic images as emerging adults walked down a narrow path. Estimates from the operation of this system suggested that a single *Wolbachia*-infected female mosquito would be released for every 900 million males, making the sex-sorting system near perfect (Crawford et al. 2020).

A trial carried out through a partnership between the Debug Project of Verily Life Sciences, MosquitoMate, and the Consolidated Mosquito Abatement District of Fresno County in 2018 demonstrated dramatic population suppression over an area nine times larger than that of the Guangzhou study. A total of more than 14 million *Wolbachia*-infected males were released as part of the study (an average of more than 78,000 per day), which led to a 96% reduction in the wild adult mosquito population; however, despite the large size of the releases, elimination was not achieved, likely due to inward migration of wild mosquitoes from neighboring untreated areas (Crawford et al. 2020). A public information campaign was conducted around the trial; however, formal documentation of this campaign is not yet available. A similar project is currently underway in Singapore through a partnership between Verily Life Sciences and the National Environment Agency of Singapore.

While neither a transgenic nor invasive technology, these IIT releases do provide lessons regarding the scale of releases that can be achieved when investment is made into automated rearing and sex-sorting facilities. Release requirements for low-threshold gene drive mosquitoes will be orders of magnitude lower than those for sterile male releases, and hence a facility capable of producing tens of millions of mosquitoes, such as the one designed by Verily Life Sciences, would be capable of achieving control over a much greater spatial scale than for IIT. The technological capacity for sex sorting is also encouraging given that male mosquitoes don't bite or transmit diseases to humans and hence may also be preferable for gene drive mosquito releases. The IIT releases enjoyed much less resistance from communities and regulatory agencies than GM sterile male releases, despite acting through a similar mechanism, highlighting issues that trials of gene drive mosquitoes will likely also face and must invest in.

2.4 Lessons from *Wolbachia*-based Population Replacement

A second approach to the use of *Wolbachia* to control mosquito-borne disease transmission is to intentionally include *Wolbachia*-infected females in a release. In IIT, care is taken to only release *Wolbachia*-infected males, as CI causes matings between *Wolbachia*-infected males and wild females to be sterile; however, CI-induced sterility, combined with the fact that *Wolbachia* is maternally inherited, provides an inheritance bias in favor of *Wolbachia* when *Wolbachia*-infected females are also present (Turelli and Hoffmann 1991) (Fig. 2.1). For *Wolbachia* strains that also block pathogen transmission, this can be used to drive the pathogen-blocking trait into the mosquito population (Moreira et al. 2009). This strategy has advanced significantly over the last decade (Hoffmann et al. 2011) and, like IIT, has faced much less resistance than GM strategies. It serves as an interesting case study for potential releases of transgenic population replacement technologies, as it has faced many of the non-GM issues that future gene drive programs will face.

The first *Wolbachia* population replacement program was carried out by the Eliminate Dengue project (now known as the World Mosquito Program) in the communities of Yorkeys Knob and Gordonvale in Queensland, Australia (Hoffmann et al. 2011). In this program, *Ae. aegypti*, the main vector of dengue and other arboviruses in Queensland, was transfected with the *w*Mel strain of *Wolbachia* from *Drosophila melanogaster*, a strain that has been shown to (1) block dengue transmission, (2) have a small associated fitness cost, and (3) be capable of driving into a small field cage (Walker et al. 2011). *Wolbachia* displays threshold properties in the presence of a fitness cost such that releases above a certain population frequency tend to spread, while releases below that frequency tend to be eliminated. The exact value of the threshold is determined by the point at which the inheritance bias induced by CI outweighs the fitness cost associated with the infection and has been estimated at ~20–30% for the *Wolbachia* strain used in this release (Hoffmann et al. 2011; Hancock et al. 2019).

The releases in Yorkeys Knob and Gordonvale were a clear success—after 10 weekly releases of 11,000–22,000 *Wolbachia*-infected *Ae. aegypti* per week, the *Wolbachia* infection reached near fixation in both populations within 3 months, despite a tropical storm postponing one of the releases in Gordonvale (Hoffmann et al. 2011) (Fig. 2.2). The finer details of this program provide an excellent example of how gene drive systems may be successfully trialed in the future. To begin, they highlight the importance of a detailed monitoring effort and adaptive release

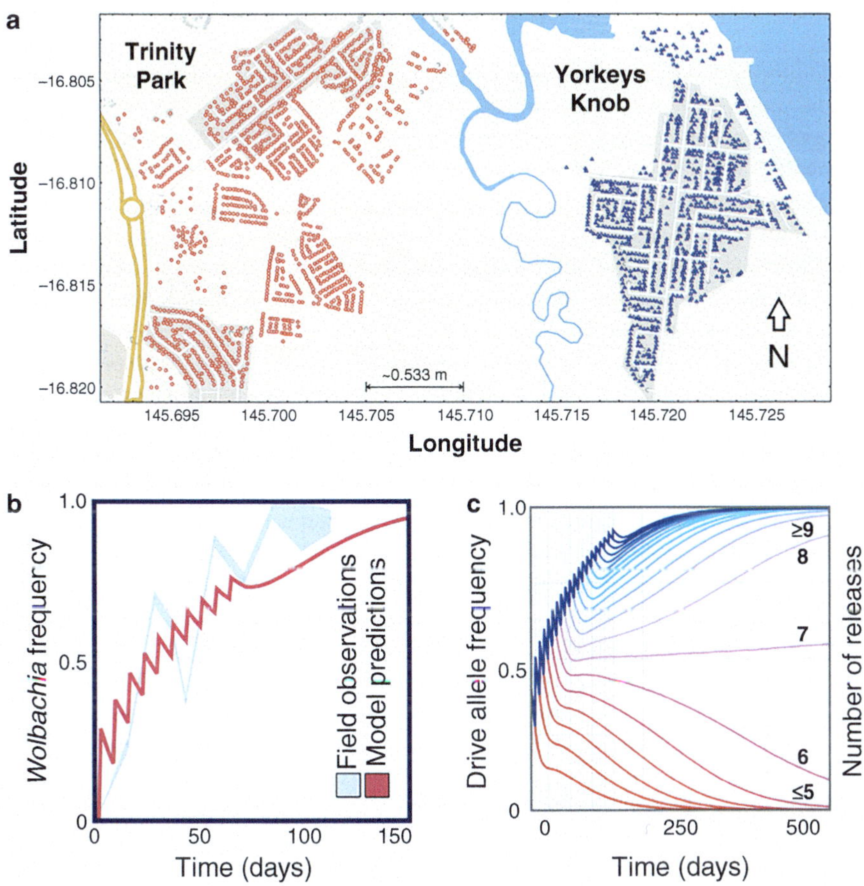

Fig. 2.2 (**a**) Landscape of Yorkeys Knob and Trinity Park in Queensland, Australia, where field trials of *Wolbachia*-based population replacement for *Aedes aegypti* were carried out and where trials of reciprocal chromosomal translocations were simulated. (**b**) Blue lines depict data for *Wolbachia* frequency over time from the *Wolbachia* population replacement field trial conducted in Yorkeys Knob in 2011 (Hoffmann et al. 2011), with line thickness representing 95% binomial confidence intervals around observed proportions. Red lines depict simulated data for an analogous release scheme consisting of 20 *Wolbachia*-infected mosquitoes per household at a coverage of 30% over 10 weeks, demonstrating good agreement with field data (Sánchez et al. 2020). (**c**) Translocation frequency over time for a given number of weekly releases of 20 adult male *Ae. aegypti* mosquitoes homozygous for the translocation per household with the intent of population replacement in Yorkeys Knob. Results are depicted for a coverage of 50%, at which seven or more releases result in the translocation being driven into the population (Sánchez et al. 2020). Due to the 50% threshold property of translocations, the same release scheme for wild types can be used to remediate translocations from the population

protocol. The releases in Yorkeys Knob and Gordonvale were accompanied by a network of 29 Biogents Sentinel mosquito traps that monitored *Wolbachia* infection frequency at the block level. Heterogeneity in *Wolbachia* infection frequency was observed, and releases were supplemented in areas where *Wolbachia* frequency was low.

Monitoring for unintended spread outside the study area was also conducted, and this did indeed reveal limited long-distance spread into a neighboring suburb from Yorkeys Knob and across a freeway from Gordonvale (Hoffmann et al. 2011). Although these migrants were expected to be lost due to being present at subthreshold levels, continued monitoring was important to confirm this. Continued monitoring was also conducted at the trials sites to confirm enduring intervention efficacy, and while the *Wolbachia* infection remained at near fixation for several years following the release, a low frequency of uninfected mosquitoes has also persisted, likely due to immigration (Hoffmann et al. 2014).

The releases in Yorkeys Knob and Gordonvale also highlight the importance of preparing for unexpected events. In addition to the tropical storm that affected both release sites and postponed one of the releases in Gordonvale, releases in a portion of Yorkeys Knob ceased two-thirds of the way into the intervention following a reported dengue case (Hoffmann et al. 2011). Although this dengue case likely originated elsewhere, a reactive insecticide intervention was carried out in surrounding households in agreement with local disease control protocols. Trials of mosquitoes with gene drive systems should make allowances for events such as these. Encouragingly, the *Wolbachia* infection continued to spread through the Yorkeys Knob *Ae. aegypti* population despite this, and no secondary dengue cases were documented following the reported case.

The Yorkeys Knob and Gordonvale releases provide an example of a successful community and regulatory engagement process. Community engagement was carried out over 2 years leading up to the releases and consisted of informal interviews, semi-structured in-depth interviews, qualitative and quantitative surveys, focus groups, historical research, and face-to-face presentations at community meetings (Hoffmann et al. 2011; McNaughton 2012). Issues explored through these activities included the sociopolitical context, lay knowledge of dengue fever and biological control programs, and acceptability of the project. Community members did raise concerns about a previous local biological control program—the introduction of the cane toad near Gordonvale in the 1930s. Largely seen as a failed biological control program, this was raised as a cautionary tale indicating the limits of scientific knowledge and the unpredictability of ecological interventions (McNaughton 2012).

The Queensland releases enjoyed substantial community support, with 85% of respondents viewing *Wolbachia* as an acceptable dengue prevention strategy in a March 2010 telephone survey (ahead of insecticides, at 66% acceptance) and 84% of respondents stating they would support a release that they were informed and updated about, that had regulatory oversight, and that was shown to be safe for people and the environment by a risk assessment carried out by Australia's Commonwealth Scientific and Industrial Research Organisation (CSIRO) (McNaughton 2012). The releases were ultimately approved by the Australian Pesticides and Veterinary Medicines Authority (APVMA) following risk assessments by CSIRO (Murphy et al. 2010) and the APVMA with support from the Federal Commonwealth Government's Department of the Environment, Water, Heritage and the Arts

(Marshall 2011). The World Mosquito Program is now exploring application of their technology beyond Australia, with active collaborations throughout Latin America, Asia, and Oceania.

In summary, key lessons from the *Wolbachia*-based population replacement strategy include the importance of (1) a detailed monitoring protocol to assess heterogeneity of spread at the field site, (2) an adaptive release scheme to supplement releases in areas of low *Wolbachia* frequency, (3) additional monitoring to assess levels of unintended spread to neighboring areas, and (4) preparing for unexpected events. The fact that *Wolbachia* infection has the potential to persist in the mosquito population for extended periods, and perhaps indefinitely, also emphasizes the need for a long-term, comprehensive, and multifaceted community engagement program.

2.5 Considerations for Trials with Reciprocal Chromosomal Translocations

Lessons from field trials of *Wolbachia*-based population replacement systems apply most closely to threshold-dependent gene drive systems, which are also expected to spread if released above a certain threshold frequency and to be eliminated if present below that frequency. One of the first of these systems to be proposed (Serebrovskii 1940; Curtis 1968), and perhaps currently one of the most promising (Sánchez et al. 2020), is reciprocal chromosomal translocations. These result from a mutual exchange between terminal segments of two nonhomologous chromosomes and produce a heterozygote reproductive disadvantage because, when translocation heterozygotes mate, several crosses result in unbalanced genotypes and hence unviable offspring. This produces a threshold frequency of 50%, which increases in the presence of a fitness cost (Curtis 1968). Early attempts to generate translocations through radiation-induced mutagenesis were abandoned due to high associated fitness costs (Laven et al. 1972; Lorimer et al. 1972); however, interest has been reignited as site-specific translocations have recently been generated using CRISPR (Lekomtsev et al. 2016; Jiang et al. 2016), and translocations generated in *D. melanogaster* using endonucleases were recently shown to drive in laboratory experiments with a threshold frequency of ~50% (Buchman et al. 2018).

A recent modeling study suggests that translocations represent one of the best systems to implement in field trials due to their symmetrical threshold properties and strong confinement potential. A key advantage of translocations is that releases required to introduce them into a population are of a similar magnitude to wild-type releases required to eliminate them once they have been introduced (Sánchez et al. 2020). Population replacement and reversion were modeled at the household level in the suburb of Yorkeys Knob, the site of the *Wolbachia* population replacement study, with low levels of migration modeled to the neighboring suburb of Trinity Park in Queensland, Australia (Fig. 2.2). Population replacement could be achieved in simulations with seven or more weekly releases of 20 *Ae. aegypti* males homozygous for the translocations per household per week (a similar magnitude to that used in the *Wolbachia* population replacement trial at the same site) and for a coverage of 50% of the households in the community. Elimination could be achieved for the same release scheme using wild *Ae. aegypti* mosquitoes.

One benefit of translocations, and other underdominant systems that have a threshold in the absence of a fitness cost, is that their release threshold is more robust than that for *Wolbachia*, which only arises in the presence of a fitness cost. This property leads to translocations being more robustly confineable to a field site than a *Wolbachia* infection, since they are unlikely to exceed the release threshold in a neighboring population purely through migration, even if they spread to near fixation at the trial site. In the translocation modeling study in Yorkeys Knob and Trinity Park (Sánchez et al. 2020), it was considered unlikely that *Ae. aegypti* mosquitoes would travel from one suburb to another by their own flight, especially in numbers sufficient to exceed the release threshold there, and so "batch migration" was instead considered, in which several mosquitoes are carried, perhaps by a vehicle, from one suburb to another at once. Batch migration events were modeled as occurring between randomly chosen neighborhoods, and the number of daily migration events and effective number of adults carried per event were varied. Results from this modeling study made a strong case for the potential to confine translocations to the release site, as the number of daily migration events required for the translocation to exceed the threshold in the neighboring suburb exceeded those inferred from field data. Specifically, 3–4 daily migration events consisting of batches of ten adults were required for translocations to spillover to the neighboring suburb in simulations (Sánchez et al. 2020), while field data suggested 1–2 daily migration events consisting of batches of less than five adult females (Hoffmann et al. 2011).

Collectively, these modeling results for translocations are encouraging for the potential to conduct field trials of *Ae. aegypti* mosquitoes with translocations because (1) translocations could be introduced on a suburban scale and remediated through releases of non-disease-transmitting male mosquitoes with release sizes on the scale of what has been previously implemented and (2) spillover of translocations into neighboring suburbs is unlikely. Lessons for the conduct of field trials with translocations may be drawn from the field trials previously described in this chapter—most importantly, for *Wolbachia*-based population replacement. These lessons highlight the importance of a detailed monitoring effort, including outside the study area, and of an adaptive release protocol that can respond to heterogeneities in spread at the trial site. They also highlight the importance of preparing for unexpected events and for conducting a long-term and comprehensive community engagement program, given that translocations have the potential to persist in the environment long term. A comparison of the RIDL and IIT releases suggests that community engagement and regulatory requirements for translocations may be stricter than for those for *Wolbachia* due to the fact that mosquitoes with translocations, generated using CRISPR or other endonucleases, will be considered GM organisms. Finally, regarding the release protocol, including a range finder release phase may help to refine fitness cost estimates and release requirements for translocations, as per a lesson from the RIDL field trial in Brazil.

2.6 Considerations for Trials with CRISPR-based Gene Drive Systems

Finally, lessons from the field trials discussed here have implications for the spectrum of CRISPR-based gene drives, from those that are nonlocalized to those that are self-limiting. Recent attention has focused on CRISPR-based homing gene drives, for their ability to spread widely and their potential to control vector-borne diseases on a wide scale (Gantz et al. 2015; Kyrou et al. 2018); however, there are also threshold-dependent gene drive systems that can now be engineered using CRISPR, such as chromosomal translocations (Buchman et al. 2018) and various forms of underdominance (Akbari et al. 2013), as well as temporally self-limiting gene drive systems, such as split drive (Li et al. 2020), which display transient drive activity before being eliminated by virtue of a fitness cost. The CRISPR revolution has also enabled gene drive countermeasures to be engineered, such as homing-based drive remediation systems, ERACR (element for the reversal of the autocatalytic chain reaction) and e-CHACR (erasing construct hitchhiking on the autocatalytic chain reaction) (Gantz and Bier 2016; Xu et al. 2020).

CRISPR-based homing gene drive systems bias inheritance in their favor by cleaving a highly specific target sequence in the host genome and copying themselves to the cut chromosome through a mechanism known as homology-directed repair (Gantz and Bier 2015; Champer et al. 2016). For high homing efficiencies and low-to-moderate fitness costs, these systems are capable of driving into populations from arbitrarily low initial frequencies. This property allows them to spread widely, and hence they are considered "nonlocalized." For these gene drive systems, while we may learn from field trials of *Wolbachia*-based population replacement systems, the scale of their potential spread and impact leads to additional and unique challenges that we must carefully consider.

One way to manage the risks associated with the potential wide-scale spread of homing-based gene drive systems is for testing to proceed iteratively through multiple phases, with each phase involving a larger spatial scale and a higher degree of human or environmental exposure (James et al. 2018) (Fig. 2.3). In this phased release pathway, initial studies are to be conducted in contained laboratories and

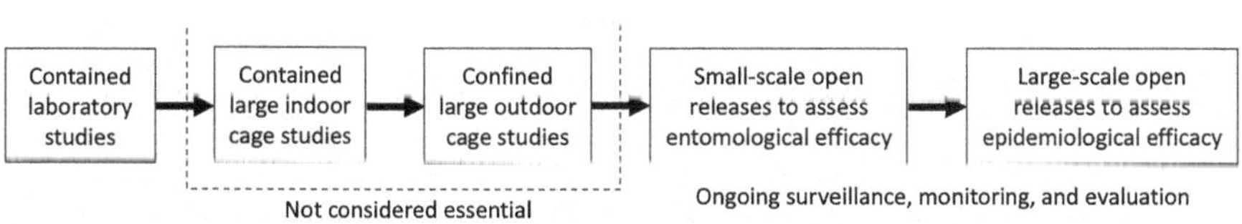

Fig. 2.3 Phased release pathway for CRISPR-based homing gene drive systems

insectaries, where product efficacy and safety are studied. Entering field-testing is a big decision, given the anticipated difficulty of remediating a homing-based gene drive system that is capable of spreading widely. Large outdoor cages present one option for moving beyond the laboratory; however, this is not considered essential since some mosquito behaviors, such as mating, and parameters, such as fitness, can only be meaningfully studied in the field. Furthermore, studies in outdoor cages must anticipate the possibility of escape occurring, and hence similar safety and efficacy criteria must be met before either outdoor cage studies or small-scale isolated releases are performed. Initial outdoor testing should be conducted at field sites within which the gene drive system is expected to be contained, for instance, on oceanic islands, following which open releases would be conducted on iteratively larger spatial scales (James et al. 2018).

Another consideration for trials of nonlocalized gene drive systems is that regulators are likely to require that a remediation plan be in place prior to field-testing (James et al. 2018). The chosen remediation strategy will depend on a number of factors, including the mode of action of the drive system and the scale and geography of the field site. A default remediation plan would be a large-scale insecticide-based campaign to eliminate the vector population at the field site. This would require an assessment of insecticide resistance in the local vector population prior to the gene drive trial. Failing this, releases of non-disease-transmitting male mosquitoes carrying a drive-resistant allele that restores the function of the gene targeted by the drive system are an attractive option, especially if the drive-resistant allele is sourced from a wild population.

Gene drive countermeasures such as ERACR and e-CHACR are another option for remediation. The ERACR system consists of a second homing system with a target site corresponding to the original drive system, essentially removing the original system as it homes into it, while utilizing the Cas9 of the original system and thus removing that as well (Gantz and Bier 2016; Xu et al. 2020). The e-CHACR system uses the Cas9 from the original homing system to home into a second site in the genome in addition to the site of the original drive system, thus driving itself into the population while removing the original system and its Cas9 in the process (Gantz and Bier 2016, Xu et al. 2020). Both of these systems hold promise; but they may not be the first choice for remediation efforts as they introduce additional transgenes into populations from which transgenes are trying to be removed.

Another potential phased release pathway is to precede the release of a nonlocalized gene drive system with a self-limiting one. Ideally, such a release would provide insights into the expected behavior of the nonlocalized system, and hence there should be strong resemblance between the two systems, to the extent possible. For a CRISPR-based homing gene drive system, one possibility is to begin with a trial of a split drive system, in which the Cas9 and guide RNA components are separated at different loci (Li et al. 2020). In the split drive system, transient drive activity occurs at the guide RNA locus when the Cas9 and guide RNA alleles co-occur in an organism; however, the Cas9 allele is gradually eliminated from the population due to its fitness cost, followed by the guide RNA if it also has a fitness cost. This transient drive activity also leads to spatial confinement, since a gene can only disperse so far in a limited number of generations. Intermediate technologies also exist for other systems. For instance, a driving Y chromosome that spreads by cleaving the X chromosome at multiple sites during spermatogenesis is expected to spread on a wide scale (Galizi et al. 2014); however, if linked to an autosome, it is self-limiting, providing an opportunity for intermediate study in the field.

For self-limiting CRISPR-based gene drive systems that could be used as an intermediate system in a field trial, similar field trial considerations apply as for chromosomal translocations. Namely, the ability to confine the release to the trial site, and to remediate transgenes from the environment as needed, is a great strength. Furthermore, it is important to combine a detailed monitoring effort, both in and outside the trial site, with an adaptive release protocol to respond to heterogeneities in spread, and to make allowances for unexpected events. A range finder release phase may help to refine fitness cost estimates and release requirements.

For nonlocalized CRISPR-based gene drive systems, the potentially wide scale of spread and difficulty of remediation emphasize the need to monitor for the gene drive system both in and outside the field trial area. Additionally, a range finder release phase may help to predict release schemes capable of achieving population control within the desired timeframe. Finally, as the spatial scale of the release grows, lessons may be learned from the experience of the Fresno IIT trial regarding automated rearing and sex sorting of mosquitoes. Knowledge of the potential scale of mosquito production will allow us to set expectations for wide-scale vector-borne disease control.

As for all of the systems discussed in this chapter, effective community and regulatory engagement is essential prior to field trials of mosquitoes engineered with CRISPR-based gene drive systems; however, this is especially important for trials of nonlocalized gene drive systems. Mosquitoes engineered with these systems are GM organisms capable of spreading widely, potentially across international borders, and are often developed in the Global North for application in the Global South. Their potential to spread across international borders highlights the desirability of a multicountry or regional agreement on their release, especially when a country that shares a border with another is being considered for field trials. Indeed, such agreements may be required by the Cartagena Protocol on Biosafety, which governs the safe transfer, handling, and use of GM organisms (referred to as "living modified organisms" in the protocol), including their transboundary spread (Secretariat of the Convention on Biological Diversity 2000; Marshall 2010).

2.7 Conclusion

The limitations of traditional insecticide based strategies to control mosquito populations, and, in particular, the widespread emergence of insecticide resistance, have spurred interest in a variety of novel biological and genetics-based vector control strategies, including SIT, IIT, RIDL, *Wolbachia* based population replacement, and CRISPR-based gene drive (Benelli et al. 2016). Trials of RIDL, IIT, and

Table 2.2 Significant field trials of novel biological and genetics-based mosquito control strategies

Method	Species	Location	Year	Outcome	Reference
SIT	*Anopheles quadrimaculatus*	Florida, USA	1962	Poor mating competitiveness	Weidhaas et al. (1962)
IIT	*Culex pipiens fatigans*	Burma (now Myanmar)	1967	Successful suppression	Laven (1967)
SIT	*Culex quinquefasciatus*	India	1971–1975	Modest suppression	Singh et al. (1975)
SIT	*Anopheles albimanus*	El Salvador	1971–1979	Significant suppression	Lowe et al. (1980)
RIDL	*Aedes aegypti*	Cayman Islands	2009	Small-scale suppression	Harris et al. (2011, 2012)
IIT	*Aedes polynesiensis*	French Polynesia	2009–2010	Demonstration of efficacy	O'Connor et al. (2012)
Wolbachia population replacement	*Ae. aegypti*	Queensland, Australia	2011	Successful population replacement	Hoffmann et al. (2011)
RIDL	*Ae. aegypti*	Juazeiro, Brazil	2012–2013	Community-scale suppression	Carvalho et al. (2015)
RIDL	*Ae. aegypti*	Jacobina, Brazil	2013	Suppression and resurgence	Garziera et al. (2017)
IIT	*Aedes albopictus*	Kentucky, USA	2014	Significant suppression	Mains et al. (2016)
IIT-SIT	*Ae. albopictus*	Guangzhou, China	2016–2018	Community-scale suppression	Zheng et al. (2019)
IIT	*Ae. aegypti*	California, USA	2018–2019	Community-scale suppression	Crawford et al. (2020)

Wolbachia over the last decade provide a series of case studies from which we may learn in preparing for field trials of CRISPR-based gene drive systems (Table 2.2).

There are challenges associated with gene drive technologies—notably, the controversies surrounding GM organisms and the potential for spread across international borders. However, these challenges are also a reason for promise as half of the world's population is at risk of vector-borne diseases, and genetic engineering provides new opportunities to interfere with pathogen transmission. In learning from recent field trials, we seek to move these technologies forward carefully and responsibly toward the eventual goal of global vector-borne disease control.

References

Akbari OS, Matzen KD, Marshall JM, Huang H, Ward CM et al (2013) A synthetic gene drive system for local, reversible modification and suppression of insect populations. Curr Biol 23:671–677

Alphey L (2002) Re-engineering the sterile insect technique. Insect Biochem Mol Biol 32:1243–1247

Benedict MQ, Robinson AS (2003) The first releases of transgenic mosquitoes: an argument for the sterile insect technique. Trends Parasitol 19:349–355

Benelli G, Jeffries CL, Walker T (2016) Biological control of mosquito vectors: past, present and future. Insects 7:52

Black WC, Alphey L, James AA (2011) Why RIDL is not SIT. Trends Parasitol 27:362–370

Buchman A, Ivy T, Marshall JM, Akbari OS, Hay BA (2018) Engineered reciprocal chromosome translocations drive high threshold, reversible population replacement in Drosophila. ACS Synth Biol 7:1359–1370

Carvalho DO, McKemey AR, Garziera L, Lacroix R, Donnelly CA et al (2015) Suppression of a field population of *Aedes aegypti* in Brazil by sustained release of transgenic male mosquitoes. PLoS Negl Trop Dis 9:e0003864

Champer J, Buchman A, Akbari OS (2016) Cheating evolution: engineering gene drives to manipulate the fate of wild populations. Nat Rev Genet 17:146–159

Crawford JE, Clarke DW, Criswell V, Desnoyer M, Cornel D et al (2020) Efficient production of male *Wolbachia*-infected *Aedes aegypti* mosquitoes enables large-scale suppression of wild mosquitoes. Nat Biotechnol 38:482–492

Curtis CF (1968) Possible use of translocations to fix desirable genes in insect pest populations. Nature 218:368–369

de Campos A, Hartley S, de Koning C, Lezaun J, Velho L (2017) Responsible innovation and political accountability: genetically modified mosquitoes in Brazil. J Respons Innov 4:5–23

Doudna JA, Charpentier E (2014) Genome editing. The new frontier of genome engineering with CRISPR-Cas9. Science 346:1258096

Enserink M (2010) GM mosquito trial alarms opponents, strains ties in Gates-funded project. Science 330:1030–1031

Enserink M (2011) GM mosquito release in Malaysia surprises opponents and scientists—again. Science Insider. Available online at: https://www.sciencemag.org/news/2011/01/gm-mosquito-release-malaysia-surprises-opponents-and-scientists-again

Fu G, Lees RS, Nimmo D, Aw D, Jin L et al (2010) Female-specific flightless phenotype for mosquito control. Proc Natl Acad Sci U S A 107:4550–4554

Galizi R, Doyle LA, Menichelli M, Bernardini F, Deredec A et al (2014) A synthetic sex ratio distortion system for the control of the human malaria mosquito. Nat Commun 5:3977

Gantz VM, Bier E (2015) The mutagenic chain reaction: a method for converting heterozygous to homozygous mutations. Science 348:442–444

Gantz VM, Bier E (2016) The dawn of active genetics. BioEssays 38:50–63

Gantz VM, Jasinskiene N, Tatarenkova O, Fazekas A, Macias VM et al (2015) Highly efficient Cas9-mediated gene drive for population modification of the malaria vector mosquito *Anopheles stephensi*. Proc Natl Acad Sci U S A 112:E6736–E6743

Garziera L, Pedrosa MC, de Souza FA, Gómez M, Moreira MB et al (2017) Effect of interruption of over-flooding releases of transgenic mosquitoes over wild populations of *Aedes aegypti*: two case studies in Brazil. Entomol Exp Appl 164:327–339

Gould F, Huang Y, Legros M, Lloyd AL (2008) A killer-rescue system for self-limiting gene drive of anti-pathogen constructs. Proc Biol Sci 275:2823–2829

Hammond A, Galizi R, Kyrou K, Simoni A, Siniscalchi C et al (2016) A CRISPR Cas9 gene drive system targeting female reproduction in the malaria mosquito vector *Anopheles gambiae*. Nat Biotechnol 34:78–83

Hancock PA, Ritchie SA, Koenraadt CJM, Scott TW, Hoffmann AA et al (2019) Predicting the spatial dynamics of *Wolbachia* infections in *Aedes aegypti* arbovirus vector populations in heterogeneous landscapes. J Appl Ecol 56:1674–1686

Harris AF, Nimmo DD, McKemey AR, Kelly N, Scaife S et al (2011) Field performance of engineered male mosquitoes. Nat Biotechnol 29:1034–1037

Harris AF, McKemey AR, Nimmo DD, Curtis Z, Black I et al (2012) Successful suppression of a field mosquito population by sustained release of engineered male mosquitoes. Nat Biotechnol 30:828–830

Hendrichs J, Robinson A (2009) Sterile insect technique. In: Resh VH, Carde RT (eds) Encyclopedia of insects, 2nd edn. Academic Press, London

Hoffmann AA, Montgomery BL, Popovici J, Iturbe-Ormaetxe I, Johnson PH et al (2011) Successful establishment of *Wolbachia* in *Aedes* populations to suppress dengue transmission. Nature 476:454–457

Hoffmann AA et al (2014) Stability of the wMel *Wolbachia* infection following invasion into *Aedes aegypti* populations. PLoS Negl Trop Dis 8:e3115

James S, Collins FH, Welkhoff PA, Emerson C, Godfray HCJ et al (2018) Pathway to deployment of gene drive mosquitoes as a potential biocontrol tool for elimination of malaria in sub-Saharan Africa: recommendations of a scientific working group. Am J Trop Med Hyg 98:1–49

Jiang J, Zhang L, Zhou X, Chen X, Huang G et al (2016) Induction of site-specific chromosomal translocations in embryonic stem cells by CRISPR/Cas9. Sci Rep 6:21918

Klassen W, Curtis CF (2005) History of the sterile insect technique. In: Dyck VA, Hendrichs J (eds) Sterile insect technique: principles and practice in area-wide integrated pest management. IAEA, Geneva, pp 3–36

Knipling EF (1955) Possibilities of insect control or eradication through the use of sexually sterile males. J Econ Entomol 48:459–462

Knipling EF (1968) The potential role of sterility for pest control. In: LeBrecque GC, Smith CN (eds) Principles of insect chemosterilization. Appleton-Century-Crofts, New York, pp 7–40

Kyrou K, Hammond AM, Galizi R, Kranjc N, Burt A et al (2018) A CRISPR-Cas9 gene drive targeting doublesex causes complete population suppression in caged Anopheles gambiae mosquitoes. Nat Biotechnol 36:1062–1066

Laven H (1967) Eradication of *Culex pipiens fatigans* through cytoplasmic incompatibility. Nature 216:383–384

Laven H, Cousserans J, Guille G (1972) Eradicating mosquitoes using translocations: a first field experiment. Nature 236:456–457

Lekomtsev S, Aligianni S, Lapao A, Bürckstümmer T (2016) Efficient generation and reversion of chromosomal translocations using CRISPR/Cas technology. BMC Genomics 17:739

LePage DP, Metcalf JA, Bordenstein SR, On J, Perlmutter JI et al (2017) Prophage WO genes recapitulate and enhance *Wolbachia*-induced cytoplasmic incompatibility. Nature 543:243–247

Li M, Yang T, Kandul NP, Bui M, Gamez S, Raban R, Bennett JB, Sánchez HM, Lanzaro GC, Schmidt H, Lee Y, Marshall JM, Akbari OS (2020) Development of a confinable gene-drive system in the human disease vector, *Aedes aegypti*. eLife 9:e51701

Lorimer N, Hallinan E, Rai KS (1972) Translocation homozygotes in the yellow fever mosquito, *Aedes aegypti*. J Hered 63:159–166

Lowe RE, Bailey DL, Dame DA, Savage KE, Kaiser PE (1980) Efficiency of techniques for the mass release of sterile male *Anopheles albimanus*. Am J Trop Med Hyg 29:695–703

Mains JW, Brelsfoard CL, Rose RI, Dobson SL (2016) Female adult *Aedes albopictus* suppression by *Wolbachia*-infected male mosquitoes. Sci Rep 6:33846

Marshall JM (2010) The Cartagena protocol and genetically modified mosquitoes. Nat Biotechnol 28:896–897

Marshall JM (2011) The Cartagena protocol in the context of recent releases of transgenic and Wolbachia-infected mosquitoes. Asia-Pacific J Mol Biol Biotechnol 19:93–100

Marshall JM, Akbari OS (2018) Can CRISPR-based gene drive be confined in the wild? A question for molecular and population biology. ACS Chem Biol 13:424–430

McNaughton D (2012) The importance of long-term social research in enabling participation and developing engagement strategies for new dengue control technologies. PLoS Negl Trop Dis 8: e1785

Moreira LA, Iturbe-Ormaetxe I, Jeffery JA, Lu G, Pyke AT et al (2009) A *Wolbachia* symbiont in *Aedes aegypti* limits infection with dengue, chikungunya, and plasmodium. Cell 139:1268–1278

Murphy B, Jansen C, Murray J, De Barro P (2010) Risk analysis on the Australian release of *Aedes aegypti* (L.) (Diptera: Culicidae) containing *Wolbachia*. CSIRO, Canberra

Nature (1975) Oh New Delhi, Oh Geneva (editorial). Nature 256:355–357

Nightingale K (2010) GM mosquito wild release takes campaigners by surprise. SciDev.Net. Available online at: https://www.scidev.net/global/policy/news/gm-mosquito-wild-release-takes-campaigners-by-surprise.html

Noble C, Adlam B, Church GM, Esvelt KM, Nowak MA (2018) Current CRISPR gene drive systems are likely to be highly invasive in wild populations. eLife. https://doi.org/10.7554/eLife.33423

O'Connor L, Pilchart C, Sang AC, Brelsfoard CL, Bossin HC, Dobson SL (2012) Open release of male mosquitoes infected with a *Wolbachia* biopesticide: field performance and infection containment. PLoS Negl Trop Dis 6:e1797

Robinson AS (2002) Mutations and their use in insect control. Rev Mutat Res 511:113–132

Sánchez HM, Bennett JB, Wu SL, Rašić G, Akbari OS et al (2020) Confinement and reversibility of threshold-dependent gene drive systems in spatially-explicit Aedes aegypti populations. BMC Biol 18:50

Secretariat of the Convention on Biological Diversity (2000) Cartagena protocol on biosafety to the convention on biological diversity. World Trade Center, Montreal

Serebrovskii AS (1940) On the possibility of a new method for the control of insect pests. Zool Zhurnal 19:618–630

Singh KRP, Patterson RS, LaBrecque GC, Razdan RK (1975) Mass rearing of *Culex fatigans*. J Commun Dis 7:31–53

Thomas DD, Donnelly CA, Wood RJ, Alphey LS (2000) Insect population control using a dominant, repressible, lethal genetic system. Science 287:2474–2476

Turelli M, Hoffmann AA (1991) Rapid spread of an inherited incompatibility factor in California *Drosophila*. Nature 353:440–442

Walker T, Johnson PH, Moreira LA, Iturbe-Ormaetxe I, Frentiu FD et al (2011) The *wMel Wolbachia* strain blocks dengue and invades caged *Aedes aegypti* populations. Nature 476:450–453

Weidhaas DE, Schmidt CH, Seabrook EL (1962) Field studies on the release of sterile males for the control of *Anopheles quadrimaculatus*. Mosq News 22:283–291

Wyss JH (2000) Screw-worm eradication in the Americas. In: Tan KH (ed) Area-wide control of fruit flies and other insect pests. Penerbit Universiti Sains Malaysia, Penang, pp 79–86

Xu XS, Bulger EA, Gantz VM, Klanseck C, Heimler SR et al (2020) Active genetic neutralizing elements for halting or deleting gene drives. Mol Cell 80:246–262

Zheng X, Zhang D, Li Y, Yang C, Wu Y et al (2019) Incompatible and sterile insect techniques combined eliminate mosquitoes. Nature 572:56–61

Chapter 3
Genetic Improvements to the Sterile Insect Technique (SIT) for the Control of Mosquito Population

P. V. D. Dilani, Y. I. N. S. Gunawardene, and R. S. Dassanayake

Abstract Mosquito-borne diseases are becoming a major health problem worldwide. At present, the principal method of controlling these diseases entirely depends on the mosquito vector control strategies. However, traditional control methods which are focussed on reducing mosquito populations through environmental management and the application of insecticides are largely ineffective. Hence, various control methods, including the release of sterile insect technique (SIT), have been proposed for the reduction of the mosquito population. As a species-specific control strategy, SIT offers considerable environmental benefits and a chemical-free option for insect control. However, the application of the SIT to mosquito control consistently suffered from lack of efficient sexing system, high fitness cost and operational difficulty in ionizing radiation, density-dependent nature of the target mosquito population and various other technical issues. The intervention of genetic engineering has led to several improvements in the operation or security of SIT programmes. The advent of mosquito transgenesis has paved the way for novel approaches in mosquito control. One possibility is a release of insects carrying dominant lethal (RIDL) strategy by engineering self-limiting gene, which offers solutions for many drawbacks of traditional SIT by providing genetic sterilization, genetic sexing, genetic containment and provision of genetic markers while maintaining its environmentally benign and species-specific utility. The success of this strategy often depends on how genetic modification affects the fitness of the mosquitoes. With several improvements and modifications allowing minimum fitness load, RIDL is now available for a wide range of mosquitoes such as *Aedes aegypti*, *Aedes albopictus* and *Anopheles stephensi* with field-testing possibilities. However, with solid epidemiological evidence and community support, widespread implementation

P. V. D. Dilani · R. S. Dassanayake (✉)
Department of Chemistry, Faculty of Science, University of Colombo, Colombo, Sri Lanka
e-mail: desha@chem.cmb.ac.lk; rsdassanayake@chem.cmb.ac.lk

Y. I. N. S. Gunawardene
Molecular Medicine Unit, Faculty of Medicine, University of Kelaniya, Ragama, Sri Lanka
e-mail: nilminisg@kln.ac.lk

B. K. Tyagi (ed.), *Genetically Modified and other Innovative Vector Control Technologies*, https://doi.org/10.1007/978-981-16-2964-8_3

of these strategies might reverse the current alarming global mosquito vector-borne diseases.

Keywords Mosquito · SIT · Transgenic · RIDL

3.1 Introduction

Mosquitoes (Diptera: Culicidae) act as vectors for the transmission of various disease-causing pathogens to humans. The first discovery that the mosquito transmits the disease was revealed by Dr. Patrick Manson in 1877; the disease was known as human lymphatic filariasis, more commonly also the elephantiasis caused by a tiny filarial worm, *Wuchereria bancrofti*, and transmitted by the mosquitoes belonging to the genus *Culex* (Chernin 1983). In 1897, a groundbreaking work done by Dr. Ronald Ross demonstrated that *Anopheles* mosquitoes transmitted malaria parasites (Ross 1897). Several types of other mosquito-borne diseases such as dengue, chikungunya, yellow fever and West Nile viral diseases have been identified from then on (Cuervo-Parra et al. 2016). Dengue and malaria, among other mosquito-borne diseases, are now, more than ever, major public health problems the world over. At present, the principal method of controlling these diseases entirely depends on the mosquito vector control (Wilson et al. 2020).

During the life cycle, the mosquito goes through four life stages: egg, larva, pupa and adult, of which the first three stages require water for the development process. Adult mosquitoes have a pair of long and narrow wings and three pairs of long slender legs. The adult female mosquito is equipped with proboscis which can pierce the skin of the host to feed on blood (Cuervo-Parra et al. 2016).

Prior to the use of insecticides, vector control was relied on environmental management with the understanding of vector behavioural ecology and tailored environmental control strategies (Wilson et al. 2020). Following the groundbreaking discovery of Ross, the vector control was primarily focussed on removing the potential breeding sites of mosquitoes. In 1899–1900, the first trial of malaria intervention was carried out by Angelo Celli among railway workers in Italy (Celli 1901a, b). In his controlled intervention study, the combined intervention of house screening, whitewashing internal walls, burning special powders (probably pyrethrum) and protective clothing has been highly successful (Ferroni et al. 2012). Another example of environmental management for malaria control was offered by Sir Malcolm Watson who led vector control efforts based on an understanding of the ecology of local vectors, *Anopheles umbrosus* and *An. maculatus*, and controlled these in Malaysia in the early 1900s through draining marshes, subsoil drainage, filling waterbodies, tree clearing and relocating housing (Walton 1922). The control of yellow fever and malaria in Cuba was heavily reliant on environmental management from 1901 to 1912 (Le Prince 1916). Although these methods were often effective, the success of this approach was often forgotten in search of more easily achievable vector control strategies.

The advent of insecticides such as dichlorodiphenyltrichloroethane (DDT) and Paris green gained favour owing to less labour-intensive nature to implement and rapid results than environmental strategies. In 1950–1960, the official vector control programmes of many countries used chemical strategies such as DDT and Paris green which successfully eradicated malaria in south-east USA and Italy (Williams 1963; Marchi and Munstermann 1987). The subsequent outbreak of dengue epidemics from the 1980s to date has been caused by the reinfestation of *Ae. aegypti* and the poor control programme implementation. The same control strategies which have been used in the 1950s using insecticides were unsuccessful (Axtell and Arends 1990; Wilson et al. 2020). This may be attributed to many reasons including increased ecological consequences, effect on non-target organisms and development of insecticide-resistant strains and, therefore, has not been sustainable in the long term which justifies the urgent need for a novel effective control strategy.

3.2 The Sterile Insect Technique

The sterile insect technique (SIT) is a species-specific, environmentally benign tactic for insect control which has been around for decades and widely used to control agricultural pests (Bushland et al. 1955). The SIT relies on the mass rearing, sterilization and release of a large number of sterile males into target areas where they compete for females with the wild male population (Fig. 3.1). Mating between wild females and sterilized males does not produce viable eggs, thereby reducing the reproductive potential of the wild population (Thomé et al. 2010). If sufficient sterile males can be released over a period of time, this will reduce the number of individuals in the target population and eventually lead to a crash of the entire population (Thomé et al. 2010).

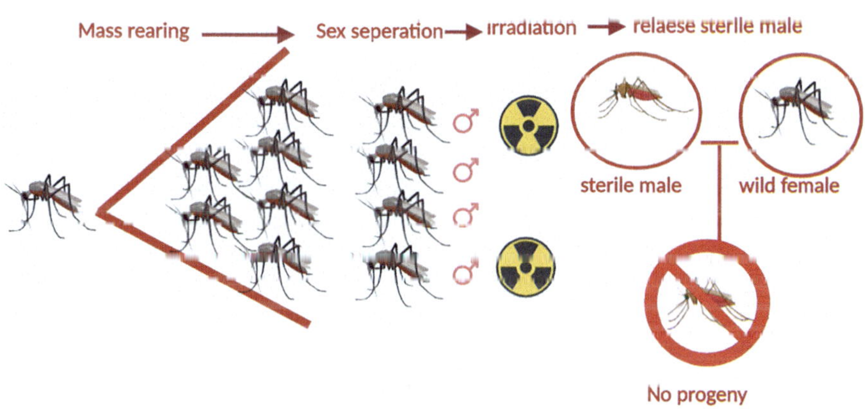

Fig. 3.1 Conventional SIT process: mass rearing of mosquitoes followed by manual sex separation and further males are sterilized by ionizing radiation. Then the sterile males are released to mate with wild females resulting in no progeny

The SIT was first conceived by Bushland and Knipling (Bushland et al. 1955) and implemented successfully in 1958 in Florida to control *Cochliomyia hominivorax* (screw worm fly) (Knipling 1979). Since then, SIT has been successfully used against several insect pest species, (1) to complete elimination of *Cochliomyia hominivorax* (Coquerel 1858), the New World screw worm fly from southern USA, Mexico and all of Central America (Wyss 2000); (2) to eradicate *Ceratitis capitata* (Wiedemann), the Mediterranean fruit fly, from Central America and Mexico (Hendrichs et al. 2002; Juan-Blasco et al. 2014); and (3) to eradicate *Glossina austeni* (Newstead 1912), the tsetse fly, from the island of Zanzibar in Tanzania (Vreysen et al. 2000).

3.2.1 The Sterile Insect Technique to Control Mosquitoes

Mosquitoes are another ideal candidate for SIT. The early success with agricultural pests led to several trials to control mosquitoes. Several mosquito releases have been performed for the purpose related to SIT in the 1960s, 1970s and early 1980s for *Aedes aegypti*, *Ae. albopictus*, *Culex pipiens*, *Cx. quinquefasciatus*, *Cx. tarsalis*, *Cx. tritaeniorhynchus*, *Anopheles albimanus*, *An. culicifacies*, *An. gambiae* and *An. quadrimaculatus* (Benedict and Robinson 2003). Different sterilization strategies have been tested during these trials for mosquitoes, including gamma irradiation, chemosterilization, cytoplasmic incompatibility and other chromosomal rearrangements. Among these trials, some have shown suppression in the mosquito population.

However, only a few cases achieved eradication; even in these cases, the programmes were not sufficient to be effective in non-isolated areas (Benedict and Robinson 2003). The first major success was achieved against *Cx. pipiens* to eradicate an isolated population from a village in Myanmar (Laven 1967). This programme used sterilized mosquitoes through cytoplasmic incompatibility, the phenomenon which can now be explained by the existence of different strains of *Wolbachia* in different mosquito populations (Benedict and Robinson 2003). The first successful programme against anopheline mosquitoes was the elimination of isolated *An. albimanus* population in a 15 km^2 area in EI Salvador where chemosterilization was used as an alternative approach to induce sterility in mosquitoes (99.8% sterility in males) with promising results, though it was not acceptable for large-scale and long-term use due to chemical leftover. Unfortunately, the project was interrupted in the mid-1970s due to political upheavals (Benedict and Robinson 2003). One of the most ambitious projects was undertaken jointly with the World Health Organization (WHO) and the Indian Council of Medical Research (ICMR) against *Cx. quinquefasciatus*, *Ae. aegypti* and *An. stephensi* in India during the 1970s. In this project, they developed a mass-rearing method to get more sterile male pupae, methods to package, mark, transport and distribute adults. Unfortunately, this trial did not get off the ground just 2 days before it was to begin due to Indian media and politicians who accused that the trial was a cover for a biological warfare (Curtis 2005). Recently, an open-field release of sterile *Ae. aegypti* males developed from a combined SIT and incompatible insect technique (IIT) approach was successfully carried out to suppress the natural populations of *Ae. aegypti* in semi-rural village settings in Thailand. However, further trials have been recommended to assure the effectiveness of the strategy, despite the combined SIT/IIT technology that has shown the potential for long-term control of *Ae. aegypti* and *Ae. albopictus* mosquitoes (Kittayapong et al. 2019).

3.2.2 Limitation of Sterile Insect Technique in Controlling Mosquitoes

SIT was a proven cost-effective strategy for elimination or eradication of most agricultural pests. However, the utility of strategy is yet to be fully assessed for mosquitoes. Despite its attractive benefits for pest control, SIT has not been operationally used against mosquitoes other than trials conducted in the 1970s and 1980s (Benedict and Robinson 2003; Alphey et al. 2010). The major difficulty for mosquito SIT was the sterilization. Ionizing radiation was the major sterilizing principle used by all recent SIT programmes though other principles such as cytoplasmic incompatibility and inter-species hybrid sterility have also been used in early trials (Benedict and Robinson 2003; Alphey et al. 2006). The major drawback of ionizing radiation is the reduction in mating competitiveness of irradiated insects (Helinski et al. 2009). Irradiation caused a dramatic loss of competitive mating ability and reduced the lifespan relative to the wild type. This contributed to the failure of SIT programmes against *An. quadrimaculatus* and *Ae. aegypti* in Florida ((Helinski et al. 2009). For a successful SIT programme, it is essential to minimize the somatic damages that occur during irradiation. Somatic damage can be reduced when irradiated later in the development of the insect as the number of cells undergoing division declines (Helinski et al. 2009). Irradiation of adults is less damaging but operationally far more difficult to perform (Phuc et al. 2007). At the same time, it is considered that the released male should be completely sterile to be applicable for a SIT programme. These factors can be partly overcome by increasing the ratio of sterile-to-wild insects (Helinski et al. 2009). In addition to the fitness cost, the operational cost is also high to implement radiation facilities. Some trials have used chemicals as an alternative to radiation to induce sterility which were proven to be effective for sterilization, but trace contamination of the chemical residues such as thiotepa and tepa has been reported in some trials (Labrecque et al. 1972).

Another prominent cause for the failure of several mosquito control programmes was the insufficient production of sterile mosquitoes due to the absence of a proper sex separation system (Benedict and Robinson 2003). Male-only releases have proven to be more efficient and less expensive than bisexual releases as only sterile males can transfer sterile sperms to wild females; otherwise, sterile females would distract sterile males from inseminating wild females (Alphey et al. 2010).

Moreover, the male-only release is desirable for mosquito SIT owing to the biting nuisance and potential disease transmission caused by adult females (Fu et al. 2010). Mosquito sexing for SIT release programmes used sexual dimorphism based on differential pupal weight, size and time of emergence by manual or mechanical means (Breeland et al. 1974; Dame et al. 1974; Kittayapong et al. 2018). Generally, female pupae are larger than male pupae in size. However, the extent of dimorphism is species specific. Though these methods have had varying degrees of success, they rarely yield a truly single sex population (Kittayapong et al. 2018). A stringent sexing system is, therefore, required for mosquito SIT for the more efficient removal of females.

Another problem of mosquito SIT is associated with the population biology of mosquitoes. The density-dependent effect can reduce, eliminate or even reverse the success of SIT programme (Alphey et al. 2010). SIT programmes have generally been targeted against agricultural pests which were not resource limited, at least not limited by the availability of larval food. However, for the mosquitoes, the density dependence occurs at the larval stage. Early death of some offspring (progeny of a mating between wild females and sterilized males) or the embryonic lethality occurring early in the embryogenesis would leave more resources (foods and availability of oviposition sites) for remaining larvae leading to less competition and higher survival rates which can ultimately counter the SIT control programme (Alphey et al. 2010). Therefore, it is required to introduce an alternative strategy, for example, late-acting lethal action which occurs after the density-dependent stage for an effective mosquito control programme (Phuc et al. 2007).

In addition to all above prominent issues, various other technical causes contributed to the failure of several SIT programmes, for example, (1) the efficiency of SIT reduced due to unexpected migration of mated females into the release area (Benedict and Robinson 2003) and (2) required production was higher than expected owing to the negative impact of sterilization, improper handling during rearing, sexing during transportation and distribution (Balestrino et al. 2017).

3.3 Improvement of SIT Through Genetic Engineering

Classical genetics have been used for decades to improve SIT programmes. Several efforts to improve mosquito SIT programmes have been performed using the mechanisms of control such as mutagen-induced dominant lethality, sex ratio distortion and chromosomal translocations. These mechanisms enhance the SIT by either improving mass rearing or removing the need for sterilization (Bruno Wilke et al. 2009). The advent of modern biotechnology has led to a resurgence in the interest of using molecular tools to enhance SIT programmes. The intervention of genetic engineering has led to several improvements in the operation or security of SIT programmes (Alphey 2007). These include:

1. Identification of released insect (sterile insect) by introducing genetic markers to distinguish from the wild target population.
2. Removing the need for radiation sterilization by providing genetic sterilization.
3. Providing stringent sex separation method: genetic sexing for efficient removal of females from the released population.
4. Mitigating the consequences of accidental release of non-irradiated insects from the mass-rearing facility through genetic containment where the insects require an artificially provided condition for survival.
5. Enabling the release of eggs instead of adults.

None of these benefits can be achieved via classical genetics, whereas genetic engineering allows these benefits to be obtained more efficiently within a short period of time and transferred readily to the next generation in contrast to the classical methods.

3.3.1 Genetic Markers

It is essential to be able to monitor the dispersal of mass-reared insects from the release site, the survival in the field and the effectiveness in competing with wild population for mates and to be able to monitor the size of the wild population. These can be facilitated by introducing markers which can distinguish between released individuals and wild population (Alphey 2002). Before the intervention of genetic engineering, the marking of the insects has been done by adding a dye to their food before release or dusting the pupae with a fluorescent powder which can be transferred to the adult upon emergence (Hagler and Jackson 2001). However, insects often groom themselves and wear off the dye over time. Moreover, the powder may be transferred to non-released insects, especially during mating, and can only be used to identify released insect not their progeny (Hagler and Jackson 2001; Alphey 2002). Genetic markers can overcome these issues and prevent ambiguous identification of released mosquitoes in SIT programmes.

Marker genes can be integrated into strains with recombinant DNA technology which result in a new phenotype that would reduce the amount of handling required and human errors (Alphey 2007). Inheritance of a marker gene may be dominant, co dominant, recessive or polygenic (control by more than one gene). Dominant acting genes are generally used for the mosquito SIT programmes. The dominant markers require only one allele for the phenotype to be present (heterozygotes). Therefore, the F1 progeny resulting from the mating of released individual and wild mosquitoes also possess the same phenotype (Hagler and Jackson 2001). Therefore, fluorescent protein genes have been introduced as genetic markers due to the easy discrimination between released and wild-type individuals. Green fluorescent protein (GFP) gene has been used as a biological-marker gene in a wide variety of studies. This gene was isolated from jellyfish *Aequorea victoria* (Prasher et al. 1992). GFP has been used as a genetic marker for gene expression in *Drosophila melanogaster* and various other insects (Plautz et al. 1996). In addition, the use of GFP as a reporter gene in mosquitoes infected with *Sindbis virus* indicated the possibility of using GFP as an effective genetic marker in transgenic mosquitoes (Higgs et al. 1996; Pinkerton et al. 2000). Enhanced green fluorescent protein (EGFP) is a more soluble and red-shifted variant of GFP and more suitable marker for screening transgenic mosquitoes (Higgs and Lewis 2000). Later, different fluorescent proteins have been developed with a wavelength different from that of GFP. These include blue-, cyan- and yellow-shifted mutants of GFP and newly isolated red-fluorescent protein (DsRed) from a coral of the *Discosoma striata* (Heikal et al. 2000; Rodrigues et al. 2001). These marker systems have now been widely used in diverse array of mosquitoes, *Ae. aegypti* (Kokoza et al. 2001; Fu et al. 2010), *Ae. fluviatilis* (Rodrigues et al. 2006), *Ae. albopictus* (Labbé et al. 2012), *An. albimanus* (Perera et al. 2002), *An. gambiae* (Grossman et al. 2001) and *An. stephensi* (Ito et al. 2002; Nolan et al. 2002).

A variety of promoter systems have been used to express the fluorescent marker genes in mosquitoes. This includes either tissue-specific or ubiquitous promoters. The artificial 3xP3 promoter has been designed to express in eye rhodopsin cells of insects (Berghammer et al. 1999). The tissue-specific nature of the promoter helps to distinguish marker expression from background fluorescence. Moreover, the promoters such as Hr5-IE1 (baculovirus IE1 promoter and Hr5 enhancer), which express in all tissues throughout the development, are more visible and easy to screen, especially for the insects whose eye pigment masks the expression of 3xP3-driven markers (Rodems and Friesen 1993). In that context, the expression of markers all over the body is preferable since some body parts may be lost in trapped insects (heads, legs, wings and antenna) (Handler 2002).

3.3.2 Integration of Genetic Sterilization, Genetic Sexing and Genetic Containment: Release of Insects Carrying Dominant Lethal (RIDL)

The advent of insect transgenesis has paved the way for novel approaches in insect control. One possibility is a genetic-based control strategy modelled on the traditional SIT by engineering self-limiting gene, which offers solutions to many drawbacks of traditional SIT with genetic sterilization, genetic sexing and genetic containment while maintaining its environmentally benign and species-specific utility. The release of insects carrying dominant lethal system (RIDL system) is such a control strategy introduced by Oxitec (Abington, UK) (Thomas et al. 2000; Qsim et al. 2017). Using this strategy, the genetic sterility and the genetic sexing can be engineered in insects using repressible dominant sex-specific lethal gene where the permissive condition can be created only in the laboratory but never be encountered by the wild. Then the lethal effect would be suppressed in the mass-rearing facility by a dietary additive like tetracycline. RIDL insects then would be reared and released into the wild without irradiation (Thomas et al. 2000; Alphey et al. 2006). Upon the release, the engineered males, which are homozygous for this RIDL construct, would pass one copy of the dominant lethal to their progeny by Mendelian inheritance. Lacking the antidote, offspring would subsequently die in the wild as same as the irradiated males but without the need for irradiation. This approach would thus mitigate the financial cost, direct and indirect damages to the insect caused by the irradiation (Alphey et al. 2006; Alphey 2007).

Initially, the RIDL was developed in a way which allows the death of all individuals (both male and female); later it has been realized that killing female only is sufficient and more effective than killing both sexes (Schliekelman and Gould 2000). The RIDL can be combined with genetic sexing for female killing. In doing so, insects can be engineered with female-specific RIDL construct which can then be mass-reared in the absence of the antidote to bring about the female-specific lethality through the expression of the lethal gene. This allows the easy and reliable sex separation for the insect control programme, and further, the survived males can pass on their RIDL construct to some of their next generation. Therefore, this strategy is more efficient than conventional SIT (Alphey et al. 2006).

The conventional SIT could unintentionally release unirradiated insects together with mass-reared irradiated insects, whereas the RIDL approach has the potential to provide satisfactory solution known as genetic containment (Alphey 2007). In RIDL, on accidental release, mass-reared insects would often die without the antidote in the environment so that these insects would not survive by their own in the wild. In addition, late-acting RIDL insects can be released at any stage of their life cycle, and eggs can also be released to the environment to widen the range of distribution and to reduce the cost involved compared to adults. Further, these eggs can be stored for several months and easy to distribute. The distribution of eggs to communities and public bodies would allow much greater community involvement than typical SIT programmes (Fu et al. 2010).

3.3.2.1 RIDL Principle

The RIDL systems so far developed for the insects are based on the tetracycline-dependent repressible systems (Heinrich and Scott 2000; Gong et al. 2005; Morrison et al. 2012; Tan et al. 2013). There are two types of tetracycline-repressible systems (Tet off), known as two-component system (bipartite expression system) and one component system (Fig. 3.2). The two-component system contains two sub-constructs known as the driver construct and the effector construct. The driver construct mainly contains tetracycline-repressible transactivator (tTA) under the control of a specific promoter. The effector construct contains the lethal gene/effector placed under the control of a minimal promoter and tTA response elements (TREs) upstream of the promoter (Gossen and Bujard 1992; Gong et al. 2005; Haghighat-Khah et al. 2019) which contain several tetO sequences. The presence of tetracycline represses the lethal system by binding to tTA, followed by a conformational change in the tTA, which prevents the binding of tTA to the TRE site. This

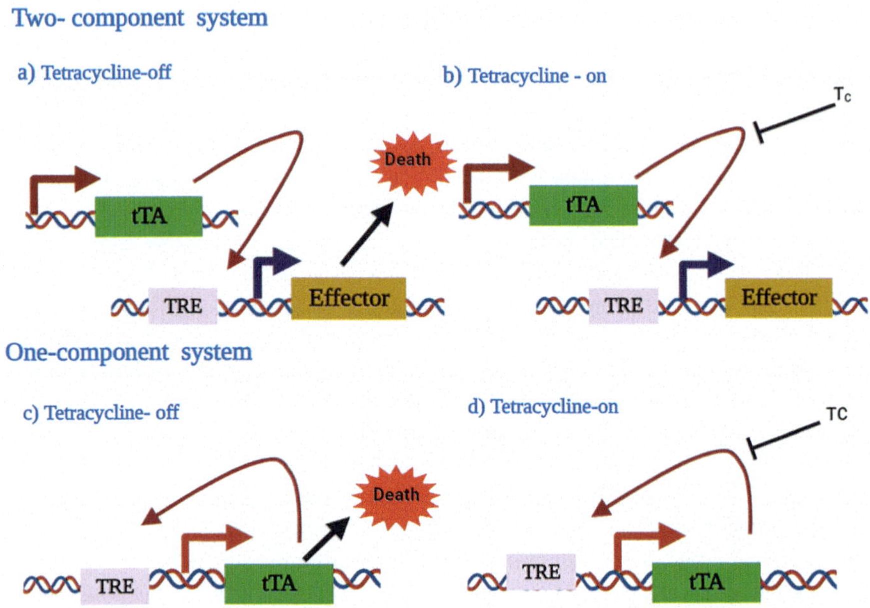

Fig. 3.2 Schematic of tetracycline-repressible lethal systems. (**a**, **b**) Two-component tetracycline-repressible lethal systems; in the absence of tetracycline (T_c), tTA binds TRE and drives the expression of effector gene, leading to death. In the presence of tetracycline, tTA binds tetracycline and changes the conformation of tTA leading to the repression of effector expression to inactivate the lethal system. (**c**, **d**) One-component positive feedback system; in the absence of tetracycline, low levels of basally expressed tTA bind TRE to increase the expression of tTA to cause the death. In the presence of tetracycline, tTA preferentially bind to tetracycline to repress the expression of tTA to prevent the lethality

allows the development of insects normally on a dietary tetracycline supplement for mass rearing at the laboratory condition. In the absence of the tetracycline in the wild, tTA binds to the TRE and drives the expression of an effector gene and leads to death in the case of a lethal effector. The simplified variant of this system is one-component positive feedback circuit. This system consists of a minimal promoter adjacent to TRE sites which drives the expression of tTA. In the absence of tetracycline, low levels of basally expressed tTA bind TRE and drive further expression of tTA in a positive feedback manner. When the expression of tTA becomes high, it acts as a lethal effector molecule and leads to death (Fu et al. 2007).

In one-component systems, tTA act as both the transactivator and the effector. In the presence of tetracycline, the feedback is suppressed, and the expression of tTA remains at the low levels established by the basal promoter. tTA is a fusion of tetracycline-binding domain (tetR) and activating domain of virion protein 16 (VP16) of herpes simplex virus (Gossen and Bujard 1992). Two-component systems rely on the expression of the effector gene to induce lethality (Schetelig et al. 2009, 2016; Schetelig and Handler 2012). Unlike the two-component Tet systems, the exact reason for the lethality of one-component positive feedback systems remains unknown. It has been proposed that the lethality is potentially due to tTA toxicity, change in expression of critical transcripts, transcriptional squelching or interference with ubiquitin-dependent proteolysis pathways due to high protein production (Gong et al. 2005; Alphey and Oxitec Ltd. 2015; Bryk et al. 2017). However, the study performed on positive feedback circuit in *D. melanogaster* has shown that the lethality is not caused by aforementioned mechanisms; instead they proposed the lethality might be due to the results of integration site-based stochastic differential gene expression due to the tissue-specific expression patterns (Bryk et al. 2017).

3.3.2.2 Bi-sex and Female-Specific RIDL Systems

A prototype of highly efficient two-component tetracycline-dependent system was first demonstrated for *D. melanogaster* using tTA. In this study, it has shown successful usage of two systems, known as bi-sex RIDL (non-female-specific) and female-specific RIDL (fsRIDL) (Thomas et al. 2000). In RIDL control programme, the released insects are homozygous for the RIDL construct, and they transfer one copy of the dominant lethal system to their offspring after mating with wild counterpart. In the bi-sex RIDL procedure, all offspring would die in the field due to the expression of the effector gene in the absence of tetracycline on the contrary to the female-specific RIDL approach which targets only the females. In another study to achieve the female specificity of the *D. melanogaster*, the female-specific fat body enhancer yolk protein 1 (*YP1*) or yolk polypeptide 3 (*YP3*) has been used to drive the expression of lethal gene hid, Ras64B^{V12} and female-specific lethal gene, msl-2^{NOPU} (Heinrich and Scott 2000; Thomas et al. 2000).

These two systems and their outcomes were later used for more economically important insects. In doing so, bi-sex positive feedback RIDL system (one component) was constructed for *C. capitata* and achieved the lethality in the early developmental stages of heterozygous progeny (Gong et al. 2005). Later, the first transgenic embryonic lethality system was developed based on two-component tetracycline-repressible lethal system for *C. capitata*, showing complete reproductive sterility (Schetelig et al. 2009). Furthermore, the bi-sex RIDL system has been used to achieve repressible lethality to control another agriculturally important lepidopteran pest, *Pectinophora gossypiella* (pink bollworm) (Morrison et al. 2012). Moreover, sex-specific alternative splicing has been used to engineer insects such as *C. capitata* with female-specific autocidal genetic systems. In this system, the insertion of the cassette exon from the *C. capitata* transformer gene to disrupt the tTA transcript during male splicing but not in the female, in turn, induces the same female-specific lethal phenotype (Fu et al. 2007). Later, the sex-specific alternative splicing was used to develop female-specific lethality in *Bombyx mori* (silkworm), and more recently, genetically engineered *Plutella xylostella* (diamondback moth) were released to the field as the first released RIDL agricultural pest (Shelton et al. 2020).

3.3.2.3 Application of RIDL for Mosquito Control

The RIDL-based genetic systems have been developed for several species of mosquitoes including *Ae. aegypti*, *Ae. albopictus* and *An. stephensi.* The development of late-acting RIDL strain of *Ae. aegypti* was first reported in 2007 (Phuc et al. 2007). This study has recovered four transgenic lines of *Ae. aegypti* carrying tetracycline-repressible, lethal positive feedback systems to evaluate the tetracycline-repressible lethality. Of these four lines, three have shown highly penetrant (95–100%) lethality which has been caused at early larval stage (L1–L2) for two lines, and late-acting lethality has shown for one line (LA513A) during the transition from late larvae to early pupae (Phuc et al. 2007). Later, Oxitec developed the latter line to produce OX513A RIDL bi-sex strain, harbouring a genetic construct, LA513A, which is the most successful strain developed to date and currently available for field use. This construct mainly consists of a marker gene, *DsRed2*, and the variant of *tTA* gene (*tTAV*), TRE comprising multiple copies of tetO sequence to which tTAV binds, together with a minimal promoter from *Drosophila* heat-shock protein (Hsp70) and the 3′ untranslated region (UTR) sequence of the gene *Dmelfs(1)K10* of *Drosophila* (Fig. 3.3). Mosquitoes harbouring the RIDL construct can be readily detected under fluorescent light due to the expression of DsRed2.

When the OX513A males are released into the wild and they mate with wild-type females, their offspring inherit the lethality trait followed by late larval death in the absence of tetracycline in their diet. Prior to release of OX513A strain into the field, a study had been carried out to compare life history characteristics such as mortality, developmental rate, adult size and longevity with respect to its wild counterparts under the controlled laboratory conditions and found lower larval survival and reduced adult longevity in OX513A strain. The outcomes of this study suggested that the performance issues associated with the RIDL line can generally be alleviated using optimized rearing methods and laboratory-based cage trials as well as field trials (Bargielowski et al. 2011). Later, several open-field trials have been performed to solve critical issues concerning whether genetic modification itself and/or rearing the mosquito in laboratory condition might compromise the mating competitiveness and penetrance of the lethal phenotype. The first open-field trial demonstrated that

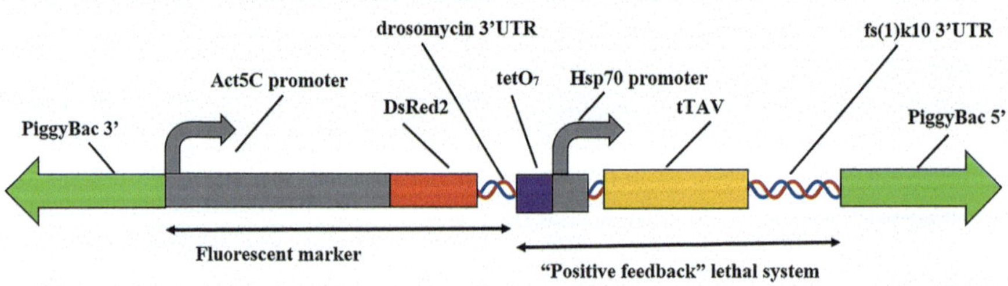

Fig. 3.3 Schematic representation of transposon LA513. LA513 is an 8.4 kb non-autonomous piggyBac-based transposon. Act5C promoter drives the expression of the marker gene DsRed2. tTAV is controlled by its binding site tetO, the minimal promoter of *Drosophila* Hsp70 and a 3′ UTR sequence from *Drosophila* fs(1)K10

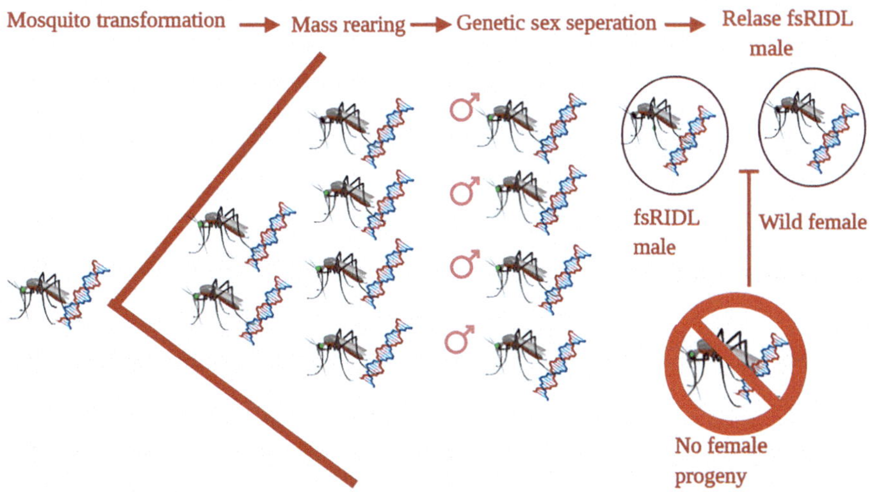

Fig. 3.4 fsRIDL mosquito control strategy: mosquito is developed by the transformation of RIDL gene construct and inserting marker gene into the insect genome followed by the mass rearing. Genetic sexing allows the male-only release to mate with wild female resulting in no female progeny

genetically engineered OX513A mosquitoes released across 10 ha for a 4-week period at the selected field site on Grand Cayman—the largest of the Cayman Islands, a British overseas territory in the Caribbean—mated successfully with wild females and fertilized their eggs. The results suggested that OX513A males can compete well with wild male in the field. The penetrance of the lethal phenotype of the transgene in F1 hybrid between OX513A homozygous male and wild female was tested, and 96.5% mortality has been observed in another preliminary study (Harris et al. 2011). Furthermore, sustained releases of OX513A *Ae. aegypti* males led to 80% relative reduction of the target wild population in the Cayman (Harris et al. 2012). Sustained series of field releases of OX513A males in Brazil have been resulted in 95% suppression of local *Ae. aegypti* population, based on adult trapping data, and 81% suppression based on egg trapping data over the course of 1 year suggesting the feasibility of this product to the control of field population of *Ae. aegypti* (Carvalho et al. 2015).

Even though OX513A bi-sex RIDL *Ae. aegypti* strain was highly effective in controlling the target mosquito population, it is still lacking an efficient genetic system to eliminate disease-transmitting female mosquitoes. Therefore, attempts were made to develop a fsRIDL system that would be highly effective for mosquito control than that of bi-sex strain. Since the latter approach could provide a genetic sexing mechanism (Fig. 3.4), it could be used more efficiently to control *Ae. aegypti* with male-only release approach. Moreover, this approach is highly desirable for the insects for which the reliable physical sex separation methods are not available and avoid the labour-intensive steps required in sex separation. Recently, an attempt was made to develop genetic sexing strain (GSS) of *Ae. aegypti* engineered to have a

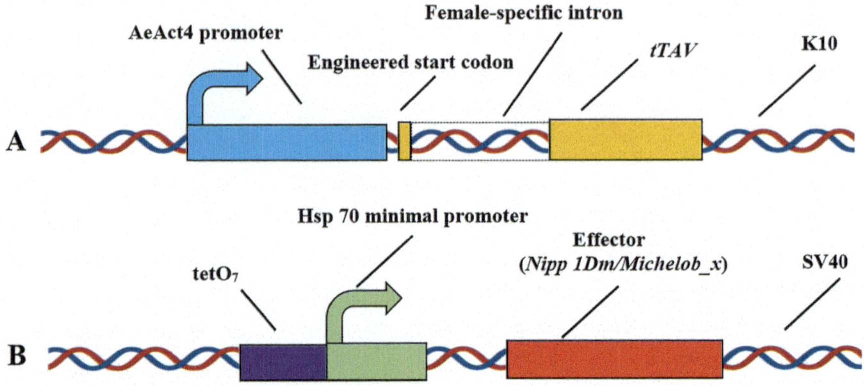

Fig. 3.5 Schematic representation of a two-component fsRIDL system. (**a**) Basic elements of the driver construct containing AeAct4 promoter and an extra engineered start codon (ATG) at 5′-end leading to the alternatively spliced intron to be within the tTAV open reading frame (ORF). (**b**) Basic elements of the effector construct containing TREs (heptamerized tetO sequences), hsp 70 minimal promoter and effector gene (*Nipp1 DM* or *Michelob-x*). 5′ and 3′ PiggyBac ends have not been included

repressible female-specific flightless phenotype using either single transgene (positive feedback system) or two separate transgenes (two-component system) based on the promoter sequences of *Ae. aegypti* Actin-4 gene (*AeAct-4*), specific to the indirect flight muscle (IFM) of females with expression starting at L4 larvae (Fu et al. 2010).

Thereafter, a mosquito line having an *AeAct-4 –tTA* driver construct has been crossed with two lethal effector components containing *Nipp1Dm* and *micelob_x* under control of the TRE and has shown that 69.8–98.3% of females were to be flightless phenotype (Fig. 3.5). Further, some flightless phenotypes have been observed even without the presence of an effector gene suggesting its suitability of developing as positive feedback RIDL system. The leakiness of the female specificity of the *AeAct-4* promoter has shown to be an issue in developing the mosquito lines as an effective mosquito control method. Since the analysis of the *AeAct-4* gene has shown the low abundance of transcripts in males, the strength of the female specificity of the promoter sequence was evaluated by crossing *AeAct-4 –tTA* diver construct with a mosquito line carrying reporter gene DsRed, driven under the control of TRE together with a minimal promoter (TRE-DsRed2). Progeny of these two lines had a strong red fluorescence in developing IFM tissue of late L4 female larvae, pupae and adult female mosquito in the absence of tetracycline. However, weak expression of DsRed in other female-specific tissues in a few lines of female and in one line of male has been observed possibly due to the position effect as the extent of expression of the transgene is sensitive to its chromosomal context (Fu et al. 2010). The female specificity has been therefore refined further by exploiting the sex-specific alternative splicing nature of the *AeAct-4* gene. The driver component was modified with a start codon (ATG) at the 5′-end of the male-specific exon, and the strain was designated as OX3604C. Then in the female-specific variant, ATG was in frame with tTA leading to functional protein while it interposes a frameshift with multiple stop codons in the male-specific transcript. Further, OX3604C was observed to have strong DsRed expression exclusively in the IFM tissues of female pupae and adults as well (Fu et al. 2010). Later, this approach was used in developing an ideal combination of female-specific late-acting RIDL system to make flightless phenotype rather than lethality. However, in the field, inability to fly is likely to be lethal since adult mosquitoes are unable to escape from the water upon emergence, unable to access the food resource and escape from predation as well as unable to mate and fly to locate blood meal (Fu et al. 2010). Another important feature was that the lethal effect or the flightless phenotype exhibits only at the adult stage. Modelling has shown if the lethal phenotype occurs after density-dependent phase, mosquito suppression can be significantly enhanced since the mosquito population dynamics are regulated by negative density dependence (Phuc et al. 2007).

In a trial study carried out by introducing the fsRIDL strain, OX3604C males into small and large laboratory cage showed the successful elimination of the targeted population (De Valdez et al. 2011). However, such effectiveness was not observed as anticipated when tested in large outdoor field cages perhaps due to fitness costs associated with mating competitiveness (Facchinelli et al. 2013). Later, Oxitec was successful in introducing a new second-generation self-limiting *Ae. aegypti* fsRIDL system called OX5034, based on the improvement of first-generation *Ae. aegypti* OX513A with an additional performance and operational benefits including genetic sexing, greater cost-effectiveness and potentially higher rates of mosquito control. A recent study which dealt with releasing of OX5034 fsRIDL mosquitoes has shown 96% suppression of wild *Ae. aegypti* populations in the city of Indaiatuba, Brazil, with disappearance expected of OX5034 males from the environment within ten generations after releasing was stopped (Oxitec 2019).

It was later shown the same technology works in a related species such as *Ae. albopictus* (Labbé et al. 2012). In a study, a segment of *Ae. albopictus* Actin-4 (*AealbAct-4*) gene characterized and compared with *Ae. aegypti* homologue has shown sequence, and the functional similarities to *AeAct-4* and hence both *Ae. aegypti* and *Ae. albopictus* were engineered with *AealbAct-4* promoter to drive a dominant lethal gene in the IFM, and this resulted in a female-specific flightless phenotype. Although this was the first report of engineering *Ae. albopictus*, it has not been tried out yet for population suppression requiring assessment of its suitability in suppressing the wild population ideally by creating homozygous *Ae. albopictus* strains.

The technological advances achieved with regard to *Aedes* mosquitoes were later adapted to the human malaria vector, *An. stephensi* to control the transmission of malaria parasites. Towards this goal, *An. stephensi* was engineered to confer repressible female-specific phenotype using similar driver construct, *AeAct-4 –tTA*, and effector construct (*teto-Nipp1Dm*), previously used for the genetic transformation of *Ae. aegypti* (Fu et al. 2010; Marinotti et al. 2013).

Further experiments by replacing the *AeAct-4* 5′-end sequence with the orthologous sequence from *An. stephensi* Actin-4 (AsAct-4) displayed promising results necessitating the further development of flightless female-specific phenotype for application in malaria control programmes (Marinotti et al. 2013).

Even though a variety of repressible lethal systems have been tested in mosquitoes using the Tet-off, two-component expression systems or one-component positive feedback systems, experimented systems so far have entirely relied on the use of cell-autonomous effectors of which the action is restricted only to the tissue in which they are expressed (Heinrich and Scott 2000; Phuc et al. 2007; Fu et al. 2010). Therefore, there is a significant limitation in developing more complex and flexible lethal systems as the cell type expressing effector must simultaneously be sensitive to its lethal effect, preventing the use of either useful effector or useful regulatory sequences. Dissociation of the temporal and spatial expression pattern of the effector can potentially overcome this limitation. A more recent study conducted using non-cell-autonomous system capable of dissociating the temporal and spatial expression pattern of the effector in *Ae. aegypti* was designed with a novel synthetic effector to be secreted out from the adult fat body following a blood meal by a female mosquito and to affect the functioning of motor neurons to cause paralysis (Haghighat-Khah et al. 2019). Even though the penetrance of the phenotype was not as high as expected, the outcomes of the study demonstrated the feasibility of using non-cell-autonomous effectors in developing more flexible effector constructs for repressible lethal systems.

3.4 Fitness Cost of Mosquito Transgenesis

Genetic modification is receiving increasing attention in controlling the target mosquito population. The success of this strategy often depends on how genetic modification affects the fitness of mosquito. Transgenic mosquito with minimal fitness load is a prerequisite for the success of the control strategy, and the realistic estimates of transgene fitness costs are essential for planning and subsequent implementation of control strategies (Marrelli et al. 2006). The fitness cost of genetically modified mosquitoes (Catteruccia et al. 2003; Marrelli et al. 2006) depends on:

1. The expression exogenous gene and the negative effect of the protein product
2. Mutations that may cause during the integration of the transgene
3. Hitch-hiking effect and Inbreeding involved in the establishment of the homozygous line
4. Position effect

The burden of the transgene products is a major factor affecting the fitness of the transformed insects. Transgenic insects typically express multiple transgenes, for example, the marker gene and the effector gene. In addition to these genes, the RIDL-based transgenic mosquitoes carry a repressible transactivator protein (tTA) which together with other transgene products can impose huge fitness cost on the transgenic mosquitoes (Irvin et al. 2004; Fu et al. 2007; Phuc et al. 2007; Labbé et al. 2012). Expression of foreign proteins more often becomes toxic to the cells in which they are expressed. This toxicological effect is more of a concern with the RIDL where the system has been designed to selectively kill female mosquitoes and, therefore, it is essential to avoid the possible harm to the non-targeted population, i.e. males. Selection of a female-specific strong promoter with low leaky basal expression and a suitable effector protein is of paramount importance in this regard. The toxic effect of the transgenes can be partly reduced by engineering an effector gene to be expressed in tissue-specific and stage-specific manner by selecting female tissue-specific and stage-specific promoters. For example, expression of tTA effector gene in IFM of females has been regulated by Actin-4 gene, and 3XP3 promoter sequence has been used in eye-specific fluorescence marker expression in mosquitoes (Fu et al. 2010; Labbé et al. 2012), and this fluorescence marker expression in the eye is less likely to have an impact on fitness compared to the markers expressed in all over the body (Marrelli et al. 2006). Moreover, the toxicity can be varied based on the places where it is expressed. In some cases, the accumulation of foreign proteins in cytoplasm will increase the concentration than secreted proteins which may have a greater impact on fitness (Marrelli et al. 2006).

Insertion of the transgene may sometime disrupt the transcriptionally active gene which in turn can prevent the expression of a functional gene. This phenomenon is known as insertional mutagenesis. However, more often, mutagenesis is recessive and would not be detected in heterozygotes. Most often, insertions do not affect fitness due to the fact that they either integrate into a region that does not encode functional gene or do not significantly disrupt the native gene function (Lyman et al. 1996; Marrelli et al. 2006). Apart from the negative effect of transgene products and the insertional mutagenesis, hitch-hiking effect is another important factor that influences the fitness but affects only at the homozygous state of the transgenic mosquitoes. Many organisms often carry recessive mutations that decrease fitness. Since the transposition of the transgene is random, the transgene may integrate with a certain probability in the vicinity of such recessive mutations, and when the transgene is made homozygous via inbreeding, any nearby recessive mutation also becomes homozygous. This phenomenon is known as hitch-hiking effect. Fixation of such recessive mutation will cause inbreeding depression (Fig. 3.6).

Finally, the strength of the transgene expression can often be influenced by the chromatin surrounding the site of insertion which is known as the position effect. Transgenic mosquitoes generated by transposable elements often poorly express transgenes due to these position effects. Chromatin domain insulators are one potential strategy to overcome this issue. Another approach exploits the site-directed recombination system of bacteriophage *ΦC31* to avoid these effects (Franz et al. 2011).

Several studies have evaluated the impact of the transgene on the fitness of transgenic mosquitoes. Of those, one study has investigated the factors influencing fitness in a cage experiment with four lines of transgenic *An. stephensi* and found the sharp reduction of transgene allele frequency in all four lines until extinction. Furthermore, the analysis of transgene integration sites of two lines (having same

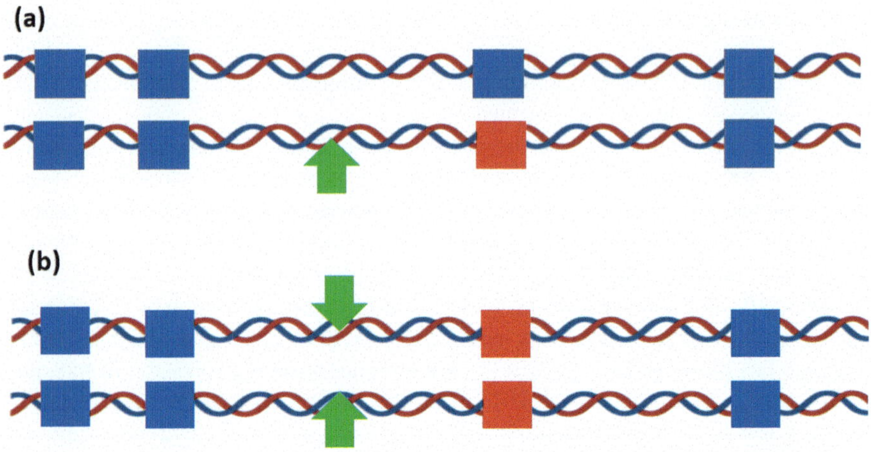

Fig. 3.6 The hitch-hiking effect: (**a**) genomic region hemizygous for the transgene insert (green arrow) and (**b**) genomic region homozygous for the transgene insert. Red boxes represent the recessive mutation. Blue boxes represent genes with no fitness load. Since the transposition of transgene is random, the transgene may integrate with a certain probability in the vicinity of the recessive mutation, and when the transgene is made homozygous, any nearby recessive gene also becomes homozygous

transposon construct) revealed that the transposon had disrupted a coding sequence of a non-essential gene in one of the lines which showed significant reduction in fitness compared to the line in which the insertion site is different and that did not disrupt an open reading frame (Catteruccia et al. 2003). Another study which examined the impact of transgenesis on the reproductive and development fitness of transgenic *Ae. aegypti* has revealed a considerable reduction in survivorship, fecundity and adult longevity in transgenic mosquitoes relative to the untransformed laboratory strain (Irvin et al. 2004). Since the aforementioned studies have maintained the transgenic lines as homozygotes, reduction in fitness for homozygous mosquitoes can be attributed to the consequences of (1) either negative effect of transgene product or insertional mutagenesis during transposition and (2) hitch-hiking effect during the establishment of homozygous lines. Since two possibilities cannot be distinguished from each other, it is often accepted to use heterozygous lines to assess the fitness cost in transgenesis (Marrelli et al. 2006). Moreover, the cost will require inundating release of mosquitoes in controlling mosquitoes in case of SIT or RIDL control programmes or employment of stronger genetic drive mechanisms when the aim is to fix pathogen refectory transgene in a natural vector population (Catteruccia et al. 2003).

3.5 Conclusion

The control of vector population is one of the effective ways to reduce transmission of many mosquito-borne diseases in spite of traditional mosquito control methods that are currently in place for this purpose seems to be ineffective. Although SIT was a proven cost effective strategy for elimination or eradication of most agricultural pests, the use of this technique for mosquito control is yet to be fully assessed due to the absence of efficient sexing system, high fitness cost and operational difficulty of radiation used for ionization, density-dependent nature of the target mosquito population and various other technical issues associated with this technique. With the possibility of genetic enhancements for the SIT, this can effectively be used for mosquito population control with genetic sterilization, genetic sexing, genetic containment and provision of genetic markers while maintaining its environmentally benign and species-specific utility. With various improvements and modifications to minimize the fitness load, RIDL strategy is now applicable for a wide range of mosquitoes with field-testing possibilities. Together with solid epidemiological evidence and community support, widespread implementation of these novel RIDL-based approaches can reverse the current alarming global mosquito vector-borne diseases.

Acknowledgement We acknowledge the Accelerating Higher Education Expansion and Development (AHEAD) operation for financial support to carry out research studies on high-tech approaches to control dengue by genetic engineering mosquito vector *Ae. aegypti.*

References

Alphey L (2002) Re-engineering the sterile insect technique. Insect Biochem Mol Biol 32 (10):1243–1247

Alphey LS (2007) Engineering insects for the sterile insect technique. In: Area-wide control of insect pests: from research to field implementation. Springer, Dordrecht, pp 51–60

Alphey L, Oxitec Ltd. (2015) Expression system for insect pest control. US patent no. 9121036

Alphey L, Baker P, Burton RS et al (2006) Genetic technologies to enhance the sterile insect technique (SIT). In: Fruit flies of economic importance: from basic to applied knowledge. Proc. 7th intl. symp. fruit flies of economic importance, pp 10–15

Alphey L, Benedict M, Bellini R et al (2010) Sterile insect methods for control of mosquito-borne diseases: an analysis. Vector-Borne Zoonotic Dis 10(3):295–311

Axtell RC, Arends JJ (1990) Ecology and management of arthropod pests of poultry. Annu Rev Entomol 35(1):101–126

Balestrino F, Puggioli A, Carrieri M et al (2017) Quality control methods for Aedes albopictus sterile male production. PLoS Negl Trop Dis 11(9):e0005881

Bargielowski I, Nimmo D, Alphey L, Koella JC et al (2011) Comparison of life history characteristics of the genetically modified OX513A line and a wild type strain of *Aedes aegypti*. PLoS One 6(6):e20699

Benedict MQ, Robinson AS (2003) The first releases of transgenic mosquitoes: an argument for the sterile insect technique. Trends Parasitol 19(8):349–355

Berghammer AJ, Klingler M, Wimmer EA (1999) A universal marker for transgenic insects. Nature 402(6760):370–371

Breeland SG, Jeffery GM, Lofgren CS, Weidhaas DE (1974) Release of chemosterilized males for the control of Anopheles albimanus in El Salvador. I. Characteristics of the test site and the natural population. Am J Trop Med Hyg 23(2):74–281

Bruno Wilke AB et al (2009) Mini-review: genetic enhancements to the sterile insect technique to control mosquito population. Asia Pac J Mol Biol Biotechnol 17(3):65–74

Bryk J, Reeves RG, Reed FA, Denton JA (2017) Transcriptional effects of a positive feedback circuit in *Drosophila melanogaster*. BMC Genomics 18(1):1–14

Bushland RC, Lindquist AW, Knipling EF (1955) Eradication of screw-worms through. Release of sterilized males. Science 122:287–288

Carvalho DO, McKemey AR, Garziera L (2015) Suppression of a field population of *Aedes aegypti* in Brazil by sustained release of transgenic male mosquitoes. PLoS Negl Trop Dis 9(7): e0003864

Catteruccia F, Godfray HC, Crisanti A (2003) Impact of genetic manipulation on the fitness of *Anopheles stephensi* mosquitoes. Science 299(5610):1225–1227

Celli A (1901a) Epidemiologia della malaria secondo le recenti vedute biologiche. Annali dell'Istituto di Igiene Sperimentale 11:44–95

Celli A (1901b) Sulla nuova profilassi della malaria. Annali dell'Istituto di Igiene Sperimentale 11:97–141

Chernin E (1983) Sir Patrick Manson's studies on the transmission and biology of filariasis. J Infect Dis 5(1):148–166

Coquerel C (1858) Note sur des larves appartenant à une espece nouvelle de diptere, (*Lucilia hominivorax*) développée dans les sinus frontaux de l'homme à Cayenne. Annls Soc enl Fr 6 (3):171–176

Cuervo-Parra JA, Cortés TR, Ramirez-Lepe M (2016) Mosquito-borne diseases, pesticides used for mosquito control and development of resistance to insecticides. In: Trdan S (ed) Insecticides resistance. IntechOpen, Rijeka, pp 111–134

Curtis CF (2005) Review of previous applications of genetics to vector control. Frontis, 33–43

Dame DA, Lofgren CS, Ford HR, Boston MD, Baldwin KF, Jeffery GM (1974) Release of chemosterilized males for the control of Anopheles albimanus in El Salvador. II. Methods of rearing, sterilization, and distribution. Am J Trop Med Hyg 23(2):282–287

De Valdez MR, Nimmo D, Betz J, Gong HF, James AA, Alphey L et al (2011) Genetic elimination of dengue vector mosquitoes. Proc Natl Acad Sci 108(12):4772–4775

Facchinelli L, Valerio L, Ramsey JM, Gould F, Walsh RK, Bond G et al (2013) Field cage studies and progressive evaluation of genetically-engineered mosquitoes. PLoS Negl Trop Dis 7(1): e2001

Ferroni E, Jefferson T, Gachelin G (2012) Angelo Celli and research on the prevention of malaria in Italy a century ago. J R Soc Med 105(1):35–40

Franz AW, Jasinskiene N, Sanchez-Vargas I, Isaacs AT, Smith MR, Khoo CC et al (2011) Comparison of transgene expression in *Aedes aegypti* generated by mariner Mos1 transposition and φc31 site-directed recombination. Insect Mol Biol 20(5):587–598

Fu G, Condon KC, Epton MJ, Gong P, Jin L, Condon GC et al (2007) Female-specific insect lethality engineered using alternative splicing. Nat Biotechnol 25(3):353–357

Fu G, Lees RS, Nimmo D, Aw D, Jin L, Gray P (2010) Female-specific flightless phenotype for mosquito control. Proc Natl Acad Sci U S A 107(10):4550–4554

Gong P, Epton MJ, Fu G, Scaife S, Hiscox A, Condon KC et al (2005) A dominant lethal genetic system for autocidal control of the Mediterranean fruitfly. Nat Biotechnol 23(4):453–456

Gossen M, Bujard H (1992) Tight control of gene expression in mammalian cells by tetracycline-responsive promoters. Proc Natl Acad Sci U S A 89(12):5547–5551

Grossman GL, Rafferty CS, Clayton JR, Stevens TK, Mukabayire O, Benedict MQ (2001) Germline transformation of the malaria vector, *Anopheles gambiae*, with the piggyBac transposable element. Insect Mol Biol 10(6):597–604

Haghighat-Khah RE, Harvey-Samuel T, Basu S, StJohn O, Scaife S, Verkuijl S et al (2019) Engineered action at a distance: blood-meal-inducible paralysis in *Aedes aegypti*. PLoS Negl Trop Dis 13(9):e0007579

Hagler JR, Jackson CG (2001) Methods for marking insects: current techniques and future prospects. Annu Rev Entomol 46(1):511–543

Handler AM (2002) Prospects for using genetic transformation for improved SIT and new biocontrol methods. Genetica 116(1):137–149

Harris AF, Nimmo D, McKemey AR, Kelly N, Scaife S, Donnelly CA et al (2011) Field performance of engineered male mosquitoes. Nat Biotechnol 29(11):1034–1037

Harris AF, McKemey AR, Nimmo D, Curtis Z, Black I, Morgan SA et al (2012) Successful suppression of a field mosquito population by sustained release of engineered male mosquitoes. Nat Biotechnol 30(9):9–11

Heikal AA, Hess ST, Baird GS, Tsien RY, Webb WW (2000) Molecular spectroscopy and dynamics of intrinsically fluorescent proteins: coral red (dsRed) and yellow (citrine). Proc Natl Acad Sci U S A 97(22):11996–12001

Heinrich JC, Scott MJ (2000) A repressible female-specific lethal genetic system for making transgenic insect strains suitable for a sterile-release program. Proc Natl Acad Sci 97 (15):8229–8232

Helinski MEH, Parke AG, Knols BG (2009) Radiation biology of mosquitoes. Malar J 8(S2):S6

Hendrichs J, Robinson AS, Cayol JP, Enkerlin W (2002) Medfly areawide sterile insect technique programmes for prevention, suppression or eradication: the importance of mating behavior studies. Fla Entomol 85(1):1–13

Higgs S, Lewis DL (2000) Green fluorescent protein (GFP) as a marker for transgenic insects. In: Insect transgenesis. CRC Press, Boca Raton, FL, pp 101–120

Higgs S, Traul D, Davis BS, Kamrud KI, Wilcox CL, Beaty BJ (1996) Green fluorescent protein expressed in living mosquitoes—without the requirement of transformation. BioTechniques 21 (4):660–664

Irvin N, Hoddle MS, O'Brochta DA, Carey B, Atkinson PW (2004) Assessing fitness costs for transgenic Aedes aegypti expressing the GFP marker and transposase genes. Proc Natl Acad Sci 101(3):891–896

Ito J, Ghosh A, Moreira LA, Wimmer EA, Jacobs-Lorena M (2002) Transgenic anopheline mosquitoes impaired in transmission of a malaria parasite. Nature 417(6887):452–455

Juan-Blasco M, Sabater-Munoz B, Pla I, Argilés R, Castañera P, Jacas JA et al (2014) Estimating SIT-driven population reduction in the Mediterranean fruit fly, *Ceratitis capitata* from sterile mating. Bull Entomol Res 104(2):233–242

Kittayapong P, Kaeothaisong NO, Ninphanomchai S, Limohpasmanee W (2018) Combined sterile insect technique and incompatible insect technique: sex separation and quality of sterile *Aedes aegypti* male mosquitoes released in a pilot population suppression trial in Thailand. Parasites Vectors 11(2):73–83

Kittayapong P, Ninphanomchai S, Limohpasmanee W, Chansang C, Chansang U, Mongkalangoon P (2019) Combined sterile insect technique and incompatible insect technique: the first proof-of-concept to suppress Aedes aegypti vector populations in semi-rural settings in Thailand. PLoS Negl Trop Dis 13(10):e0007771

Knipling EF (1979) The basic principles of insect population suppression and management, Agriculture Handbook 512. US Dept. of Agriculture, Washington, DC

Kokoza V, Ahmed A, Wimmer EA, Raikhel AS (2001) Efficient transformation of the yellow fever mosquito *Aedes aegypti* using the piggyBac transposable element vector pBac[3xP3-EGFP afm]. Insect Biochem Mol 31(12):1137–1143

Labbé GM, Scaife S, Morgan SA, Curtis ZH, Alphey L (2012) Female-specific flightless (fsRIDL) phenotype for control of *Aedes albopictus*. PLoS Negl Trop Dis 6(7):e1724

LaBrecque GC, Bowman MC, Patterson RS, Seawright JA (1972) Persistence of thiotepa and tepa in pupae and adults of Culex pipiens fatigans Wiedemann. Bull World Health Organ 47 (5):675–676. http://www.ncbi.nlm.nih.gov/pubmed/4121670. Accessed 10 Apr 2020

Laven H (1967) Eradication of Culex pipiens fatigans through cytoplasmic incompatibility. Nature 216(5113):383–384

Le Prince JAA (1916) Mosquito control in Panama. The eradication of malaria and yellow fever in Cuba and Panama. JAMA LXVI:1652

Lyman RF, Lawrence F, Nuzhdin SV, Mackay TF (1996) Effects of single P-element insertions on bristle number and viability in *Drosophila melanogaster*. Genetics 143(1):277–292

Marchi A, Munstermann LE (1987) The mosquitoes of Sardinia: species records 35 years after the malaria eradication campaign. Med Vet Entomol 1(1):89–96

Marinotti O, Jasinskiene N, Fazekas A, Scaife S, Fu G, Mattingly ST (2013) Development of a population suppression strain of the human malaria vector mosquito, Anopheles stephensi. Malar J 12(1):142

Marrelli MT, Moreira CK, Kelly D, Alphey L, Jacobs-Lorena M (2006) Mosquito transgenesis: what is the fitness cost? Trends Parasitol 22(5):197–202

Morrison NI, Simmons GS, Fu G, O'Connell S, Walker AS, Dafa'alla T et al (2012) Engineered repressible lethality for controlling the pink bollworm, a lepidopteran pest of cotton. PLoS One 7(12):e50922

Nolan T, Bower TM, Brown AE, Crisanti A, Catteruccia F (2002) piggyBac-mediated germline transformation of the malaria mosquito *Anopheles stephensi* using the red fluorescent protein dsRED as a selectable marker. J Biol Chem 277(11):8759–8762

Oxitec (2019) Oxitec successfully completes first field deployment of 2nd generation friendly *Aedes aegypti* technology. Oxitec. Archived from the original on 13 Aug 2019. https://www.prnewswire.co.uk/news-releases/oxitec-successfully-completes-first-field-deployment-of-2nd-generation-friendly-tm-aedes-aegypti-technology-829366749.html. Accessed 29 Apr 2020

Perera OP, Harrell RA, Handler AM (2002) Germ-line transformation of the South American malaria vector, Anopheles albimanus, with a piggyBac/EGFP transposon vector is routine and highly efficient. Insect Mol Biol 11(4):291–297

Phuc HK, Andreasen MH, Burton RS, Vass C, Epton MJ, Pape G et al (2007) Late-acting dominant lethal genetic systems and mosquito control. BMC Biol 5(1):11

Pinkerton AC, Michel K, O'Brochta DA, Atkinson PW (2000) Green fluorescent protein as a genetic marker in transgenic *Aedes aegypti*. Insect Mol Biol 9(1):1–10

Plautz JD, Day RN, Dailey GM, Welsh SB, Hall JC, Halpain S et al (1996) Green fluorescent protein and its derivatives as versatile markers for gene expression in living *Drosophila melanogaster*, plant and mammalian cells. Gene 173(1):83–87

Prasher DC, Eckenrode VK, Ward WW, Prendergast FG, Cormier MJ (1992) Primary structure of the Aequorea victoria green-fluorescent protein. Gene 111(2):229–233

Qsim M, Ashfaq UA, Yousaf MZ, Masoud M, Rasul I, Noor N et al (2017) Genetically modified *Aedes aegypti* to control dengue: a review. Crit Rev Eukaryot Gene Expr 27:331–340

Rodems SM, Friesen PD (1993) The hr5 transcriptional enhancer stimulates early expression from the *Autographa californica* nuclear polyhedrosis virus genome but is not required for virus replication. J Virol 67(10):5776–5785

Rodrigues F, van Hemert M, Steensma HY, Côrte-Real M, Leão C (2001) Red fluorescent protein (DsRed) as a reporter in Saccharomyces cerevisiae. J Bacteriol 183(12):3791–3794

Rodrigues FG, Oliveira SB, Rocha BC, Moreira LA (2006) Germline transformation of *Aedes fluviatilis* (Diptera:Culicidae) with the piggyBac transposable element. Mem Inst Oswaldo Cruz 101(7):755–757

Ross R (1897) On some peculiar pigmented cells found in two mosquitos fed on malarial blood. Br Med J 2(1929):1786

Schetelig MF, Handler AM (2012) Strategy for enhanced transgenic strain development for embryonic conditional lethality in Anastrepha suspensa. Proc Natl Acad Sci 109 (24):9348–9353

Schetelig MF, Caceres C, Zacharopoulou A, Franz G, Wimmer EA (2009) Conditional embryonic lethality to improve the sterile insect technique in Ceratitis capitata (Diptera: Tephritidae). BMC Biol 7(1):4

Schetelig MF, Targovska A, Meza JS, Bourtzis K, Handler AM (2016) Tetracycline-suppressible female lethality and sterility in the Mexican fruit fly, *Anastrepha ludens*. Insect Mol Biol 25 (4):500–508

Schliekelman P, Gould F (2000) Pest control by the introduction of a conditional lethal trait on multiple loci: potential, limitations, and optimal strategies. J Econ Entomol 93(6):1543–1565

Shelton AM, Long SJ, Walker AS, Bolton M, Collins HL, Revuelta L (2020) First field release of a genetically engineered, self-limiting agricultural pest insect: evaluating its potential for future crop protection. Front Bioeng Biotechnol 29(7):48215

Tan A, Fu G, Jin L, Guo Q, Li Z, Niu B et al (2013) Transgene-based, female-specific lethality system for genetic sexing of the silkworm, Bombyx mori. Proc Natl Acad Sci 110 (17):6766–6770

Thomas DD, Donnelly CA, Wood RJ, Alphey LS (2000) Insect population control using a dominant, repressible, lethal genetic system. Science 287(5462):2474–2476

Thomé RCA, Yang HM, Esteva L (2010) Optimal control of *Aedes aegypti* mosquitoes by the sterile insect technique and insecticide. Math Biosci 223(1):12–23

Vreysen MJ, Saleh KM, Ali MY, Abdulla AM, Zhu ZR, Juma KG (2000) Glossina austeni (Diptera: Glossinidae) eradicated on the island of Unguja, Zanzibar, using the sterile insect technique. J Econ Entomol 93(1):123–135

Walton HJ (1922) The prevention of malaria in the federated Malay states: a record of twenty years progress. Nature 109(2733):334–335

Williams LL (1963) Malaria eradication in the United States. Am J Public Health 53(1):17–21

Wilson AL, Courtenay O, Kelly-Hope LA, Scott TW, Takken W, Torr SJ et al (2020) The importance of vector control for the control and elimination of vector-borne diseases. PLoS Negl Trop Dis 14(1):e0007831

Wyss JH (2000) Screwworm eradication in the Americas. Ann N Y Acad Sci 916(1):186–193

Chapter 4
Advances in *Aedes* Mosquito Vector Control Strategies Using CRISPR/Cas9

P. D. S. U. Wickramasinghe, G. N. Silva, Y. I. N. Silva Gunawardene, and R. S. Dassanayake

Abstract Advancements in genetic engineering have resulted in the development of mosquitoes with impaired vector competence, thereby limiting acquisition and transmission of pathogens. The main dengue (DENV) vector, *Aedes aegypti*, is an invasive species that have spread unwittingly across the world as a result of human trade and travel. The *Ae. aegypti* mosquito species has spread across tropical and subtropical regions, with higher presence in urban regions where rapid breeding patterns have shown in artificial containers. Identification of and treating an adequate number of mosquito breeding sites as a control measure have been done for the past couple of years, and yet improvement is far from the expectations, even with well-funded and well-organized initiatives. In order to stop the pathogen transmission, genetically modified mosquitoes (GMM) needs to be created and released. Despite many Aedes-related achievements, GMM creation has been challenging. The spread of particular genetic elements that impair vector competence, trigger deleterious recessive mutations, or skew a population's sex ratio can be used to prevent the spread of vector disease, or eradicate invasive organisms in a species-specific and eco-friendly manner. In recent years, genome editing strategies have evolved to make use of a variety of nucleases, ranging from sequence-specific zinc finger nucleases to modular TALENs (transcription activator-like effector nucleases) and most recently, RNA-guided nucleases adapted from bacterial adaptive immune systems, dubbed CRISPR/Cas (clustered regularly interspaced palindromic repeats/CRISPR associated systems). By combining these methods, a new era in gene editing had emerged. Generally, both of these gene editing technologies utilize sequence-specific nucleases to generate double-stranded DNA breaks (or nicks) in the target sequence, resulting in desired DNA modifications using endogenous DNA repair mechanisms. Since cells with DNA lesions are unable to divide further, the

P. D. S. U. Wickramasinghe · G. N. Silva · R. S. Dassanayake (✉)
Department of Chemistry, University of Colombo, Colombo, Sri Lanka
e-mail: sachith@chem.cmb.ac.lk; gayathris@chem.cmb.ac.lk; rsdassanayake@chem.cmb.ac.lk

Y. I. N. Silva Gunawardene
Molecular Medicine Unit, University of Kelaniya, Ragama, Sri Lanka
e-mail: nilminisg@kln.ac.lk

B. K. Tyagi (ed.), *Genetically Modified and other Innovative Vector Control Technologies*, https://doi.org/10.1007/978-981-16-2964-8_4

nuclease-generated strand breaks must be rapidly repaired by the cell to maintain the viability. CRISPR/Cas has been widely accepted for use in a variety of organisms, including insect species, with only minor optimization steps needed thus far. CRISPR/Cas9 technology transformed the process of engineering nucleases capable of cleaving complex genomic sequences. A complementary guide RNA (gRNA) directs the Cas9 endonuclease's operation to the specific DNA target site, enabling the editing of virtually any DNA sequence without complex protein engineering and selection procedures. Apart from genome editing, the specificity and flexibility of the CRISPR/Cas9 method enables unprecedented rapid development of genetically modified organisms with mutation systems for disease vector insect control. The stability and expression of the gene construct generated by CRISPR/Cas9 or any other method must be addressed before GMM are released, in order to make sure that pathogen transmission and formulation are interrupted robustly and completely. Spreading foreign antipathogen genes through gene drive strategies among wild mosquito populations strengthens the case for a more streamlined approach. Major fields that must be adequately assessed include risk evaluation and management, conducting studies to ensure human and environmental protection, developing effective control strategies built on comprehensive gene-driving systems, and adequately addressing the ethical, legal, and social consequences of GMM release. Although GMM is theoretically feasible as a disease control method, field releases should be made only when strong scientific evidence of human and environmental protection and effectiveness are presented, and public acceptance is addressed appropriately. This chapter discusses the diverse technological advances in generating Ae. aegypti mosquitoes which are resistant to dengue virus (DENV) and other diseases, as well as the biosafety and risk assessment of these procedures. Additionally, the chapter outlines a convincing path forward for developing successful genetic-based DENV control strategies based on CRISPR/Cas9, which could be expanded to control other arboviruses while maintaining biosafety.

Keywords *Aedes* · Mosquito · Control · CRISPER/CAS9

In order to stop the pathogen transmission, genetically modified mosquitoes (GMMs) need to be created and released. Despite many *Aedes*-related achievements, GMM creation has been challenging. The spread of particular genetic elements that impair vector competence, trigger deleterious recessive mutations, or skew a population's sex ratio can be used to prevent the spread of vector disease or eradicate invasive organisms in a species-specific and eco-friendly manner (Gantz and Akbari 2018).

In recent years, genome-editing strategies have evolved to make use of a variety of nucleases, ranging from sequence-specific zinc finger nucleases to modular TALENs (transcription activator-like effector nucleases) and, most recently, RNA-guided nucleases adapted from bacterial adaptive immune systems, dubbed CRISPR/Cas (clustered regularly interspaced palindromic repeats/CRISPR-associated systems) (Gantz and Akbari 2018). By combining these methods, a new era in gene editing had emerged. Generally, both of these gene-editing technologies utilize sequence-specific nucleases to generate double-stranded DNA breaks (or nicks) in the target sequence, resulting in desired DNA modifications using endogenous DNA repair mechanisms. Since cells with DNA lesions are unable to divide further, the nuclease-generated strand breaks must be rapidly repaired by the cell to maintain the viability.

CRISPR/Cas has been widely accepted for use in a variety of organisms, including insect species, with only minor optimization steps needed thus far. CRISPR/Cas9 technology transformed the process of engineering nucleases capable of cleaving complex genomic sequences. A complementary guide RNA (gRNA) directs the Cas9 endonuclease's operation to the specific DNA target site, enabling the editing of virtually any DNA sequence without complex protein engineering and selection procedures. Apart from genome editing, the specificity and flexibility of the CRISPR/Cas9 method enable unprecedented rapid development of genetically modified organisms with mutation systems for disease vector insect control.

The stability and expression of the gene construct generated by CRISPR/Cas9 or any other method must be addressed before GMM is released, in order to make sure that pathogen transmission and formulation are interrupted robustly and completely. Spreading foreign antipathogen genes through gene drive strategies among wild mosquito populations strengthens the case for a more streamlined approach. Major fields that must be adequately assessed include risk evaluation and management, conducting studies to ensure human and environmental protection, developing effective control strategies built on comprehensive gene-driving systems, and adequately addressing the ethical, legal, and social consequences of GMM release. Although GMM is theoretically feasible as a disease control method, field releases should be made only when strong scientific evidence of human and environmental protection and effectiveness are presented, and public acceptance is addressed appropriately (Touré et al. 2003).

This chapter discusses the diverse technological advances in generating *Ae. aegypti* mosquitoes which are resistant to dengue virus (DENV) and other diseases, as well as the biosafety and risk assessment of these procedures. Additionally, the chapter outlines a convincing path forward for developing successful genetic-based DENV control strategies based on CRISPR/Cas9, which could be expanded to control other arboviruses while maintaining biosafety.

4.1 Introduction

Aedes aegypti is an endophilic domesticated mosquito which has expanded its living sphere and likely continues to spread vastly over recent decades. It has been the primary vector of dengue virus (DNV) and other epidemiological important viruses such as the chikungunya (CHIKV), yellow fever (YFV), and Zika virus (ZIKV) (Buchman et al. 2020).

Containers such as buckets, birdbaths, tires, water storage jars, and flowerpots are main locations where the female mosquito lays eggs and bears larvae. The larvae metamorphose into pupae after development through four larvae stages. As the larval stage, the pupal stage too is aquatic. After 2 days, an adult mosquito fully develops and breaks through the pupae into a flying adult mosquito.

Aedes aegypti's spread has been largely attributed to the globalization of travel and transportation. Heavily populated areas with insufficient availability of safe drinking water, waste management, and sanitation have been the major habitats of *Ae. aegypti* (Honório et al. 2009). Historically, the spread of *Ae. aegypti* intercontinentally was mainly by naval transportation, and it is considered as to be carried on from continental Europe from Madeira (via ferries). A prominent suggestion is that *Ae. aegypti*, which developed domesticated behavior in West Africa, the species' birthplace, and extended colonization and dissemination in the tropical regions, has resultantly evolved in vectoring interhuman transmission of viruses such as dengue in a highly efficient manner (Weaver and Reisen 2010). Domestication may protect against environmental conditions and a range of egg-laying (oviposition) environments but may also lead to an increase in vulnerability to control measures for the removal of species. The ability of *Ae. aegypti* to survive in densely populated areas and in close proximity to humans confirms its evolutionary existence as a good vector. Male and female adult mosquitoes also feed on plant nectar; however, females blood feed exclusively on humans in order to obtain required proteins to produce eggs, and the activity is high during daytime. Since eggs can withstand desiccation for extended periods of time, they are easily spread to new locations.

Early strategies such as finding and treating/destroying mosquito breeding sites have become futile, even for the foremost well-funded organizations. Despite the failures, moderate approaches such as reducing larval habitats, the use of pesticides, and the use of insecticide-impregnated nets are still used worldwide. However, conventional interventions face obstacles, including shifts in biting times, environmental concerns pertaining insecticides, the rise of new insecticide-resistant mosquitoes, and also the complexity of distributing and reliably using bed nets (Fu et al. 2010).

Aedes aegypti's ability to rapidly develop in temperate regions is presently limited due to its intolerance of mild winters and, more specifically, the high mortality of eggs exposed to frost (Otero et al. 2006; Gould and Higgs 2009). Current habitat patterns will change in the future by means of increased northern and southern expansion due to global climate change (Weaver and Reisen 2010). Given that conventional vector management methods have fallen short of expectations in terms of eradicating the spread of mosquito-borne diseases, new approaches should be considered immediately.

4.2 Global Distribution of Vector-Borne Diseases and Need for Indigenously Developed Technologies

Vector-borne diseases account for more than 17% of the global communicable diseases and with fatalities over 700,000 each year. High rates of mobility and mortality caused by vector-borne diseases attribute to the poverty and great majority of the world's population (over 80%) living in disease-prone areas (World Health Organization 2017). These diseases curtail major economic and social development. Moreover, heavy financial losses on economies are caused through lost production and reduced working hours. In prevention of arboviral diseases, various public health strategies are being used, and sustaining those strategies in conjunction with modern technological advancements is critical for prevention of vector control of mosquito-transmitted diseases.

Biocontrol methods or more sophisticated techniques are needed to assist in reducing the duration of insecticide application, which is currently the primary method for mosquito control. As a result, the number of countries reporting insecticide resistance in *Aedes* mosquitoes has risen significantly over the last decade. Mosquito behavior is critical in vector control programs, and additional information about the chemical ecology of mate seeking, swarming landmarks, and mate selection in swarming sites is necessary to enhance control strategies. Experiments on vector competency are necessary to determine the most effective control strategy for an ecosystem. Vector competence experiments incorporate a plethora of variables into their design, including ecological considerations, the vector's health and fecundity, and so on. Scientists considered eradicating global mosquito populations using a variety of methods after being prompted by global health advocates. Elimination will be exceedingly difficult, as there are over 3500 mosquito species found in virtually every geographical area (Fang 2010).

Even if global mosquito eradication is possible, it may be undesirable due to the fact that only a few mosquito species contribute as human disease vectors, and widespread eradication could have significant ecosystem consequences.

Due to numerous obstacles in mosquito eradication, scientists are making significant strides toward the production of disease-free genetically modified mosquitoes. Theoretically, GMM would cause minimal ecological disruption because the targeted mosquito species would retain their ecological niche, not allowing the invasive species from displacing them.

As a result, novel vector control techniques, such as using genetically modified mosqutos to eradicate mosquito populations or render mosquitos incapable of transmitting pathogens, are critical. Generally, genetically dependent vector control approaches fall into two categories. The first group of procedures is intended for population control, containment, or eradication, with the aim of reducing or even eliminating particular insect species through the development of genes that are either (conditionally) lethal or capable of rendering the insects sterile. Genetic control is used to eradicate a wild population by introducing sexually transmitted factors that decrease the population's reproduction rate or introduce refractoriness. Typically,

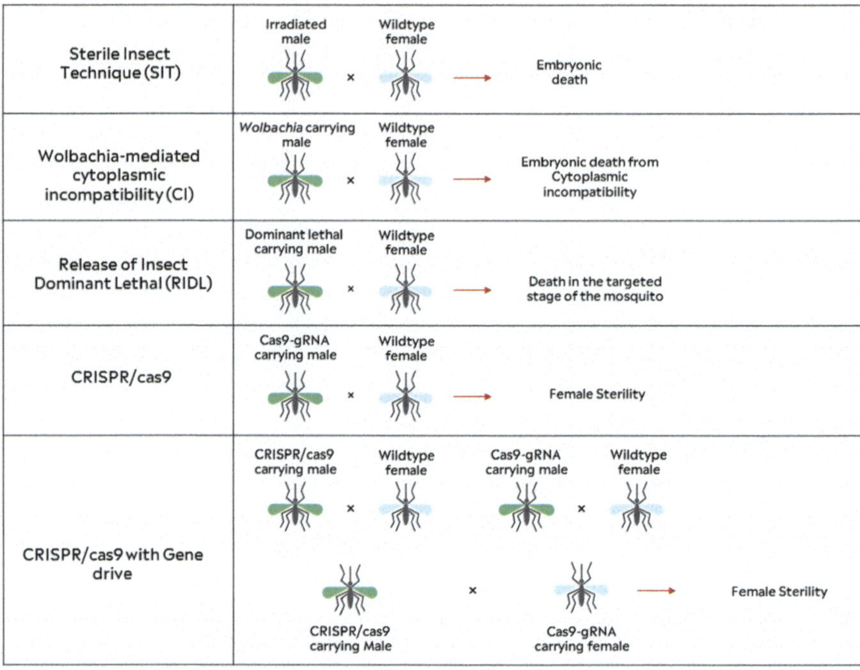

Fig. 4.1 A summary of current and potential methods for genetically modified mosquito growth. Population suppression of mosquito vectors employs sterile insect technology (SIT), release of insect carrying a dominant lethal (RIDL), and *Wolbachia*-mediated cytoplasmic incompatibility (CI). CRISPR/Cas9-mediated genetic alteration of mosquitoes results in embryonic genetic modification, which results in population transformation/replacement. Gene drives are non-Mendelian inherited genetic elements that can spread across populations. In a gene drive scheme, genetically engineered mosquitoes could be crossed with wild-type mosquitoes, and the altered genes would be preferentially transmitted by all offspring with approximately 100% likelihood, resulting in an optimal population replacement

these sexually transmitted pathogens are introduced into wild populations by males that have been mass-produced in a factory and released. Wild males compete with the released males for virgin female insemination. A released male can successfully transmit his traits on, sterility or disease resistance into the wild population (depending on the method of control). The sterile insect technique (SIT), which involves the release of insect carrying a dominant lethal (RIDL), and *Wolbachia*-mediated cytoplasmic incompatibility (CI) are available strategies for suppressing the wild population of mosquitoes (Fig. 4.1).

In SIT, a large number of males that have been sterilized with radiation will be released into a target population. These SIT males can mate with indigenous females, causing the reproductive ability of females to be reduced, which results in the eradication or suppression of the target population provided that male mosquitoes are released in adequate amounts and for an extended period of time. Reduced fecundity has a detrimental effect on both the arthropod-borne pathogen transmission and the insect population. A form of SIT is the release of insects carrying a dominant lethal (RIDL) gene/allele in which repressible female-specific lethality is induced via a transgene mechanism. Due to the absence of radiation sterilization, the technology is referred to as an evolution of SIT. The RIDL strategy involves the introduction of a dominant lethal transgene and artificial suppression of its expression to allow insects to be reared (Alphey et al. 2013). Mosquitoes have been genetically engineered to transfer a lethal gene to offspring that is genetically regulated by the tetracycline gene (Singh et al. 2018); thus, offspring of genetically modified mosquitoes need tetracycline in water to survive. The absence of tetracycline in environmental water results in larval death (von Seidlein et al. 2017).

Wolbachia is a widespread endosymbiotic bacterium of arthropods and nematodes that causes significant changes in its hosts' reproductive biology, thereby facilitating the spread of infection among uninfected populations. Cytoplasmic incompatibility (CI) is a form of postmating sterility that occurs when a spermatozoon from a person infected with a particular *Wolbachia* strain fertilizes an uninfected egg or an egg infected with another incompatible *Wolbachia* strain. CI significantly decreases the number of offspring produced by crosses between uninfected females and *Wolbachia*-infected males. Females infected with *Wolbachia*, on the other hand, will mate with infected or uninfected males without experiencing a drop in progeny. CI has the potential to drive speciation and is being aggressively sought as a method of controlling human diseases resulting from mosquito vectors.

The strategies outlined above are successful against mosquitoes only when frequent releases occur, and therefore long-term financial and administrative commitments must be met even if transmission is discontinued.

The second type of category seeks to change or eliminate populations. The objective is not to eradicate the vector but to engineer a replacement capable of adding an effector gene capable of reducing or preventing disease transmission in the wild population. Male and female mosquitoes carrying an effector gene that decreases or eliminates their ability to spread viruses such as dengue or Zika are used in the technique. When these genetically engineered mosquitoes mate with wild mosquitoes, the factor spreads throughout the population, essentially making mosquitoes incapable of transmitting the pathogen without population suppression. RNA interference (RNAi) appears to be an attractive mechanism for suppressing virus replication in arboviruses spread by *Aedes* mosquitoes.

RNAi is an invertebrate innate immunity pathway that acts as an antiviral immune response in mosquitoes, effectively controlling arbovirus replication and facilitating virus transmission (Franz et al. 2009). RNAi can, therefore, be a major determinant of the vector capacity of arbovirus mosquitoes (Franz et al. 2009), and it may inhibit RNA viral infections. In one study, the inverted-repeat (IR) sequences derived from DENV strain 2 (DENV-2) genomic RNA were genetically engineered. The double-stranded RNA evolved soon after IR-RNA expressed in the midgut of female mosquitoes after viremic blood ingestion triggering the DENV-2 endogenous RNAi pathways into the mosquito midgut and has been shown to have lower levels of vector competence for DENV-2 genetically triggering pathways. However, in the initial stage, laboratory research showed that transgenic mosquitos carrying an antivirus effector gene had lost their strength and viral resistance because of the genetic changes taking place outside of the area being targeted (Franz et al. 2009). On top of that, several other studies indicated that the expression may impose substantial fitness costs and also may impose some kind of epigenetic silence on the unusual inverted repeating structure, making the strategy ineffective after several mosquito generations. Due to the incomplete nature of the RNAi strategy (Krueger et al. 2007) and the presence of nonspecific RNAs (Jackson et al. 2005), further research was progressed in developing a method capable of directly altering/modifying a genome area without affecting mosquito fitness or fertility.

Genome engineering, alternatively referred to as genome editing, is a highly precise technique that enables making changes of an organism's genome at a specific locus. These techniques became popular tools with scientists and researchers. Initially, homologous recombinations (HRs) were used to silence the gene, but their utility is limited due to their inefficiency and tedious nature. Genome engineering technology has evolved over time to generate unique double-stranded DNA (dsDNA) or single-stranded DNA (ssDNA) breaks in a targeted region of the genome, which are then repaired by endogenous repair machinery (Rudin et al. 1989; Rouet et al. 1994). To repair DSBs (double-stranded breaks), one of two distinct repair pathways is used: homology-directed repair (HDR) or nonhomologous end joining (NHEJ) (Kass and Jasin 2010). To date, four highly effective artificial nuclease systems for inducing DSBs and enabling site-specific genome editing have been discovered. These include homing endonucleases (HEases) or mega nucleases (MNs) (Stoddard 2006), transcription activator-like effector nucleases (TALENs) (Hockemeyer et al. 2011; Reyon et al. 2012), zinc finger nucleases (ZFNs) (Sander et al. 2011a; Wood et al. 2011), and clustered regularly interspaced short palindromic repeats (CRISPR) and CRISPR/Cas9 from *Streptococcus pyogenes* (Garneau et al. 2010; Gasiunas et al. 2014).

Due to the targeted DNA error, MNs are exempt as a method for genome editing (Silva et al. 2011; Porteus 2016). Scientists were under the impression that ZFN could be a replacement for MNs. However, the ZFN domain seems to exhibit context-dependent binding (Maeder et al. 2008), and assembling the ZFN domain with the required binding specificity requires a lengthy screening process (Sander et al. 2011b). Also, high manufacturing cost of ZFNs made it less optional. TALEN itself is another programmable DNA-binding protein, as ZFN proteins. However, because TALEN is context dependent, substantial gene construction is required which is also a labor-intensive process (Juillerat et al. 2014). Nonetheless, the CRISPR/Cas9 method has been shown to be a cost-effective, rapid, and powerful genome-editing tool, although one requires considerable labor.

CRISPR/Cas9 are present in a wide variety of bacteria involved in antiviral immunity (Marraffini and Sontheimer 2008). Variation can be observed in CRISPR sequences, ranging between 30 and 40 nucleotides, and is interspersed with direct repeats. These repeats are linked to neighboring genes encoding nuclease or helicase proteins, which are referred to as CRISPR-associated, or Cas, proteins (Jansen et al.

2002; Brouns et al. 2008; Jinek et al. 2012; Hartenian and Doench 2015). Cas proteins identify, catch, and integrate invading genetic material (and, in some cases, self-DNA) into spacer-containing DNA sequences (Brouns et al. 2008; Jinek et al. 2012). CRISPR sequences are translated into short RNA sequences called CRISPR-RNA (crRNA), which then form a complex with the Cas protein (s) to target and label previously incorporated viral DNA into the bacterial genome (Jinek et al. 2012). CRISPR-RNA (crRNA) and the trans-activating crRNA (tracrRNA) complex will target and cleave exogenous DNA, thus removing the plasmid or virus that invaded the cell.

Cas proteins could be reprogrammed to cleave a variety of targets based on the sequence of the short guide RNA (sgRNA), which is often derived from the crRNA/ tracrRNA (van der Oost et al. 2014; Zetsche et al. 2015). The protospacer adjacent motif (PAM) is required for Cas nuclease to generate double-stranded breaks (DSBs) at its target site. Although the PAM is present on the viral genome, it is absent on the bacterial sequence, allowing for precise viral DNA cleavage. Cas9's PAM site can be modified to accept nucleotide sequences of different lengths (Kleinstiver et al. 2015).

After a Cas9/gRNA complex cleaves a targeted genomic sequence, one of two alternative DNA repair mechanisms will reinstate the integrity of chromosomes: NHEJ (Chiruvella et al. 2013; Aylon and Kupiec 2004), which usually results in insertions and/or deletions (indels) of a few base pairs (bp) of DNA near the gRNA cleaving site, or HDR (Aylon and Kupiec 2004; Jasin and Rothstein 2013), which can repair the cut using a homology spanning DNA template. Small insertions or deletions can occur as a result of errors made during canonical C-NHEJ repair. These indels can result in frameshift mutations within a gene's open reading frame, resulting in knockout mutants as well as truncated proteins. Additionally, it is possible to exclude parts of a gene by using several guide RNAs in tandem. By comparison, if the DSB is repaired through the HDR pathway, the break can be repaired perfectly using either the homologous chromosome or an exogenous template called single-stranded oligodeoxynucleotides (ssODN). By providing a template for HDR, the experimenter can either eliminate undesirable alleles from a population or insert foreign DNA into the population (Fig. 4.2).

A genetic modification of a mosquito species can be beneficial epidemiologically only if it is present in a significant proportion of the target population. The genetically altered mosquitoes certainly need to maintain a high level of prevalence and efficacy over several vector generations. However, the primary hurdle is the difficulty in dispersing transgenes into wild populations. This issue is explicable using fundamental Mendelian genetics. When a transgenic mosquito mates with a wild mosquito, only 50% of the offspring are transgenic. When the above progeny outcross to wild mosquitoes again, the frequency of the transgene drops to 25% in the subsequent generation, indicating that the transgene is eventually filtered out (Fig. 4.3a). If the transgene incurs a fitness cost, the probability of transgene removal could be higher than anticipated. The only other option is to introduce transgenic males at a rate exceeding that of the wild population and to continue doing so indefinitely.

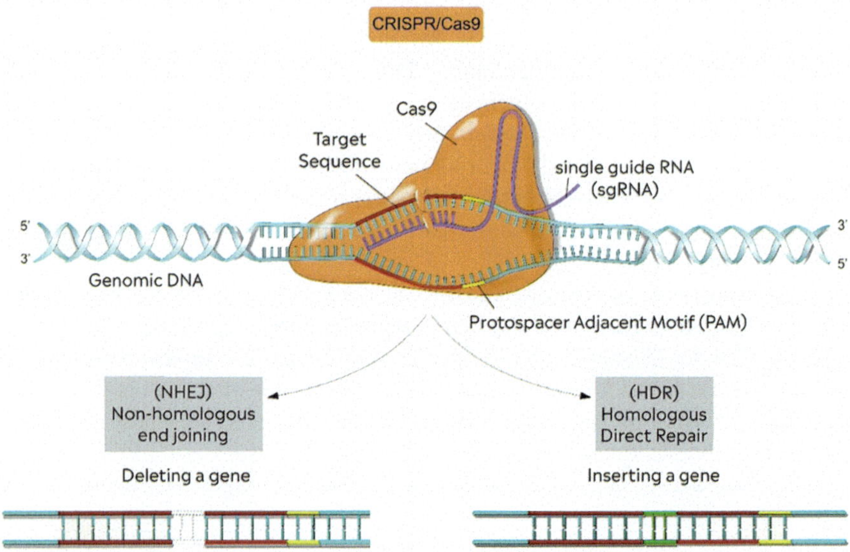

Fig. 4.2 CRISPR/Cas9 system with nonhomologous end joining (NHEJ) and homology-directed repair (HDR) mechanism. A DNA double-stranded break (DSB) can be repaired through HDR or NHEJ. In HDR, the DSB is corrected using a template and can be used to introduce a gene, a portion of a gene, or even a single point mutation precisely. In contrast to HDR, NHEJ pathway is usually error prone and can create indels (small insertions/deletions) at the DSB, resulting in gene inactivation by frameshift to generate a premature stop codon

Even though a selective disadvantage is present, a proper system is required in increasing the prevalence within the population over time and is termed as gene drive (GD) systems. Gene drives are an older idea and occur quite commonly in nature. Since at least 1887, biologists were aware of "selfish" genes. The main objective of such genes is to manipulate the reproductive process for enhanced spread. The genes might act in several ways such as triggering processes that kill off sperm that doesn't include them or alter the DNA replication mechanism for frequent copy mechanism. Numerous researchers have proposed that these "selfish genes" could serve as the foundation for "gene drives" capable of disseminating engineered traits throughout wild populations. Over a decade ago, Austin Burt proposed the first gene drives based on site-specific "homing" endonuclease genes. These genes influence inheritance by severing the homologous chromosome and inducing the cell to replicate them when the break is repaired. Numerous efforts have concentrated on defining an autosomal allele as genetically driven. The requirement is to have greater than 50% of the offspring to inherit alleles from an individual carrying a single copy of the allele. Numerous variants of genetic elements which can favorably drive their transmission at the expense of other genes were discovered in symbiotic/parasitic organisms and are frequently referred to as selfish genes/selfish genetic elements

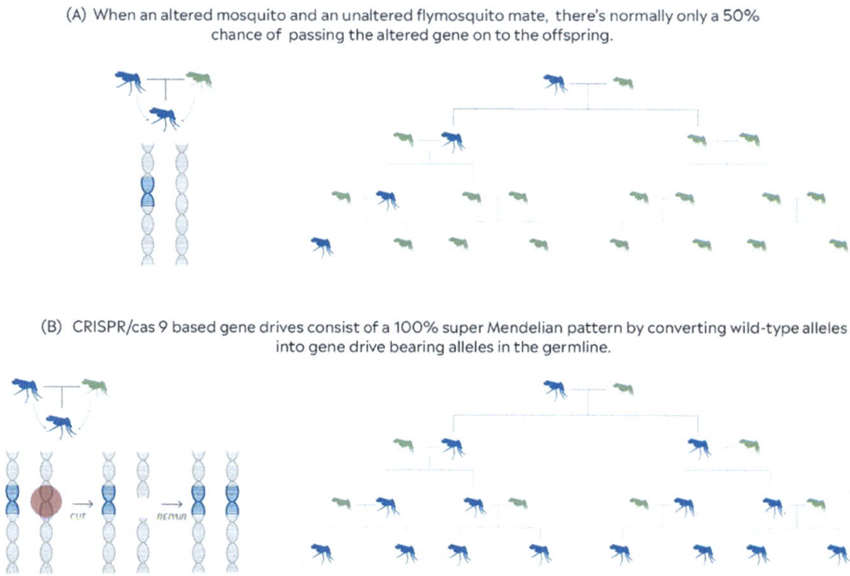

Fig. 4.3 Modification of the chromosome via gene drive. Mendelian inheritance as opposed to "super" Mendelian inheritance. (**a**) Where a transgene is not sex linked, Mendel's law of independent assortment predicts a 50% inheritance rate. Without repeated introduction, the transgene is likely to be lost due to a variety of factors, including the genetic drift and the transgene's fitness cost. (**b**) By converting wild-type alleles to gene drive alleles, gene drives based on CRISPR/Cas9 supersede Mendel's law of independent assortment in the germline. The expression of Cas9 and guide RNA results in stable integration of the genes onto a single chromosome, followed by a second subsequent modification, and propagation of the gene drive factor occurs in the target population

(James 2005; Sinkins and Gould 2006; Marshall 2009; Alphey et al. 2013; Burt 2014; Gantz and Bier 2016).

A novel system termed as gene drive (GD) is developed which has the ability to gradually increase the prevalence within the population, despite a selective disadvantage. Selfish DNA systems, which have the ability to spread while offering no benefit to the human, are the primary source of inspiration for the development of gene drive systems. Gene drives are an older concept that occur frequently in nature. Even in the nineteenth century, biologists have been aware of "selfish" genes that can exploit the reproductive process in order to increase their chances of spreading. The genes may trigger processes that exclude sperm that does not contain them, or they may alter the DNA replication process in such a way that they are duplicated more frequently than other genes. Numerous researchers have proposed that these "selfish genes" could serve as the foundation for gene drives having the capability of disseminating modified traits among wild populations (Craig et al. 1960; Sinkins and Gould 2006; Alphey and Bonsall 2014). The first gene drive was proposed by Austin Burt around a decade ago on site-specific "homing" endonuclease genes (Burt 2003). Inheritance patterns will be modified as these genes influence severely the homologous chromosome and cause the cell to replicate them when the break is repaired. Numerous efforts have concentrated on defining an autosomal allele as genetically driven if more than 50% of progeny inherit the allele from a person bearing a single copy of the allele (Curtis 1968). Well-studied examples of such selfish elements or organisms include chromosomal rearrangements (Curtis 1968), transposons (Skipper et al. 2013), Medea elements (Chen et al. 2007; Ward et al. 2011; Akbari et al. 2014), homing endonuclease genes (HEGs) (Deredec et al. 2011; Windbichler et al. 2011; Alphey and Bonsall 2014), maternal effect lethal under dominant elements (Akbari et al. 2013), bacterial endosymbiont/parasite *Wolbachia* (Rasgon 2007), and the recently developed CRISPR/Cas9 system.

CRISPR/Cas9 has been widely adopted and is now considered the de facto standard technique for genome editing using the GD feature. Numerous experiments worldwide are being conducted to generate knockout and knock-in mutants in vector mosquitoes using CRISPR/Cas9. The use of CRISPR/Cas9 to generate mutant mosquito strains such as nonpathogenic mosquitoes, disturbed mosquitoes that seek hosts, male-only mosquitoes, and wingless mosquitoes will be novel strategies for controlling/eradicating the spread of fatal diseases by vector mosquitoes A possible requirement for an effective population replacement strategy is a strong link between the genome-modifying construct and an efficient, species-specific GD system capable of rapidly repairing the transgene into the wild-type population at a rapid rate than predicted by Mendelian inheritance. The above modifications could then be used to spread via gene drive, with the goal of ultimately suppressing the mosquito population.

4.3 CRISPR-Based Gene Drive to Control Mosquito Vectors

Synthetic gene drives are used to rapidly alter the genetic makeup of wild populations in order to modify or eliminate specific characteristics. A possible application of GD would be to genetically prevent female mosquitos from reproducing, followed by the release of genetically engineered mosquitoes into an island ecosystem to mate with the wild population. Formation of the male offspring only will eventually result in the population of the targeted mosquito being reduced or eradicated.

Any organism that reproduces sexually and with a reasonable frequency may be a gene drive target. Gene drives are unique to organisms that reproduce via the meiotic process, which results in the generation of gametes. Meiosis occurs frequently in eukaryotes but is absent in prokaryotes (all bacteria and archaea). The genes that are altered by engineered drives are only found in the nucleus and not in other regions of the cell, such as mitochondrial DNA. As mentioned previously, gene drives function when an organism produces gametes via meiosis. Diploid organisms have their nucleic DNA composed of two distinct sets of genetic material, one from each parent. The DNA would be combined and halved during meiosis prior to the production of gametes (e.g., sperm or egg cells). During gametogenesis, the parents' haploid DNA is randomly redistributed and becomes diploid again after fertilization. While normal genes inherit at a 50% chance, gene drive elements have increased the percentage of transmission, and hence scientists refer to this as "super-Mendelian" inheritance (Chevalier 2001; Hammond et al. 2016; Grunwald et al. 2019). When genetically engineering an organism with special or specifically designed gene drives, the gene drive methods may force their own inheritance to a level of 80% or nearly 100% (Fig. 4.3b).

In little more than a year following the first demonstrations in human cells, CRISPR/Cas9 has been successfully functioning in many different species including *Drosophila melanogaster* (Bassett et al. 2013), the mosquitoes *Ae. aegypti* (Basu et al. 2015), *Anopheles stephensi* (Gantz and Bier 2015), and *Anopheles gambiae* (Hammond et al. 2016).

Insects such as fruit flies and mosquitoes detect the presence or absence of moisture in the air through transient receptor potential (TRP) channels, a process known as hygrosensation. Waterwitch (*wtrw*) is a subfamily of TRP channel that is critical for detecting absolute humidity, which is necessary for survival. Kistler et al. (2015) studied the efficiency of a CRISPR/Cas9-mediated gene-editing system with the *Aaeg-wtrw* (*Ae. aegypti* waterwitch) locus to generate mutations through disparate repair mechanisms and achieved various types of mutations in several *Ae. aegypti* mosquito genomic loci (Kistler et al. 2015). The study focused on site-directed mutagenesis in *Ae. aegypti* using type II CRISPR/Cas9. Kistler et al. (2015) injected 200-bp-long sgRNAs and ssODN into mosquito embryos to determine the rate of mutations associated with indel formation via two types of mechanisms, the C-NHEJ pathway and ssODN insertion via the HDR pathway. In G0 survivors, indel generation occurred at a rate of 24%, whereas ssODN insertion occurred at a rate of 0.71%, indicating a predominating position for C-NHEJ repair. Additionally, the study demonstrated an indel inheritance rate of 18.9% and a ssODN inheritance rate of 0.6% in G1 progeny, indicating that somatic mutation rates may be a viable method for directing germline experiments.

Basu et al. (2015) have effectively used CRISPR/Cas9 to target a variety of genes involved in DNA repair, RNAi, and sex determination in *Ae. aegypti*. The study made use of the multiplexed impact of CRISPR/Cas9 on six distinct genes (kmo, loqs, r2d2, ku70, lig4, and nix) (Hall et al. 2015). Additionally, taking into account the possibility that the editing rate varies across the genome, the authors built 40 additional sgRNAs and tested their editing ability in transient embryo assays, producing various forms of somatic and germline mutations in *Ae. aegypti* mosquitoes. Additionally, the study established a platform for rapid identification of sgRNAs using a transient embryo assay, which increased the efficiency of gene editing and HR-mediated repair when CRISPR/Cas9 was used.

Dong et al. (2015) were the first research group to alter the enhanced cyan fluorescent protein (ECFP) gene in an *Ae. aegypti* mosquito line. The alteration leads to expressing two distinct eye markers using the CRISPR/Cas9 system (Dong et al. 2015). Along with Cas9, two sgRNAs were used to target distinct regions of the

ECFP gene using in vitro transcribed mRNAs for germline transformation, resulting in the generation of four distinct G1 pools with a knockout efficiency of 5.5%. PCR amplification, cloning, and sequencing experiments showed indels (insertions or deletions) ranging from two to 27 nucleotides in the ECFP target gene, demonstrating that CRISPR/Cas9-mediated gene editing is possible in *Ae. aegypti*. The successful disruption of the marker gene established that the CRISPR/Cas9 system is a viable tool for targeted gene disruption in *Ae. aegypti*.

In the same year, Hall et al. (2015) demonstrated the creation of feminized genetic males by knocking out the male-determining gene (nix) using the CRISPR/Cas9 gene-editing system. The study discovered that silencing nix, a gene thought to play a critical role in regulating fruitless splicing, and doublesex results in sexual dimorphism in *Ae. aegypti*. Additionally, the study revealed that nix is sufficient to initiate male growth, paving the way for female mosquitoes to be converted to harmless males (Hall et al. 2015). This study can pave the way for the custom editing of female mosquito genomes via CRISPR/Cas9. These studies could result in more significant inventions that will aid in the fight against the spread of insect-borne deadly diseases in developing countries (Reegan et al. 2016).

Although development of gene drives and GDOs proceeded at a relatively slow pace, the introduction of CRISPR/Cas9 in 2012 (Jinek et al. 2012) accelerated development and advancements, igniting a frenzied method for implementation. CRISPR/Cas9 is typically used to cause DNA cleavage at specific sites, allowing cells to undergo HDR or NHEJ and allowing for the addition of specific elements during repair. CRISPR/Cas becomes a "drive element" when used as a homing endonuclease. In this case, the entire CRISPR/Cas build would need to be duplicated into its own target site (cf. Fig. 4.3). This is only possible if the outer boundaries of the construct contain sequences homologous to those found adjacent to the CRISPR/Cas target site. Once the construct is present on one chromosome, it generates the CRISPR/Cas molecule, an RNA-guided endonuclease that cleaves the DNA at the construct's target site in the parallel (homologous) chromosome. With the cleavage of the target site, the repair mechanism initiates and uses the homologous chromosome, in this case the CRISPR gene drive construct, as a template for repair. With the aid of homology-directed repair (HDR), the GD construct is copied into the target site, ensuring its "inheritance" and dissemination.

CRISPR/Cas9-based gene drives, which resemble a "mutagenic chain reaction (MCR)," are well known (Gantz and Bier 2015). There are two distinct strategies: population suppression (to reduce or eliminate a target species) and modification (to spread a specifically designed or desired trait). For instance, a population suppression drive could be a gene drive spreading female sterility. In theory, if passed onto each offspring and also carried and thus spread by each male, wild populations would decline rapidly and ultimately collapse. However, significant practical complications as well as unexpected and unanticipated consequences may exist.

MCRs (Gantz and Bier 2015) or related constructs (DiCarlo et al. 2015) are relatively recent entrants into this well-established domain of selfish DNA. "CRISPR/Cas9-driven autocatalysis that generates insertional mutants can be achieved by a construct consisting of three genetic elements: a central element that encodes Cas9 that can be expressed in both somatic and germline cells; a gRNA sequence encoded by an ubiquitously expressed gene, targeting the desired DNA sequence in the genome; and homology arms designed to flank the genomic sequence immediately adjacent to the DSB which must be placed flanking the Cas9/gRNA cassette" (Gantz and Bier 2016) (cf. Fig. 4.3). This tripartite construct could allow Cas9 cleavage in the chromosomal target at the site defined by the gRNA and insertion of HDR-guided flanking sequences. Cas9 and gRNA expression from the insertion allele could then result in the cleavage of the opposing allele (cf. Fig. 4.3), and the insertion of the Cas9/gRNA cassette is achieved via HDR. The construction of CRISPR/Cas9 GD included the Cas9 gene driven by a gonad-specific promoter and the homologous sequence containing sgRNAs to be guided by a pol III promoter. Insertion of the GD to the homologous chromosome is achieved by Cas9-sgRNA complex that creates a DSB at the target site. DSB is repaired by inserting the GD via HDR pathway, effectively replicating its copy number in each generation. The technique can also be modified in disrupting desired genes or in rapidly introducing a transgene into a population.

Genetic modification of the vector is not merely an alternative to traditional control techniques; it also has a number of distinct advantages. The technique relies on the released mosquito's natural actions to locate and mate with the local population. The released mosquito is not constrained by human access restrictions, and it can reach an infinite number of locations that conventional technology has been unable to locate or penetrate. The technique described above is critical in rural and urban areas, areas with insufficient transportation and infrastructure, and areas with political turmoil or military activity. Each genetic mechanism is exclusive to a single species and is unlikely to have an effect on similar mosquito species or other invertebrates. The adverse effect of such strategy is little to none.

4.4 Biosafety of the Use of Genetically Modified Mosquitoes

As a novel approach, the World Health Organization has developed guidelines to ensure that GMMs are used safely in a variety of countries and areas. Field trials should begin with the release of sterile or self-limiting modified male mosquitoes to familiarize themselves with the technology in conditions where its effects can be reversed by discontinuing releases (Benedict 2003). To date, field releases of genetically modified mosquitoes have concentrated on nonreplicating, functionally sterile males (which do not bite).

Additionally, physical confinement can be used to research modified mosquitoes incorporating gene drive (Alphey et al. 2002; Benedict et al. 2008). There should have been a test outside the laboratory for GMM to reproduce or spread the modification to wild mosquitoes. Unlike other public health products such as medications, vaccines, and insecticides, no pipeline exists to transfer candidate GMM from the laboratory to field deployment through safety and efficacy trials.

Due to the lack of a pre-existing structure, scientists had to play an important, independent role in defining the path of product development. The challenge provided potential end users and technology beneficiaries with an unprecedented opportunity to participate in products' discovery and creation from the beginning. This approach ensures that protection and efficacy criteria are developed as design characteristics into the adapted mosquito strains. The role of researchers is to identify gaps and to contribute to regulatory standards for the goods they produce. These standards cover not only legislative requirements but also a broader regulatory framework that takes the fears and requirements of each community in which the product is used into account.

The expected duration of the persistence of GMMs in the environment after the release of GMMs can specify these technologies further. Transgenic components and how they operate in the wild determine GMM's durability. With the use of "self-limiting" strategies, the prevalence of genetic modification in the mosquito population decreases gradually until eradication. GMMs have been designed to be sterile in some cases to prevent the genetic change of the offspring from becoming heritage. In other cases, GMM is designed to mate the effect in a short time into the local population of mosquitoes, but the effect would eventually filter out by crossbreeding with local mosquitoes over many generations. Only periodic re-releases of GMM, whose frequencies depend on genetic modifications, may sustain the impact of self-limiting approaches. These methods are reversible when releases are stopped, and therefore, they do not likely lead to permanent environmental changes. Nevertheless, continued development and distribution costs are involved in the need for reintroduction.

The genetic modification is to be distributed throughout the local mosquito population with "self-sustaining" methods. While these approaches can provide long-term and cost-effective protection against pathogen transmission, unintended consequences are likely to be tough to reverse than with self-restricting approaches.

Due to the diversity of genetic approaches under review and the circumstances under which they can be used, there is no standardized formula for evaluating these technologies. It will be important to assess individual needs on an individual basis, taking into account all possible benefits and risks.

Akbari et al. (2015) suggested a set of proposals to limit the accidental spreading of gene-driving organisms in nature following the successful demonstration of the Cas9-based gene drive in *Drosophila melanogaster*. The containment limitations include molecular containment, ecologic containments, reproductive containment, and barrier containment (Akbari et al. 2015). Akbari et al. further propose that all laboratory gene drive research should include no less than two of the abovementioned metrics. By adhering to the guidelines developed by Akbari et al. (2015), adverse effects of GMM release can be avoided, allowing the strategy to be one of the most effective vector control strategies.

4.5 Conclusion

The CRISPR/Cas9 system is a revolutionary tool for both prokaryotic and eukaryotic genetics. It has been preliminarily established and developed in model and non-model insects. Highly efficient knockout and knock-in experiments have been successfully conducted in model insects such as *Drosophila* and silkworm and in non-model insects such as butterfly, mosquito, and beetle. Based on CRISPR/Cas9, several well-designed systems have been developed, including gene drive and regulation and DNA/RNA tracking systems, which will have significant impact on functional studies and pest control. Researchers have improved the CRISPR/Cas9 system in insects, resulting in easier and more effective design and use. Successful modifications of CRISPR/Cas9 have been made in cells and model animals, implying that CRISPR/Cas9 has the potential for broad application and development in insects together with the use of gene drive elements, especially in case of the deadly dengue vector, *Ae. aegypti.*

References

Akbari OS, Matzen KD, Marshall JM, Huang H, Ward CM, Hay BA (2013) A synthetic gene drive system for local, reversible modification and suppression of insect populations. Curr Biol 23: 671–677. https://doi.org/10.1016/j.cub.2013.02.059

Akbari OS, Chen CH, Marshall JM, Huang H, Antoshechkin I, Hay BA (2014) Novel synthetic Medea selfish genetic elements drive population replacement in Drosophila; a theoretical exploration of Medea-dependent population suppression. ACS Synth Biol 3(12):915–928. https://doi.org/10.1021/sb300079h

Akbari OS, Bellen HJ, Bier E, Bullock SL, Burt A, Church GM, Cook KR, Duchek P, Edwards OR, Esvelt KM, Gantz VM, Golic KG, Gratz SJ, Harrison MM, Hayes KR, James AA, Kaufman TC, Knoblich J, Malik HS, Matthews KA, O'Connor-Giles KM, Parks AL, Perrimon N, Port F, Russell S, Ueda R, Wildonger J (2015) BIOSAFETY: safeguarding gene drive experiments in the laboratory. Science 349:927–929. https://doi.org/10.1126/science.aac7932

Alphey N, Bonsall MB (2014) Interplay of population genetics and dynamics in the genetic control of mosquitoes. J R Soc Interface 11:20131071. https://doi.org/10.1098/rsif.2013.1071

Alphey L, Beard CB, Billingsley P, Coetzee M, Crisanti A, Curtis C, Eggleston P, Godfray C, Hemingway J, Jacobs-Lorena M, James AA, Kafatos FC, Mukwaya LG, Paton M, Powell JR, Schneider W, Scott TW, Sina B, Sinden R, Sinkins S, Spielman A, Touré Y, Collins FH (2002) Malaria control with genetically manipulated insect vectors. Science 298:119–121. https://doi.org/10.1126/science.1078278

Alphey L, McKemey A, Nimmo D, Neira Oviedo M, Lacroix R, Matzen K, Beech C (2013) Genetic control of Aedes mosquitoes. Pathog Glob Health 107:170–179. https://doi.org/10.1179/2047773213Y.0000000095

Aylon Y, Kupiec M (2004) DSB repair: the yeast paradigm. DNA Repair (Amst) 3:797–815. https://doi.org/10.1016/j.dnarep.2004.04.013

Bassett AR, Tibbit C, Ponting CP, Liu J-L (2013) Highly efficient targeted mutagenesis of Drosophila with the CRISPR/Cas9 system. Cell Rep 4:220–228. https://doi.org/10.1016/j.celrep.2013.06.020

Basu S, Aryan A, Overcash JM, Samuel GH, Anderson MAE, Dahlem TJ, Myles KM, Adelman ZN (2015) Silencing of end-joining repair for efficient site-specific gene insertion after TALEN/

CRISPR mutagenesis in Aedes aegypti. Proc Natl Acad Sci 112:4038–4043. https://doi.org/10.1073/pnas.1502370112

Benedict M (2003) The first releases of transgenic mosquitoes: an argument for the sterile insect technique. Trends Parasitol 19:349–355. https://doi.org/10.1016/S1471-4922(03)00144-2

Benedict M, D'Abbs P, Dobson S, Gottlieb M, Harrington L, Higgs S, James A, James S, Knols B, Lavery J, O'Neill S, Scott T, Takken W, Toure Y (2008) Guidance for contained field trials of vector mosquitoes engineered to contain a gene drive system: recommendations of a scientific working group. Vector-Borne Zoonotic Dis 8:127–166. https://doi.org/10.1089/vbz.2007.0273

Brouns SJJ, Jore MM, Lundgren M, Westra ER, Slijkhuis RJH, Snijders APL, Dickman MJ, Makarova KS, Koonin EV, van der Oost J (2008) Small CRISPR RNAs guide antiviral defense in prokaryotes. Science 321:960–964. https://doi.org/10.1126/science.1159689

Buchman A, Gamez S, Li M, Antoshechkin I, Li H-H, Wang H-W, Chen C-H, Klein MJ, Duchemin J-B, Crowe JE, Paradkar PN, Akbari OS (2020) Broad dengue neutralization in mosquitoes expressing an engineered antibody. PLoS Pathog 16:e1008103. https://doi.org/10.1371/journal.ppat.1008103

Burt A (2003) Site-specific selfish genes as tools for the control and genetic engineering of natural populations. Proc R Soc Lond B Biol Sci 270:921–928. https://doi.org/10.1098/rspb.2002.2319

Burt A (2014) Heritable strategies for controlling insect vectors of disease. Philos Trans R Soc Lond Ser B Biol Sci 369:20130432. https://doi.org/10.1098/rstb.2013.0432

Chen C-H, Huang H, Ward CM, Su JT, Schaeffer LV, Guo M, Hay BA (2007) A synthetic maternal-effect selfish genetic element drives population replacement in Drosophila. Science 316:597–600

Chevalier BS (2001) Homing endonucleases: structural and functional insight into the catalysts of intron/intein mobility. Nucleic Acids Res 29:3757–3774. https://doi.org/10.1093/nar/29.18.3757

Chiruvella KK, Liang Z, Wilson TE (2013) Repair of double-Strand breaks by end joining. Cold Spring Harb Perspect Biol 5:a012757–a012757. https://doi.org/10.1101/cshperspect.a012757

Craig GB, Hickey WA, VandeHey RC (1960) An inherited male-producing factor in Aedes aegypti. Science 132:1887–1889. https://doi.org/10.1126/science.132.3443.1887

Curtis CF (1968) Possible use of translocations to fix desirable genes in insect pest populations. Nature 218:368–369. https://doi.org/10.1038/218368a0

Deredec A, Godfray HCJ, Burt A (2011) Requirements for effective malaria control with homing endonuclease genes. Proc Natl Acad Sci U S A 108:E874–E880. https://doi.org/10.1073/pnas.1110717108

DiCarlo JE, Chavez A, Dietz SL, Esvelt KM, Church GM (2015) Safeguarding CRISPR-Cas9 gene drives in yeast. Nat Biotechnol 33:1250–1255. https://doi.org/10.1038/nbt.3412

Dong S, Lin J, Held NL, Clem RJ, Passarelli AL, Franz AWE (2015) Heritable CRISPR/Cas9-mediated genome editing in the yellow fever mosquito, Aedes aegypti. PLoS One 10:e0122353. https://doi.org/10.1371/journal.pone.0122353

Fang J (2010) Ecology: a world without mosquitoes. Nature 466:432–434. https://doi.org/10.1038/466432a

Franz AWE, Sanchez-Vargas I, Piper J, Smith MR, Khoo CCH, James AA, Olson KE (2009) Stability and loss of a virus resistance phenotype over time in transgenic mosquitoes harbouring an antiviral effector gene. Insect Mol Biol 18:661–672. https://doi.org/10.1111/j.1365-2583.2009.00908.x

Fu G, Lees RS, Nimmo D, Aw D, Jin L, Gray P, Berendonk TU, White-Cooper H, Scaife S, Kim Phuc H, Marinotti O, Jasinskiene N, James AA, Alphey L (2010) Female-specific flightless phenotype for mosquito control. Proc Natl Acad Sci 107:4550–4554. https://doi.org/10.1073/pnas.1000251107

Gantz VM, Akbari OS (2018) Gene editing technologies and applications for insects. Curr Opin Insect Sci 28:66–72. https://doi.org/10.1016/j.cois.2018.05.006

Gantz VM, Bier E (2015) Genome editing. The mutagenic chain reaction: a method for converting heterozygous to homozygous mutations. Science 348:442–444. https://doi.org/10.1126/science.aaa5945

Gantz VM, Bier E (2016) The dawn of active genetics. BioEssays 38:50–63. https://doi.org/10.1002/bies.201500102

Garneau JE, Dupuis M-È, Villion M, Romero DA, Barrangou R, Boyaval P, Fremaux C, Horvath P, Magadán AH, Moineau S (2010) The CRISPR/Cas bacterial immune system cleaves bacteriophage and plasmid DNA. Nature 468:67–71. https://doi.org/10.1038/nature09523

Gasiunas G, Sinkunas T, Siksnys V (2014) Molecular mechanisms of CRISPR-mediated microbial immunity. Cell Mol Life Sci 71:449–465. https://doi.org/10.1007/s00018-013-1438-6

Gould EA, Higgs S (2009) Impact of climate change and other factors on emerging arbovirus diseases. Trans R Soc Trop Med Hyg 103:109–121. https://doi.org/10.1016/j.trstmh.2008.07.025

Grunwald HA, Gantz VM, Poplawski G, Xu X-RS, Bier E, Cooper KL (2019) Super-Mendelian inheritance mediated by CRISPR-Cas9 in the female mouse germline. Nature 566:105–109. https://doi.org/10.1038/s41586-019-0875-2

Hall AB, Basu S, Jiang X, Qi Y, Timoshevskiy VA, Biedler JK, Sharakhova MV, Elahi R, Anderson MAE, Chen X-G, Sharakhov IV, Adelman ZN, Tu Z (2015) A male-determining factor in the mosquito Aedes aegypti. Science 348:1268–1270. https://doi.org/10.1126/science.aaa2850

Hammond A, Galizi R, Kyrou K, Simoni A, Siniscalchi C, Katsanos D, Gribble M, Baker D, Marois E, Russell S, Burt A, Windbichler N, Crisanti A, Nolan T (2016) Europe PMC Funders Group Europe PMC Funders author manuscripts: A CRISPR-Cas9 gene drive system targeting female reproduction in the malaria mosquito vector Anopheles gambiae. Nat Biotechnol 34:78–83. https://doi.org/10.1038/nbt.3439

Hartenian E, Doench JG (2015) Genetic screens and functional genomics using CRISPR/Cas9 technology. FEBS J 282:1383–1393. https://doi.org/10.1111/febs.13248

Hockemeyer D, Wang H, Kiani S, Lai CS, Gao Q, Cassady JP, Cost GJ, Zhang L, Santiago Y, Miller JC, Zeitler B, Cherone JM, Meng X, Hinkley SJ, Rebar EJ, Gregory PD, Urnov FD, Jaenisch R (2011) Genetic engineering of human pluripotent cells using TALE nucleases. Nat Biotechnol 29:731–734. https://doi.org/10.1038/nbt.1927

Honório NA, Codeço CT, Alves FC, Magalhães MAFM, Lourenço-de-Oliveira R (2009) Temporal distribution of Aedes aegypti in different districts of Rio De Janeiro, Brazil, measured by two types of traps. J Med Entomol 46:1001–1014. https://doi.org/10.1603/033.046.0505

Jackson WT, Giddings TH, Taylor MP, Mulinyawe S, Rabinovitch M, Kopito RR, Kirkegaard K (2005) Subversion of cellular autophagosomal machinery by RNA viruses. PLoS Biol 3:e156. https://doi.org/10.1371/journal.pbio.0030156

James AA (2005) Gene drive systems in mosquitoes: rules of the road. Trends Parasitol 21:64–67. https://doi.org/10.1016/j.pt.2004.11.004

Jansen R, van Embden JDA, Wim G, Schouls LM (2002) Identification of genes that are associated with DNA repeats in prokaryotes. Mol Microbiol 43:1565–1575. https://doi.org/10.1046/j.1365-2958.2002.02839.x

Jasin M, Rothstein R (2013) Repair of strand breaks by homologous recombination. Cold Spring Harb Perspect Biol 5:a012740–a012740. https://doi.org/10.1101/cshperspect.a012740

Jinek M, Chylinski K, Fonfara I, Hauer M, Doudna JA, Charpentier E (2012) A programmable dual-RNA-guided DNA endonuclease in adaptive bacterial immunity. Science 337:816–821. https://doi.org/10.1126/science.1225829

Juillerat A, Dubois G, Valton J, Thomas S, Stella S, Maréchal A, Langevin S, Benomari N, Bertonati C, Silva GH, Daboussi F, Epinat J-C, Montoya G, Duclert A, Duchateau P (2014) Comprehensive analysis of the specificity of transcription activator-like effector nucleases. Nucleic Acids Res 42:5390–5402. https://doi.org/10.1093/nar/gku155

Kass EM, Jasin M (2010) Collaboration and competition between DNA double-strand break repair pathways. FEBS Lett 584:3703–3708. https://doi.org/10.1016/j.febslet.2010.07.057

Kistler KE, Vosshall LB, Matthews BJ (2015) Genome engineering with CRISPR-Cas9 in the mosquito aedes aegypti. Cell Rep 11:51–60. https://doi.org/10.1016/j.celrep.2015.03.009
Kleinstiver BP, Prew MS, Tsai SQ, Topkar VV, Nguyen NT, Zheng Z, Gonzales APW, Li Z, Peterson RT, Yeh J-RJ, Aryee MJ, Joung JK (2015) Engineered CRISPR-Cas9 nucleases with altered PAM specificities. Nature 523:481–485. https://doi.org/10.1038/nature14592
Krueger U, Bergauer T, Kaufmann B, Wolter I, Pilk S, Heider-Fabian M, Kirch S, Artz-Oppitz C, Isselhorst M, Konrad J (2007) Insights into effective RNAi gained from large-scale siRNA validation screening. Oligonucleotides 17:237–250. https://doi.org/10.1089/oli.2006.0065
Maeder ML, Thibodeau-Beganny S, Osiak A, Wright DA, Anthony RM, Eichtinger M, Jiang T, Foley JE, Winfrey RJ, Townsend JA, Unger-Wallace E, Sander JD, Müller-Lerch F, Fu F, Pearlberg J, Göbel C, Dassie JP, Pruett-Miller SM, Porteus MH, Sgroi DC, Iafrate AJ, Dobbs D, McCray PB, Cathomen T, Voytas DF, Joung JK (2008) Rapid "open-source" engineering of customized zinc-finger nucleases for highly efficient gene modification. Mol Cell 31:294–301. https://doi.org/10.1016/j.molcel.2008.06.016
Marraffini LA, Sontheimer EJ (2008) CRISPR interference limits horizontal gene transfer in staphylococci by targeting DNA. Science 322:1843–1845. https://doi.org/10.1126/science.1165771
Marshall JM (2009) The effect of gene drive on containment of transgenic mosquitoes. J Theor Biol 258:250–265. https://doi.org/10.1016/j.jtbi.2009.01.031
Otero M, Solari HG, Schweigmann N (2006) A stochastic population dynamics model for Aedes Aegypti: formulation and application to a city with temperate climate. Bull Math Biol 68:1945–1974. https://doi.org/10.1007/s11538-006-9067-y
Porteus M (2016) Genome editing: a new approach to human therapeutics. Annu Rev Pharmacol Toxicol 56:163–190. https://doi.org/10.1146/annurev-pharmtox-010814-124454
Rasgon J (2007) Population replacement strategies for controlling vector populations and the use of Wolbachia pipientis for genetic drive. J Vis Exp:225. https://doi.org/10.3791/225
Reegan AD, Ceasar SA, Paulraj MG, Ignacimuthu S, Al-Dhabi NA (2016) Current status of genome editing in vector mosquitoes: a review. Biosci Trends 10:424–432
Reyon D, Tsai SQ, Khayter C, Foden JA, Sander JD, Joung JK (2012) FLASH assembly of TALENs for high-throughput genome editing. Nat Biotechnol 30:460–465. https://doi.org/10.1038/nbt.2170
Rouet P, Smih F, Jasin M (1994) Introduction of double-strand breaks into the genome of mouse cells by expression of a rare-cutting endonuclease. Mol Cell Biol 14:8096–8106. https://doi.org/10.1128/mcb.14.12.8096
Rudin N, Sugarman E, Haber JE (1989) Genetic and physical analysis of double-strand break repair and recombination in Saccharomyces cerevisiae. Genetics 122:519–534
Sander JD, Dahlborg EJ, Goodwin MJ, Cade L, Zhang F, Cifuentes D, Curtin SJ, Blackburn JS, Thibodeau-Beganny S, Qi Y, Pierick CJ, Hoffman E, Maeder ML, Khayter C, Reyon D, Dobbs D, Langenau DM, Stupar RM, Giraldez AJ, Voytas DF, Peterson RT, Yeh J-RJ, Joung JK (2011a) Selection-free zinc-finger-nuclease engineering by context-dependent assembly (CoDA). Nat Methods 8:67–69. https://doi.org/10.1038/nmeth.1542
Sander JD, Yeh J-RJ, Peterson RT, Joung JK (2011b) Engineering zinc finger nucleases for targeted mutagenesis of zebrafish. Methods Cell Biol 104:51–58. https://doi.org/10.1016/B978-0-12-374814-0.00003-3
Silva G, Poirot L, Galetto R, Smith J, Montoya G, Duchateau P, Pâques F (2011) Meganucleases and other tools for targeted genome engineering: perspectives and challenges for gene therapy. Curr Gene Ther 11:11–27
Singh RK, Dhama K, Khandia R, Munjal A, Karthik K, Tiwari R, Chakraborty S, Malik YS, Bueno-Marí R (2018) Prevention and control strategies to counter Zika virus, a special focus on intervention approaches against vector mosquitoes—current updates. Front Microbiol 9:87. https://doi.org/10.3389/fmicb.2018.00087
Sinkins SP, Gould F (2006) Gene drive systems for insect disease vectors. Nat Rev Genet 7:427–435. https://doi.org/10.1038/nrg1870

Skipper K, Andersen P, Sharma N, Mikkelsen J (2013) DNA transposon-based gene vehicles—scenes from an evolutionary drive. J Biomed Sci 20:92. https://doi.org/10.1186/1423-0127-20-92

Stoddard BL (2006) Homing endonuclease structure and function. Q Rev Biophys 38:49. https://doi.org/10.1017/S0033583505004063

Touré YT, Oduola AM, Sommerfeld J, Morel CM (2003) Biosafety and risk assessment in the use of genetically modified mosquitoes for disease control. World Health Organization, pp 217–222

van der Oost J, Westra ER, Jackson RN, Wiedenheft B (2014) Unravelling the structural and mechanistic basis of CRISPR–Cas systems. Nat Rev Microbiol 12:479–492. https://doi.org/10.1038/nrmicro3279

von Seidlein L, Kekulé AS, Strickman D (2017) Novel vector control approaches: the future for prevention of Zika virus transmission? PLoS Med 14:e1002219. https://doi.org/10.1371/journal.pmed.1002219

Ward CM, Su JT, Huang Y, Lloyd AL, Gould F, Hay BA (2011) Medea selfish genetic elements as tools for altering traits of wild populations: a theoretical analysis. Evolution 65:1149–1162. https://doi.org/10.1111/j.1558-5646.2010.01186.x

Weaver SC, Reisen WK (2010) Present and future arboviral threats. Antivir Res 85:328–345. https://doi.org/10.1016/j.antiviral.2009.10.008

Windbichler N, Menichelli M, Papathanos PA, Thyme SB, Li H, Ulge UY, Hovde BT, Baker D, Monnat RJ, Burt A, Crisanti A (2011) A synthetic homing endonuclease-based gene drive system in the human malaria mosquito. Nature 473:212–215. https://doi.org/10.1038/nature09937

World Health Organization (2017) Vector-borne diseases. https://www.who.int/news-room/fact-sheets/detail/vector-borne-diseases. Accessed 15 Oct 2020

Wood AJ, Lo T-W, Zeitler B, Pickle CS, Ralston EJ, Lee AH, Amora R, Miller JC, Leung E, Meng X, Zhang L, Rebar EJ, Gregory PD, Urnov FD, Meyer BJ (2011) Targeted genome editing across species using ZFNs and TALENs. Science 333:307. https://doi.org/10.1126/science.1207773

Zetsche B, Gootenberg JS, Abudayyeh OO, Slaymaker IM, Makarova KS, Essletzbichler P, Volz SE, Joung J, van der Oost J, Regev A, Koonin EV, Zhang F (2015) Cpf1 is a single RNA-guided endonuclease of a class 2 CRISPR-Cas system. Cell 163:759–771. https://doi.org/10.1016/j.cell.2015.09.038

Chapter 5
Evaluation of Transgenic *Aedes aegypti* L. Strain in India: A Friendly Mosquito

Prabhakargouda B. Patil, Kamlesh K. Yadav, Shaibal K. Dasgupta, Usha B. Zehr, Shirish R. Barwale, and Bharat Char

Abstract *Aedes aegypti* L. is a primary vector of dengue and is also responsible for transmission of other arboviral diseases including chikungunya, yellow fever, and Zika virus worldwide. India is attributed with as much as 34% of the global burden of dengue infection as per the recent reports. It is generally acknowledged controlling the vector population in the environment is the key to effectively curb the transmission of vector-borne diseases. Existing conventional methods for controlling the vector populations are not alone sufficient for reducing populations below disease transmission thresholds and would need integration of innovative strategies. With the advancement in modern biotechnology, scientists have been trying to explore the use of genetically engineered insects for their potential use in area-wide method for vector control. One among such strains being widely tested is OX513A *Ae. aegypti* strain termed as Friendly™ mosquitoes, a genetically engineered strain with a repressible dominant homozygous lethal gene construct and DsRed2 fluorescent marker gene. Male adults of OX513A strain are intended to suppress the wild population of *Ae. aegypti* in the released environment by population suppression strategy. The lethal gene renders mating events between OX513A males and wild-type females unsuccessful by passing a copy of gene to all the

P. B. Patil (✉) · K. K. Yadav
Gangabishan Bhikulal Investment and Trading Limited (GBIT), Jalna, Maharashtra, India
e-mail: prabhakar.patil@mahyco.com; kamlesh.yadav@mahyco.com

S. K. Dasgupta
Gangabishan Bhikulal Investment and Trading Limited (GBIT), Jalna, Maharashtra, India

Tata Institute for Genetics and Society, Bengaluru, Karnataka, India
e-mail: shaibal.dasgupta@tigs.res.in

U. B. Zehr · S. R. Barwale
Gangabishan Bhikulal Investment and Trading Limited (GBIT), Jalna, Maharashtra, India

Mahyco Private Limited, Jalna, Maharashtra, India
e-mail: usha.zehr@mahyco.com; shirish.barwale@mahyco.com

B. Char
Mahyco Private Limited, Jalna, Maharashtra, India
e-mail: bharat.char@mahyco.com

B. K. Tyagi (ed.), *Genetically Modified and other Innovative Vector Control Technologies*, https://doi.org/10.1007/978-981-16-2964-8_5

progeny of mated female adults leading to the death of the progeny during development stages in the absence of repressor antidote tetracycline. This OX513A strain was reviewed and recommended for carefully planned pilot deployment and randomized control trials with epidemiological outcomes to build evidence for routine programmatic use against *Ae. aegypti*-borne diseases by World Health Organization's Vector Control Advisory Group in the year 2016. The strain is being tested globally in different countries including India for its efficacy and has been deregulated in Brazil for its use in vector control program. This chapter reviews our research investigations conducted on transgenic OX513A *Ae. aegypti* strain and its potential for use in dengue vector control in India.

Keywords *Aedes aegypti* · OX513A · Dengue · Chikungunya · Genetic engineering

5.1 Introduction

Aedes aegypti Linnaeus (1762), a cosmo-tropical vector mosquito, is the type species of the subgenus *Stegomyia* Theobald (1901) commonly known as the yellow fever mosquito and is a primary vector for transmission of dengue as well as several other incapacitating arboviral diseases such as chikungunya, yellow fever, and the newly emerging Zika virus worldwide (Brady et al. 2012; Bhardwaj et al. 2017). *Ae. aegypti* has received high attention and is in focus worldwide due to its high vectorial capacity in transmission of dengue disease—the fastest-emerging vector-borne disease in the world. Dengue, endemic in more than 100 countries, is reported to be most rapidly spreading mosquito-borne disease, with a 30-fold increase in global incidence during the past 50 years. Recent estimates suggest annual dengue infections to be 390 million, of which 96 million manifest different levels of disease severity (Bhatt et al. 2013; Matthews et al. 2018). Distribution of dengue disease globally is comparable to that of malaria with an estimation of 3.97 billion population from 128 countries that are living in areas at risk for infection with an annually reported cases of 50–100 million (Gupta et al. 2012; Shepard et al. 2014). The World Health Organization (WHO) has defined global strategies for prevention and control of dengue, aiming to reduce the burden of this disease during 2012–2020 (World Health Organization 2012).

Geographical spread of vector *Ae. aegypti* from urban to rural areas has further raised the risk of epidemic outbreaks increasing its economic burden on health-care system. This increase in alarming situation is due to global warming, population growth and movements, urbanization, and lacunae in mosquito control strategies and tools (Kay and Vu 2005). Unlike other vector species, *Ae. aegypti* has tremendously adapted to the environmental conditions to make its presence throughout the year irrespective of the seasons. The presence of this species throughout the year can be attributed to its close association with man or anthropophilic nature and ability to breed and sustain in small compact sites or containers like discarded containers, coconut shells, flowerpots, water storage containers, etc. (Harrison et al. 1972). *Aedes aegypti* is backed with the advantage of its ability of the eggs to withstand desiccation more than 6 months and hatch with the return of favorable environmental conditions.

5.1.1 Dengue in India

Dengue (DENV) disease has been considered of prime importance among mosquito-borne diseases due to not only the severe forms of the disease and fatal nature but also the lack of vaccine and specific drugs. Dengue infection is caused by any of the four antigenically similar viruses (DENV, 1–4 serotypes) and is the most important arthropod-borne viral infection globally in tropical and subtropical regions (Mutheneni et al. 2017). Infection with any of the strain is thought to confer immunity toward other variants of respective DENV serotype; however, immunity against other serotypes is partial and for a transient period of 2–3 months. It has been reported that possibility of severe form of dengue infection increases during subsequent reinfection with other DENV serotypes (Halstead 2007). The first epidemiology of dengue in India was reported in Chennai (formerly known as Madras) in 1780, and the first outbreak was in Kolkata (formerly known as Calcutta) in 1963 (Ramakrishnan et al. 1964; Chaturvedi and Nagar 2008). The total number of dengue cases has significantly increased since 2001 and has been consistently increasing for the past 10 years. The number of reported cases during the year 2017 was 188,401 (NVBDCP), it remained close to 1.5 lakh in year 2019 (Fig. 5.1). It is important to emphasize that recent study on dengue infection cases in Madurai District has shown underreporting cases to be 282 times less than the number of cases reported per year (Shepard et al. 2014).

Infection trend of the dengue disease indicates frequent dengue outbreaks in various regions and its geographical expansion from urban to rural regions including increased severity of infections (Gupta and Reddy 2013; Chakravarti et al. 2012). Several factors including unplanned urbanization and environmental changes have been attributed for the outbreak of dengue. Studies have shown the association between weather and extrinsic incubation period (EIP) of dengue virus and its

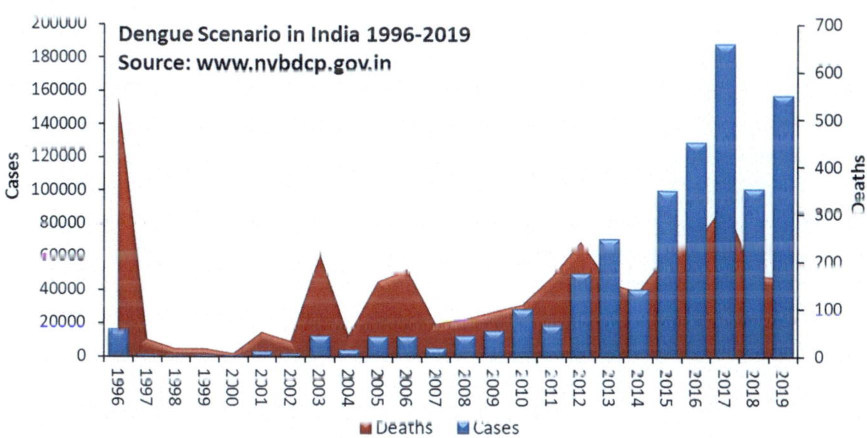

Fig. 5.1 Annual dengue cases and deaths in India during the years from 1996 to 2019. (*Source*: On public domain, https://nvbdcp.gov.in)

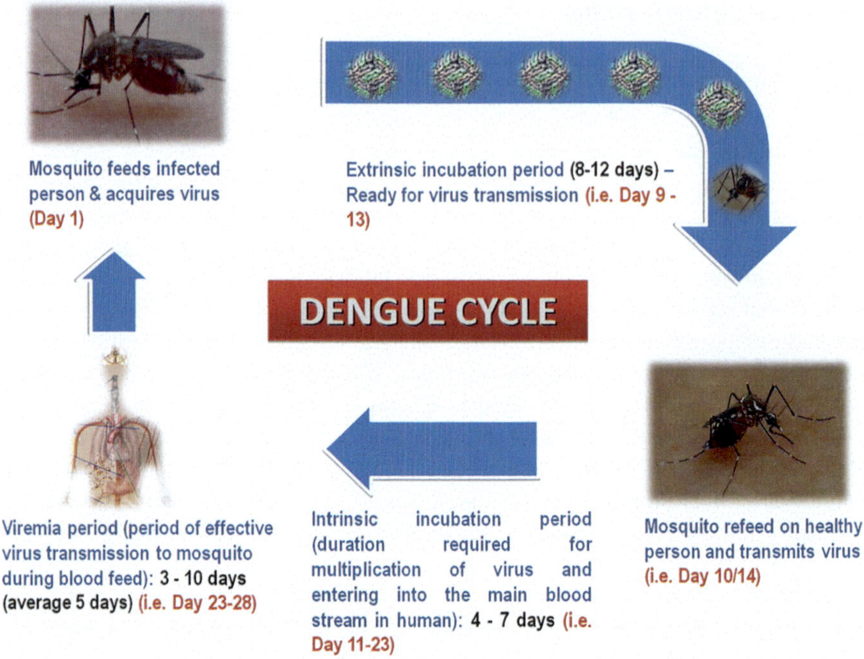

Fig. 5.2 Transmission cycle of dengue virus in hosts (human to *Ae. aegypti* to human). (*Source*: This work)

variability in various climatic zones in India. The EIP is the time period required for virus to multiply in mosquito after feeding on viremic blood meal and become infectious (Fig. 5.2). This EIP has been considered as an important factor and plays a critical role in dengue outbreaks. One of the studies on EIP estimate by monthly mean temperature has shown lowest EIP values in eastern and western coasts and higher EIP values in central, northern, and northeastern parts of India (Mutheneni et al. 2017).

The economic impact caused due to dengue disease burden has been reported to be hundreds of times greater than estimates based entirely on official reporting. As per a recent study, total annual cost of dengue was US $547 million, and the direct medical cost of dengue was reported to be $0.43 per capita in 2012 (Shepard et al. 2014). Another earlier study has shown a preliminary estimate on the immediate cost of chikungunya and dengue in Gujarat State to be INR 3.7 billion per annum and extrapolated the data to estimate the immediate cost to the whole of India, which was reported to be INR 61 billion (Mavalankar et al. 2009). In addition to the economic burden caused due to the direct cost of treatment of the disease, impact on several other aspects like emotional and long-term burden of illness and/or deaths due to these diseases, tourism, education, and per capita income is overlooked, which have a larger impact and long-term effect.

5.1.2 Chikungunya in India

Chikungunya virus (CHIKV) belongs to family *Togaviridae* and genus *Alphavirus* and has a single-stranded RNA genome, a 60–70-nm-diameter capsid, and phospholipid envelope and was first isolated in 1953 in Tanzania (Robinson 1955; Pialoux et al. 2007). The word chikungunya, termed for both the virus and the disease, means "to walk bent over" in east African languages and is due to the characteristic symptom of joint pains besides fever, rash, and incapacitating arthralgia (Enserink 2006). In India, the first major epidemic was reported in Kolkata in 1963 followed by those in Puducherry, Tamil Nadu, Andhra Pradesh, Madhya Pradesh, and Maharashtra (Sudeep and Parashar 2008). The CHIKV infection outbreak re-emerged in 2006 after a gap of 32 years from the year 1973 and was reported in 13 states infecting all age-groups. In 2006, more than 1.3 million suspected cases for CHIKV infection were reported in India, and subsequently frequent infections were reported every year (Ravi 2006; Cecilia 2014) in all states (Fig. 5.3).

5.1.3 Ae. aegypti *Vector Control*

Aedes aegypti is mainly a container breeder with only a few ounces of fresh water required to breed. Survey reports have shown breeding habitats consisting of various types including large water storage sites and medium-sized water holding sites like rock holes and tires, while highest preferred breeding sites are small containers like

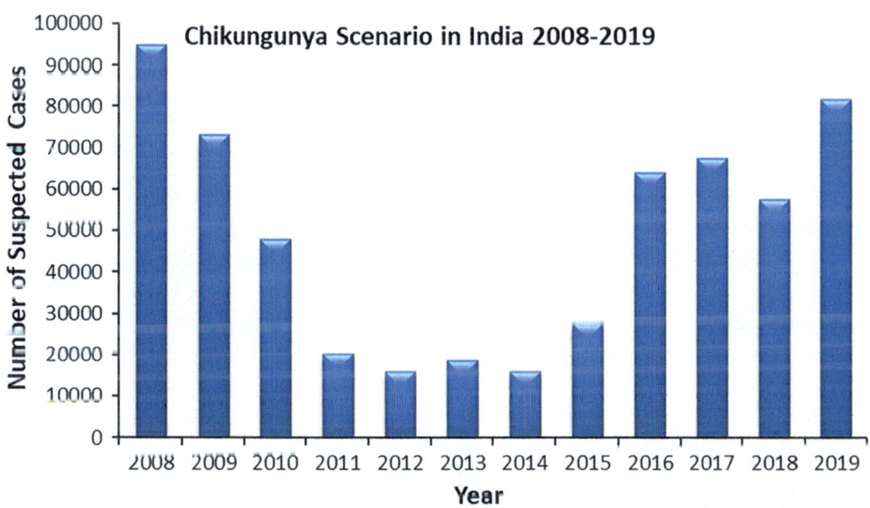

Fig. 5.3 Annual number of suspected chikungunya cases in India from 2008 to 2019. (*Source*: On public domain, https://nvbdcp.gov.in)

flowerpots, discarded cups, coconut shell, tires, etc. One of the important factors for the spread of dengue disease is the potential to transmit disease rate by *Ae. aegypti* population at a threshold level as low as two adult mosquitoes emerging per day in a locality of 100 people (Lee et al. 2008). Population density and distribution are dependent on seasonal variations due to its sensitivity to changes in temperature and humidity. Awareness among communities and proper waste management to eliminate potential breeding sites would complement the implementation of integrated vector management to curb the transmission of diseases spread by *Ae. aegypti* and other mosquito-borne diseases.

5.1.4 Genetically Modified Mosquitoes

"If major advances are to be made in coping with most of the major arthropod pest problems, then the tactics and strategies for managing such insects, ticks and mites must change. They must change from the current, limited scale, reactive, broad-spectrum measures to preventive measures that are target-pest specific and rigidly applied on an area-wide basis" (Knipling 1992). Mosquitoes will continue to challenge mankind, forcing a review of existing management tools and strategies and stimulating the development and need for new innovative control strategies. Investing in research for new innovative technologies including genetically modified (GM) insects for management of vector mosquitoes could be an integral part of strategies to combat with the frequent epidemic disease outbreaks. GM insects are one of the emerging technologies and have several advantages; however, their use and implementation in integrated management for vector control yet need public and scientific discourse. Genetically modified mosquito (GMM) strains are primarily being developed for their use in area-wide control program, in conjunction with other vector control tools, to control and reduce the rate of disease transmission. GMM has been reported to be compatible with other disease control methods, in conjunction with other disease control methods, such as the use of drugs and/or vaccines (Alphey et al. 2013; Bhatt et al. 2013). For instance, the inclusion of GMMs in integrated approach to control disease transmission could improve the protective potential of new vaccines by reducing the number of infectious mosquitoes. During the last decade, one such transgene-based sterile insect technique being tested for its effectiveness in many countries, including India, is OX513A strain, a genetically engineered *Ae. aegypti* mosquito strain. Open-field trials conducted in Brazil have shown effective suppression of the natural population of *Ae. aegypti* by sustained releases of OX513A male adults over a period of time, and the strain has been deregulated for its use in management and controlling the *Ae. aegypti* population (Harris et al. 2011, 2012; Carvalho et al. 2015).

For implementation of any such new technologies in various scientific fields including GMOs, GM crops, pharmaceutical chemicals, etc., a phased testing pathway is considered as a more appropriate way for risk assessment (UNICEF/UNDP/World Bank/WHO Special Programme for Research and Training in

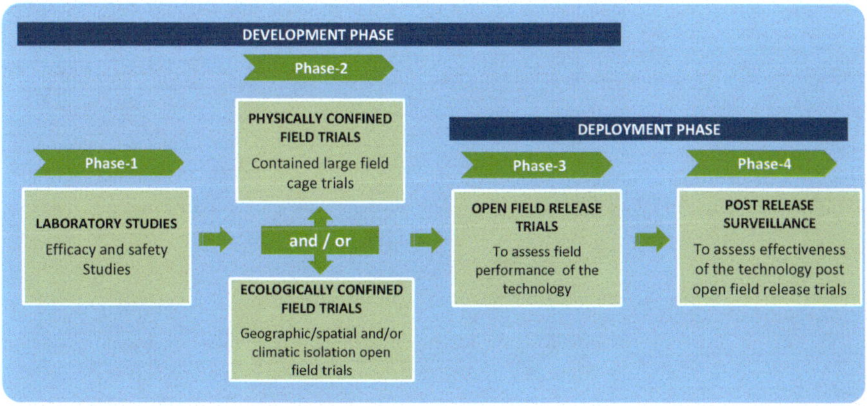

Fig. 5.4 Development and deployment phases for the implementation of genetically engineered mosquitoes as per WHO guidelines. (Adapted from WHO Special Programme for Research and Training in Tropical Diseases 2014)

Tropical Diseases 2015). The WHO has proposed guidance framework for testing and implementation of genetically modified vectors as illustrated in Fig. 5.4. The phased testing also includes and considers addressing the regulatory decisions and ethical, social, and cultural issues at each phase.

In this chapter, we discuss the studies conducted during phased evaluation of the OX513A strain under laboratory conditions as per the regulatory requirements in India.

5.2 OX513A *Aedes aegypti* L. Strain

OX513A *Ae. aegypti* strain, a genetically engineered strain, was originally described by Phuc et al. (2007), termed earlier as RIDL® insects (release of insects carrying dominant lethal gene) and also now referred as Friendly™ mosquitoes. OX513A is developed with a single insertion comprising of two dominantly inherited genes, a self-limiting repressible dominant lethal gene and fluorescent marker gene (DsRed2).

The repressible self-limiting gene component confers a lethal phenotype, in the absence of repressor antidote tetracycline, leading to mortality during the development stages; hence, immature stages fail to develop successfully into adults. Fluorescent marker gene encodes for a red fluorescence protein DsRed2, visible under a fluorescence microscope equipped with specific filters at excitation range of 510–550 nm and emission at 600 nm (Fig. 5.5).

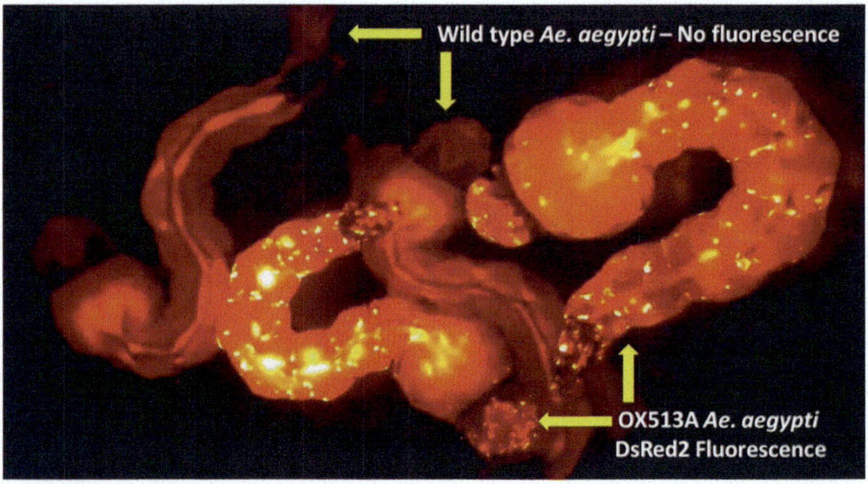

Fig. 5.5 DsRed2 fluorescence visible in third instar larvae of OX513A *Ae. aegypti* strain (Leica MZ10 F stereomicroscope with fluorescence, excitation filter ET545/30x nm—emission filter ET620/60m nm). (*Source*: This work)

5.2.1 Lethal Gene Construct

The lethal gene construct has a positive feedback system in OX513A strain (Fig. 5.6a) and is composed of a tetO binding domain (tetO), a minimal promoter, and a tTAV (tetracycline-repressible transcriptional activator). In the absence of repressor, a small amount of tTAV is produced from basal expression of the minimal promoter that binds to the tetO binding sites (Fig. 5.6b).

The binding of tTAV enhances expression of the minimal promoter producing more tTAV which in turn binds to more tetO sites. This positive feedback leading to excess production of tTAV protein in the absence of repressor (tetracycline) becomes deleterious to cell function, eventually resulting in death of the individual before adulthood.

In contrast, the presence of repressor tTAV is sequestered due to binding of tTAV with repressor (Fig. 5.6c); this precludes the positive feedback and excess production of tTAV, while only the benign, basal level of tTAV is produced. Thus, OX513A *Ae. aegypti* strain is reared in water containing 30μg/mL concentration of repressor tetracycline for suppression of lethal gene expression (Phuc et al. 2007).

5.3 Phase 1: Laboratory Studies

Eggs of *Ae. aegypti* OX513A strain were imported during the month of September in 2011, from Oxitec in accordance with the import permit issued by the Department of Biotechnology (DBT), Government of India, New Delhi. The culture of the

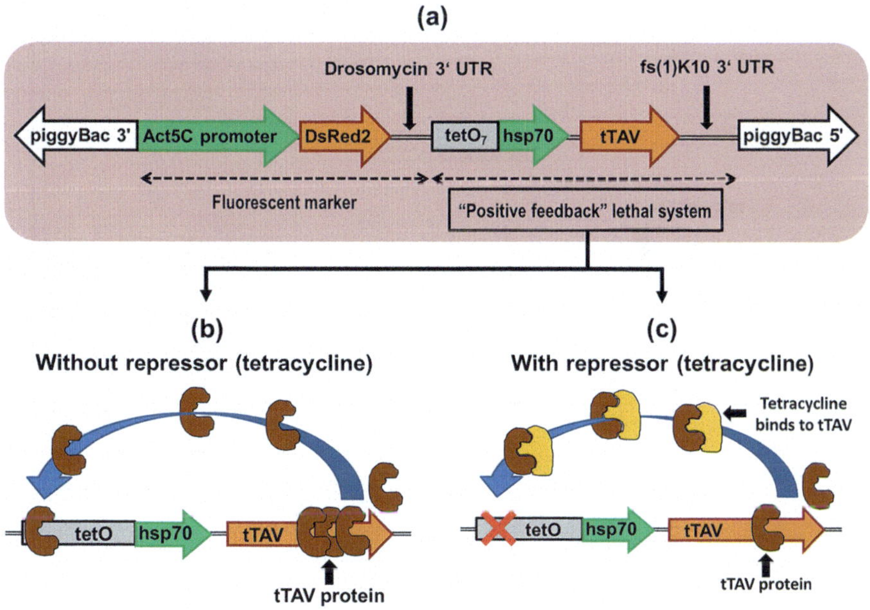

Fig. 5.6 Schematic representation of gene construct and positive feedback system in OX513A strain (Adapted from Phuc et al. 2007; Gong et al. 2005). (**a**) Transgenic OX513A are readily identified by red fluorescence visible due to expression of DsRed2. The positive feedback system is composed of tTAV, a tetracycline-repressible transcriptional activator, and is under the control of its own binding site, tetO, a minimal promoter from *Drosophila* hsp70, and a 3′ UTR sequence from *Drosophila fs(1)K10*. (**b**) In the absence of repressor (tetracycline), tTAV produced from basal expression of the minimal promoter binds to tetO binding sites and drives expression of more tTAV, which in turn binds to more tetO sites enhancing expression of tTAV in a positive feedback loop. Large quantities of tTA produced in high enough levels are deleterious to cells, possibly due to the interaction of the VP16 domain with key transcription factors. (**c**) In the presence of repressor (tetracycline), tTAV binds with tetracycline due to its high affinity toward tetracycline and does lead to expression of more tTAV. (*Source*: This work)

OX513A strain was initiated from the eggs imported and maintained under Arthropod Containment Level 2 (ACL-2) laboratory (Fig. 5.7) located at Dawalwadi, Jalna District, of Maharashtra State (Arthropod Containment Levels 2003). In addition, two wild-type strains from Indian region, a strain from Aurangabad region (AWD) and from Delhi region (DEL), were maintained under laboratory conditions and used for the laboratory studies. Following the import of the strain and as per the WHO guidelines and recommendations from the Review Committee for Genetic Manipulation (RCGM), Department of Biotechnology, Government of India, Phase 1 laboratory studies on OX513A strain were conducted.

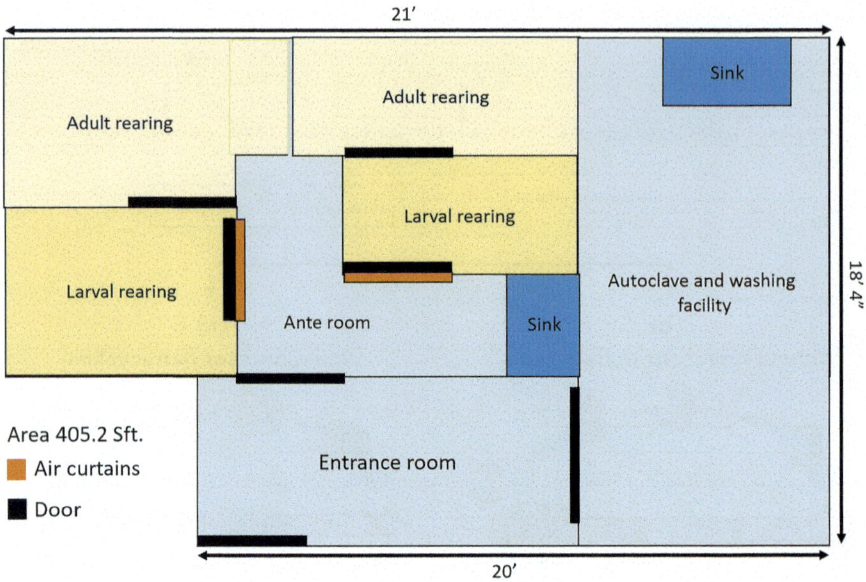

Fig. 5.7 Layout of the Arthropod Containment Level 2 laboratory facility. (*Source*: This work)

5.3.1 Evaluation of Lethal Gene Expression

This study was conducted to evaluate the lethality induced by gene expression during the developmental phases in offsprings of wild-type *Ae. aegypti* females mated with OX513A male adults in the absence of tetracycline (chlortetracycline hydrochloride), an antidote used to repress expression of the lethal gene during the development phase (Phuc et al. 2007).

For generating heterozygous eggs, 60 female adults of wild type were outcrossed with 30 male adults of OX513A strain in a cage of size 30 cm^3. Mated female adults were allowed for oviposition post-blood meal, and eggs laid were collected and used for the study. The treatment group in the experimental setup consisted of (a) homozygous OX513A strain larvae and (b) heterozygous larvae reared, both reared in the absence of tetracycline. Control groups were (1) OX513A strain larvae; (2) wild-type DEL strain larvae, both supplemented with tetracycline; and (3) wild-type DEL larvae reared without tetracycline. Each group consisted of three replicates with 200 larvae in each. Larvae in the treatment and control groups were provided with tetramin fish food ad libitum, and observations were recorded till emergence of adults.

Results demonstrated the mortality during developmental stages in homozygous and heterozygous conditions reared without tetracycline in the rearing medium (data unpublished). Homozygous offspring collapsed during the larval stages expressing 93.7% mortality during the larval stages, while pupal and adult mortalities were found to be 6.1% and 0.2%, respectively, with 99.97% total mortality (Fig. 5.8).

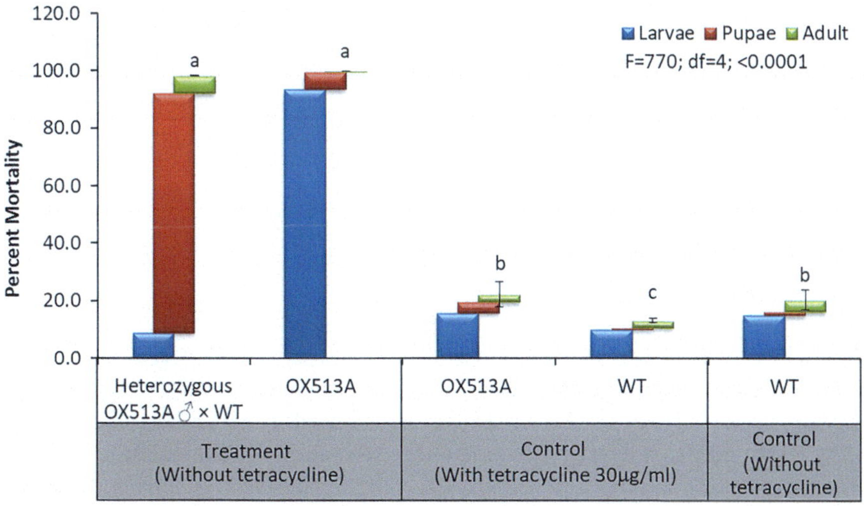

Fig. 5.8 Mortality rate (%) during developmental stages of homozygous and heterozygous (progeny of OX513A ♂ × wild type ♀) progeny of OX513A strain *Ae. aegypti* reared without repressor tetracycline. (*Source*: This work)

Heterozygous offspring mortality was observed during the pupal stage exhibiting 83.1% pupal mortality and 9.1% mortality during larval stage. The pupae formed were transferred to adult rearing cages for eclosion. Mortality during adult stage (adults dead on the surface of water immediately after eclosion and nonflying adults dead during the first day of emergence were taken into account) was found to be 6.1%. Total mortality observed was 98.3% and 1.63% adults were emerged. Adults emerged could not survive for a longer period with a median life span of 5 days, most of the female adults emerged died prior to blood meal (within 2 days), and only 0.4% of female adults survived long enough to take blood meal. In control groups, the emerged adults (homozygous reared with tetracycline and wild-type adults) survived successfully.

Although a very small percentage of the heterozygous adults survived under laboratory conditions, it may be noted that, under natural environment, the daily survival rate is found to be very low compared to adult life span under laboratory conditions due to harsh conditions in the environment. Moreover the female adults have to survive at least the extrinsic incubation period, typically 7–13 days for dengue, to have any possibility of transmitting the virus from person to person (Watts et al. 1987). Earlier report on fitness modeling indicates that for efficient use of OX513A strain of *Ae. aegypti*, the mean fitness of heterozygous offspring of the released homozygous OX513A males mated with wild-type females should be <10% that of wild type (Phuc et al. 2007).

It was concluded based on the experimental observations that homozygous more than 99% offspring collapse before reaching adult stage when reared in the absence of tetracycline and none of the offspring emerged into successful adults. In heterozygous offspring reared in the absence of tetracycline, more than 90% of the offspring collapse before reaching adult stage, and a very small percent of the emerged adults survive (<1%) with lower fitness as compared to wild-type counterpart, indicating higher efficacy for its use in wild population suppression. Furthermore, these survivors need to be fully fit and survive long enough under harsh natural open-field conditions.

5.3.2 Life Table Parameters

Uses of GMOs including mosquitoes have been demonstrated by several workers as an effective and alternative option in the combat against pest insects and disease-carrying vectors (Li et al. 2008). However, successful implementation of such technologies needs stringent evaluation under laboratory conditions for fitness and effectiveness for controlling target species in the environment (Curtis and Sinkins 1998; Sinkins and Gould 2006; Huang et al. 2007). Assessment of the life table parameters of GMOs is an initial and important step in the phased evaluation of such strains for their subsequent use. In view of the studies on fitness cost with respect to life table, reproductive parameters and population growth parameters were analyzed for transgenic OX513A strain in comparison with two wild-type DEL and AWD strains (Patil et al. 2015).

The strains were reared from eggs to adult stages, and observations were carried out throughout the developmental and reproductive period. For the experimental setup, eggs of transgenic OX513A *Ae. aegypti* strain and two non-transgenic wild-type strains of *Ae. aegypti*, DEL and AWD strains from the cyclic colony, were submerged in tap water and placed under low pressure for 20 min for synchronized hatching. In post-hatching, the larvae were transferred individually to plastic cups [8 cm (Ø) and 3.5 cm (ht)] containing 20 mL of tap water with 120 replicates for each strain. Tetracycline (30μg/mL) was added to the tap water in plastic cups holding OX513A strain larvae for repressing the lethal gene expression. And the larvae were provided with a drop of liquifry for the first day followed with tetramin fish food ad libitum throughout the larval development period. Observations on the development were recorded daily to determine developmental period of each aquatic developmental stage.

Thirty male/female pupae of each strain were introduced separately into adult rearing cages (16 cm^3) for adult emergence. Adults emerged were fed with 10% sucrose solution in cotton pads, and on the fourth day of emergence, male/female adults of each strain were paired (30 pairs), and each pair was transferred to mating cages for a period of 24 h to ensure successful insemination of female adults by male adults. In postmating period, female adults from each mating cage were separated to individual rearing adult box [6.5 cm (Ø) and 8.0 cm (ht)] holding small containers containing water and a paper strip for oviposition by female adults. Female adults were fed with animal blood meal for oviposition. Adult survivorship and fecundity were recorded daily for determining life span and reproductive parameters.

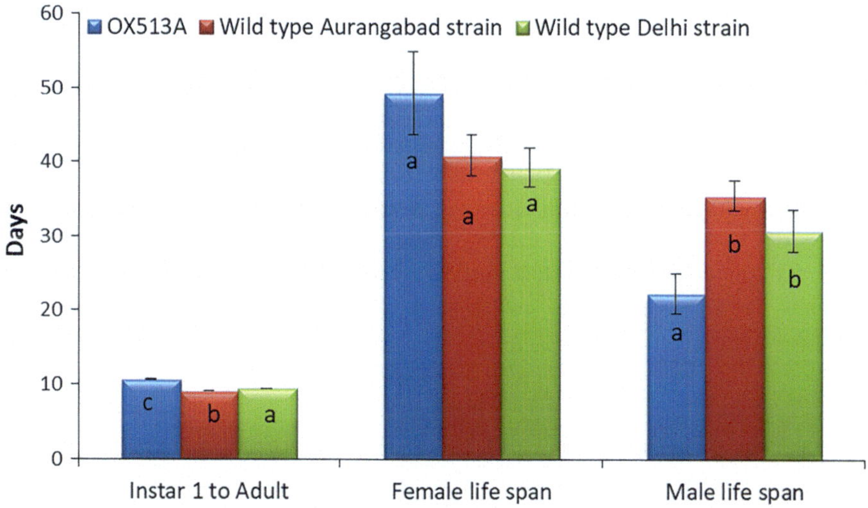

Fig. 5.9 Aquatic developmental period and male/female adult life span of OX513A and wild-type strains of *Ae. aegypti*. (*Source*: This data is reproduced with permission from author's own publication, Patil et al. 2015, © 2014 Gangabishan Bhikulal Investment and Trading Limited. Pest Management Science published by John Wiley & Sons Ltd. on behalf of the Society of Chemical Industry)

Stage-specific developmental period of transgenic and wild-type strains revealed significant variations within the larval instars. Larval developmental period (instar I to IV) for OX513A strain (8.5 days) was found to be significantly higher compared to wild-type strains of DEL and AWD with 7.4 and 7.1 days, respectively ($p < 0.0001$). Pupal period varied between 2.1 and 2.2 for wild-type and transgenic strains ($p < 0.05$). The developmental period from instar I to adult emergence for OX513A strain was found to be 10.7 days and differed significantly from DEL to AWD strains exhibiting developmental period of 9.4 and 9.1 days ($p < 0.0001$), respectively (Fig. 5.9). Observations on the stage-specific mortality revealed 18% mortality for OX513A strain during second instar and 8–10% during fourth instar for DEL and AWD strains.

Average life span of female adults of transgenic OX513A strain was 49.3 days demonstrating nonsignificant difference compared to wild type strains with a survival period of 39.3 and 40.9 days for DEL and AWD strains, respectively. Life span of male adults was found to be significantly less compared to wild type strains; however, there was no significant difference observed between the wild-type strains (Fig. 5.9). Adult female survival following each oviposition decreased consistently, and survival was found to be extended greater than 80 days for some of the females in all the strains.

Demographic parameters (net reproductive rate, mean generation time, and doubling time) of the transgenic strain showed similarity and were comparable with the wild-type strains.

5.3.2.1 Net Reproductive Rate [$R_0 = \Sigma l_x m_x$]

R_0 represents the average number of offspring that a female reproduces during her lifetime termed as net reproductive rate. l_x is the proportion of surviving adult females at each age (x), and m_x is the average number of offsprings reproduced to the surviving adult females at each age. Sum of the products ($l_x m_x$) of all the age-groups will give the growth rate per generation of the population under defined conditions. Value of $R_0 > 1.0$ indicates increase in the population, while value of $R_0 = 1.0$ indicates that a population is neither increasing nor decreasing but replacing its number exactly and if value of $R_0 < 1.0$ indicates decrease in population.

5.3.2.2 Mean Generation Time [$T_c = \Sigma x l_x m_x / R_0$]

T_c represents average interval between the birth of an individual and the birth of its offspring during each generation.

5.3.2.3 Doubling Time [$T_d = \text{Ln } 2/r_m$]

T_d is the duration required for a population growing at a specified rate to double in size of the population.

Fitness of the transgenic OX513A strain in comparison with wild-type strains was clear indicating similarity when demographic parameters were assessed. The population growth parameters assessed (net reproductive rate, mean generation time, intrinsic rate of increase, and doubling time) were significantly similar except doubling time for AWD strain was slightly but significantly high ($F = 5.72$, $p < 0.005$) compared to doubling time of OX513A and DEL strains (Fig. 5.10). These demographic growth parameters indicate similarity of fitness of transgenic OX513A strain in comparison with wild-type DEL and AWD strains. There are several reports that development of transgenic insects comes at a substantial cost of fitness (Irvin et al. 2004). Results demonstrated that the OX513A strain behaves similarly in comparison with wild-type strains in terms of development, reproductive, and population growth parameters under laboratory conditions. A similar study on life table parameters for OX513A strain in comparison with wild-type Malaysian strain has reported similarity for several parameters involving preoviposition period, lifetime fecundity, offspring sex ratio, and female sterility (Lee et al. 2009).

The study showed the life span of male adults to be significantly lower compared to wild-type strains. It must be emphasized here that OX513A strain is developed and intended for use outside the laboratory to suppress the target population in the field; therefore, the transgenic is neither required nor expected to persist in the environment for a longer time, imposing no risk in the environment.

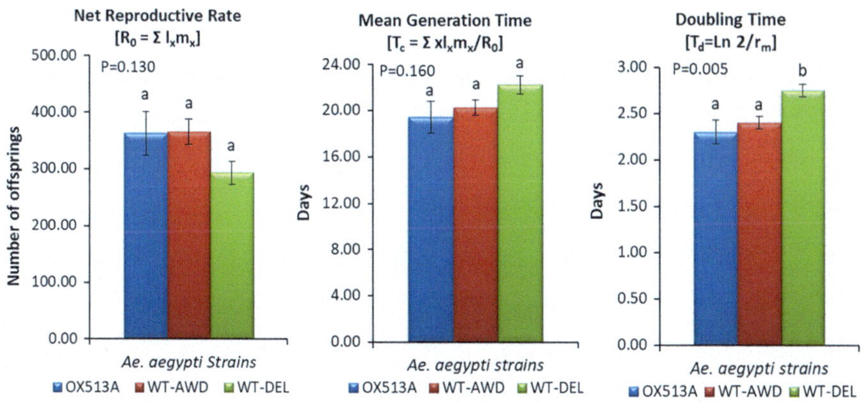

Fig. 5.10 Demographic parameters of OX513A and wild-type strains of *Ae. aegypti*. (*Source*: This data is reproduced from author's own publication with permission from Patil et al. 2015, © 2014 Gangabishan Bhikulal Investment and Trading Limited. Pest Management Science published by John Wiley & Sons Ltd. on behalf of the Society of Chemical Industry)

5.3.3 Mating Competitiveness

Mating competitiveness experiments were performed under defined laboratory conditions in adult rearing cages (30 cm^3) between OX513A and wild-type DEL strains. Three-day-old adults were used for the study, and the experiments were conducted by introducing OX513A males, wild-type Delhi strain females, and wild-type Delhi strain males in the ratio of 5:5:5, respectively (Patil et al. 2015). In each cage, five wild-type female adults were introduced and allowed for acclimatization for 1 h, followed by introducing male adults of OX513A and wild-type DEL strain together (5:5) to ensure equal chances of mating for both strains. A total of 57 replicates were performed.

Adults were allowed for 24-h mating period followed by separation of female adults to individual rearing boxes [6.5 cm (Ø) and 8.0 cm (ht)] and subsequently provided with blood meal for oviposition. Eggs laid by individual females were allowed for embryonic development and to be hatched for fluorescence screening during the third instar larval stage under fluorescence microscope (*Nikon* fluorescence microscope) under the filter with an excitation/emission range of 540 nm/ 600 nm. The presence of fluorescence in the larvae indicated females mated with OX513A males and the absence of fluorescence in case of females mated by wild-type males, and the presence of both fluorescence and nonfluorescence larvae indicated double mating with males of both strains.

Data of females mating against OX513A strain, DEL strain male adults, and double mating (OX513A and DEL male adults) revealed 42.8%, 39.6%, and 9.1% mating, respectively (Fig. 5.11). The mean number of adult females mated with respective to males including double mating (mated with both OX513A and DEL males) was significantly similar within the values for single paternity of OX513

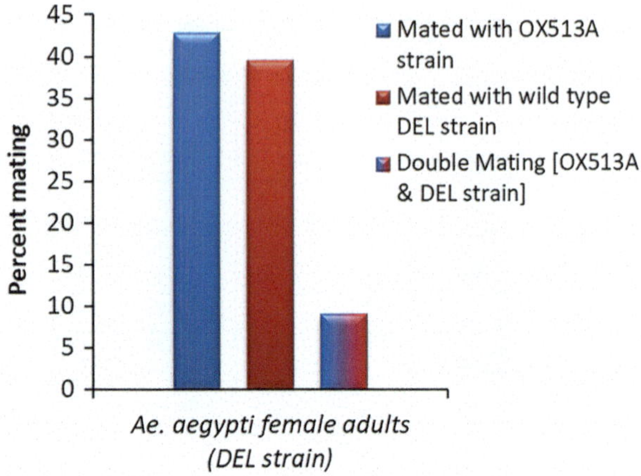

Fig. 5.11 Percent mating of wild-type female adults with male adults of OX513A and wild-type strains. (*Source*: This data is reproduced from author's own publication with permission from Patil et al. 2015, © 2014 Gangabishan Bhikulal Investment and Trading Limited. Pest Management Science published by John Wiley & Sons Ltd. on behalf of the Society of Chemical Industry)

males/DEL males and differed significantly for double-mated females ($F = 40.2$, $p = 0.0001$). The analysis of mating proportion 122:113:26 by chi-square test against expected mating proportion of 130.5:130.5:0.0 (mated with OX513A strain males/mated with DEL strain males/double mated with both OX513A and DEL strain males) revealed nonsignificant deviation at 0.05 level ($X^2 = 0.012$, $p = 0.9942$). Although a small percent (9.1%) of double mating was observed, the chi-square test revealed that the deviation of relative mating proportion for double mating is nonsignificant indicating the observed double mating to be negligible (~0). The expected mating proportion was calculated by equal distribution of the total number of observed mated female adults against single mating with OX513A/DEL strain and nil for double mating. And the calculation of expected proportion was based on consideration of the fact that the female adult is monogamous and double mating is of rare occurrence in the natural environment.

Double mating in the laboratory cage is attributable to the restriction of movement for mated female adults in the cage which could have led for forced mating for the second time by sex-starved male adults. This is not the case in the environment as the female once inseminated by a male will rarely mate for the second time. This phenomenon of single mating is due to the factor called "matrone" transferred in semen by male adults rendering female adults unreceptive to males for second mating and refractory to further copulation (Craig 1967; Helinski and Harrington 2011). In conclusion, the results of the mating experiments demonstrated OX513A strain to be equally competitive and fit in comparison with wild-type males under laboratory conditions.

5.3.4 *Insecticidal Susceptibility Tests*

One of the safety aspects when considering GMM releases, in addition to any inserted genes, is possibility of carrying potentially deleterious genes in the strain

Table 5.1 Susceptibility status of the OX513A and two wild-type *Ae. aegypti* strains tested against insecticides

		Susceptibility status (mortality)	
Insecticides	Strains	Male adults	Female adults
DDT (4%)	OX	Resistance likely	Resistant (<90%)
	AWD	Resistant (<90%)	Resistant (<90%)
	DEL	Resistant (<90%)	Resistant (<90%)
Malathion (5%)	OX	Susceptible (100%)	Susceptible (100%)
	AWD	Susceptible (100%)	Susceptible (100%)
	DEL	Susceptible (100%)	Susceptible (98%)
Permethrin (0.75%)	OX	Susceptible (100%)	Susceptible (100%)
	AWD	Susceptible (100%)	Susceptible (100%)
	DEL	Susceptible (100%)	Susceptible (100%)
Deltamethrin (0.05%)	OX	Susceptible (100%)	Susceptible (100%)
	AWD	Susceptible (100%)	Susceptible (100%)
	DEL	Susceptible (100%)	Susceptible (100%)

Source: This data is reproduced from author's own publication with permission from Patil et al. 2018, Copyright © 2018 Prabhakargouda B. Patil et al.

such as those encoding for insecticide resistance. Even for cases where GMM strains are unable to persist and/or the introgression of background genes into wild populations is not of concern due to self-limiting genes, their inherent susceptibility to insecticides may have implications for risk mitigation, where an attempt to clear them from a specific environment is expected. In the view, the transgenic OX513A strain along with wild-type strains was tested for insecticidal susceptibility (Patil et al. 2018).

Adults of all three strains were tested using WHO insecticide susceptibility testing kits, with insecticide-impregnated papers (Vector Control Research Centre, Universiti Sains, Malaysia). The insecticides tested were (1) 4% dichlorodiphenyltrichloroethane, DDT (organochlorine), (2) 5% malathion (organophosphate), (3) 0.75% permethrin (pyrethroid), and (4) 0.05% deltamethrin (pyrethroid). Insecticide tests were carried out as per the WHO bioassay procedure against 4–5-day-old adult male/female fed with sugar (World Health Organization 2016). Male and female adult mosquitoes were exposed separately to insecticide-impregnated papers for a period of 60 min with cumulative knockdown time recorded at an interval of 5 min. Following the WHO guidelines, mosquitoes exposed to insecticide were considered "knocked down" if a mosquito was unable to stand or fly in a coordinated way or observed without flight, fallen to the bottom of the exposed tube. Assessment criteria for susceptibility tests were as follows: 98–100% indicated full susceptibility, <98% suggested resistance was likely, and <90% confirmed the presence of resistance (World Health Organization 2016).

Mortality data of male and female adults of OX513A and wild-type strains indicated susceptibility to malathion, deltamethrin, and permethrin insecticides (Table 5.1). Knockdown time for male and female adults of all the strains against

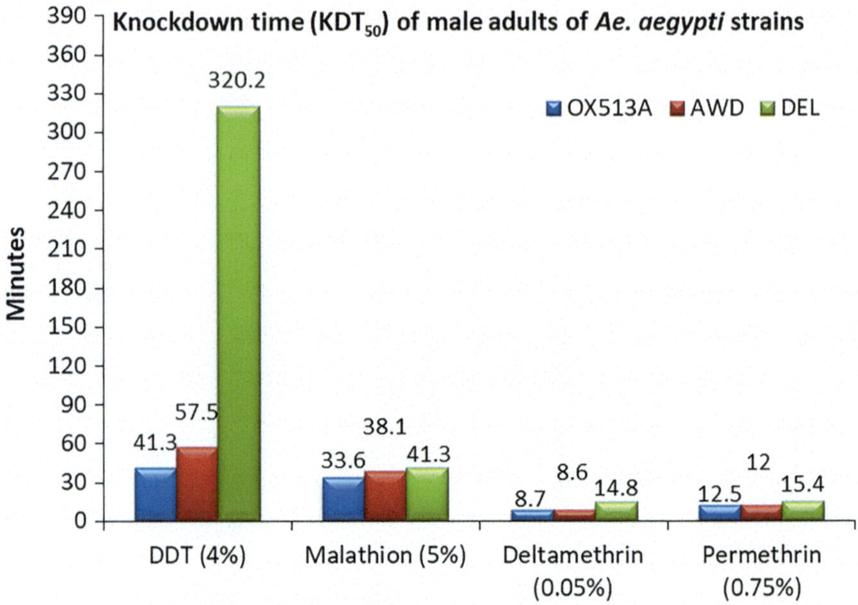

Fig. 5.12 Knockdown times (KDT) for male adults of OX513A and wild-type *Ae. aegypti* strains against the four insecticides. (*Source*: This data is reproduced from author's publication with permission from Patil et al. 2018, Copyright © 2018 Prabhakargouda B. Patil et al.)

insecticides revealed OX513A strain to be of equally and slightly higher susceptibility to insecticide tested in comparison with wild-type strains (Figs. 5.12 and 5.13).

Knockdown times (KDT_{50}) for male and female adults of transgenic OX513A strain against DDT were 41.3 and 50 min, respectively, and mortality within 24 h observed for female adults (70.1%) was significantly lower than observed for males (90.9%). Assessment of the susceptibility tests against the WHO interpretation criteria confirmed the presence of resistance in all three strains. While comparing mortality rates for DDT, OX513A were less resistant than either wild-type strains; indeed, the male of OX513A strain demonstrated mortality of >90% categorizing them as "resistance likely" (Table 5.1) in contrast to male adults of either wild-type strains scoring them as "resistance confirmed." Earlier numerous reports have shown the presence and increase of resistance to DDT in *Ae. aegypti* population as a worldwide phenomenon, likely due to usage of DDT in vector control programs for a prolonged period (Dhiman et al. 2014). These susceptibility test results were consistent with those previously reported for transgenic OX513A along with Malaysian *Ae. aegypti* strain (Nazni et al. 2009). The study demonstrated that OX513A strain poses very low risks, due to the presence of self-limiting gene, and is susceptible to insecticides, if required to eliminate them from the released environment.

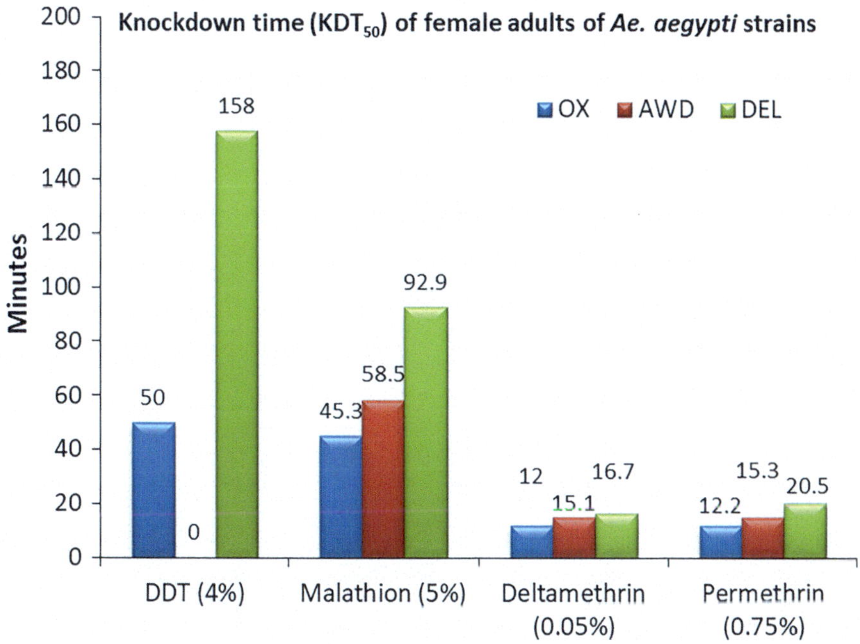

Fig. 5.13 Knockdown times (KDT) for female adults of OX513A (OX) and wild-type *Ae. aegypti* strains against the four insecticides. (*Source*: This data is reproduced from author's own publication with permission from Patil et al. 2018, Copyright © 2018 Prabhakargouda B. Patil et al.)

5.3.5 Analysis of Mitochondrial Cytochrome Oxidase I

Understanding the genetic deviation in *Ae. aegypti* to track the evolution and molecular ecology is of crucial importance for implementing measures to control the vector mosquitoes more effectively. Analysis of the mitochondrial DNA regions is one of the standardized tools for understanding the genetic variations in the vector population. Mitochondrial DNA (mtDNA) represents only a tiny fraction of organismal genome size, yet it is by far the most popular molecular diversity marker in animals (Galtier et al. 2009). Adoption of mtDNA as marker of choice is well known as mtDNA is relatively easy to amplify and it appears in multiple copies in the cell. Mitochondrial DNA is highly variable in populations in the natural environment because of its elevated mutation rate, which can shed insight about population history over short time frames (Avise et al. 1987; Moritz et al. 1987; Ratnasingham and Hebert 2007).

A study on the mtDNA cytochrome oxidase 1 (CO1) gene as a genetic marker was carried out to understand the variations in geographically closely related natural populations of *Ae. aegypti* and the laboratory strains (OX513A, wild-type AWD, and DEL strains) of *Ae. aegypti* (Gandhi et al. 2020). Field samples of *Ae. aegypti* mosquito were collected by ovitrap method from nine villages of Badnapur Taluka,

Table 5.2 Haplotypes, collection sites of isolates, and accession number for each isolate

Haplotypes	Laboratory and field isolates (code)—accession no.
H1	• Laboratory wild-type strain of Aurangabad region (AWD)—MK805533 • Akola village (AK)—MK805532 • Deogaon village (DG)—MK805534 • Haldola village (HA)—MK805537
H2	• Doksal village (DK)—MK805535 • Dawalwadi village (DW)—MK805536 • Khadgaon village (KH)—MK805538 • Matarewadi village (MA)—MK805539 • Rajewadi village (RJ)—MK805541
H3	• Ramkheda village (RA)—MK805540
H4	• Laboratory wild-type strain of Delhi region (DEL)—MK805542
H5	• Transgenic OX513A strain (OX)—MK805543

Source: This data is reproduced with permission from Gandhi et al. 2020,

Maharashtra State, India. Eggs collected were hatched and reared till adult emergence for identification of *Ae. aegypti* species following morphological identification keys described by Barraud (1934). Sample species identified were preserved in container under −80 °C for further analysis.

Analysis of the COI sequences of field and laboratory strains of *Ae. aegypti* revealed the presence of five haplotypes (H1 to H5). The haplotype H2 was found in five populations, followed by haplotype H1 observed in four populations, whereas H3, H4, and H5 represented only single population (Table 5.2 and Fig. 5.14). Wild-type AWD strain with H1 haplotype showed similarity with three village isolates AK, DG, and HA, while isolates from villages DK, DW, KH, MA, and RJ had a single nucleotide variation representing H2 haplotype. Field sample from village RA belonged to the H3 haplotype separating from the other observed haplotypes by six mutational events. Wild-type laboratory strain DEL and transgenic OX513A strain had a total of 9 and 12 nucleotide variations among all isolates representing haplotypes H4 and H5, respectively. Multiple sequence analysis showed a total of 20 polymorphic sites (36.8%) in all the five haplotype populations identified. It was observed among analyzed field populations that RA belonging to H3 haplotype was separated from other haplotypes demonstrating six mutational events and formed a monophyletic cluster.

Multiple sequence alignment analysis was performed with the COI sequences of isolates of this study and sequences retrieved from the NCBI reported from India and worldwide. And phylogenetic relationship was examined among COI gene sequences of all field and laboratory samples along with 30 Indian isolates using the Bayesian method. The transgenic OX513A strain exhibited the maximum similarity with the isolates from Rajasthan, Andhra Pradesh, and Tamil Nadu in India and with the worldwide isolates from Singapore, Thailand, and Martinique regions. Although the OX513A isolate exhibited genetic divergence among other

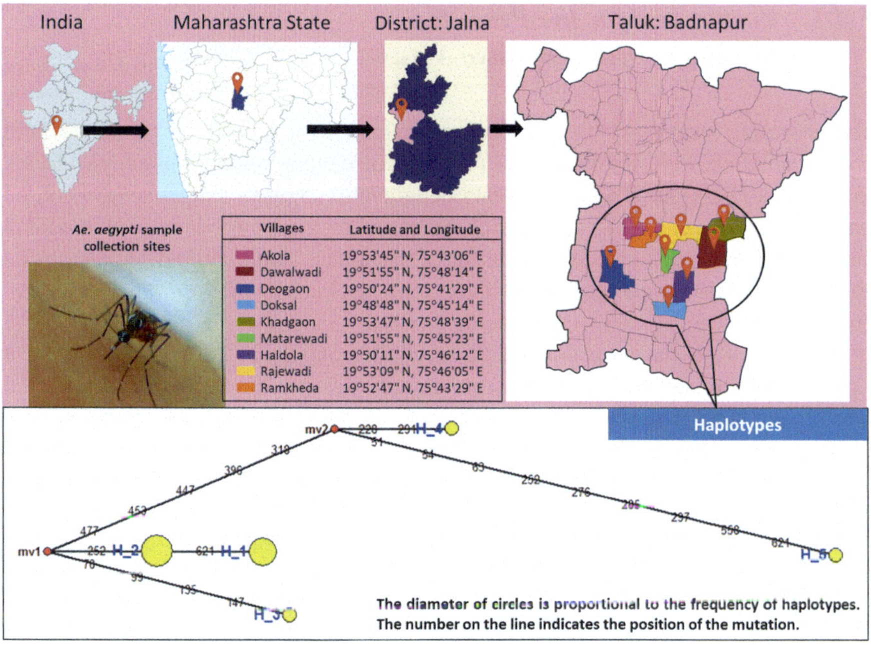

Fig. 5.14 Median-joining haplotype network of mtDNA COI sequences of *Ae. aegypti* isolates. Diameter of circles is proportional to the frequency of haplotype. Number on the line indicates the position of the mutation on the analyzed sequences. (*Source*: This figure is reproduced with permission from Gandhi et al. 2020, © 2019 Korean Society of Applied Entomology. Published by Elsevier B.V. All rights reserved)

field and laboratory isolates, the OX513A isolate has shown similarity with other isolates of Indian region and worldwide (Gandhi et al. 2020).

5.4 Phase 2: Physically Contained Field Cage Studies

As mentioned earlier, the evaluation of genetically engineered mosquitoes for its potential use in public health tools involves phased testing including contained trials with natural environmental settings. Phase 2 trials involve physically contained field cage studies for testing the strains for efficacy studies intended for field releases. The transgenic OX513A strain was evaluated under physically contained large field cages with natural environmental settings as per the recommendations from Indian regulatory board, RCGM, Department of Biotechnology, Government of India.

5.4.1 *Vector Test for the Presence of Dengue/Chikungunya Virus*

The mosquito strains AWD and OX513A planned for physically contained field cage evaluation were tested for the presence of dengue and chikungunya viruses using VectorTest® kit. The VectorTest® dengue/chikungunya antigen assay is based on an immuno-chromatographic wicking assay method, and a dual monoclonal antibody "sandwich" principle is used in the assay. Antibodies detecting virus-specific antigens have been immobilized on a wicking test strip (www.vectortest.com).

Test zones on the strip contain the second monoclonal antibody and form a "sandwich" (first antibody-antigen-second antibody) and can be visualized as a red line on the strip if the antigen is present. Unbound antibodies are captured in a control zone and form a second red line, confirming that the test has been carried out correctly (Fig. 5.15). The test does not cross-react with other flaviviruses (for dengue kits) and alphaviruses (for chikungunya kits).

The test results showed the presence of only a control line on the dipstick, a negative test result for both strains tested for dengue and chikungunya virus, indicating dengue- and chikungunya virus-free strains (Fig. 5.15).

5.4.2 *Physically Contained Field Cage Studies*

Phase 1 laboratory studies on mating competitiveness, life table parameters, and efficacy of the OX513A strain showed promising results confirming the strain to be fit and equally compatible with wild-type strains of *Ae. aegypti* (Patil et al. 2015), while the efficacy studies under contained natural environment to understand and envisage field performance of the OX513A strain are desirable for open-field release trials and field implementation of the technology in the future.

The physically contained facility was developed at Dawalwadi, Jalna District, Maharashtra State, India, with a built-up area of 580 m^2 (29 m × 20 m) consisting of six field cages fastened on a wooden platform elevated 1 m above the ground which was supported on legs made out of steel (Fig. 5.16). Experiments were conducted in these physically large field cages to evaluate transgenic OX513A males for mating competitiveness with wild-type AWD strain of *Ae. aegypti* and to demonstrate suppression of established target wild-type (AWD) population by sustained release of OX513A male adults.

These experiments performed under physically contained field cage studies have shown OX513A strain to be equally competent with wild-type strain (data under communication). Sustained release of the OX513A strain male adults has been demonstrated with excellent efficacy and capable of suppressing the target population to the point of eradication within a period between 10 and 16 weeks (data under communication). Results have shown unequivocally that in these contained

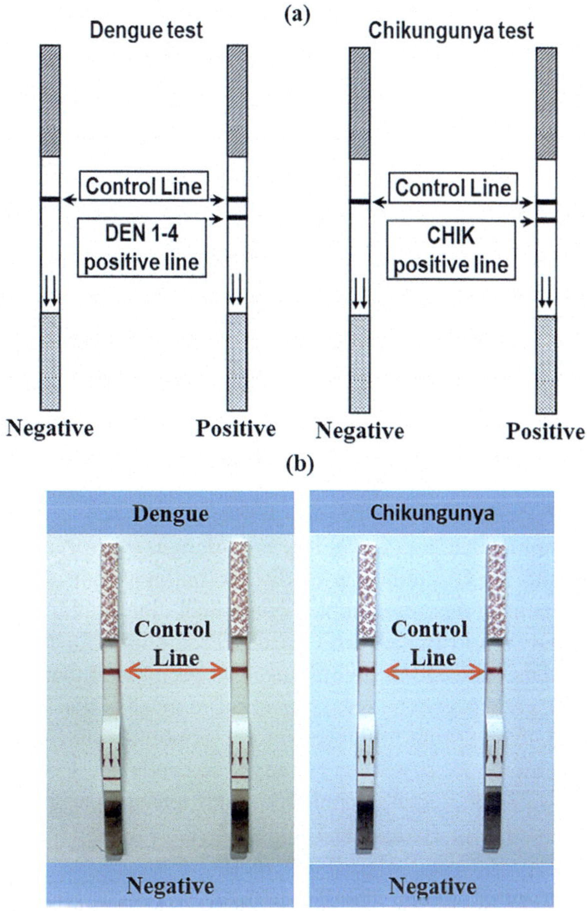

Fig. 5.15 Testing of OX513A strain *Aedes aegypti* L. for the presence of dengue and chikungunya virus by VectorTest® kit, (**a**) reference for interpretation of results for the presence of dengue/chikungunya antigen assay, (**b**) test results of AWD/OX513A *Ae. aegypti* strain for the presence of dengue/chikungunya. (*Source*: This work)

environments, OX513A mosquitoes are compatible and effective tool for the reduction of wild Indian *Ae. aegypti* populations and fully support the progression to Phase 3 outdoor open-field release trials.

5.5 *Ae. aegypti* Field Surveillance

Surveillance is an important component of vector control program for management of the mosquito population effectively well below the disease transmission threshold level. Also vector surveillance provides information for developing risk assessment that in turn could be used to qualitatively or quantitatively predict the occurrence of vector-borne disease or pest outbreaks (Beech et al. 2009). Among various surveillance methods, ovitraping is one among cost-effective and convenient tools

Fig. 5.16 Physically contained field cage facility for evaluation of transgenic OX513A strain. (*Source*: This work)

recommended for surveillance in dengue vector and can be used not only for monitoring or prediction of disease outbreak but also for assessing the impact of any control measures implemented including implementation of GMMs (Focks and UNDP/World Bank/WHO Special Programme for Research and Training in Tropical Diseases 2004). Ovitrap use has been recommended for differentiating infestation levels between areas, making it more sensitive than the larval survey currently used, even during low levels of vector population (Azil et al. 2010).

An ovitrap surveillance study was conducted to collect baseline information on *Ae. aegypti* vector population and seasonable abundance in the ten villages (Fig. 5.17) of Badnapur Taluka, Maharashtra State, India (Patil et al. 2020). The ovitrap consisted of a black plastic container of 300 mL volume, with a base diameter of 6.5 cm, opening diameter of 7.8 cm, and 9.0 cm in height. The ovitraps are placed containing water with a strip of germination sheet surrounding the inner side of the container half immersed in the water. Ovitraps were placed indoors and outdoors in randomly selected houses scattered over each study area. Thirty to 40 ovitraps were placed both indoors and outdoors monthly and were collected after 6 days. Paper strips from the ovitraps were collected separately and brought to the laboratory for further analysis. The paper strips containing eggs were immersed in water for hatching and reared till adult emergence for identification of the species.

The ovitrap index was calculated as follows (Lim et al. 2010):

$$\text{Ovitrap index (OI)} = \frac{\text{No. of ovitraps positive for oviposition}}{\text{No. of total traps recovered}} \times 100$$

A total of 2236 and 2074 ovitraps were recovered from indoor and outdoor sites, respectively, in the study villages throughout the surveillance period spanning over 14 months. Surveillance revealed the presence of *Ae. aegypti* mosquito species throughout the year with slightly lower ovitrap index during the period from

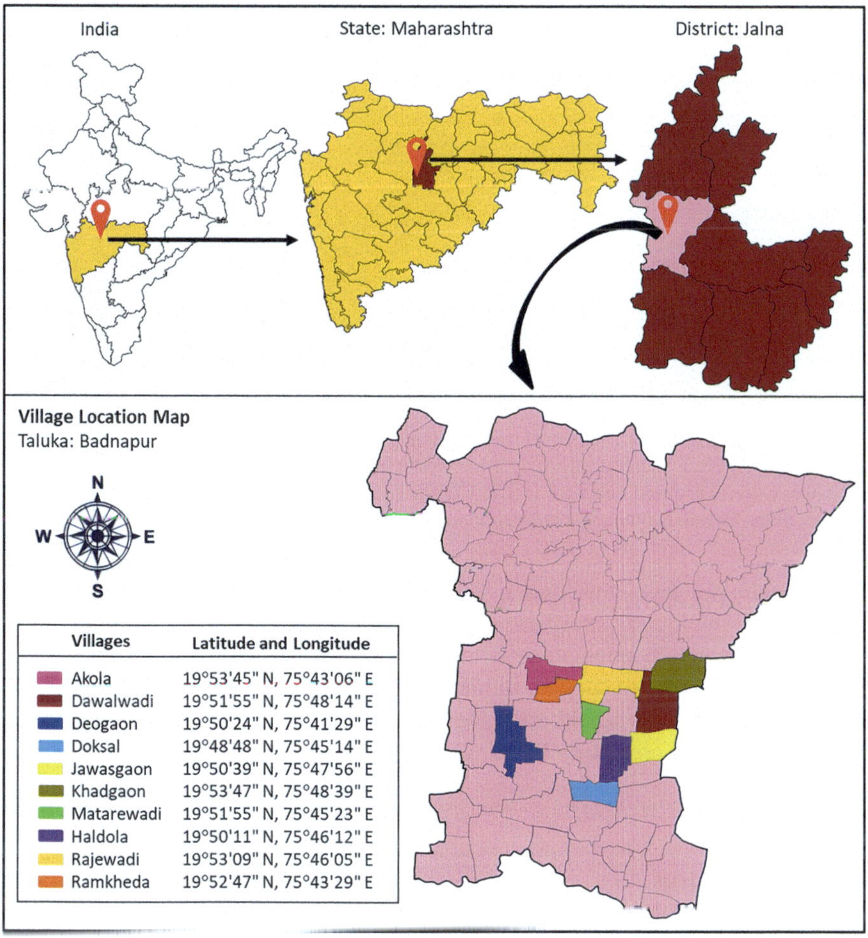

Fig. 5.17 Map location of villages selected for ovitrap surveillance of *Ae. aegypti* in Maharashtra State, India. (*Source*: This figure is reproduced from author's own publication with permission from Patil et al. 2020, this work is licensed under a Creative Commons Attribution-NonCommercial 3.0 Unported License. https://creativecommons.org/licenses/by-nc/3.0/legalcode)

March to July and higher ovitrap index during the survey period between August and October (Fig. 5.18). Ovitrap index for *Ae. aegypti* was found to be in the range of 7.4–53.0 indicating increase in ovitrap index post-monsoon season. Independent *t*-test between average OI of indoor and outdoor ovitraps surveyed in the villages over a period of 14 months revealed nonsignificant difference ($p > 0.05$) revealing equal preference for indoor and outdoor breeding of *Ae. aegypti*. Lee (1990) has reported that *Ae. aegypti* had equal preference for both indoor and outdoor containers. *Ae. aegypti* bites and rests indoors. Studies suggest that an ovitrap index above 10% for *Aedes* species in an area may indicate a possible risk for outbreak of dengue (Lee 1992; Norzahira et al. 2011).

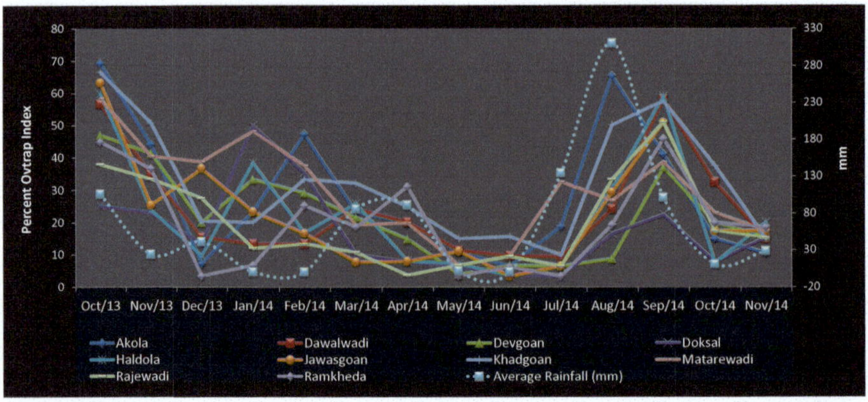

Fig. 5.18 Ovitrap index of *Ae. aegypti* in ten villages of Jalna District, Maharashtra State, India. (*Source*: This figure is reproduced from author's own publication with permission from Patil et al. 2020, this work is licensed under a Creative Commons Attribution-NonCommercial 3.0 Unported License. https://creativecommons.org/licenses/by-nc/3.0/legalcode)

Data acquired from this ovitrap surveillance study in these villages provide important entomological information for effective vector control program to control *Ae. aegypti* mosquito population in these areas. Further this study provides data pertinent for selection of sites for future studies to evaluate transgenic OX513A *Ae. aegypti* strain under open-field conditions for its efficacy within Indian environments with approval from regulatory bodies in the future.

5.6 Conclusion

Assessment of the OX513A strain under laboratory conditions and physically contained facility has shown positive results indicating its potential use in integrated vector control program to control *Ae. aegypti* population and to test the strain further under open-field conditions for suppression of wild *Ae. aegypti* population under natural environment. This novel, species-specific self-limiting technology can deliver long-term protection while minimizing the ecological footprint associated with pest management practices. This is due to the fact that the mosquitoes inheriting the gene from OX513A *Ae. aegypti* do not survive in nature due to the lack of tetracycline and the released males die a natural death, thus making it a self-limiting and with no environmental footprint. There is no toxin involved; there are no adverse effects on human or environmental health and limited prospect of resistance development—providing true sustainability. Long-term reductions in *Ae. aegypti* prevalence would provide universal benefits to public health and social well-being and significantly reduce the economic burden being imposed upon national health-care services.

The World Health Organization, in their 2012 Global Strategy for Dengue Prevention and Control, states "effective vector control measures are critical to achieving and sustaining reduction of morbidity attributable to dengue and decreasing the incidence of infection and preventing outbreaks of the disease." In the year 2016, World Health Organization's Vector Control Advisory Group has reviewed and recommended for carefully planned pilot deployment and randomized control trials with epidemiological outcomes to build evidence for routine programmatic use of OX513A against *Ae. aegypti*-borne diseases (http://www.who.int/neglected_diseases/news/mosquito_vector_control_response/en/). In Brazil, OX513A *Ae. aegypti* strain received approval for "commercial release" throughout the country from the National Technical Commission of Biosafety (CTNBio) in 2014. The OX513A strain for its use to control *Ae. aegypti* population has been proven globally at locations in different countries and has the potential to succeed at a national level in India.

The studies conducted were a stepwise phased evaluation of the Friendly mosquito OX513A *Ae. aegypti* strain under laboratory conditions and physically contained field cages with natural environmental settings for efficacy of the strain. Investigations have shown the strain as a potential tool for suppression of wild population of *Ae. aegypti* and support to proceed for further evaluation of the strain under open-field conditions. These studies on OX513A reviewed in this chapter would underpin the implementation of a groundbreaking approach to insect control within the Indian subcontinent.

References

Alphey L, McKemey A, Nimmo D et al (2013) Genetic control of *Aedes* mosquitoes. Pathog Glob Health 107(4):170–179. https://doi.org/10.1179/2047773213Y.0000000095

Arthropod Containment Levels (2003) Vector Borne Zoonotic Dis 3:75–90

Avise JC, Arnold J, Ball RM et al (1987) Intraspecific phylogeography: the mitochondrial DNA bridge between population genetics and systematics. Annu Rev Ecol Syst 18:489–522. https://doi.org/10.1146/annurev.es.18.110187.002421

Azil AH, Long SA, Ritchie SA et al (2010) The development of predictive tools for preemptive dengue vector control: a study of *Aedes aegypti* abundance and meteorological variables in North Queensland, Australia. Tropical Med Int Health 15:1190–1197. https://doi.org/10.1111/j.1365-3156.2010.02592.x

Barraud PJ (1934) The Fauna of British India, including Ceylon and Burma. In: Diptera. Family Culicidae. Tribes Megarhinini and Culicini, vol 5. Taylor & Francis, London, pp 1–463

Beech CJ, Nagaraju J, Vasan SS et al (2009) Risk analysis of a hypothetical open field release of a self-limiting transgenic *Aedes aegypti* mosquito strain to combat dengue. Asia Pac J Mol Biol Biotechnol 17(3):99–111

Bhardwaj S, Gokhale MD, Mourya DT (2017) Zika virus: current concerns in India. Indian J Med Res 146(5):572–575. https://doi.org/10.4103/ijmr.ijmr_1160_17

Bhatt S, Gething PW, Brady OJ et al (2013) The global distribution and burden of dengue. Nature 496:504–507. https://doi.org/10.1038/nature12060

Brady OJ, Gething PW, Bhatt S et al (2012) Refining the global spatial limits of dengue virus transmission by evidence-based consensus. PLoS Negl Trop Dis 6:e1760. https://doi.org/10.1371/journal.pntd.0001760

Carvalho DO, Mckemey AR, Garziera L et al (2015) Suppression of a field population of *Aedes aegypti* in Brazil by sustained release of transgenic male mosquitoes. PLoS Negl Trop Dis 9(7): e0003864. https://doi.org/10.1371/journal.pntd.0003864

Cecilia D (2014) Current status of dengue and chikungunya in India. WHO South East Asia J Public Health 3(1):22–26. https://doi.org/10.4103/2224-3151.206879

Chakravarti A, Arora R, Luxemburger C (2012) Fifty years of dengue in India. Trans R Soc Trop Med Hyg 106:273–282. https://doi.org/10.1016/j.trstmh.2011.12.007

Chaturvedi UC, Nagar R (2008) Dengue and dengue haemorrhagic fever: Indian perspective. J Biosci 33:429–441. https://doi.org/10.1007/s12038-008-0062-3

Craig GB Jr (1967) Mosquitoes: female monogamy induced by male accessory gland substance. Science 156(781):1499–1501. https://doi.org/10.1126/science.156.3781.1499

Curtis CF, Sinkins SP (1998) *Wolbachia* as a possible means of driving genes into populations. Parasitology 116(Suppl):S111–S115. https://doi.org/10.1017/s0031182000084997

Dhiman S, Rabha B, Yadav K et al (2014) Insecticide susceptibility and dengue vector status of wild *Stegomyia albopicta* in a strategically important area of Assam, India. Parasites Vectors 7 (295):1–5. https://doi.org/10.1186/1756-3305-7-295

Enserink M (2006) Infectious diseases. Massive outbreak draws fresh attention to little-known virus. Science 311:1085. https://doi.org/10.1126/science.311.5764.1085a

Focks DA, UNDP/World Bank/WHO Special Programme for Research and Training in Tropical Diseases (2004) A review of entomological sampling methods and indicators for dengue vectors. World Health Organization, Geneva. https://apps.who.int/iris/handle/10665/68575

Galtier N, Nabholz B, Giemin S et al (2009) Mitochondrial DNA as a marker of molecular diversity: a reappraisal. Mol Ecol 18:4541–4550. https://doi.org/10.1111/j.1365-294X.2009.04380.x

Gandhi R, Yadav KK, Patil PB et al (2020) Molecular analysis of mitochondrial cytochrome oxidase I gene of *Aedes aegypti* L. mosquitoes. J Asia Pac Entomol 23:51–59. https://doi.org/10.1016/j.aspen.2019.10.006

Gong P, Epton M, Fu G et al (2005) A dominant lethal genetic system for autocidal control of the Mediterranean fruitfly. Nat Biotechnol 23:453–456. https://doi.org/10.1038/nbt1071

Gupta B, Reddy BP (2013) Fight against dengue in India: progresses and challenges. Parasitol Res 112:1367–1378. https://doi.org/10.1007/s00436-013-3342-2

Gupta N, Srivastava S, Jain A, Chaturvedi UC (2012) Dengue in India. Indian J Med Res 136:373–390. https://www.ijmr.org.in/text.asp?2012/136/3/373/101976

Halstead SB (2007) Dengue. Lancet 370:1644–1652. https://doi.org/10.1016/s0140-6736(07)61687-0

Harris AF, Nimmo D, McKemey AR et al (2011) Field performance of engineered male mosquitoes. Nat Biotechnol 29(11):1034–1037. https://doi.org/10.1038/nbt.2019

Harris AF, Mckemey AR, Nimmo D et al (2012) Successful suppression of a field mosquito population by sustained release of engineered male mosquitoes. Nat Biotechnol 30 (9):828–830. https://doi.org/10.1038/nbt.2350

Harrison BA, Boonyakanist P, Mongkolpanya K (1972) Biological observations on *Aedes seatoi* Huang in Thailand with notes on rural *Aedes aegypti* (L.) and other *Stegomyia* populations. J Med Entomol 9(1):1–6. https://doi.org/10.1093/jmedent/9.1.1

Helinski ME, Harrington J (2011) Male mating history and body size influence female fecundity and longevity of the dengue vector *Aedes aegypti*. J Med Entomol 48(2):202–211. https://doi.org/10.1603/me10071

Huang Y, Magori K, Lloyd AL et al (2007) Introducing desirable transgenes into insect populations using Y-linked meiotic drive—a theoretical assessment. Evolution 61:717–726. https://doi.org/10.1111/j.1558-5646.2007.00075.x

Irvin N, Hoddle MS, O'Brochta DA et al (2004) Assessing fitness costs for transgenic *Aedes aegypti* expressing the GFP marker and transposase genes. Proc Natl Acad Sci U S A 101(3):891–896. https://doi.org/10.1073/pnas.0305511101

Kay B, Vu SN (2005) New strategy against *Aedes aegypti* in Vietnam. Lancet 365(9459):613–617. https://doi.org/10.1016/S0140-6736(05)17913-6

Knipling EF (1992) Brief talk on the occasion of receiving the world food prize on 12 Oct 1992 at Des Moines, Iowa, USA. Unpublished document

Lee HL (1990) Breeding habitats and factors affecting breeding of *Aedes* larvae in urban towns of Peninsular Malaysia. J Biosci 1(2):107–112

Lee HL (1992) *Aedes* ovitrap and larval survey in several suburban communities in Selangor, Malaysia. Trop Biomed 9:29–34

Lee H, Vasan S, Nazni W et al (2008) Scientific report on the innovative application of *Aedes aegypti* RIDL-sterile insect technique to combat dengue and chikungunya in Malaysia. Institute of Medical Research, Malaysia, Kuala Lumpur

Lee HL, Joko H, Nazni WA et al (2009) Comparative life parameters of transgenic and wild strain of *Aedes aegypti* in the laboratory. Dengue Bull 33:103–114

Li C, Marrelli MT, Yan G et al (2008) Fitness of transgenic *Anopheles stephensi* mosquitoes expressing the SM1 peptide under the control of a vitellogenin promoter. J Hered 99 (3):275–282. https://doi.org/10.1093/jhered/esn004

Lim KW, Sit NW, Norzahira R et al (2010) Dengue vector surveillance in insular settlements of Pulau Ketam, Selangor, Malaysia. Trop Biomed 27(2):185–192

Matthews BJ, Dudchenko O, Kingan SB et al (2018) Improved reference genome of *Aedes aegypti* informs arbovirus vector control. Nature 563(7732):501–507. https://doi.org/10.1038/s41586-018-0692-z

Mavalankar DV, Puwar TI, Govil D et al (2009) A preliminary estimate of immediate cost of chikungunya and dengue to Gujarat, India. IIMA Working Papers WP2009-01-01, Indian Institute of Management, Ahmedabad, India

Moritz C, Dowling TE, Brown WM (1987) Evolution of animal mitochondrial DNA: relevance for population biology and systematics. Annu Rev Ecol Syst 18:269–292. https://doi.org/10.1146/annurev.es.18.110187.001413

Mutheneni SR, Morse AP, Caminade C et al (2017) Dengue burden in India: recent trends and importance of climatic parameters. Emerg Microbes Infect 6(8):e70. https://doi.org/10.1038/emi.2017.57

Nazni WA, Selvi S, Lee HL et al (2009) Susceptibility status of transgenic *Aedes aegypti* (L.) against insecticides. Dengue Bull 33:124–129. https://apps.who.int/iris/handle/10665/170922

Norzahira R, Hidayatulfathi O, Wong HM et al (2011) Ovitrap surveillance of the dengue vectors, *Aedes* (Stegomyia) *aegypti* (L.) and *Aedes* (Stegomyia) *albopictus* Skuse in selected areas in Bentong, Pahang, Malaysia. Trop Biomed 28(1):48–54

Patil PB, Reddy BPN, Gorman K et al (2015) Mating competitiveness and life-table comparisons between transgenic and Indian wild-type *Aedes aegypti* L. Pest Manag Sci 71:957–965. https://doi.org/10.1002/ps.3873

Patil PB, Gorman KJ, Dasgupta SK (2018) Self-limiting OX513A *Aedes aegypti* demonstrate full susceptibility to currently used insecticidal chemistries as compared to Indian wild type *Aedes aegypti*. Psyche J Entomol 2018. https://doi.org/10.1155/2018/7814643

Patil PB, Dasgupta SK, Seshu Reddy KV et al (2020) Surveillance of *Aedes* mosquito species in villages of Jalna district, Maharashtra. India J Entomol Zool Stud 8(2):669–677

Phuc H, Andreasen MH, Burton RS et al (2007) Late-acting dominant lethal genetic systems and mosquito control. BMC Biol 5:1–11. https://doi.org/10.1186/1741-7007-5-11

Pialoux G, Gaüzère BA, Jauréguiberry S et al (2007) Chikungunya, an epidemic arbovirosis. Lancet Infect Dis 7(5):319–327. https://doi.org/10.1016/s1473-3099(07)70107-x

Ramakrishnan SP, Geljand HM, Bose PN et al (1964) The epidemic of acute haemorrhagic fever, Calcutta, 1963; epidemiological inquiry. Indian J Med Res 52:633–650

Ratnasingham S, Hebert PDN (2007) Bold: the barcode of life data system (http://www.barcodinglife.org). Mol Ecol Notes 7(3):355–364. https://doi.org/10.1111/j.1471-8286.2007.01678.x

Ravi V (2006) Re-emergence of chikungunya virus in India. Indian J Med Microbiol 24:83–84. https://doi.org/10.4103/0255-0857.25175

Robinson MC (1955) An epidemic of virus disease in Southern Province, Tanganyika Territory, in 1952–53. I. Clinical features. Trans R Soc Trop Med Hyg 49:28–32. https://doi.org/10.1016/0035-9203(55)90080-8

Shepard DS, Halasa YA, Tyagi BK et al (2014) Economic and disease burden of dengue illness in India. Am J Trop Med Hyg 91:1235–1242. https://doi.org/10.4269/ajtmh.14-0002

Sinkins SP, Gould F (2006) Gene drive systems for insect disease vectors. Nat Rev Genet 7:427–435. https://doi.org/10.1038/nrg1870

Sudeep AB, Parashar D (2008) Chikungunya: an overview. J Biosci 33:443–449. https://doi.org/10.1007/s12038-008-0063-2

UNICEF/UNDP/World Bank/WHO Special Programme for Research and Training in Tropical Diseases (2015) Biosafety for human health and the environment in the context of the potential use of genetically modified mosquitoes (GMMs): a tool for biosafety training based on courses in Africa, Asia and Latin America, 2008–2011. World Health Organization, Geneva. https://apps.who.int/iris/handle/10665/180388

Watts DM, Burke DS, Harrison BA et al (1987) Effect of temperature on the vector efficiency of *Aedes aegypti* for dengue 2 virus. Am J Trop Med Hyg 36:143–152. https://doi.org/10.4269/ajtmh.1987.36.143

WHO Special Programme for Research and Training in Tropical Diseases (2014) World Health Organization, guidance framework for testing of genetically modified mosquitoes. World Health Organization, Geneva. https://apps.who.int/iris/handle/10665/127889

World Health Organization (2012) Global strategy for dengue prevention and control, 2012–2020. World Health Organization, Geneva. https://apps.who.int/iris/handle/10665/75303

World Health Organization (2016) Test procedures for insecticide resistance monitoring in malaria vector mosquitoes, 2nd edn. World Health Organization, Geneva. https://apps.who.int/iris/handle/10665/250677

Chapter 6
Three Decades of Malaria Vector Control in Sudan: The Plausible Role of Sterile Insect Technique (SIT)

Arwa Elaagip and Adeogun Adedapo

Abstract In Northern State, Sudan, a feasibility study for sterile insect technique (SIT) in an area-wide integrated pest management was established for the first time in an African country. The aim of the study was to see whether it is feasible, from a technical, an economical and a biological perspective, to use sterile male mosquitoes to control mosquito populations in designated areas in the African context. The project was focussed on *Anopheles arabiensis*, one of the major malaria vectors. Meteorological data, larval surveillance and population genetic studies were carried out on the disease vectors. The first phase of the study focussed on the development of an efficient sex-separation system, development of dose-sterility curves for the pupal and adult stages and testing of a range of doses in competition experiments to determine effective sterility dose. This stage was followed by a semi-field phase that monitored their swarming and mating behaviours, effectiveness of irradiated males in competitive experiments with wild males and insemination rates. Information regarding irradiation and transportation of irradiated males were also obtained during the study. Unfortunately, the SIT study was terminated in 2017 before starting field release of irradiated males. In spite of the challenges, such investment need not be totally abandoned as valuable experience has been gained and capacity built, which are of high value to malaria control program in Sudan.

Keywords SIT · Malaria · *Anopheles arabiensis* · Dongola · Sudan

A. Elaagip (✉)
Department of Parasitology and Medical Entomology, Faculty of Medical Laboratory Sciences, University of Khartoum, Khartoum, Sudan

A. Adedapo
Public Health and Epidemiology Department, Nigerian Institute of Medical Research, Lagos, Nigeria

Department of Biological Sciences, Faculty of Basic Medical and Applied Sciences, Lead City University, Ibadan, Oyo State, Nigeria

B. K. Tyagi (ed.), *Genetically Modified and other Innovative Vector Control Technologies*, https://doi.org/10.1007/978-981-16-2964-8_6

6.1 Introduction

Malaria vector control in Africa is ever becoming more challenging. Though still inadequate, investments in malaria vector control have not yielded significant results and strongly characterized by failures and abandonments (WHO 2020). The increased global burden of malaria, particularly in sub-Saharan Africa, and the development and distribution of insecticide resistance among mosquito vectors emphasize the urgent demand of implementation of new vector control technologies that are applicable at regional and subregional levels (Lees et al. 2015). Some of these new vector control tools have been proposed, but of importance is the sterile insect technique (SIT) which has been successful in Africa (Wilke et al. 2009) and has given impetus to the utilization of this method for suppression or elimination of mosquito vectors in some regions of Africa in area-wide integrated pest management (AWIPM) programmes (Benedict and Robinson 2003), including Sudan (El Sayed et al. 2009), South Africa (Munhenga et al. 2011) and Mauritius (Lees et al. 2015).

The nineteenth century witnessed significant improvements in the use of modern biotechnological techniques in vector control with accumulation of novel molecular research on mosquito vectors (Grossman et al. 2001; Catteruccia et al. 2005) but also witnessed its shared challenges and technical barriers that contributed to their limitations in integrated vector control strategies including the SIT programme. For instance, inadequate understanding of mosquito life history and gene flow in nature, as well as the fitness and competitiveness (Poda et al. 2018), and capacity building and strengthening of national institutions to support and regulate activities of SIT projects (Mshinda et al. 2004) suffered significant setbacks. There were several technical prerequisites needed to be addressed for the successful implementation of any SIT project in Africa; these include the mass production of mosquito populations with reasonable cost to provide sufficient numbers for each release; the fitness and competitiveness of the sterile male compared with wild males; low population density of the target species in the wild during the time of release; detailed information about spatial and temporal dynamics of the male population in the target area; a good background about mating behaviour, breeding sites, and flight range; natural or artificial barriers to exclude immigration; and that only one species should be dominant in the target area (Vreysen 1995; Klassen and Curtis 2005; Helinski et al. 2006a; Damiens et al. 2013).

Lessons learnt from dedicated research projects conducted on mosquito sterilization between 1950 and 1980 in the Americas and Asia (Lofgren et al. 1974; Dame et al. 1981; Benedict and Robinson 2003) on some principal points, such as the immigration of females from outside the target area, and method of sterilization (Helinski et al. 2006a) gave high hopes to the success of the SIT programme for malaria vector control in Sudan. We, therefore, review the SIT method for malaria vector control in Sudan over the past 30 years.

6.2 The Major Malaria Vector Control Interventions in Sudan

Malaria vector control interventions in Sudan date back to 1900, with the use of larval source management (LSM) strategies (Balfour 1904; El Gadal et al. 1985); however, indoor residual spray (IRS) was initiated in the 1950s with the introduction of BHC (benzene hexachloride) and later dichlorodiphenyltrichloroethane (DDT) (El Gadal et al. 1985; Haridi 1972). The IRS programme was quickly challenged with significant increase in insecticide resistance in the disease vectors. This led to the switch from organochlorines to organophosphates in the late 1970s (El Gadal et al. 1985). Pyrethroid insecticides were later introduced in IRS in the 1990s (Malik et al. 2006). Massive distribution of long-lasting insecticide-treated nets (LLINs) started in the year 2013 with the distribution of PermaNet 2.0 in 13 states (Ismail et al. 2018). The emergence of resistance to pyrethroids also challenged these vector control interventions, and as such, bendiocarb insecticide was introduced in IRS (Sudan IVM Strategic Plan, 2014–2018). Up till date, LLINs, IRS and LSM are the main malaria vector control interventions in Sudan; however, other methods have been introduced, including SIT, as part of IVM.

6.3 The First SIT Programme in Africa: Sudan

The first SIT feasibility study in Africa was supported by the International Atomic Energy Agency (IAEA) and the Islamic Development Bank (IDB) to eliminate the population of *Anopheles arabiensis* Patton (Diptera: Culicidae), the major malaria vector in Sudan (Helinski et al. 2006a). The project received full backing from the Ministry of Science and Technology, Ministry of Health, National Malaria Control Programme and Tropical Medicine Research Institute in Sudan (Robinson et al. 2009). The study was designed to assess the feasibility (from a technical, an economical and a biological perspective) of using sterile male mosquitoes to control mosquito populations in designated areas in the African context (Malcolm et al. 2007).

The project focussed on *An. arabiensis*, the main malaria vector in Sudan (Helinski et al. 2006a), which belongs to the *An. gambiae* species complex (White 1974). The field site of the pilot project was situated in Northern State, Sudan, and contained along a narrow strip on either side of the River Nile extending from Egypt in the north up to the Fourth Cataract in the south. This strip was surrounded by the Great Desert on the eastern and western sides (Malcolm et al. 2007; El Sayed et al. 2009) (Fig. 6.1). Several ethical, legal and social issues associated with this project were considered and evaluated before the commencement of the project (El Sayed et al. 2009). Larval surveillance in random and static sites was performed in the two selected localities, Merowe and Dongola on a monthly basis, in northern Sudan for 2 years (Robinson et al. 2009), and positions of the characterized breeding sites were

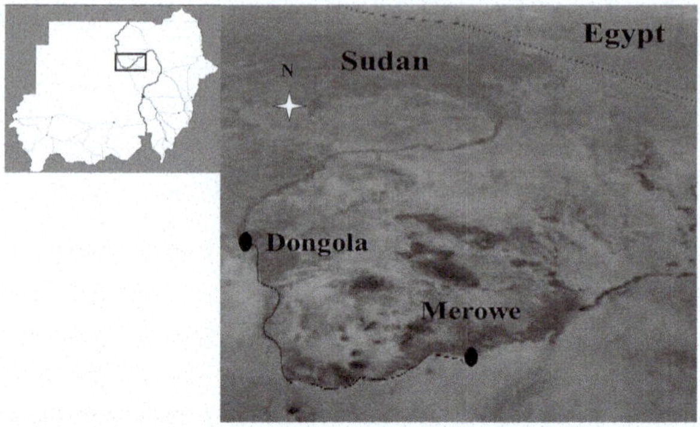

Fig. 6.1 Satellite image of the SIT project area along the Nile, situated between Dongola and Merowe, in Northern State, Sudan. (Adapted from Helinski et al. 2006a)

fixed and linked to a Global Positioning System (GPS). Moreover, potential breeding sites were identified using high-resolution satellite images of the project area. Meteorological data were collected using two automated weather stations located in the two sites (Helinski et al. 2006a). Studies on population genetics of *An. arabiensis* vectors found across the two project sites were performed, and the results indicated that there were no chromosomal or molecular differences between local mosquito populations (Azrag et al. 2016).

A rearing facility in Khartoum State was attached to the project to produce 1 million sterile male *An. arabiensis* per day, after the first releases of sterilized male mosquitoes produced from genetic sexing strain in 2007 (Robinson et al. 2009). Mosquito colony populations were originally from Dongola site (colonized in 2004 from specimens collected near Dongola, Northern State, Sudan) and have been maintained on powdered Koi Floating Blend fish food since its origination. Larvae and adults were cultured in a climate-controlled room at 27 ± 1 °C, $60 \pm 10\%$ RH and a photoperiod of 12:12 (L/D) h, including dusk (1 h) and dawn (1 h). Adults were kept in plastic cages ($30 \times 30 \times 30$ cm) with constant access to a 10% sucrose plus 0.2% methylparaben solution and a weekly mechanically defibrinated cattle blood meal (Damiens et al. 2012).

The first phase of the study focussed on the development of an efficient sex-separation system (Malcolm et al. 2007) and dose-sterility curves for the pupae and adult stages (Helinski et al. 2006a) followed by competition experiments with a range of doses. Initial experiment of competitiveness of sterile males was performed in the laboratory to gain insight into the level of competitiveness among them. Thereafter, a comparison of competitiveness between sterile males and unirradiated laboratory-reared males was designated in large cages. Competition experiments took place in a semi-field setting "greenhouse", in which irradiated males were made to compete with wild males for the wild females (Scolari et al. 2010; Damiens et al. 2013) followed by small-scale releases (Lees et al. 2015). The implementation of this programme was divided into two phases: the laboratory phase and the semi-field studies.

6.3.1 The Laboratory Phase of SIT in Sudan

The laboratory phase of the project was focussed on three main categories: mass production, sexing and sterilization (Damiens et al. 2013).

6.3.1.1 Mass Production

The target of mass rearing was to reach the production of 1 million good-quality sterile males of *An. arabiensis*. To achieve this aim, larval diets were re-evaluated and reconstituted (Damiens et al. 2012). Bovine liver powder and tuna meal were found to be the best for larval survival, developmental rate and adult size. The addition of a vitamin mixture further improved the diet (Damiens et al. 2012). Mostly, water source had a direct link to mosquito size and developmental time with mineral water and powder feed used for the optimal quality of reared mosquitoes (Akpodiete et al. 2019).

The effects of colonization and irradiation on important reproductive traits of *An. arabiensis* were evaluated. Two strains of *An. arabiensis* were used: (1) F5 from a laboratory reared and (2) late larval instars from field collection. Pupal irradiation was performed using 95 Gy; then several adult reproductive traits including "insemination rate, fecundity, oviposition behaviour, fertility and male survival" were evaluated in different mating combinations. The results indicated that irradiation and colonization alter reproductive traits (Poda et al. 2018). To explore the effect of mass rearing (irradiation, chilling, packing and release time on longevity of irradiated male mosquito), a study was conducted on mass-reared immature stages of *An. arabiensis* and *Aedes aegypti* from Khartoum, Sudan, and Juazeiro, Brazil, respectively, using a rack and tray system. The study results showed that irradiated mass-reared males have reduced lifespan in comparison to laboratory-maintained controls under simulated field conditions and that *An. arabiensis* showed to be more sensitive to the handling process and release time than *Ae. aegypti* (Culbert et al. 2018).

In order to maximize egg collections, effects of texture, shade, height and shape of the ovipositional containers on female *An. arabiensis* behaviour were investigated. Based on this study, a strong association was found between preferences for oviposition on humid substrates over free-standing water. In addition, oviposition sites with square shape were preferred, and a majority of the eggs were laid at corners.

6.3.1.2 Sexing

A previous report from Curtis (1978) explained that the gene for dieldrin resistance has been artificially male linked among sibling species of *An. gambiae* complex; with this, females can be selectively killed with dieldrin, and only batches of sterile males can be reared. A genetic sexing strain (GSS) was found in *An. arabiensis*

(Dongola strain) based on a dieldrin-resistance mutation (Yamada et al. 2012). The GSS "Ano IPCL1" was reared in the laboratory and used for rearing and production of sterile males. The sterile males were produced by treating Ano IPCL1 eggs with dieldrin to allow high elimination of female L1 larvae and high production of males needed with a significant reduction in cost, space and labour requirements (Yamada et al. 2012). The sex separation of Ano IPCL1 has been achieved by translocating a dieldrin-resistant allele to the Y chromosome. The presence of the Y-autosome translocation primarily affected the natural fertility of the males, and this required amplification steps during mass rearing (Oliva et al. 2012). Few information is available on the effects of the genetic modification or the dieldrin and irradiation treatments on the reproductive traits of GSS males in Sudan (Oliva et al. 2012). In particular, the effect of the genetic modification for the sexing strain, dieldrin treatment and 70-Gy irradiation on sperm production was assessed (Damiens et al. 2013). However, a more recent sex-separation study of SIT programme for elimination of *Ae. aegypti* and *Ae. albopictus* mosquitoes targeted at different stages of the mosquitoes has been attempted using mechanical and behavioural techniques instead of dieldrin and irradiation treatments. With 100% separation, the use of a spiked blood meal with 8 ppm of ivermectin was identified as the most effective way as compared with the mechanical methods (Gunathilaka et al. 2019).

6.3.1.3 Sterilization

Dose-sterility curves were determined after exposing the male mosquitoes to gamma rays at the immature stages. Before emergence, pupae were irradiated briefly at 22–26 h old. The tested doses ranged between 0 and 100 Gy, and the effects on adult emergence, male survival, induced sterility and insemination capability were evaluated (Helinski et al. 2006b). Results demonstrated that longer X-ray wavelengths have significant effect on the outcome of the sterile males' longevity as well as an increase on the efficacy of sterilization at low doses (Urquidi et al. 2015). Another study done in South Africa concluded that the optimal irradiation dose induces male sterility without compromising mating vigour at 75 Gy when using Cobalt-60 irradiation doses ranging from 70 to 100 Gy (Munhenga et al. 2016). The best stage of development for mass-scale mosquito irradiation is pupa, but pupa irradiation has led to a lower rate of insemination at the highest dose compared to adult irradiation. However, as irradiation alone considerably led to either a reduction in sperm count or prevention of the production of new sperm, dieldrin treatment appears to have unexpected radioprotectant effect (Damiens et al. 2013).

Little is known about the quality and quantity of sperm as determinants of male reproductive success. Data largely suggests that irradiation has no effect on sperm quality except for a study by Helinski and Knols (2009), who reported that testes of pupal irradiated males produced fewer and smaller sperm in comparison to unirradiated ones. Multiple insemination was also studied in *An. arabiensis*. To evaluate the occurrence of multiple insemination, novel labelling system with the stable isotopes 15N and 13C was used to assess the incidence of multiple insemination when males were irradiated at the pupal stage and emerged radiated adults were left to compete with unirradiated males. Overall, the data indicated that only 25% of inseminated females were inseminated with both irradiated and non-irradiated males (Helinski et al. 2008b).

6.3.2 *The Semi-Field Phase of SIT*

One of the techniques that characterized this stage was mark-release and recapture. This method has been successful with *An. arabiensis* and *An. gambiae* s.s. (Touré et al. 1998). The swarming and mating behaviours of *An. arabiensis* mosquitoes were conducted at two field sites in Dongola, northern Sudan. Observations were drawn in the field sites and in a contained semi-field enclosure. Furthermore, the swarming behaviour of irradiated adult males with non-irradiated ones was studied. The result of the experiment showed that the irradiated-marked males joined the natural swarms regularly, indicating their probable competitiveness with other wild males (Hassan et al. 2014).

The mark-release recapture technique was also used to evaluate the capacity of released irradiated *An. arabiensis* males in survival, dispersal and participation in swarms that are occurring at varying distances from the release site using irradiated male mosquitoes. Mean distance travelled, daily probability of survival and estimated population size were calculated from the recapture data. Findings showed that the released irradiated *An. arabiensis* mosquitoes were able to find and participate in swarms and possibly initiate swarms; however, the likelihood of survival decreased with the age, but participation in the swarm and distance travelled by older males appeared to be higher than by younger males (Ageep et al. 2014). As for the monitoring strategy, five types of traps (CDC light trap, BG trap, big and small pet shelters, vertical and horizontal pots and Waleed trap, "local trap named after the designer") were evaluated. The traps were set at different land use/land cover areas. Findings showed that the horizontal pot traps along with the backpack aspirator serve as efficient tools for SIT monitoring in Sudan (Ali et al. 2018).

In addition, a field cage was developed for experimental purposes by establishing resting sites under favourable conditions. Mosquitoes were irradiated in Khartoum and transported as adults by air to the site designated for future discharges (400 km from the laboratory). Mating and survival studies of irradiated laboratory males and field-sampled males were conducted in the field cage in two small scale competition experiments. Few challenges were observed from the irradiated mosquitoes, mostly associated with the absence of rearing facilities in close proximity to the irradiation source. Experimental studies in the field cage showed that mating occurred at high frequencies and that high irradiated males were able to inseminate wild females at rates similar to those of wild males. Survival of mosquitoes from the cage, based on recapture after mating, was satisfactory and approximately 60% of the mosquitoes were recaptured after one night. A few information about male competitiveness were also obtained. Studies concluded that although conditions were challenging, there were no major obstacles associated with the small-scale irradiation and transportation of mosquitoes making the field cage suitable for experiments (Helinski et al. 2008a).

6.4 Lessons Learnt

The SIT programme has been shown to be an effective and sustainable genetic approach to control populations of selected insects (Robinson et al. 2009). The feasibility study of SIT in Sudan aimed to introduce genetic sterility in females of *An. arabiensis* populations in the field following their mating with released sterile males (Helinski et al. 2006a). This process resulted in population reduction or elimination caused by dominant lethal mutations induced in sperm of the released males (Robinson et al. 2009). This project was largely successful in Sudan until it was terminated in 2017 before the field release of irradiated males.

Several SIT field trials have been carried out on mosquitoes with varying degrees of success (Robinson et al. 2009). In spite of the termination of the feasibility study in Sudan, which emphasizes the crucial role of government in malaria control in Sudan, essential information have been gained regarding mass-rearing, mosquito-mating behaviour, population dynamics and regulation, dispersal, level of reproductive isolation, genetic sex-separation strain, effective sterilization doses and procedures and its implications on effectiveness of sterilized males in the field.

In response to the increasing interest and demand for the development and application of SIT as stand-alone or part of integrated vector control intervention, it is important to have strong political will to drive the process. Hence, this emphasizes the role of regulatory authorities in the entire SIT intervention planning, implementation and sustenance. It is hoped that the effectiveness of this intervention can be demonstrated in the near future as this serves as a potential powerful tool to the limited arsenal of interventions available for use against mosquito-borne diseases in Sudan.

References

Ageep T, Damiens D, Alsharif B, Ahmed A, Salih E, Ahmed F, Diabaté A, Lees R, Gilles J, El Sayed B (2014) Participation of irradiated *Anopheles arabiensis* males in swarms following field release in Sudan. Malar J 13:484. http://www.malariajournal.com/content/13/1/484

Akpodiete N, Diabate A, Tripet F (2019) Effect of water source and feed regime on development and phenotypic quality in *Anopheles gambiae* (s.l.): prospects for improved mass-rearing techniques towards release programmes. Parasite Vector 12:210. https://doi.org/10.1186/s13071-019-3465-0

Ali B, Ageep T, Ahmed A, Salih E, Tagelsir F, El Sayed B (2018) Field evaluation of novel candidate traps for surveillance of mosquitoes in area for sterile insect technique trial in Sudan. Poster

Azrag R, Ibrahim K, Malcolm C, El Rayah E, El-Sayed B (2016) Laboratory rearing of *Anopheles arabiensis*: impact on genetic variability and implications for Sterile Insect Technique (SIT) based mosquito control in northern Sudan. Malar J 15:432. https://doi.org/10.1186/s12936-016-1484-2

Balfour A (1904) First report of the Welcome Tropical Research Laboratories at the Gordon Memorial College Department of Education. Gordon Medical College, Khartoum, p 12

Benedict M, Robinson A (2003) The first releases of transgenic mosquitoes: an argument for the sterile insect technique. Trends Parasitol 19:349–355

Catteruccia F, Benton J, Crisanti A (2005) An *Anopheles* transgenic sexing strain for vector control. Nat Biotechnol 23:1414–1417

Culbert N, Maiga H, Somda N, Gilles J, Bouyer J, Mamai W (2018) Longevity of mass-reared, irradiated and packed male *Anopheles arabiensis* and *Aedes aegypti* under simulated environmental field conditions. Parasite Vector 11:603. https://doi.org/10.1186/s13071-018-3191-z

Curtis C (1978) Genetic sex separation in *Anopheles arabiensis* and the production of sterile hybrids. Bull World Health Organ 56(3):453–454

Dame D, Lowe R, Williamson D (1981) Assessment of released sterile *Anopheles albimanus* and *Glossina morsitans morsitans*. In: Kitzmiller JB, Kanda T (eds) Cytogenetics and genetics of vectors. Elsevier Biomedical, New York, pp 231–248

Damiens D, Benedict M, Wille M, Gilles J (2012) An inexpensive and effective larval diet for *Anopheles arabiensis* (Diptera: Culicidae): eat like a horse, a bird, or a fish? J Med Entomol 49 (5):1001–1011. https://doi.org/10.1603/ME11289

Damiens D, Vreysen M, Gilles J (2013) *Anopheles arabiensis* sperm production after genetic manipulation, dieldrin treatment, and irradiation. J Med Entomol 50(2):314–316. https://doi.org/10.1603/ME12058

El Gadal AA, Haridi AM, Hassan F, Hussein H (1985) Malaria control in the Gezira-Managil irrigated scheme of the Sudan. J Trop Med Hyg 88:153–159

El Sayed B, Malcolm C, Babiker A, Malik E, El Tayeb M, Saeed N, Nugud A, Knols B (2009) Ethical, legal and social aspects of the approach in Sudan. Malar J 8(Suppl. 2):S3. https://doi.org/10.1186/1475-2875-8-S2-S3

Grossman G, Rafferty C, Clayton J, Stevens T, Mukabayire O, Benedict M (2001) Germline transformation of the malaria vector, *Anopheles gambiae*, with the *piggyBac* transposable element. Insect Mol Biol 10:597–604

Gunathilaka N, Ranathunge T, Udayanga L, Wijegunawardena A, Gilles J, Abeyewickreme W (2019) Use of mechanical and behavioural methods to eliminate female *Aedes aegypti* and *Aedes albopictus* for sterile insect technique and incompatible insect technique applications. Parasit Vectors 12:148. https://doi.org/10.1186/s13071-019-3398-7

Haridi AM (1972) Partial exophily of *Anopheles gambiae* species B in Khashm Elgirba area in eastern Sudan. Bull World Health Organ 46:39–46

Hassan M, Zain H, Basheer M, Elhaj H, El-Sayed B (2014) Swarming and mating behavior of male *Anopheles arabiensis* Patton (Diptera: Culicidae) in an area of the Sterile Insect Technique Project in Dongola, northern Sudan. Acta Trop 132S:S64–S69

Helinski M, Knols B (2009) Sperm quantity and size variation in un-irradiated and irradiated males of the malaria mosquito *Anopheles arabiensis* Patton. Acta Trop 109:64–69

Helinski M, El-Sayed B, Knols B (2006a) The Sterile Insect Technique: can established technology beat malaria? Entomol Berichten 66(1):13–20

Helinski M, Parker A, Knols B (2006b) Radiation-induced sterility for pupal and adult stages of the malaria mosquito *Anopheles arabiensis*. Malar J 5:41. https://doi.org/10.1186/1475-2875-5-41

Helinski M, Hassan M, El-Motasim W, Malcolm C, Knols B, El-Sayed B (2008a) Towards a sterile insect technique field release of *Anopheles arabiensis* mosquitoes in Sudan: irradiation, transportation, and field cage experimentation. Malar J 7:65. https://doi.org/10.1186/1475-2875-7-65

Helinski M, Hood R, Knols B (2008b) A stable isotope dual-labelling approach to detect multiple insemination in un-irradiated and irradiated *Anopheles arabiensis* mosquitoes. Parasit Vectors 1:9. https://doi.org/10.1186/1756-3305-1-9

Ismail BA, Kafy HT, Suleiman JE, Subramaniam K, Thomas B, Kassim NF, Ahmad AH, Knox TB, Kleinschmidt I, Donelly MJ (2018) Temporal and spatial trends in insecticide resistance in *Anopheles arabiensis* in Sudan: outcomes from an evaluation of implications of insecticide for malaria vector control. Parasit Vectors 11:122

Klassen W, Curtis C (2005) History of the sterile insect technique. In: Dyck VA, Hendrichs J, Robinson AS (eds) Sterile insect technique: principles and practices in area-wide integrated pest management. Springer, Dordrecht, pp 39–68

Lees R, Gilles J, Hendrichs J, Vreysen M, Bourtzis K (2015) Back to the future: the sterile insect technique against mosquito disease vectors. Curr Opin Insect Sci 10:156–162

Lofgren C, Dame D, Breeland S, Weidhaas D, Jeffery G, Kaiser R, Ford H, Boston M, Baldwin K (1974) Release of chemosterilized males for the control of *Anopheles albimanus* in El Salvador III. Field methods and population control. Am J Trop Med Hyg 23:288–297

Malcolm C, Welsby D, El Sayed B (2007) SIT for malaria vector *Anopheles arabiensis* in northern state, Sudan: an historical review of the field site. In: Vreysen MJB, Robinson AS, Hendrichs J (eds) Area-wide control of insect pests. Springer, Dordrecht, pp 361–372

Malik EM, Ali E, Mohamed TA (2006) Efforts to control malaria in Sudan—case study of the National Malaria Control Programme, 2001–2005. SIMET 11:77–85

Mshinda H, Killeen G, Mukabana W, Mathenge E, Mboera L, Knols B (2004) Development of genetically modified mosquitoes in Africa. Lancet Infect Dis 4:264–265. http://infection.thelancet.com

Munhenga G, Brooke B, Chirwa T, Hunt R, Coetzee M, Govender D, Koekemoer L (2011) Evaluating the potential of the sterile insect technique for malaria control: relative fitness and mating compatibility between laboratory colonized and a wild population of *Anopheles arabiensis* from the Kruger National Park, South Africa. Parasit Vectors 4:208. http://www.parasitesandvectors.com/content/4/1/208

Munhenga G, Brooke B, Gilles J, Slabbert K, Kemp A, Dandalo L, Wood O, Lobb L, Govender D, Renke M, Koekemoer L (2016) Mating competitiveness of sterile genetic sexing strain males (GAMA) under laboratory and semi-field conditions: steps towards the use of the Sterile Insect Technique to control the major malaria vector *Anopheles arabiensis* in South Africa. Parasit Vectors 9:122. https://doi.org/10.1186/s13071-016-1385-9

Oliva C, Benedict M, Soliban S, Lemperiere G, Balestrino F, Gilles J (2012) Comparisons of life-history characteristics of a genetic sexing strain with laboratory strains of *Anopheles arabiensis* (Diptera: Culicidae) from northern Sudan. J Med Entomol 49(5):1045–1051. https://doi.org/10.1603/ME11292

Poda S, Guissou E, Maïga H, Bimbile-Somda S, Gilles J, Rayaisse J, Lefèvre T, Roux O, Dabiré R (2018) Impact of irradiation on the reproductive traits of field and laboratory *An. arabiensis* mosquitoes. Parasit Vectors 11:641. https://doi.org/10.1186/s13071-018-3228-3

Robinson A, Knols B, Voigt G, Hendrichs J (2009) Conceptual framework and rationale. Malar J 8 (Suppl 2):S1. https://doi.org/10.1186/1475-2875-8-S2-S1

Scolari F, Siciliano P, Gomulski P, Bonomi A, Malacrida G (2010) Safe and fit genetically modified insects for pest control: from lab to field applications. Genetica. https://doi.org/10.1007/s10709-010-9483-7

Touré Y, Dolo G, Petrarca V, Traoré S, Bouaré M, Dao A, Carnahan J, Taylor C (1998) Mark–release–recapture experiments with *Anopheles gambiae s.l.* in Banambani Village, Mali, to determine population size and structure. Med Vet Entomol 12:74–83

Urquidi J, Brar R, Rodriguez S, Hansen I (2015) The development of new radiation protocols for insect sterilization using long wavelength X-rays. Radiat Phys AIP Conf Proc 1671:020010-1–020010-7. https://doi.org/10.1063/1.4927187

Vreysen M (1995) Radiation induced sterility to control Tsetse flies. PhD thesis, Wageningen Agricultural University

White G (1974) *Anopheles gambiae* complex and disease transmission in Africa. Trans Roy Soc Trop Med Hyg 68:278–301

WHO (2020) World malaria report: 20 years of global progress and challenges. World Health Organization, Geneva. ISBN: 978-92-4-001579-1. https://www.who.int/publications/i/item/9789240015791

Wilke A, Nimmo D, John O, Kojin B, Capurro M, Marrelli M (2009) Mini-review: genetic enhancements to the sterile insect technique to control mosquito populations. Asia Pac J Mol Biol Biotechnol 17(3):65–74

Yamada H, Benedict M, Malcolm C, Oliva C, Soliban S, Gilles J (2012) Genetic sex separation of the malaria vector, *Anopheles arabiensis,* by exposing eggs to dieldrin. Malar J 11:208–219. https://doi.org/10.1186/1475-2875-11-208

Chapter 7
Perspectives into Genetic Manipulations for Control of Dengue Vector (*Aedes aegypti* Linnaeus, 1762) with Reference to Progress in Indian Experiments

Ritwika Chatterjee, Sajal Bhattacharya, and Brij Kishore Tyagi

Abstract Vector-borne diseases like malaria, dengue, chikungunya, Japanese encephalitis, Zika and others claim millions of lives across the globe annually, and as such their control has become an ardent necessity. Past attempts over the decades have introduced vector control through chemical, biological and environmental means. However, these measures, already in place, failed to completely bring down the mortality rates from vector-borne diseases, most of which lack a vaccine to prevent epidemics or even a specific antidote to treat patients. The modern development of technologies such as the release of insects carrying a dominant lethal (RIDL) gene system, an example of transgenesis; the *Wolbachia*-based cytoplasmic incompatibility inducing infertility in female insects, an example of paratransgenesis; and the revolutionary gene drive (CRISPR/Cas9) technology, has their roots in the sterile insect technology (SIT), which worked by creating sterilized males through irradiation to compete with their wild counterparts and subsequently mate with females in nature to produce infertile eggs; a technology meant to gradually and finally exterminate the vector population in nature. These technologies have shown great promise, albeit many imperfections, particularly regarding acceptance by the concerned societies. As far as vector control is concerned, we have attempted to simplify their definitions for the common man so that the intricate scientific jargon about these technologies do not instill any fear or doubts to the end users.

Keywords Transgenesis · Paratransgenesis · Gene editing · CRISPR/Cas9 · Vector control

---

R. Chatterjee (✉) · S. Bhattacharya
Post-Graduate Department of Zoology, Asutosh College, University of Calcutta, Kolkata, West Bengal, India

B. K. Tyagi
SpoRIC, VIT University, Vellore, Tamil Nadu, India

B. K. Tyagi (ed.), *Genetically Modified and other Innovative Vector Control Technologies*, https://doi.org/10.1007/978-981-16-2964-8_7

7.1 Introduction

Vectors play a pivotal role in dissemination of a large number of deadly and debilitating diseases such as malaria, filariasis, dengue, chikungunya, Zika virus, onchocerciasis, Chagas' disease, West Nile virus, Japanese encephalitis, etc. that, under the impact of changing climate, are prevalent widely across the tropical and subtropical countries (Gould and Higgs 2009). The control of vector for the global elimination of several diseases like dengue and Zika virus has taken a greater degree of importance due to lack of a specific vaccine or a drug to deal with the infections. Conventionally, insecticides, repellants, biological agents and environmental tools have been applied to keep the vectors under control or to deter them from making contact with human beings, but all the methodologies had several pitfalls in their operation (administrative, operational and storage), and the control of vector was never really achieved to evade infection from mosquito bites (Tyagi 2020). The modern biotechnological advancement has offered several new technologies which are largely sound, economic and practically applicable. While past technologies such as the release of insects having dominant lethal (RIDL) gene system (transgenesis) and *Wolbachia*-based introduction of cytoplasmic incompatibility in the host body of vector and pathogen to bring about infertility in female insects (paratransgenesis) have been successful in attracting the world's attention, notwithstanding social, regulatory and operational bottlenecks, the application of recently developed gene drive technologies, encompassing CRISPR/Cas9 (*c*lustered *r*egularly *i*nterspaced *s*hort *p*alindromic repeats/*C*RISPR-*as*sociated protein *9*), in vector control has rekindled an altogether new hope in vector control and the diseases they transmit.

7.2 Tools of Genetics: A Potential Game Changer in Vector Management

The introduction of genetic engineering tools in the manipulation of both vectors and disease pathogens has put forth a rejuvenated enterprise globally (Marshall 2010, 2011). Transgenesis involves the direct alteration in the vector through the introduction of a transgene or modified gene, whereas paratransgenesis deals with genetic manipulation of a symbiotic microbe whose protein product interferes with the infective stages of the pathogen. Both these genetic approaches have been preliminarily exploited and have provided considerable insight towards an era of highly promising vector control technologies.

All of the above technologies are, however, developed through highly complicated processes using genetics, genomics, biotechnology and biochemistry as their basic disciplines. In spite of their immense effectiveness, these technologies have left an indelible imprint of a kind of enigmatic maze or jargon of terminologies on the mind of a common man, whereby the end user of the benefits of these technologies finds himself sceptical to accept these tools. We have in this chapter tried to define these technologies in a very simple manner with examples so that even an early student of science can see through the scientific intricacies of these processes and understand, in fine, their practical utility in consonance with the prevailing socioecological considerations. In essence, we are saying, "What are they?"

7.3 Transgenesis

Transgenesis, which literally means transfer of a foreign gene in the hosts body, actually involves the direct alteration in the vector through the introduction of a transgene or modified gene. By introducing a small sequence of foreign DNA into a fertilized egg or developing embryo, animals can be made transgenic. Genetically modified mosquitoes have been created aiming at suppressing target mosquito populations by releasing transgenic males carrying a lethal dominant gene. As described above, an experimentally introduced DNA segment within the genome of a host animal is called a transgene, and it is the transgene which takes the centre stage of host's biological sense of functionality. A gene of the host at the insertion site can be altered or disrupted and in many cases, a transgene will do both. For example, while on one hand, it will disrupt an endogenous gene expressing a new gene product, it can also impact direct genetic modifications in vector mosquitoes as a novel initiative towards disease control. In Cayman Islands and Brazil, the use of OX513A has been successfully demonstrated to suppress the vector populations of *Aedes aegypti*, the primary vector of the deadly dengue infection (Harris et al. 2012; Carvalho et al. 2015).

In India, two institutions, e.g. the International Institute of Biotechnology and Toxicology (IIBAT) at Padappai, Tamil Nadu, and Maharashtra Hybrid Seeds Co. (MAHYCO) at Jalna, Maharashtra State, took up laboratory and simulated field studies in the early 2000s, with an objective to release transgenic mosquito, *Aedes aegypti* OX513A in the field, following approval from the regulators and community acceptance.

7.3.1 International Institute of Biotechnology and Toxicology (IIBAT)

The Coromandel Indag Research Centre, a farmed non-profit research trust/society recognized by the Ministry of Science and Technology, Govt. of India under the DSIR scheme, was established in 1978, and it was subsequently rechristened as the Fredrick Institute of Plant Protection and Toxicology (FIPPAT) in 1985 and eventually the International Institute of Biotechnology and Toxicology (IIBAT) in 2002. It earned the reputation of the first ever India-owned successful institutional venture of pesticide manufacture and evaluation. In 2004, the IIBAT rose to the level of a

GLP-certified contract research organization delivering contract services in the areas of toxicology, analytical chemistry, ecotoxicology, genetic toxicology, radioisotope studies, etc.

The IIBAT was permitted officially to import the transgenic strain of *Aedes aegypti*, OX513A, from Oxitec (Oxford Insect Technologies) Limited, a UK-based company and a spin-off from Oxford University, in 2008, and they were mandated to carry out studies on single objective: "Mating competitiveness of an Asian outcrossed RIDL® strain of mosquito Aedes aegypti under total containment". In order to carry out experiments, the IIBAT constructed an Arthropod Containment Level II (ACL II) laboratory with all necessary facilities (Fig. 7.1) to conduct experiments with rearing, sexing out, mating, etc. on both the GMM (OX513A) and wild populations of the mosquito, *Aedes aegypti*.

A lot of biological parameters including mating competitiveness between the RIDL and wild *Aedes aegypti* populations were found to be satisfactorily matching. However, the IIBAT suddenly aborted the project after a successful duration of nearly 3 years!

7.3.2 Gangabishan Bhikulal Investment and Trading Limited (GBIT)

The laboratory and other paraphernalia on genetically modified *Aedes aegypti*, imported as eggs from the Oxitec, UK, are located at the site of Maharashtra Hybrid Seeds Co. (MAHYCO), founded in 1964. The MAHYCO is an agricultural Indian company based at Dawalwadi in Jalna (Maharashtra State) and is one of the major seed producers in the country. Its offshoot, Gangabishan Bhikulal Investment and Trading Limited (GBIT), established in 1996, established research and development wing alluding to studies on the biological compatibilities between OX513A *Aedes aegypti* and wild strain of the mosquito, eventually to utilize the technology for the control of *Ae. aegypti*—the deadly dengue vector mosquito worldwide (Patil et al. 2015). Based on extensive investigations on the Friendly™ OX513A *Aedes aegypti* in various situations, including field in Cayman Islands, the following salient features were understood:

1. Friendly™ OX513A is a self-limiting, efficient and cost-effective tool for mosquito control.
2. Friendly™ OX513A is species specific, with no direct impact on non-target organisms.
3. Friendly™ OX513A is environmentally friendly with no adverse effect on human or environmental health.
4. Release in field is made of male mosquitoes which do not bite and cannot transmit disease.
5. Released males seek out wild females and mate; eggs laid by females die before reaching adulthood and so the pest population declines.

(a) (b) (c) (d)

Fig. 7.1 Arthropod Containment Laboratory essentials for biological experiments on RIDL and wild *Aedes aegypti* populations, (**a**) mating chamber, (**b**) anteroom, (**c**) wild larval section, (**d**) RIDL larval section, (**e**) double curtain, (**f**) air curtain and (**g**) AC with net covering. (*Source*. This work)

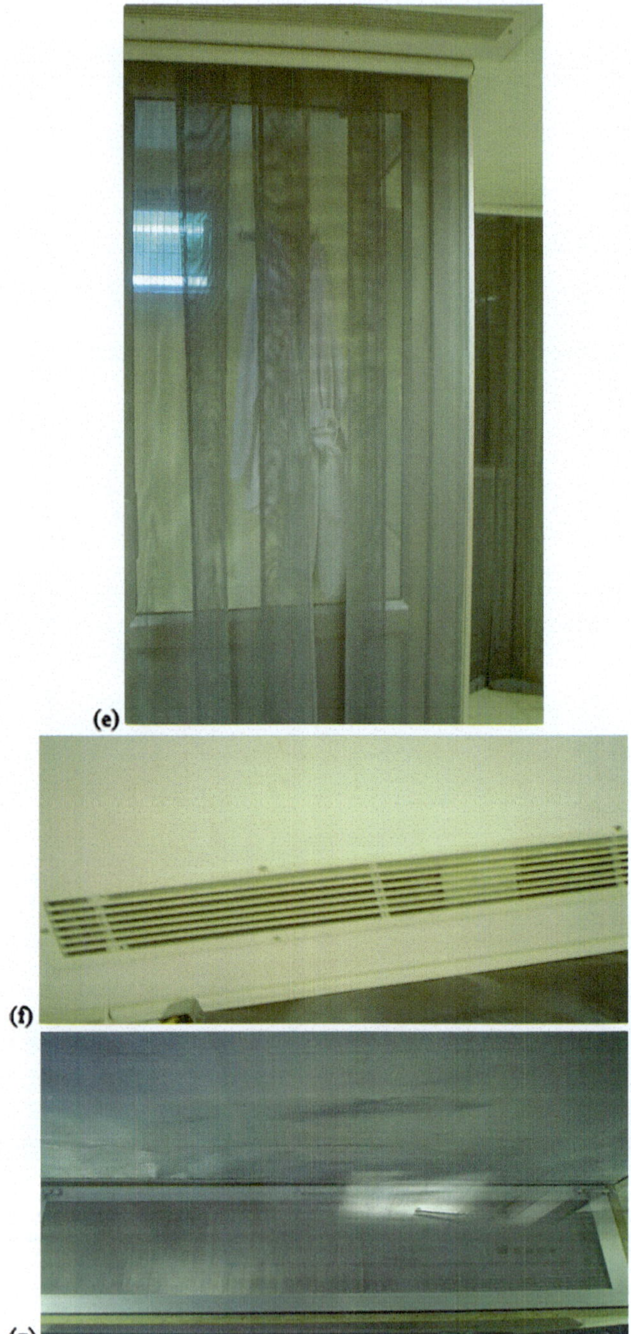

Fig. 7.1 (continued)

Fig. 7.2 Experiment with the field release of *Wolbachia*-transformed *Aedes albopictus* in an island village in Guangzhou, China. (*Source*: This work; Dr. B.K. Tyagi, *personal archive)*

6. Released adult mosquitoes do not establish in the environment due to the absence of tetracycline in nature.

With such a prolific background in genetic and hybridization research, the institute had taken up to study at laboratory and simulated field situations the suitability of application in nature after getting green signal from the concerned regulatory authority in the country.

GBIT built a state-of-art Arthropod Containment Level II laboratory facility for phase I research activities on OX513A *Ae. aegypti* strain and a physically contained facility for conducting phase II field cage trials (Fig. 7.2). They have been working on the Friendly™ OX513A strain since 2011 in collaboration with Oxitec representative in India. Evaluation of the Friendly™ OX513A strain under laboratory conditions during phase I studies and physically contained large field cage during phase II studies simulating natural environmental conditions has shown encouraging results (Patil et al. 2018). They aimed at successful suppression of the wild-type *Ae. aegypti* population in the field cage trial and are trying to push their studies to stage III, i.e. field release, subject to mandatory regulatory approval. Whenever executed, this experiment will offer a vastly useful database.

7.4 Paratransgenesis

Paratransgenesis deals with genetic manipulation of an intracellular symbiotic microbe whose protein product interferes with the infective stages of the pathogen and/or genome of the host vector. For successful implication of paratransgenesis, the scope of cultivation of the symbiotic microorganisms should be plausible, and their amenability to genetic manipulation within the host body is also necessary. Moreover, they are required to be easily propagated in insects such as mosquitoes to facilitate the propagation of chosen traits. Of several bacteria assayed so far, *Wolbachia* (many species), naturally prevalent in a large number of arthropod species including mosquitoes though quaintly enough entirely lacking in *Aedes aegypti*, has shown tremendous potential to bring about control of certain vector mosquitoes. For instance, in China, *Wolbachia*-induced cytoplasmic incompatibility is being deployed to suppress the *Aedes albopictus* population at an island village in Guangzhou (Fig 7.2).

Paratransgenesis is a technique applied to replace the vector population and/or exterminate pathogens via introduction of transgenic symbiont bacteria into the vectors. Correct identification of proteins that are capable of preventing the transmission of pathogens within the vector species is the necessary first step. Thereafter, the genes encoding the identified proteins are introduced into the symbiont, so that they can be ultimately expressed within the vector. The final step involves the introduction of the transgenic variants of symbionts into the natural vector populations. In India, the Indian Council of Medical Research's Vector Control Research Centre, Puducherry, has undertaken a research project on *Wolbachia*-based control of *Aedes aegypti* in collaboration with Monash University, Australia.

7.5 Gene Drives and CRISPR

CRISPR, pronounced as "crisper" for convenience of conversation, is among the latest technologies being applied successfully, to a large extent, in agriculture and, in infancy or trial stage, in certain health-related matters such as human disease vectors and animal pests (Marshall and Akbari 2018). Specialized CRISPR-associated protein or Cas proteins like "Cas-9" snip foreign DNA into small fragments and paste them into CRISPR sequence to generate CRISPR RNAs (crRNAs). These crRNAs guide the subsequent recognition and processing of exogenous genetic material by Cas nucleases. A crRNA, or guide RNA (gRNA), is able to cut along specific parts of the DNA sequence, allowing us to edit genes more easily, cheaply and quickly than ever before. The CAS-9 clips off DNA at a pre-decided locus with a precision of a phlebotomist.

The CRISPR is now available in many formats, and CRISPR/Cas-9 is one highly specific moiety or design whose one potential use is gene drives. It can be simply understood from the Mendelian formula that a parent's traits are usually inherited among 50% of their offspring, but a gene drive can dictate a trait to be inherited greater than 50% of the time. This implies that through gene drive, DNA is manipulable resulting into kinds of individuals and populations which express traits as per designed requirement of science.

Gene drive technologies aim to propagate a trait faster among infective mosquito species and can lead to vector population control or an imminent break to disease transmission. It is a technology that exploits genetic engineering to propagate a particular set of genes throughout a natural vector population by altering the probability of inheritance, such that a specific allele will be transmitted to offspring

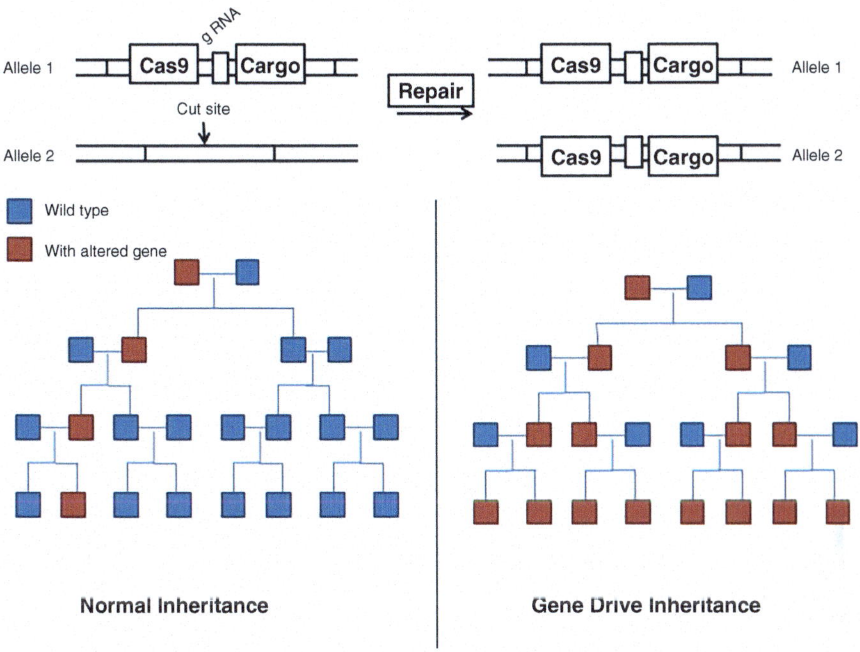

Fig. 7.3 Representation of the general principle of gene drive inheritance vs. normal inheritance. (*Source*: This work)

(instead of the Mendelian 50% probability). Gene drives can be brought about by a variety of mechanisms. Gene drives are expected to provide an effective means of genetically modifying specific populations and/or entire species. The technique can employ adding, deleting, disrupting or modifying genes (Fig. 7.3).

In contrast to their available technology, gene drives have the greatest advantage in effecting manipulation in genome from within the same host DNA where the new trait is finally obtained. Thus, there is no need at all for transferring a gene from outside host, generally a bacterium, or other microbes, and therefore also there are no undue fears of the unseen dangers due to introduction of a foreign gene. In fact, gene drives can be used for a variety of purposes with varied goals in the realms of health, for example, suppression as well as replacement of disease mosquito vector or pest population (Fig. 7.4).

7.6 Conclusion

Vector-borne diseases, notably those spread by mosquitoes, have caused massive havoc globally, with millions of lives being affected and leaving economies grappling with challenges even in the face of slowly emerging solutions. Over the years

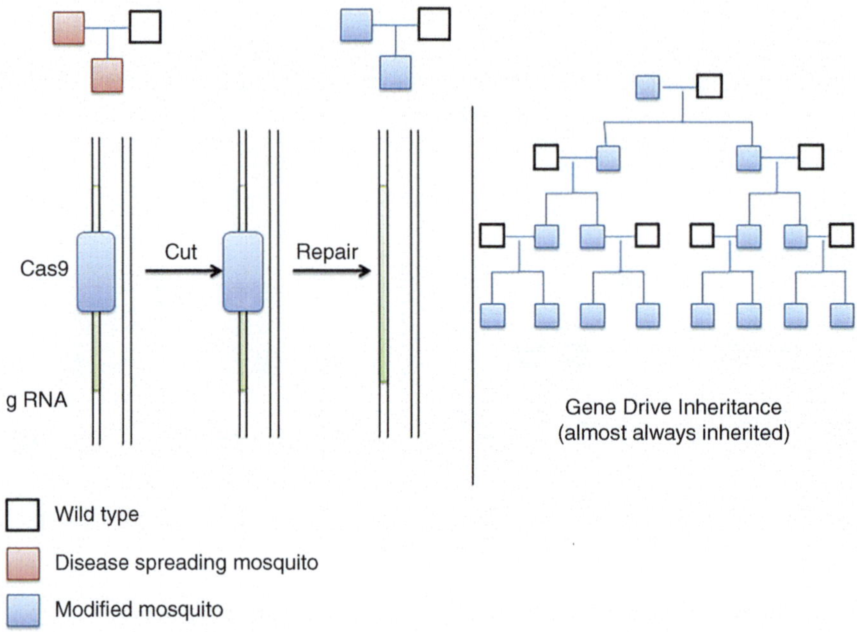

Fig. 7.4 Gene drive inheritance of the trait causes it to spread in the population. (*Source*: This work)

with innumerate policies in place and programmes encompassing biological, chemical, environmental control of vectors and diseases they transmit put into action, the problem is yet far from over. The novel concepts of genetic modifications in symbiotic gut microbiome such as *Wolbachia* and *Asaia*, or identification of innate immune response factors like FREPs, to block pathogenic transmission, are en route to becoming successful models for a one-of-a-kind disease intervention technique in vector biology. The implementation of CRISPR/Cas9 and RNAi-based editing tools has further propagated the trend towards a more holistic control programme. This not only allowed to interpret and intercept vectors and their disease transmission cycles but opened up possibilities to dissect the molecular interactions that occur during the course of infections. Methods such as gene drive, which is a natural process and technology of genome manipulation that propagates a particular suite of genes throughout a population by altering the probability that a specific allele will be transmitted to offspring and adds in the suppression/replacement of pest or vector population, have further opened new vistas of vector control technologies. In fact, both transgenesis and paratransgenesis are two powerful tools of vector control, and several labs across India are involved in experimental implementation of these techniques at various stages of vector and disease control research in animals and plants. One such instance is illustrated through the OX513A *Aedes aegypti* strain developed as a control measure against dengue. The search for novel ways to curb vector-host interactions and subsequent disease manifestations is, however, multidimensional and has led scientists to take a closer look at the mosquito olfactory receptors. Among them, ionotropic receptors (IRs) are understood to have an important effect on host detection. To verify this, CRISPR/Cas9-based editing has been employed to modify Ir8 receptors in *Aedes aegypti*. These findings can be applied to generate transgenic mosquitoes with a potential to reduce, suppress or replace the vector population, making way for newer vector control strategies. In essence, the introduction of genetics as a combat tool against the widely prevalent vector-borne diseases has given a directionality in an otherwise bleak disease control scenario, plagued by drug resistance, climate-change-induced behavioural modifications, increasing demographics and mutating viral load.

References

Carvalho DO, McKemey AR, Garziera L et al (2015) Suppression of a field population of Aedes aegypti in Brazil by sustained release of transgenic male mosquitoes. PLoS Negl Trop Dis 9(7): e0003864

Gould EA, Higgs S (2009) Impact of climate change and other factors on emerging arbovirus diseases. Trans R Soc Trop Med Hyg 103(2):109–121

Harris AF, McKemey AR, Nimmo D et al (2012) Successful suppression of a field mosquito population by sustained release of engineered male mosquitoes. Nat Biotechnol 30(9):828–830

Marshall JM (2010) The Cartagena protocol and genetically modified mosquitoes. Nat Biotechnol 28:896–897

Marshall JM (2011) The Cartagena protocol in the context of recent releases of transgenic and Wolbachia-infected mosquitoes. Asia Pac J Mol Biol Biotechnol 19:93–100

Marshall JM, Akbari OS (2018) Can CRISPR-based gene drive be confined in the wild? A question for molecular and population biology. ACS Chem Biol 13:424–430

Patil PB, Niranjan Reddy BP, Gorman K et al (2015) Mating competitiveness and life-table comparisons between transgenic and Indian wild-type *Aedes aegypti* L. Pest Manag Sci 71 (7):957–965

Patil PB, Gorman Kevin J, Dasgupta Shaibal K, Seshu Reddy KV, Barwale Shirish R, Zehr Usha B (2018) Self-limiting OX513A *Aedes aegypti* demonstrate full susceptibility to currently used insecticidal chemistries as compared to Indian wild-type *Aedes aegypti*. Psyche:7814643, 7 p. https://doi.org/10.1155/2018/7814643

Tyagi BK (2020) Mosquito hunters: a history of hostilities against man's deadliest foe—the mosquito—since 1881. Scientific Publishers, Jodhpur. 474 pp

Chapter 8
Aedes Control Using Sterile Insect Technique (SIT) in Malaysia

Wasi Ahmad Nazni, Guat-Ney Teoh, Shaikh Ismail Shaikh Norman Hakimi, Mohd Azam Muhammad Arif, Maheswaran Tanusshni, Mohd Adnan Nuradila, Achim Nurfarahin Hanini, Irfan Ahmad Shazia, Aik-Meng Tan, Hamzah Rabizah, Mohamad Dzomir Ahmad Zainuri, Asim Hasnor Hadi, Yoon-Ling Cheong, Ahmad Norazah, Hamidou Maiga, Rosemary Susan Lees, and Lee Han-Lim

Abstract The continued occurrence of massive outbreaks of *Aedes*-borne viral diseases of dengue, chikungunya, Zika and yellow fever, in spite of intensive and

---

W. A. Nazni (✉) · G. N. Teoh · S. I. Shaikh Norman Hakimi · M. A. Muhammad Arif · M. Tanusshni · M. A. Nuradila · A. Nurfarahin Hanini · I. A. Shazia · A.-M. Tan · L. Han-Lim
Medical Entomology Unit, Institute for Medical Research, Kuala Lumpur, Malaysia
e-mail: nazni@moh.gov.my

H. Rabizah
Entomology and Pest Unit, Melaka State Health Department, Melaka, Malaysia

M. D. Ahmad Zainuri
Division of Agrotechnology and Biosciences, Malaysian Nuclear Agency, Bangi, Kajang, Selangor, Malaysia
e-mail: zainuri@nuclearmalaysia.gov.my

A. Hasnor Hadi
Health Education Unit, Melaka State Health Department, Melaka, Malaysia

Y.-L. Cheong
Biomedical Museum Unit, Special Resource Centre, Institute for Medical Research, Kuala Lumpur, Malaysia
e-mail: cheongyl@moh.gov.my

A. Norazah
Infectious Disease Research Centre, Institute for Medical Research, Kuala Lumpur, Malaysia
e-mail: norazah@moh.gov.my

H. Maiga
Insect Pest Control Laboratory, Joint FAO/IAEA Division of Nuclear Techniques in Food and Agriculture, International Atomic Energy Agency, Vienna International Centre, Vienna, Austria
e-mail: H.Maiga@iaea.org

R. S. Lees
LITE (Liverpool Insect Testing Establishment), Vector Biology Department, Liverpool School of Tropical Medicine, Liverpool, England, UK
e-mail: rosemary.lees@lstmed.ac.uk

B. K. Tyagi (ed.), *Genetically Modified and other Innovative Vector Control Technologies*, https://doi.org/10.1007/978-981-16-2964-8_8

extensive application of conventional control measures, necessitates application of new tools, such as sterile insect technique (SIT), to stem the tide. Sterile insect technique is a form of biological control method, whereby sterile male insects are released in overwhelming numbers in the wild. These sterile males compete with the wild males to mate with the wild females. The females mated with sterile males will produce sterile eggs that will not hatch. Sustained release of sterile males over a period of time will lead to suppression or elimination of the natural population. Sterility is induced by gamma ray from the radioisotope of cobalt-60 and cesium-137 or X-ray. SIT is safe, cost-effective and environmentally non-polluting, and insects are unable to develop resistance to this method. SIT has a strong track record of success in elimination of agricultural pests, and this led to increased interest in using SIT against mosquitoes of public health importance. Studies and trials against *Aedes* were conducted in the 1960s and, more recently, against *Ae. albopictus* with promising results. Attempts were also made to apply SIT for the control of *Aedes* in several countries. Malaysia's first experience with SIT was in the 1990s, when the Malaysian Nuclear Agency teamed with the MARDI and local universities to sterilise an agricultural pest, the diamondback moth. To prepare for the possible threat of introducing the Old World screw worm (*Chrysomya bezziana*) into Australia from neighbouring countries, from 1995 to 2000, Australia and Malaysia undertook a collaborative myiasis control research project located at the Institut Haiwan, Kluang, Malaysia. The project assisted in suppression trials of the screw worm in Malaysia and supported research that developed and evaluated improved Old World screw worm suppression and eradication techniques. In 2014, in collaboration with the Malaysian Nuclear Agency, the IMR conducted preliminary studies to determine the optimum sterilising dose of gamma irradiation against *Ae. aegypti*. In addition, the impact of sterilisation on the biological parameters of *Ae. aegypti* was also determined. The most effective sterilising dose that did not adversely affect the male was determined to be 55 Gy. Subsequently, a new initiative of field release of sterile *Aedes aegypti* males for the control of dengue was initiated in 2019. This 2-year programme will aim to release gamma ray-sterilised *Ae. aegypti* males in three trial sites to reduce the natural mosquito population to a level below the threshold required for dengue transmission. To ensure public acceptance, public engagement, a prerequisite for a successful release programme, will be conducted intensively prior to the release.

Keywords *Aedes aegypti* · Irradiation · Suppression · Cesium-137

8.1 Introduction

Sterile insect technique (SIT) is a species-specific, non-polluting insect control method that relies on the release of large numbers of sterile males. Mating of released sterile males with native females reduces the females' reproductive potential, and, eventually, if males are released in sufficient numbers over a long enough period of time, elimination or suppression can be achieved in local target population.

SIT is typically used against mosquitos via gamma irradiation due to its high energy and penetration. Cobalt-60 (60Co) and radioisotopes of cesium-137 (137Cs) are the most commonly used gamma ray sources for this purpose, as both have a long half-life and emit high-energy gamma rays. Insects that have been radiation sterilised are considered radioactively free because there are no residual radioactivity remains on the insects (Whitten and Mahon 2005).

8.1.1 Previous Successes of the SIT

Sterile insect technique is an effective biological vector control method that has a proven track record of success against a variety of agricultural pest insects, beginning with the successful elimination of the New World screw worm *Cochliomyia hominivorax* in the USA, Central America and Libya. For more than five decades, SIT has been a major focus of the Joint Food and Agriculture Organization/International Atomic Energy Agency (FAO/IAEA) Programme on Nuclear Techniques in Food and Agriculture. It entails both applied research to control and eliminate pest insects and the transfer of the SIT package to member states via field projects, so that they can benefit from improved plant, animal and human health, as well as enhanced development of economics. SIT has been used to control a variety of agricultural and animal pests, including fruit flies, moths, tsetse flies and screw worm flies. Farmers suffer economic losses as a result of these pests, which are difficult to control or eliminate using conventional methods. The use of SIT has successfully controlled or eliminated some of these pests and thereby boosts agricultural and animal production for food. So far, ten agricultural and animal pests have been successfully eradicated or suppressed (Table 8.1). The use of the sterile insect technique (SIT) in New World screw worm *Cochliomyia hominivorax* eradication programme has been successfully demonstrated. As a result of a 45-year area-wide campaign, suppression and eradication have been achieved in the USA, Mexico, Belize, Guatemala, Honduras, El Salvador, Nicaragua, Costa Rica, Panama north of the Canal and some Caribbean Islands and in the outbreak in Libya (North Africa).

The success of agricultural pest efforts has led to increased interest in the use of SIT against public health pests, particularly the mosquitoes, including *Anopheles*, *Culex* and *Aedes* species (Table 8.2). The earliest field trial of SIT against *Ae. aegypti* was conducted in the USA (Morlan et al. 1962). The results were, however, inconclusive primarily due to lack of understanding of the biology of the sterile males, especially their mating competitiveness which was not evaluated prior to release. With escalation of *Aedes*-borne diseases of dengue, Zika and chikungunya, and the continued absence of antivirals and effective vaccine, the IAEA called upon member countries to use SIT and new other control tools to stem the tide by suppressing mosquito vectors (Kostas 2016).

Table 8.1 List of agricultural and animal pests successfully eradicated or suppressed (Dyck et al. 2005)

- The screw worm fly (*Cochliomyia hominivorax*) was eradicated from the USA, Mexico, Central America, Puerto Rico and Libya.
- The Mexican fruit fly (*Anastrepha ludens*, Loew) was eradicated from most of northern Mexico.
- The tsetse fly was eradicated from Zanzibar.
- The Mediterranean fruit fly (medfly, *Ceratitis capitata*, Wiedemann) was eradicated from the northern part of Chile and southern parts of Argentina, Peru and Mexico. It is being suppressed by SIT in fruit-producing areas of Croatia, Israel, South Africa and Spain.
- The codling moth (*Cydia pomonella*) is being effectively suppressed in parts of British Columbia, Canada [6].
- The pink bollworm (*Pectinophora gossypiella*) was eradicated from south-western USA and north-western Mexico.
- The false codling moth (*Thaumatotibia leucotreta*) is being effectively suppressed in parts of South Africa.
- The cactus moth (*Cactoblastis cactorum*) was eradicated from an outbreak in Yucatán, Mexico.
- The melon fly (*Bactrocera cucurbitae*, Coquillett) was eradicated from Okinawa.
- The onion fly (*Delia antiqua*) was managed in onion production areas in the Netherlands

8.1.2 Successful SIT Pilot Trials Against Aedes *Mosquitoes*

In order to resurrect the use of the SIT against *Aedes* mosquitos, an Italian group of researchers released 896–1590 sterile males of *Aedes albopictus* per hectare per week between 2005 and 2009, inducing up to 70–80% sterility in the target populations in three pilot sites ranging in size from 16 to 45 ha. The egg density was also reduced by 70–80%, resulting in a decrease in the wild population of *Ae. albopictus*. Activities are ongoing in a number of FAO/IAEA-supported projects in FAO and IAEA member states to develop the technical capacity and entomological knowledge required to test methods in *Aedes* control pilot trials with SIT component (Bouyer et al. 2020). In Mauritius, pilot sites have been identified (Iyaloo et al. 2014), and long-term surveillance data have given a good understanding of the dynamics of the target *Ae. albopictus* populations. Attempts are also being made by various researchers to control *Aedes* using SIT alone or in combination with other measures in China (*Ae. albopictus*) (Li 2016), Thailand (*Ae. aegypti*) (Kittayapong et al. 2018, 2019), Sweden (*Ae. sticticus*) (Lundstrom 2016) and La Reunion island (*Ae. albopictus*) (Gouagna 2016).

Pilot studies are ongoing in many countries against *Ae. aegypti* and *Ae. albopictus*, including the USA, Greece, Spain, Mauritius, French Polynesia and Singapore (Bouyer et al. 2020). The most successful SIT releases so far have been in Guangzhou, China, where between 2016 and 2017 suppression efficiency of >95% of *Ae. albopictus* was achieved in two sites ~30 ha in area. For more information about how the SIT can be used to control mosquito populations, readers are advised to see the video available at the following link, https://www.iaea.org/newscenter/multimedia/videos/using-nuclear-science-to-control-mosquitoes.

Table 8.2 Summary of SIT release trials against mosquito vectors

Diseases transmitted	Target species	Location	Sterilisation method	Outcome	References
Dengue, chikungunya, Zika and yellow fever	*Aedes aegypti*	Delhi, India	Thiotepa sterilisation or sex-ratio distorter plus translocations	High female competitiveness	Ansari et al. (1977), Suguna et al. (1977), Grover et al. (1976b), Curtis (1976)
Japanese encephalitis, West Nile and filariasis	*Culex quinquefasciatus*	Myanmar	Cytoplasmic incompatibility	Eradication of small village population	Laven (1967)
	Culex quinquefasciatus	Florida, USA	Chemosterilisation with thiotepa	Moderate competitiveness	Patterson et al. (1970)
	Culex pipiens L.	Montpellier, France	Chromosome translocations	Semi-sterility in wild population	Laven et al. (1972), Cousserans (1974)
	Culex quinquefasciatus	Delhi, India	Thiotepa sterilisation or cytoplasmic incompatibility plus translocations	Moderate competitiveness	Singh et al. (1975), Sharma et al. (1972), Grover et al. (1976a), Yasuno et al. (1978), Curtis et al. (1982), Curtis (1976)
	Culex tritaeniorhynchus Giles	Village near Lahore, Pakistan	Pericentric inversion plus translocation	Moderate competitiveness	Baker et al. (1978, 1979)
	Culex tarsalis Coquillet	California, USA	Adult irradiation	Reduced competitiveness for wild females but super-competitiveness for colony females	Reisen et al. (1982)
Malaria	*Anopheles quadrimaculatus* Say	Florida, USA	Pupal irradiation and adult release	Poor competitiveness of colonised males	Weidhaas et al. (1962), Dame et al. (1964)
	Anopheles gambiae s.s. Giles	Burkina Faso	Sterile hybrid males	Poor competitiveness of hybrid males	Davidson et al. (1970)

(continued)

Table 8.2 (continued)

Diseases transmitted	Target species	Location	Sterilisation method	Outcome	References
	Anopheles albimanus	El Salvador	Chemosterilisation of pupae with bisazir	100% sterility in wild population and population undetectable after 5 months	Lofgren et al. (1974)
	Anopheles albimanus	Salvador	Bisazir sterilisation	Successfully suppressed natural population	Seawright et al. (1978), Dame et al. (1981)
	Anopheles culicifacies Giles	Lahore, Pakistan	Bisazir sterilisation	Suboptimal mating competitiveness	Reisen et al. (1981), Baker et al. (1980, 1981)

Source: This work

8.1.3 Benefits and Safety of SIT on Environment

SIT is a species-specific biocontrol method that is environmentally friendly and has minimal off-target effects on other species. Under the International Standards for Phytosanitary Measures, the International Plant Protection Convention (IPCC) recognises sterile insects produced for SIT as beneficial organisms (ISPMs). As a result, all SIT applications against fruit flies, moths, tsetse flies or screw worms have raised no safety, health or environmental concerns, necessitating no special regulation (ISPMs 2005). The main dengue vector, *Aedes aegypti*, originated in Africa and spread to the rest of the world by the 1930s. As a result, the species is no longer native to Malaysia but is now endemic to the country. The domestic or peri-domestic environment, such as human dwellings, shop houses, public buildings and so on, is thus the target ecosystem that could potentially be exposed by the release of sterile mosquitoes. As a result, the target ecosystem is likely to include, but is not limited to, other invertebrate species like spiders, as well as vertebrates like reptiles, small domestic animals and subsistence farming animals like chickens. None of the threatened or endangered species on Malaysia's International Union for Conservation of Nature and Natural Resources (IUCN) red list occupy the same habitat or ecosystem as *Aedes aegypti*, namely, the domestic or peri domestic environment.

Aedes aegypti typically prefers sheltered spaces within houses and apartments, laying their eggs in relatively clear, stagnant water typically found in containers in the urban environment. Non-target organisms such as spiders, amphibians and reptiles do not feed exclusively on one species of mosquito. A study conducted in Germany showed that all mosquitoes together comprised only 0.16% of the total diet of an amphibian species in the Rhine Valley (Blum et al. 1997). *Aedes aegypti* is not a pollinator species of important crops in Malaysia, such as the oil palm. Mosquitoes are only known to exclusively pollinate one type of orchid that is only present in northern cold climates, such as Canada, Norway and Alaska (Thien 1969). Nor can it interbreed with pollinators of such crops. We are not talking about a world without mosquitoes; our objective is only to reduce *selectively* the population of one of the 3500+ species of mosquitoes to a level below which it cannot transmit dreaded diseases such as dengue, Zika and chikungunya.

In the presence of sterile insects in the environment, there is little to no danger (Nagel and Peveling 2005). They are biodegradable, act in a species-specific manner and do not persist after release. As a result, the sterile insect technique is regarded as one of the most environmentally friendly insect pest control methods. SIT emits no toxic chemicals or chemical residues into the environment. As long as the reproduction is sexual, target insects cannot develop resistance to SIT. There is currently no evidence that SIT resistance has developed (Robinson and Hendrichs 2005).

SIT is also a cost-effective tool. A 2010 report estimated that, based on 2008 values, the cost to produce 1 million sterile males of *Ae. aegypti*, *An. albimanus* and *Cx. quinquefasciatus* was USD 228, USD 389 and USD 196, respectively (Alphey et al. 2010). The average cost per capita for conventional dengue control was estimated to be USD 0.06 which is equivalent to RM 0.24 (Packierisamy et al.

2015). In SIT for Malaysia, the cost to produce 1 million sterile *Ae. aegypti* was estimated at RM 912.00. Therefore, the cost of 1 sterile *Ae. aegypti* is RM 0.00091. The release rate for sterile *Ae. aegypti* would be 2.5 per person, which is equivalent to a cost per capita of RM 0.0022.

Another advantage of SIT, though not only limited to it, is that the suppression of a single species would impact on transmission of multiple diseases by the same species, e.g. dengue, Zika and chikungunya transmitted by *Ae. aegypti.*

8.1.4 Challenges of SIT Application

Although species specificity is highly desirable from an environmental standpoint, it may be a limitation when controlling multiple vector species at the same time, because sterile males of each species must be mass reared and released, because each species does not cross-mate with each other. SIT is thus best suited to areas where there is only one vector for a specific disease, such as dengue, where both *Aedes aegypti* and *Aedes albopictus* are vectors. Furthermore, mass rearing of the target mosquito vector and male sex sorting are frequently cited as major challenges in SIT application. However, with the development of innovative automatic technologies for pupae and adult sex separation, these challenges are surmountable.

Another challenge is the isolation of the release sites, without which fertile females are expected to migrate. As a mitigation strategy, buffer zones around the release sites are recommended.

8.1.5 Public Perception of SIT

Historically, sterile release methods for pest insect control in agriculture have been well accepted in comparison to alternatives (e.g. insecticides) due to their (1) low environmental impact and (2) relatively unobtrusive deployment. This has been especially noticeable in cities. There are also numerous arguments in favour of introducing SIT with community participation, because SIT (1) is safe for the environment and leaves no toxic residues, (2) is cost-effective, (3) is species specific and (4) poses no risk of resistance.

All current and proposed sterile insect release programmes use cobalt-60 (60Co), cesium-137 (137Cs) or X-rays as an irradiation source. This fear is dispelled by research into the irradiation of foodstuffs, where irradiation levels are much higher. All of the studies on irradiated foods have concluded that there are no levels of residual radiation to be concerned about. Because the same sources of irradiation are used, the facts about food irradiation can be extended to insect sterilisation procedures. Furthermore, insect sterilisation for sterile insect releases necessitates a much lower radiation dose (35–100 Gy, depending on mosquito species) than food irradiation (Yamada et al. 2019; IAEA 2002).

Given the low levels of irradiation for insects, the small biomass of released insects and their widespread dispersal, there are no plausible reasons to be concerned about residual radioactivity or radiation-induced toxins in released insects. On the other hand, one could argue that the SIT is more environmentally friendly than some alternative suppression methods, such as synthetic pesticides. Pesticide contamination of foodstuffs, worker exposure and environmental damage are frequently very real issues, especially in developing countries (EJF 2002, 2003). One of the SIT's advantages over chemical control is the absence of any collateral damage caused directly by the release of radiation-sterilised insects into the environment.

The public's perception of these risks is frequently distorted, owing to a lack of familiarity with the actual risk levels involved. There is a need for accurate information on radiation risks to be disseminated to the general public. Increasing public knowledge is obviously one way scientists can reduce public confusion about radiation. To do so, we must (1) define risk concepts in lay terms, (2) explain radiation using a simple vocabulary, (3) use a simple vocabulary to explain radiation and what it is and is not and (4) educate the public and the media about radiation's benefits and risks.

8.2 *Wolbachia* and SIT Strategy

The question could be asked as to why we should consider population suppression approaches and not simply use the *Wolbachia*-based population replacement strategy. There is no doubt that *Wolbachia*-based population replacement approach is a promising method to spread a transinfected *Ae. aegypti* strain to provide protection against dengue, chikungunya and, perhaps, other pathogens (Bourtzis et al. 2014). Recent studies suggest that the blocking efficiency is positively associated with *Wolbachia* infection levels (Martinez et al. 2014; Nazni et al. 2019). In turn, high infection densities may have severe negative effects on host fitness, reducing invasive strength dynamics, as observed in nature with the wMel-infected *Aedes aegypti* strain (Hoffmann et al. 2014). On the other hand, if the invasiveness is reduced, one of the primary benefits of the population replacement strategy is lost, since the introduced population would not spread and the same level of mass rearing and multiple area-wide releases would now be required as for population suppression strategies, the extent of which would depend on how much the infection can spread in nature.

There is no single superior approach as there is no "magic bullet" for management of all insect vector populations or all vector-borne diseases. All available tools and strategies are potentially useful and should be considered, not as stand-alone methods but as part of an integrated approach. The SIT has a successful track record for fruit flies, Lepidoptera, screw worms and tsetse flies and some pilot applications for mosquitoes. It also has the added advantage and value of being largely immune to the development of resistance. The advantages and disadvantages of each population control method should be considered, and all potentially useful approaches should be evaluated and compared using standardised operating procedures and quality control protocols (Scott et al. 2006).

8.3 SIT in Malaysia

Malaysia's first experience with SIT occurred between 1992 and 1996, when the Malaysian Nuclear Agency collaborated with MARDI and local universities to try to sterilise the diamondback moth, an agricultural pest of cabbage. They successfully sterilised the pest using cobalt-60 at 150–500 Gy, but the method was not used to control the pest due to the high cost of mass rearing the moth and the lack of mass-rearing facilities in Malaysia (Omar and Mamat 1996). In contrast to the use of SIT against the diamondback moth, mass rearing of mosquitos such as *Aedes aegypti* has long been established. Methodologies for rearing several mosquito species chosen in response to renewed interest in SIT are becoming standardised (cf. Benedict et al. 2009), such as *Ae. aegypti* (Carvalho et al. 2014) and *Aedes albopictus* (Carvalho et al. 2014; Medici et al. 2011). Protocols and equipment have been developed at the FAO/IAEA Insect Pest Control Laboratory (IPCL) in collaboration with numerous scientists around the world to allow mass rearing of both *Anopheles* and *Aedes* species, including low-cost and effective diets (Damiens et al. 2012; Puggioli et al. 2013; Somda et al. 2019), a tray and rack system for aqueous stages (Balestrino et al. 2012, 2014a), adult oviposition cages (Balestrino et al. 2010a, 2014b; Maïga et al. 2019, 2020) and a vortex system for separation of anopheline larvae from pupae (Balestrino et al. 2011). Methods for storing *Aedes aegypti* and *Aedes albopictus* eggs (Zheng et al. 2015a) and quantifying *Ae. aegypti* and *Ae. albopictus* eggs (Zheng et al. 2015a) and *Anopheles arabiensis* eggs have been developed (Maiga et al. 2016). Furthermore, methods for optimising the hatching of stored eggs have been developed, allowing the loading of trays with the desired number of eggs to produce a predictable density of larvae (Zheng et al. 2015b). Furthermore, pupae can be easily quantified volumetrically, allowing adult cages to be loaded with the desired number of adults. In addition to advancements in mosquito rearing on a larger scale, protocols for irradiating a number of species (Yamada et al. 2019) with a dose sufficient to induce complete sterility without significantly reducing male performance are well established for *Aedes albopictus* (Balestrino et al. 2010b; Bellini et al. 2013), *An. stephensi* (Akram and Aslamkhan 1975), *An. pharoensis* (Wakid et al. 1976) and *An. arabiensis* (Helinski Meh Parker and Knols 2006).

In another development, to prepare for the possibility of introducing the Old World screw worm (*Chrysomya bezziana*) into Australia from neighbouring countries, Australia and Malaysia collaborated on a myiasis control research project based at the Institut Haiwan in Kluang, Malaysia, from 1995 to 2000. The project supported screw worm suppression trials in Malaysia, as well as research that developed and evaluated improved Old World screw worm suppression and eradication techniques. A demonstration was made to show that mass-reared and sterilised screw worms were fit and competitive in the field to confirm and provide confidence in the SIT's efficacy (R. J. Mahon, unpublished data). This species was successfully reared in hydroponics. A majority of the Old World screw worm-rearing methods were based on the techniques developed by the USDA for the New World screw worm (Wyss 2002), but innovations to mass-reared larvae were developed. At the Institut Haiwan, a small pilot mass-rearing facility was built, and novel production engineering methods were used to rear this species (Mahon and Ahmad 2000). In its current configuration, it has the potential to produce approximately 6 million sterile flies per week. Unfortunately, the output of Old World screw worm larvae from a given volume of diet is significantly lower than that of New World screw worm larvae. There is significant room for improvement in the efficiency of mass rearing the Old World screw worm.

In 2014, in collaboration with the Malaysian Nuclear Agency, the IMR conducted the preliminary studies to determine the optimum sterilising dose of gamma irradiation against *Ae. aegypti* that exerts low adverse effects on mating competitiveness of irradiated males compared to wild-type males. In addition, mass rearing, sex sorting and the impact of sterilisation on the biological parameters of *Ae. aegypti* such as longevity, sterility, effects on dengue virus development in sterile females, etc. were also determined. The most effective sterilising dose that did not adversely affect the male mating competitiveness was determined to be 55 Gy. With the completion of these baseline studies, field evaluation of SIT to control *Ae. aegypti* was proposed to be conducted in dengue hotspots. Subsequently, on 3 May 2016, the IMR team briefed the Honourable Minister of Health on two new *Aedes* control strategies, SIT and *Wolbachia*. The meeting agreed that SIT should be conducted in dengue hotspots out of the Klang Valley, to avoid overlapping with *Wolbachia* release. As a result, the new initiative of field release of sterile *Ae. aegypti* males commenced in 2018 with establishing the mass-rearing capacity.

Seeing the interest Malaysian was having to embark on the SIT initiative, in September 2016, the IMR and Malaysian Nuclear Agency collaborated with the IAEA to conduct a workshop on application of SIT to control disease vectors. The workshop was successfully conducted with researchers from 44 countries attended the workshop. From the presentations and discussion, it was clear that SIT is being actively researched in many countries facing the scourges of vector-borne diseases.

The WHO has spearheaded a new strategic approach to reprioritise vector control under the Global Vector Control Response (GVCR), 2017–2030. The new vector control responses are seen as game changer. The Vector Control Advisory Group set up by the World Health Organization (WHO) has reviewed and recommended vector control technologies including the sterile insect technique (SIT).

The time is opportune for SIT to be applied in mosquito control, in view of the vast experience and voluminous data acquired from research conducted in dozen of countries over the past several years. The continued occurrence of massive outbreaks of *Aedes*-borne viral diseases of dengue, chikungunya, Zika and yellow fever, in spite of intensive and extensive applications of conventional control measures, necessitates application of new tools, such as SIT to stem the tide.

8.4 The Current Proof-of-Concept Study in Melaka, Malaysia

In 2014, in collaboration with the Malaysian Nuclear Agency, the IMR conducted preliminary studies to determine the optimum sterilising dose of gamma irradiation, 55 Gy, against *Ae. aegypti*, with the minimum impact on the biological parameters of male mosquitoes. Subsequently, a new initiative of field release of sterile *Ae. aegypti* males for the control of dengue was proposed to be initiated in 2018, entitled "Field Evaluation of Sterile Insect Technique for *Ae. aegypti* Suppression" funded under the Malaysian National Institutes of Health (NIH) grant. This 2-year programme aimed to release gamma ray-sterilised *Ae. aegypti* males in a trial site to reduce the natural mosquito population to a level below the threshold required for dengue transmission; a control site was included for comparison.

8.4.1 Methodology

The current pilot field trial in Melaka was reviewed and approved by the Medical Research Ethical Committee (NMRR-17-2652-39099 S1R2), Ministry of Health. The Biosafety Department in the Ministry of Natural Resources has stated that no regulatory approval is required for the release of irradiated insects.

Two apartment sites were selected based on the density of *Ae. aegypti*: Pangsapuri Kota Laksamana, which is the treated site, and Pangsapuri Taman Tasik Utama, the untreated site. To ensure public acceptance and engagement, a prerequisite for a successful release programme, workshops were conducted intensively prior to the first release (Fig. 8.1).

Mosquitoes were mass reared in the IMR insectarium, and after sorting, male pupae were sterilised by a gamma ray irradiator at a dosage of 55 Gy. The sterile males were released in the trial site weekly to obtain a suppression of more than 90%. The number of males to be released was determined by the natural population level of *Ae. aegypti*. The field population of the mosquito was monitored using ovitraps weekly to measure the level of suppression, compared with a control site, where only conventional dengue control activities were conducted (Figs. 8.2 and 8.3).

8.4.2 Preliminary Results

On 26 December 2018, a meeting was held with the Melaka State Health Department/Jabatan Kesihatan Negeri (JKN) Melaka and Village Community Management Council Kota Laksamana/Majlis Pengurusan Komuniti Kampung (MPKK) Kota

Fig. 8.1 Education materials distributed in Kota Laksamana residents to provide information about the SIT. (*Source*: This work)

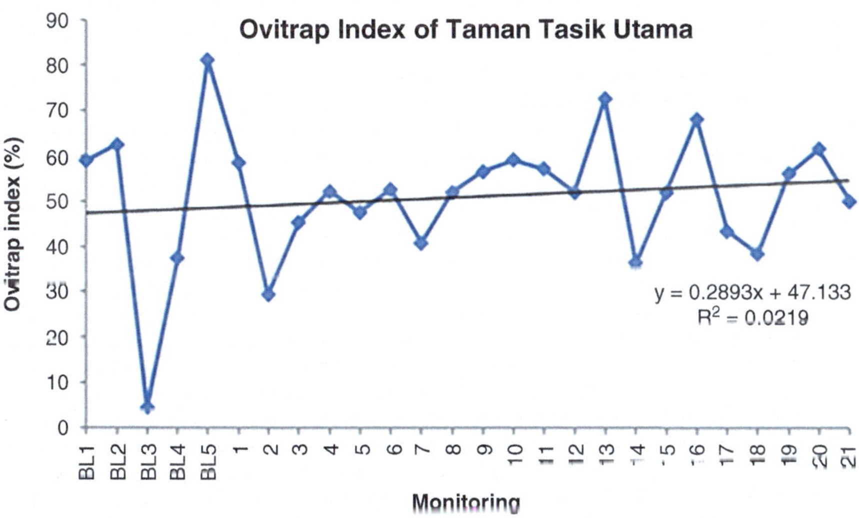

Fig. 8.2 Ovitrap index measured for 5 weeks of baseline monitoring (BL1–5) and 21 weeks of releases in Taman Tasik Utama, the control site where no sterile male *Aedes aegypti* were being released. The trend line shows a slight increase in the percentage of ovitraps that were positive for larvae over time. (*Source*: This work)

Laksamana on 21 February 2019. Banners and posters were posted in Kota Laksamana in March 2019, and educational material on SIT was distributed to provide information to the community, and on 10 March 2019, a briefing was held

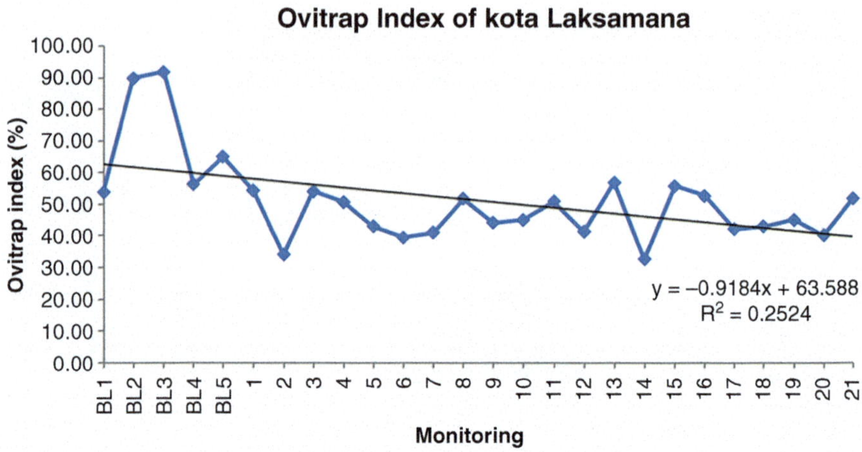

Fig. 8.3 Ovitrap index measured for 5 weeks of baseline monitoring (BL1–5) and 21 weeks of releases in Kota Laksamana, the treatment site where sterile male *Aedes aegypti* were released every week. The trend line shows a decrease in the percentage of ovitraps that were positive for larvae over time. (*Source*: This work)

with local residents. Everybody who attended these briefings had a positive response to the planned releases. A post-release survey of residents was conducted by the Melaka Health Education Unit in due course.

Although the study is still ongoing, the evidence for suppression of *Ae. aegypti* in Kota Laksamana is promising. The evaluation of the effectiveness of this approach is based on the reduction of *Ae. aegypti* population in the release site.

To date, we have conducted 64 releases of sterile male *Ae. aegypti* in Kota Laksamana. The monitoring to evaluate the suppression is still under progress, and more analysis is required. Initial trends show a reduction in ovitrap index and larval density compared to the untreated control sites where sterile *Ae. aegypti* mosquitoes were not released (Fig. 8.4).

The data exhibited (Fig. 8.5) that there is a suppression in comparing the larvae per trap in the treated sites within individual blocks ($p = 0.0001$) when compared with the baseline data. However, in the control site Taman Tasik Utama, there was no suppression observed within the individual blocks ($p = 0.72$). There was a significant difference in larvae per trap when comparing the treated site to control site ($p = 009$).

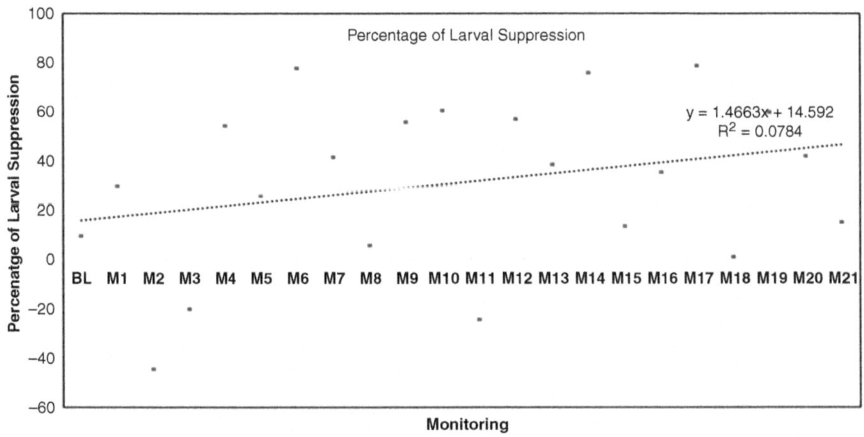

Fig. 8.4 Larval suppression during the first 21 weeks of releases of sterile male *Aedes aegypti*. The points show the percentage reduction in larval density in the release site, Kota Laksamana, compared to the untreated control site Taman Tasik Utama each week. The trend line shows an increase in the percentage suppression over time. (*Source*: This work)

8.5 Conclusion

In the present trial in Melaka, since the release of sterile male *Ae. aegypti* was initiated, trends in mosquito population data are promising. Further analysis is required to determine whether the apparent suppression is significant and to determine whether the release numbers are sufficient to effectively control the wild *Ae. aegypti* population. The release of sterile male *Ae. aegypti* has been expanded to two more localities as requested by the stakeholders. The control locality Taman Tasik Utama has been converted to treated site as there was an upsurge of dengue cases in this locality. Another release locality was situated in Johor. The most important criterion to be noted in any mosquito release programme is to ensure that the release locality has good boundaries and barrier to minimise invasion of wild mosquitoes from outside. This is because improper boundaries will lead to failure of mosquito release programmes. The trial will proceed as planned in order to collect more field data which are essential for making a correct interpretation and conclusive outcomes. It is hoped that with some further optimisation, the *Ae. aegypti* population can be reduced to below the threshold required to transmit dengue. If the mosquito population can be reduced enough, the SIT releases will stop local transmission and prevent outbreaks in the mosquito release localities in Melaka and Johor, Malaysia.

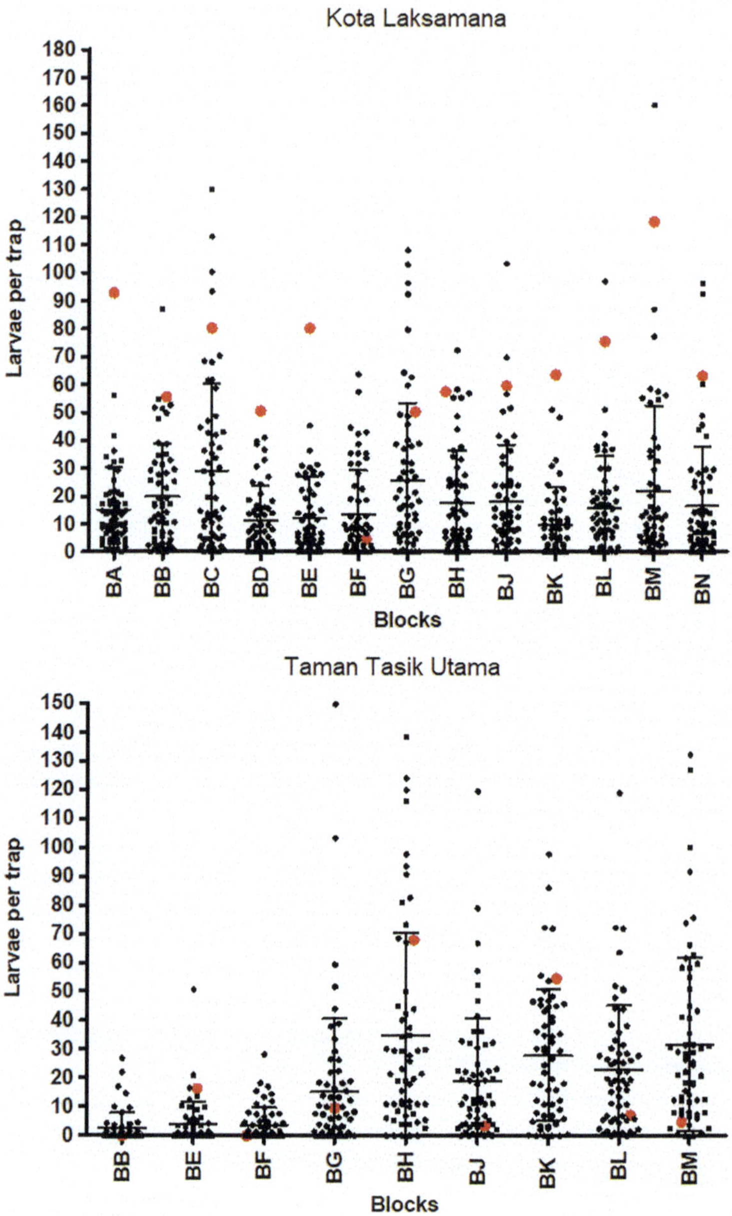

Fig. 8.5 The result is based on a point which indicates larvae per trap per monitoring in individual blocks. The red dot represents the larvae per trap prior to release of sterile male mosquitoes. Kota Laksamana is the release site and Taman Tasik Utama is the control site. (*Source*: This work)

References

Akram M, Aslamkhan M (1975) Production of dominant lethal mutations by gamma irradiation in the malaria mosquito, *Anopheles stephensi*. Pak J Zool 7:177–184

Alphey L, Benedict M, Bellini R, Clark GG, Dame DA, Service MW, Dobson S (2010) Sterile-insect methods for control of mosquito-borne diseases: an analysis. Vector Borne Zoonotic Dis 10:295–311

Ansari MA, Singh KR, Brooks GD, Malhotra PR, Vaidyanathan V (1977) The development of procedures and techniques for mass rearing of *Aedes aegypti*. Indian J Med Res 65:91–99

Baker RH, Sakai RK, Saifuddin UT (1978) Genetic sexing technique for a mosquito sterile male release. Nature 274(5668):253–255

Baker RH, Reisen WK, Sakai RK, Hayes CG, Aslamkhan M, Saifuddin UT et al (1979) Field assessment of mating competitiveness of male *Culex tritaeniorhynchus* carrying a complex chromosomal aberration. Ann Entomol Soc Am 72(6):751–758

Baker RH, Reisen WK, Sakai RK, Rathor HR, Raana K, Azra K, Niaz S (1980) *Anopheles culicifacies*: mating behavior and competitiveness in nature of males carrying a complex chromosomal aberration. Ann Entomol Soc Am 73(5):581–588

Baker RH, Sakai RK, Raana K (1981) Genetic sexing for a mosquito sterile male release. J Hered 72:216–218

Balestrino F, Soliban SM, Gilles J, Oliva C, Benedict MQ (2010a) Ovipositional behavior in the context of mass rearing of *Anopheles arabiensis*. J Am Mosq Control Assoc 26:365–372

Balestrino F, Medici A, Candini G, Carrieri M, Maccagnani B, Calvitti M, Maini S, Bellini R (2010b) Gamma ray dosimetry and mating capacity studies in the laboratory on *Aedes albopictus* males. J Med Entomol 47:581–591

Balestrino F, Giles JRL, Soliban SM, Nirschl A, Benedict QE, Benedict MQ (2011) Mosquito mass rearing technology: a cold-water vortex device for continuous unattended separation of *Anopheles arabiensis* pupae from larvae. J Am Mosq Control Assoc 27:227–235

Balestrino F, Benedict MQ, Gilles JR (2012) A new larval tray and rack system for improved mosquito mass rearing. J Med Entomol 49:595–605

Balestrino F, Puggioli A, Gilles Jrl Bellini R (2014a) Validation of a new larval rearing unit for *Aedes albopictus* (Diptera: Culicidae) mass rearing. PLoS One 9:e91914

Balestrino F, Puggioli A, Bellini R, Petric D, Gilles JRL (2014b) Mass production cage for *Aedes albopictus* (Diptera: Culicidae). J Med Entomol 51:155–163

Bellini R, Medici A, Puggioli A, Balestrino F, Carrieri M (2013) Pilot field trials with *Aedes albopictus* irradiated sterile males in Italian urban areas. J Med Entomol 50(2):317–325

Benedict MQ, Knols Bgj Bossin HC, Howell PI, Mialhe E, Caceres C, Robinson AS (2009) Colonisation and mass rearing: learning from others. Malar J 8:S4

Blum S et al (1997) Culicidae (Diptera) in diet of predatory stage of anurans (Amphibia) in humid biotopes of the Rhine Valley. J Vector Ecol 22(1):23–29

Bourtzis K, Dobson SL, Xi Z, Rasgon JL, Calvitti M, Moreira LA, Bossin H, Moretti R, Baton LA, Hughes GL, Mavingui P, Gilles J (2014) Harnessing mosquito-Wolbachia symbiosis for vector and disease control. Acta Trop 132(Suppl):150–163

Bouyer J, Culbert NJ, Dicko AH, Gomez Pacheco M, Virginio J, Pedrosa MC, Garziera L, Macedo Pinto AT, Klaptocz A, Germann J, Wallner T, Salvador-Herranz G, Argiles Herrero R, Yamada H, Balestrino F, MJB V (2020) Field performance of sterile male mosquitoes released from an uncrewed aerial vehicle. Sci Robot 5:eaba6251

Carvalho DO, Nimmo D, Naish N, McKemey AR, Gray P, Wilke AB, Marrelli MT, Virginio JF, Alphey L, Capurro ML (2014) Mass production of genetically modified *Aedes aegypti* for field releases in Brazil. J Vis Exp 83:e3579

Cousserans J (1974) Expérience de lutte génétique contre *Culex pipiens* dans la région de Montpellier. Synthèse de quatre années d'observations

Curtis CF (1976) Testing systems for the genetic control of mosquitoes. In: Proc XV Int Congr Ent. Washington, DC, pp 106–116

Curtis CF, Brookes GD, Grover KK, Krishnamurthy BS, Laven H, Rajagopalan PK et al (1982) A field trial on genetic control of *Culex p. fatigans* by release of the integrated strain IS-31B. Entomologia Experimentalis et Applicata 31:181–190

Dame DA, Woodard DB, Ford HR, Weidhaas DE (1964) Field behavior of sexually sterile *Anopheles quadrimaculatus* males. Mosq News 24(1):6–14

Dame DA, Lowe RE, Williamson DL (1981) Assessment of released sterile *Anopheles albimanus* and *Glossina morsitans morsitans*. In: Pal R, Kitzmiller JB, Kanda T (eds) Cytogenetics and genetics of vectors: proceedings of a symposium of the XVIth International Congress of Entomology. Kodansha, Elsevier Biomedical Press, Tokyo, New York

Damiens D, Benedict MQ, Wille M, Gilles JRL (2012) An inexpensive and effective larval diet for *Anopheles arabiensis* (Diptera: Culicidae): eat like a horse, a bird or a fish? J Med Entomol 10: 1–1011

Davidson G, Odetoyinbo JA, Colussa B, Coz J (1970) A field attempt to assess the mating competitiveness of sterile males produced by crossing two member species of the *Anopheles gambiae* complex. Bull World Health Organ 42:55–67

Dyck VA, Hendrichs J, Robinson AS (2005) Chapter 1.1: History of sterile insect technique. In: Klassen W, Curtis CF (eds) Sterile insect technique principles and practise in area-wide integrated pest management. Springer, Dordrecht, 799pp, pp 4–28

Environmental Justice Foundation (EJF) (2002) Death in small doses: Cambodia's pesticide problems and solutions. Environmental Justice Foundation, London. http://www.ejfoundation.org/

Environmental Justice Foundation (EJF) (2003) What's your poison? Health threats posed by pesticides in developing countries. Environmental Justice Foundation, London. http://www.ejfoundation.org/poison.html

Gouagna LC (2016) Mosquito SIT in France and French territories. In: FAO/IAEA workshop on sterile insect technique-based approaches to control populations of mosquito disease vectors: with special reference to dengue, chikungunya & Zika vectors, 5–9 Sept, Kuala Lumpur

Grover KK, Curtis CF, Sharma VP, Singh KRP, Dietz K, Agarwal HV, Razdan RK, Vaidyanathan V (1976a) Competitiveness of chemosterilized males and cytoplasmically incompatible-translocated (IS-31B) males of Culex pipiens fatigans in the field. Bull Entomol Res 66:469–480

Grover KK, Suguna SG, Uppal DK, Singh KRP, Ansari MA, Curtis CF et al (1976b) Field experiments on the competitiveness of males carrying genetic control systems for *Aedes aegypti*. Entomologia Experimentalis et Applicata 20(1):8–18

Helinski Meh Parker AG, Knols BG (2006) Radiation-induced sterility for pupal and adult stages of the malaria mosquito *Anopheles arabiensis*. Malar J 5:41

Hoffmann AA, Iturbe-Ormaetxe I, Callahan AG, Phillips BL, Billington K, Axford JK, Montgomery B, Turley AP, O'Neill SL (2014) Stability of the wMel *Wolbachia* infection following invasion into *Aedes aegypti* populations. PLoS Negl Trop Dis 8:e3115

IAEA (2002) Technical reports series no. 409. Dosimetry for food irradiation. International Atomic Energy Agency, Vienna, p 161

International Standards for Phytosanitary Measures (2005) Guidelines for the export, shipment, import and release of biological control agents and other beneficial organisms. https://www.evira.fi/files/attachments/fi/kasvit/torjuntaeliot/1146657660135_ispm3.pdf

Iyaloo DP, Elahee KB, Bheecarry A, Lees RS (2014) Guidelines to site selection for population surveillance and mosquito control trials: a case study from Mauritius. Acta Trop 132:S140–S149

Kittayapong P, Kaeothaisong NO, Ninphanomchai S, Limohpasmanee W (2018) Combined sterile insect technique and incompatible insect technique: sex separation and quality of sterile Aedes aegypti male mosquitoes released in a pilot population suppression trial in Thailand. Parasit Vectors 11:657

Kittayapong P, Ninphanomchai S, Limohpasmanee W, Chansang C, Chansang U, Mongkalangoon P (2019) Combined sterile insect technique and incompatible insect technique: the first proof-of-concept to suppress Aedes aegypti vector populations in semi-rural settings in Thailand. PLoS Negl Trop Dis 13:e0007771

Kostas B (2016) Integrated vector management approaches with a sterile insect technique (SIT) component to control *Aedes* mosquito populations. In: FAO/IAEA workshop on sterile insect technique-based approaches to control populations of mosquito disease vectors: with special reference to dengue, chikungunya & Zika vectors, 5–9 Sept, Kuala Lumpur

Laven H (1967) Eradication of *Culex pipiens fatigans* through cytoplasmic incompatibility. Nature 216(5113):383–384

Laven H, Cousserans J, Guille G (1972) Eradicating mosquitoes using translocations: a first field experiment. Nature 236(5348):456–457. https://doi.org/10.1038/236456a0

Li Y (2016) Updated progress of field trial to suppress *Aedes albopictus* population using the combined IIT/SIT in China. In: FAO/IAEA workshop on sterile insect technique-based approaches to control populations of mosquito disease vectors: with special reference to dengue, chikungunya & Zika vectors, 5–9 Sept, Kuala Lumpur

Lofgren CS, Dame DA, Breeland SG, Weidhaas DE, Jeffery G, Kaiser R, Ford R, Boston MD, Baldwin K (1974) Release of chemosterilized males for the control of *Anopheles albimanus* in El Salvador. III. Field methods and population control. Am J Trop Med Hyg 23:288–297

Lundstrom J (2016) Toward SIT-based control of *Aedes sticticus* populations in Sweden. In: FAO/IAEA workshop on sterile insect technique-based approaches to control populations of mosquito disease vectors: with special reference to dengue, chikungunya & Zika vectors, 5–9 Sept, Kuala Lumpur

Mahon RJ, Ahmad H (2000) Mass rearing the old world screw-worm fly, *Chrysomya bezziana*. In: Tan KH (ed) Proceedings: area-wide control of fruit flies and other insect pests. International conference on area-wide control of insect pests, and the 5th international symposium on fruit flies of economic importance, 28 May–5 June 1998, Penang, Malaysia. Penerbit Universiti Sains Malaysia, Pulau Pinang, pp 325–328

Maiga H, David Damiens D, Diabaté A, Dabiré RK, Ouédraogo GA, Lees RS, Gilles JRL (2016) Large-scale Anopheles arabiensis egg quantification methods for mass-rearing operations. Malar J 15:72. https://doi.org/10.1186/s12936-016-1119-7

Maïga H, Mamai W, Bimbilé Somda NS, Konczal A, Wallner T, Herranz GS et al (2019) Reducing the cost and assessing the performance of a novel adult mass-rearing cage for the dengue, chikungunya, yellow fever and Zika vector, *Aedes aegypti* (Linnaeus). PLoS Negl Trop Dis 13(9):e0007775

Maïga H, Mamai W, Bimbilé Somda NS, Wallner T, Poda BS, Salvador-Herranz G et al (2020) Assessment of a novel adult mass-rearing cage for *Aedes Albopictus* (Skuse) and Anopheles Arabiensis (Patton). Insects 11(11):801

Martinez J, Longdon B, Bauer S, Chan YS, Miller WJ, Bourtzis K, Teixeira L, Jiggins FM (2014) Symbionts commonly provide broad spectrum resistance to viruses in insects: a comparative analysis of Wolbachia strains. PLoS Pathog 10:e1004369

Medici A, Carrieri M, Scholte E-J, MacCagnani B, Dindo ML, Bellini R (2011) Studies on *Aedes albopictus* larval mass-rearing optimization. J Econ Entomol 104:266–273

Morlan HB, McCray EM, Kilpatrick JW (1962) Field tests with sexually sterile males for control of *Aedes aegypti*. Mosq News 22:295–300

Nagel P, Peveling R (2005) Environment and the sterile insect technique. In: Dyck VA, Hendrichs J, Robinson AS (eds) Sterile insect technique, Principles and practice in area-wide integrated pest management, Springer, Dordrecht, pp 499–524

Nazni WA, Hoffmann AA, NoorAfizah A, Cheong YL, Mancini MV, Golding N et al (2019) Establishment of Wolbachia strain wAlbB in Malaysian populations of *Aedes aegypti* for dengue control. Curr Biol 29(24):4241–4248

Omar D, Mamat MJ (1996) The feasibility of using sterile insect technique for the control of diamondback moth on cabbage in Cameron Highlands. In: Proceedings of the 3rd international workshop on "The management of diamondback moth & other crucifer insects", Kuala Lumpur, 29 Oct–1 Nov 1996

Packierisamy PR, Ng C-W, Dahlui M, Inbaraj J, Balan VK, Halasa YA, Shepard DS (2015) Cost of dengue vector control activities in Malaysia. Am J Trop Med Hyg 93(5):1020–1027. https://doi.org/10.4269/ajtmh.14-0667

Patterson RS, Weidhaas DE, Ford HR, Lofgren CS (1970) Suppression and elimination of an island population of *Culex pipiens quinquefasciatus* with sterile males. Science 168(3937):1368–1369

Puggioli A, Balestrino F, Damiens D, Lees RS, Soliban SM, Madakacherry O, Dindo ML, Bellini R, Gilles JRL (2013) Efficiency of three diets for larval development in mass rearing *Aedes albopictus* (Diptera: Culicidae). J Med Entomol 50:819–825

Reisen WK, Baker RH, Sakai RK, Mahmood F, Rathor HR, Raana K, Toqir G (1981) *Anopheles culicifacies* Giles: mating behavior and competitiveness in nature of chemosterilized males carrying a genetic sexing system. Ann Entomol Soc Am 74(4):395–401

Reisen WK, Milby MM, Asman SM, Bock ME, Meyer RP, McDonald PT, Reeves WC (1982) Attempted suppression of a semi-isolated *Culex tarsalis* population by the release of irradiated males: a second experiment using males from a recently colonized strain. Mosq News 42(4):565–575

Robinson AS, Henderichs J (2005) Prospects for the future development & application of the SIT. In: Dyck VA, Hendrichs J, Robinson AS (eds) Area-wide integrated pest management, sterile insect technique principles and practice. Springer, Dordrecht, pp 727–760

Scott TW, Rasgon J, Black W, Gould F (2006) Fitness studies: developing a consensus methodology. In: Knols BGJ, Louis C (eds) Bridging laboratory and field research for genetic control of disease vectors. Springer, Dordrecht, pp 171–181

Seawright JA, Kaiser PE, Dame DA, Lofgren CS (1978) Genetic method for the preferential elimination of females of *Anopheles albimanus*. Science 200(4347):1303–1304

Sharma VP, Patterson RS, Ford HR (1972) A device for the rapid separation of male and female pupae. Bull World Health Organ 47:429–432

Singh KRP, Patterson RS, LaBrecque GC, Razdan RK (1975) Mass rearing of *Culex pipiens fatigans* Wied. J Commun Dis 7(1):31–53

Somda NSB, Maïga H, Mamai W, Yamada H, Ali A, Konczal A et al (2019) Insects to feed insects-feeding Aedes mosquitoes with flies for laboratory rearing. Sci Rep 9(1):1–13

Suguna SG, Curtis CF, Kazmi SJ, Singh KRP, Razdan RK, Sharma VP (1977) Distorter-double translocation heterozygote systems in *Aedes aegypti*. Reprinted from Genetica 47(2):117–123

Thien LB (1969) Mosquito pollination of *Habenaria obtusata* (Orchidaceae). Am J Bot 56:232–237

Wakid AM, Tantawy AO, Abdel-Malek AA, El-Gazzar LM (1976) Irradiation of the immature stages of the mosquito, *Anopheles pharoensis* Theob., with ^{60}Co. J Appl Entomol 80:311–316

Weidhaas DE, Schmidt CH, Seabrook EL (1962) Field studies on the release of sterile males for the control of *Anopheles quadrimaculatus*. Mosq News 22(3):283–291

Whitten M, Mahon R (2005) Misconceptions and constraints. In: Dyck VA, Hendrichs J, Robinson AS (eds) Area-wide integrated pest management, sterile insect technique principles and practice. Springer, Dordrecht, pp 601–626

World Health Organisation (2017) Global vector control response 2017–2030, WHO/HTM/GVCR/2017.01

Wyss JH (2002) Overview of the sterile insect technique in screw-worm fly eradication. In: Proceedings of the screw-worm fly emergency preparedness conference, 12–15 Nov 2001, Canberra, Australia. Agriculture, Fisheries and Forestry, Canberra, pp 176–181

Yamada H, Maiga H, Juarez J, Carvalho DDO, Mamai W, Ali A et al (2019) Identification of critical factors that significantly affect the doseresponse in mosquitoes irradiated as pupae. Parasit Vectors 12(1):1–13

Yasuno M, Macdonald WW, Curtis CF, Grover KK, Rajagopalan PK, Sharwa LS et al (1978) A control experiment with chemosterilized male *Culex pipiens fatigans* Wied. in a village near Delhi surrounded by a breeding-free zone. Med Entomol Zool 29(4):325–343

Zheng ML, Zhang DJ, Damiens DD, Yamada H, Gilles JRL (2015a) Standard operating procedures for standardized mass rearing of the dengue and chikungunya vectors *Aedes aegypti* and *Aedes albopictus* (Diptera: Culicidae)—I—egg quantification. Parasit Vectors 8:42

Zheng ML, Zhang DJ, Damiens DD, Lees RL, Gilles JRL (2015b) Standard operating procedures for standardized mass rearing of the dengue and chikungunya vectors *Aedes aegypti* and *Aedes albopictus* (Diptera: Culicidae)—II—egg storage and hatching. Parasit Vectors 8:348

Chapter 9
Integrated Management of Malaria Vectors in Africa

Ruth Mbabazi, Karim Maredia, Badria Babiker El-Sayed, Agapitus Kato Babumba, Moussa Savadogo, and Olalekan Akinbo

Abstract Malaria disease is a major public health burden in Africa. The control of malaria vectors is a critical component for prevention, management, and eradication of malaria disease. This chapter presents information on the current status of malaria vector control in Africa with emphasis on integrated vector management (IVM) programs. The chapter highlights innovative and emerging technologies such as sterile insect technique, gene drive, *Wolbachia*-based biological control, and other technologies for malaria vector control in Africa which can be integrated into IVM programs. The chapter also provides global resources on malaria vector management programs.

Keywords Management · Malaria · Vectors · Africa

9.1 Malaria Disease Burden in Africa

The vector-borne diseases remain a major public health burden across sub-Saharan Africa (AUDA-NEPAD 2020). More than 1 million people die worldwide every year from malaria disease, and nearly 80% of these victims are children from Africa. Malaria deaths in Africa alone account for 90% of all malaria deaths worldwide

R. Mbabazi (✉) · K. Maredia (✉)
Michigan State University, East Lansing, MI, USA
e-mail: mbabazi@msu.edu; kmaredia@msu.edu

B. B. El-Sayed
Tropical Medicine Research Institute, National Centre for Research, Khartoum, Sudan

A. K. Babumba
Uganda Virus Research Institute, Entebbe, Uganda
e-mail: akato@uvri.go.ug

M. Savadogo · O. Akinbo
African Union Development Agency New Partnership for Africa's Development, Ouagadougou, Burkina Faso
e-mail: MoussaS@nepad.org; OlalekanA@nepad.org

B. K. Tyagi (ed.), *Genetically Modified and other Innovative Vector Control Technologies*, https://doi.org/10.1007/978-981-16-2964-8_9

(WHO 2018). Population groups at higher risk of malaria infection and developing severe disease are infants, children under 5 years of age, pregnant women and patients with HIV/AIDS, displaced populations, and travelers.

According to the 2020 World Malaria Report, 29 countries accounted for 95% of malaria cases globally, 51% of these cases being reported from five African nations only, viz., Nigeria (27%), the Democratic Republic of the Congo (12%), Uganda (5%), Mozambique (4%), and Niger (3%) (WHO 2020a). In 2019, Nigeria (23%), the Democratic Republic of the Congo (11%), the United Republic of Tanzania (5%), Mozambique (4%), Niger (4%), and Burkina Faso (4%) were the top five countries with the highest record of malaria deaths globally (WHO 2020b). Environmental factors including the emerging threats associated with climate change such as rising temperatures and changing precipitation patterns can alter the geographic and seasonal distributions of existing vectors and vector-borne diseases and potentially worsen and explode the malaria vector populations.

The control of vector-borne diseases represents one of the greatest global public health challenges of the twenty-first century. Malaria and other vector-borne diseases contribute substantially to the global burden of diseases and disproportionately affect poor and underserved populations living in tropical and subtropical regions of the world (Benelli and Beier 2017). In the absence of effective control measures, these vector-borne diseases have a major impact on public health and livelihoods. Sub-Saharan Africa has a high burden of vector-borne diseases, and many diseases occur concurrently in the same geographical location and timeframe. According to WHO Malaria Report (2020b), of the 82 malaria endemic countries that provided data for 2010–2019, 28 countries had detected resistance to all most commonly used insecticide classes by at least one malaria vector from one or several collection sites, and 73 countries had detected resistance to at least one insecticide class.

This chapter presents information on the status of malaria vector control in Africa with emphasis on integrated vector management (IVM) programs. In addition, the chapter also highlights emerging technologies such as SIT (sterile insect technique), gene drive, *Wolbachia*-based biological control, and other innovations for malaria vector control in African continent which can be integrated into the IVM programs. Box 9.1 enlists the malaria vectors (both primary and secondary) and parasites prevalent in Africa.

Box 9.1: Malaria Vectors and Malarial Parasites in Africa

The *Anopheles* species of mosquitoes that serve as vectors of malarial parasite in Africa include:

- *Anopheles arabiensis*
- *Anopheles gambiae*
- *Anopheles funestus*

Africa also has secondary *Anopheles* species. They include:

(continued)

Box 9.1 (continued)

- *Anopheles moucheti*
- *Anopheles nili*
- *Anopheles revulorum*

Malaria causing plasmodium parasites common in Africa include:

- *Plasmodium falciparum* (most prevalent malaria parasite)
- *Plasmodium vivax*

9.2 Current Practices for Malaria Vector Control

There are many malaria vector programs and practices currently implemented in Africa and other parts of the world (Raghavendra et al. 2011; Brooke et al. 2013). The methods used for mosquito control include:

- Insecticide-treated mosquito nets
- Indoor residual spraying of chemical insecticides
- Larval source management (Yaw et al. 2016)
- Fogging in the areas surrounding houses and vegetation
- Neem leaves smoke in the houses and other traditional methods

There are a number of other methods and approaches that are recommended and/or under development. These include house improvement for preventing entry of adult mosquitoes (Carter and Karunaweera 2020; Furnival-Adams et al. 2020), sugar bait for the control of adult mosquitoes, mass administration of antimalarial drugs, swamp sprays to control mosquito breeding sites, and use of special repellents. Transgenic mosquitoes and effective malaria vaccine are still being developed. The majority of the national vector control programs in Africa lacks an integrated approach. For effective management of malaria vectors, integrated approach should be promoted and practiced.

9.3 Integrated Management of Malaria Vectors (IVM) in Africa

The policy makers and leaders in Africa are working toward the elimination of malaria by 2030 (ALMA 2016). An integrated vector management (IVM) has been increasingly recommended as a sustainable strategy for malaria vector control (Duchet et al. 2013; WHO 2012, 2016a, b; Mafra-Neto and Dekker 2019). The IVM is not a new concept, and the basic principles of IVM have been used over the past century in many parts of the world for mosquito control. The IVM includes the use of a combination of all available methods, tools, and tactics including cultural,

Table 9.1 Key elements of integrated vector management for malaria control (Mutero et al. 2012)

S. no.	Element	Description
1.	Advocacy, social mobilization, and legislation	Promotion of IVM principles in designing policies in all relevant agencies and organizations
2.	Collaboration within the health sector and with other sectors	Consideration of all options for collaboration within and between public and private sectors
3.	Integrated approach	Ensure rational use of available resources by integrating nonchemical and chemical vector control methods and other disease control methods
4.	Evidence-based decision-making	Adaptation of strategies and interventions to local ecology, epidemiology, and resources guided by research evidence and routine disease surveillance programs
5.	Capacity building	Provision of the essential material infrastructure and human resources to manage IVM strategies

biological, physical, chemical, environmental/habitat management and genetical and other innovative approaches for effective control and management of vectors (Beier et al. 2008; Lobo et al. 2018; Killeen et al. 2017; Sougoufara et al. 2020).

With the increased development of resistance of malaria vectors to insecticides, the IVM is gaining increasing importance in Africa and worldwide for malaria vector control. According to WHO Malaria Report (2020b), of the 82 malaria endemic countries that provided data for 2010–2019, 28 have detected resistance to all four of the most used insecticide classes in at least one malaria vector and one collection site, and 73 have detected resistance to at least one insecticide class (WHO 2020b). In order for the IVM programs to succeed, more efforts are required to create an enabling policy environment for effective and successful deployment of various approaches and technologies and public-private partnerships.

All the tactics of standard vector control are included in the IVM approach. The IVM approach goes beyond the integration of traditional control measures that emphasize strategies to make vector control programs compatible with national health systems. IVM also incorporates decision-making based on human and institutional resources and engages community participation (Mutero et al. 2012). To promote success, IVM requires continual monitoring, evaluation, and legislation, linked with a strong commitment and concerted action by governments and other international organizations. The surveillance of vectors and effective tracking of the disease are critical factors for the success of IVM programs. The community-based approach encompassing health education, awareness, and prevention is highly recommended. In addition, an inter-sectoral collaboration and community-based health extension workers and programs are needed at the grassroots level (Table 9.1).

The key factors that should be considered in choosing appropriate vector control tools for IVM programs include the following:

- Effectiveness in reducing the disease or infection
- Efficacy or effectiveness
- Vector characteristics including insecticide resistance
- Human and environmental safety
- Affordability/cost-effectiveness
- Acceptability and community participation
- Easy implementation/delivery of the intervention

9.4 Country Case Studies of Integrated Vector Management (IVM) Program in Africa

Four case studies are offered below for analyzing the results of various SIT programs being undertaken in different countries in Africa. These case studies are unique because some of them have been carried out almost to the level of final stage of release of the treated mosquitoes in the nature for a demonstrable impact on the density of the vector population of a given mosquito species.

9.4.1 Case Study # 1: Integrated Vector Management in Sudan

Malaria is widely endemic all over Sudan with various levels of endemicity depending on rainfall. Some foci in central Sudan, being adjacent to irrigated areas, are hyperendemic. Malaria is unstable in the semiarid Savannah of central and northern Sudan. The probability for transmission after the rains is small. The great majority of infective bites takes place in September and October, immediately prior to the seasonal peak of malaria cases. *Anopheles arabiensis* is the main malaria vector in the whole country.

Malaria remains a major public health problem in Sudan. The total population is at risk of malaria, and about 86.7% of them are at high risk. Malaria in Sudan constituted about 36% of the estimated malaria cases in the WHO Eastern Mediterranean Region. In 2019, malaria breached the epidemic threshold when over 1.8 million cases and a mortality rate of 13 per 10,000 were reported from across the country. There was 30% increase compared to the same period in 2018 (WHO 2018). With this sharp increase, malaria accounted for 12.4% of all diseases reported in 2019.

The disease is considered a priority in the "National Health Policy 2017–2030" and in the National Health Recovery and Reform Policy and Strategic Plan, 2020–2022 (NHRRP-SP). The documents consider progress in malaria control as an important landmark for poverty reduction and achievement of Sustainable Development Goals (SDGs).

9.4.1.1 Integrated Management of Malaria Vector Anopheles Arabiensis

Integrated vector management (IVM) is "a rational decision-making process for optimal use of resources for vector control". The WHO Member States in the Eastern Mediterranean (EM) Region, including Sudan, are encouraged to adopt and implement IVM through a Regional Committee Resolution, EMRC52.R6 (2005), and another Regional Committee Resolution, EM/RC58/R.10 (2011), that call for improvement of management of pesticides that are being heavily used to combat the increasing burden of vector-borne diseases (Sudan Federal Ministry of Health 2020).

The foundation for two IVM strategic plans developed in Sudan for the periods 2007–2012 and 2014–2018 was to address the challenges, fill in the gaps, and translate the vision and the desired changes for the Federal Ministry of Health into actions for sustainable vector control through the use of cost-effective, environmentally friendly, and sustainable interventions that minimize the risks associated with the use of pesticides for humans, animals, and environment (Sudan Federal Ministry of Health 2020). Although Sudan is endemic with many other vector-borne diseases such as dengue, lymphatic filariasis, rift valley fever, yellow fever, West Nile virus, leishmaniasis, and onchocerciasis, the main focus of the integrated management is on the malaria vector due to its heavy burden.

9.4.1.2 Malaria Vector Control Activities and Interventions

All vector control activities are performed by Integrated Vector Management Department (IVM), Environmental Health and Food Control Directorate, Primary Health Care General Directorate of the Federal Ministry of Health. IVM department uses chemicals and LLINs (long-lasting insecticide-treated nets) for vector control and prevention as follows:

1. *Chemical control.* The key strategies of the WHO for the control of VBDs (vector-borne diseases) are vector control, personal protection, and community participation. Vector control is mainly directed toward reduction and/or interruption of malaria transmission by preventing human contact with malaria-bearing mosquitoes, reducing the longevity of adult mosquitoes, eliminating breeding sites, or killing the mosquito larvae. The main control methods are effective indoor residual spraying, community-wide coverage, and use of LLINs, LSM (larval source management) interventions, repellents, etc. The use of IRS (indoor residual spraying) and LLINs remains the mainstream malaria vector control tools.

   Recent estimates show that the widespread deployment of insecticide-based interventions has been, overwhelmingly, the driver of the reduction in malaria in Africa. Of the 663 million clinical cases estimated to have been averted due to malaria control interventions since 2000, 78% were attributable to insecticide-treated nets or indoor residual spraying (IRS). These two interventions have been massively scaled up since 2000 (WHO 2015a; World Malaria Report 2015b). They have been mainly used in Sudan beside larval control.

2. *Indoor residual spraying (IRS).* The use of IRS in Sudan is restricted to irrigated areas which have high intensity of transmission and is usually done during the transmission season. It is done by conducting residual house spraying two times per year using bendiocarb insecticide. During the execution of IVM strategic plan for the period 2014–2018, seven states with a total population of 5 million were targeted for IRS. Only two of them, Gezira and Sennar which represent about 74% of the targeted area, were covered. The coverage rate was 95.7%.
3. *Larval source management (LSM).* LSM is usually done in urban areas using temephos larvicide. During the same abovementioned period, LSM was being performed every week in 110 urban settings distributed in 18 states. Environmental management and biological control were also used on a limited scale.
4. *Long-lasting insecticide-treated nets (LLINs).* LLINs are used for the control of malaria in rural areas where transmission is seasonal. Twelve states with a total population of 26 million were targeted for coverage. The nets are distributed to the residents every 3 years. The overall coverage rate in these states was 96.6% (LLIN for every two persons) by 2018. A tracking system was used for following up of usage, ownership, durability, and bio-efficacy of the nets.

   For malaria in pregnancy, there is currently a policy of free distribution of one LLIN to pregnant women through antenatal care (ANC) services and to infants through Expanded Programme on Immunization (EPI).

9.4.1.3 Monitoring and Evaluation of Vector Control

1. *Vector surveillance system.* About 106 sentinel sites were identified for routine surveillance of mosquitoes and other vectors. Out of these, 73 sites were used for monitoring of indoor resting mosquitoes. The data is used for evaluation of the impact of control measures in the implemented areas.
2. *Monitoring of insecticides resistance.* Resistance to pyrethroids is extensive throughout sub Saharan Africa, while resistance to the three non-pyrethroid chemical classes used for IRS is simultaneously emerging in many regions. It was, therefore, necessary for Sudan to establish a strong system for monitoring of resistance. Seventy three sites were used for monitoring of insecticide resistance, 49 monitored annually and 24 every 2 years. In 2017, susceptibility testing was performed in 60 sites in 17 states. A high level of pyrethroids resistance was detected in the malaria vector *An. arabiensis*. Resistance to pyrethroids was also reported in many parts of Central and Eastern Sudan. This is annoying, as pyrethroids are the only insecticides currently recommended by the WHO for use in LLINs and among the cheapest, longest-lasting insecticides for IRS. Development of this resistance is a big challenge for malaria control in Sudan.

9.4.1.4 Malaria Case Management: Diagnosis and Treatment

Malaria case management activities are coordinated and performed by the Disease Control Directorate (CDC), Primary Health Care General Directorate of the Federal Ministry of Health. The CDC is responsible for coordination and performance of malaria case management by diagnosis and treatment. The current policy for malaria diagnosis in Sudan is to examine all suspected malaria cases by microscopy or Rapid Diagnostic Tests (RDTs) (Sudan Federal Ministry of Health 2020). The national treatment protocol includes artemether-lumefantrine as first-line treatment for *P. falciparum* malaria, parenteral artesunate for severe malaria, artesunate suppositories as pre-referral treatment, primaquine for 14 days as anti-relapse treatment for *P. vivax*, and a single-dose of primaquine as gametocytocidal for *P. falciparum* malaria in northern states, where the prevalence is low (targeted for elimination) (Sudan Ministry of Health 2017). There is functional subprogram for integrated community case management (iCCM) in collaboration with maternity and child health. Over 7000 communities have been identified as preliminary target.

9.4.1.5 Impact of Integrated Vector Management on Malaria Burden in Sudan

Despite the great effort in implementing this integrated malaria vector management, *Anopheles arabiensis* is still there. This is indicated by the continuous prevalence of malaria disease in the targeted states during the period 2015–2017 and 2019. The annual prevalence was on an average of 49/1000 and the range was 29–81/1000. The lowest was reported in 2015, and the highest was in 2019 indicating that malaria burden is increasing. Although suitable control methods were selected for each setting and carefully implemented and monitored, no specific impact was associated with any of these interventions (Sudan Ministry of Health 2015–2017, 2019). The problem was most probably increased by the low proportion which was not covered by the interventions in the presence of the very efficient malaria vector *Anopheles arabiensis*.

Novel approaches are required for integration to strengthen the impact and to achieve effective protection against malaria vector. The good collaboration and links between IVM department and research institutions and the enhanced capacity of IVM department in infrastructure and human resources are encouraged for considering joint implementation of new validated control methods in the future. The sterile insect technique could be a suitable option for IVM programs.

9.4.2 *Case Study # 2: Integrated Vector Management in Uganda*

9.4.2.1 Background: Malaria as a Vector-Borne Disease in Uganda

Malaria is endemic in 95% of Uganda, affecting approximately 90% of 35 million people as per the Uganda Malaria Reduction Strategic Plan. Uganda's position on the Equator makes it an environment that is conducive for malaria and other VBD vectors to thrive. The country has the third highest number of annual deaths from malaria in Africa, as well as some of the highest reported malaria transmission rates in the world, with approximately 16 million cases reported in 2013 and over 10,500 deaths annually (UMRSP 2014–2020). The whole population of Uganda is at risk of malaria. According to the Uganda Malaria Indicator Survey 2014–2015, malaria prevalence in children under 5 years old ranges from <1% in Kampala to 48% in Karamoja in the northeast region. Malaria accounts for 30–50% of outpatient consultations, 15–20% of inpatient admissions, and 9–14% of inpatient deaths and is responsible for nearly half of inpatient pediatric deaths (MOH 2014). Malaria has an indirect impact on the economy and development in general. The socioeconomic impact of malaria includes out-of-pocket expenditure for consultation fees, drugs, and transport.

9.4.2.2 The Malaria Parasites

Malaria parasite rate ranges from 1% in Kampala to >60% in the northern region (WHO et al. 2015b; MIS 2015). Seasonal peaks follow the two rainy seasons per year (May–July and November–January). The four species of malaria parasites exist in Uganda. *Plasmodium falciparum*, the predominant malaria parasite in the country, accounts for 98% of the malaria parasites, while *P. malariae*, *P. ovale*, and *P. vivax* account for 6%, 1%, and < 1%, respectively. Coinfections with different species were demonstrated, with a regional variation from as low as 1.7% coinfection in the mid-eastern region to as high as 10.7% in Karamoja in northeast region (Uganda Malaria Indicator Survey (MIS) 2014–2015).

9.4.2.3 The Malaria Vectors

The main malaria vectors are *Anopheles gambiae* sensu stricto (s.s.), *Anopheles funestus*, and *An. arabiensis*. *An. gambiae* is the dominant species in most places, while *An. funestus* is generally found at higher altitudes and during the short dry seasons of September to November. The vectors coexist in the country. *Anopheles gambiae* s.s. is highly anthropophilic, endophagic, and endophilic and is considered to be a more efficient malaria vector than the other two species. *Anopheles funestus* shows fairly consistent behavior of generally anthropophilic and endophilic activity throughout its range and breeds more frequently in water bodies with vegetation such as rice fields and ponds. In the absence of insecticide use, its endophilic resting behavior combines with a relatively high longevity to make it an efficient vector. *Anopheles arabiensis* is zoophilic, exophagic, and exophilic but may be endophagic. The behavior allows these mosquitoes to adapt quickly to counter long-lasting insecticide-treated nets (LLINs) and indoor residual spraying (IRS) through avoidance of treated surfaces. Recent effective IRS and LLIN interventions have changed the profile of vector distribution and composition in Uganda. In some areas of northern Uganda, the entomological inoculation rates were among the highest recorded in the world (EIR range from 397 to about 1586).

9.4.2.4 Malaria Control Efforts in Uganda

Starting as early as 1900, Uganda has over 100 years of experience in malaria control. The national malaria programming in Uganda has been evolving over the past years adapting to international policy commitments. Uganda aligned its health agenda to the United Nations global development frameworks, first to the Millennium Development Goals (MDGs) 2000–2015 and to the current Sustainable Development Goals (SDGs) 2016–2030, especially the targets under Goal 3 "Ensure healthy lives and promote well-being for all at all ages," specifically target 3—"By 2030, end the epidemics of AIDS, tuberculosis, malaria and neglected tropical diseases and combat hepatitis, water-borne diseases and other communicable diseases." At regional level, Uganda affirmed its commitment to malaria prevention and control by joining the African Union Heads of States as signatory to the Abuja Declaration on "Universal Access to HIV/AIDS, Tuberculosis and Malaria Services by 2010" in 2000. In response to the United Nations secretary-general's call for 100% coverage of malaria control interventions and its elimination in 2008, Uganda adapted to universal coverage of malaria control interventions.

Further commitment has been made by adapting all national malaria control policies and strategic plans to international policies and strategies of the World Health Organization (WHO), Roll Back Malaria (RBM), and Global Fund for AIDS, Tuberculosis and Malaria (GFATM) guidelines. The National Malaria Control Program (NMCP) policy guidelines are up to date with various WHO policy guidelines including those on selection and procurement of malaria rapid diagnostic tests (mRDTs), preserving the effectiveness of modern malaria vector control and case management besides overall alignment to the WHO's Global Technical Strategy (GTS). In line with the WHO's High Burden to High Impact (HBHI), Uganda adopted the targeted response to malaria strategy and the Mass Action Against Malaria (MAAM) strategy for mobilization of partners, other public sectors, private sector, and the entire population down to every household with strong leadership of the Ministry of Health.

9.4.2.5 Past Successes and Failures in Malaria Control and Elimination in the Country and Lessons Learned

The malaria control program has insidiously trudged through a journey of both success and failure since its inception as a unit in 1995. To date, Uganda has successfully developed and implemented five merozoite surface protein (MSP) that focused on increasing rollout of interventions to achieve universal coverage and access that were guided by three malaria control policies. Implementation has been impactful: malaria incidence reduced from 460/1000 population in 2013 to 282/1000 population in 2018, while malaria mortalities reduced from 320 deaths per day in 2009 to less than 20 in 2018). On program stewards, the engagement of the parliamentary which created Parliamentary Forum on Malaria (UPFM) for policy advocacy and the adoption of MAAM 2 years ago for mobilization of multi-sectoral approach to malaria programming have been the major successes. The development of community systems from community medicine distributors (CMDs) to now under the village health team (VHT) who can test and treat malaria has been a major avenue for extending service delivery at household level. This is implemented through the (integrated community case management (iCCM) strategy.

Despite the many successes recorded, malaria control in Uganda has also had some challenges, some of which include the following:

- Increased resistance to insecticides and medicines including pyrethroid resistance in a large part of the country and countrywide resistance to chloroquine and sulfadoxine-pyrimethamine leading to change to high-cost products, artemisinin-based combination therapies (ACTs) and new-generation LLINs and IRS insecticides, affecting access to these preventive and lifesaving commodities.
- Inadequate coverage of interventions among vulnerable refugee populations arising from political strife in the region.
- Inadequate coverage with effective interventions due to inadequate funding and delayed access to available funds.

Whereas the program was recently elevated to a division, the centralization of administrative systems does not make it operationally effective to access funds in real time for implementation

The key lessons learned include:

- Until recently, the opportunity to engage the private sector which treats over 50% of malaria cases, as partners in malaria control and elimination strategies, was missed.
- The influx of unprecedented numbers of refuges needs to be considered in the management of malaria interventions over above the local populace in the area where refugee camps get established.
- Failure to put in place resources for rapid response for epidemics costs the country timely and cost-effective interventions.
- Engagement and promotion of co payments and full medical cover could ease access and early management of malaria cases.

- Delay in implementation of iCCM in hard-to-reach areas hampers timely management of malaria cases.

9.4.2.6 Progress Toward Integrated Vector Management (IVM) in Uganda

The past decade has seen renewed global emphasis on malaria vector control. Support from the Global Fund to Fight AIDS, Tuberculosis and Malaria, the World Bank, the President's Malaria Initiative (PMI), the Department for International Development (DFID), and others has achieved significant long-lasting insecticide-treated nets (LLINs) and indoor residual spraying (IRS) coverage in Uganda, leading to substantial reductions in the prevalence of malaria, even where the disease was highly endemic (WHO 2013). The Ministry of Health (MOH) in Uganda is vigorously promoting and scaling up LLINs and IRS with support from the Global Fund and several donors such as the USAID/PMI and the DFID-UK, which supports the Abt Associates-led Uganda IRS Project Phase II.

However, despite efforts to control and reduce the burden of malaria and other VBDs, they remained the major causes of morbidity and mortality in Uganda. Subsequently, a comprehensive vector control need assessment was conducted in August 2014 (MOH 2014), which revealed several factors that undermine the effectiveness of vector control in the country. These include among others:

- Inadequate capacity for implementing evidence-based vector control strategies at all levels—national, regional, district, and community—often resulting in a suboptimal choice or improper timing of interventions and subsequent waste of valuable resources.
- The NMCP and the Vector Control Division (VCD) program focus on a single disease, and often it is not fully integrated into health systems, raising concern about sustainability.
- The patterns of most vector-borne diseases (VBDs), including malaria, are affected by climate change, environmental degradation, and urbanization, pointing to the need for an adaptive management approach to vector control based on local evidence.
- Other sector ministries, such as agriculture, animal industry, and fisheries; ministry of works and transport, environment, finance, culture, and tourism; as well as communities are often not well informed and/or involved in vector control. As a result, they are often unaware of the consequences of their actions on the incidence of VBDs.
- LLINs and IRS rely heavily on the use of a limited choice of insecticides. Thus, the development of resistance could undermine control efforts without additional mitigation measures.

The UMRSP 2014–2020 had four strategic interventions on malaria prevention, namely, scale up and sustain IRS in 50 districts, sustain universal access to LLINs, build capacity for larval source management (LSM), and strengthen capacity in

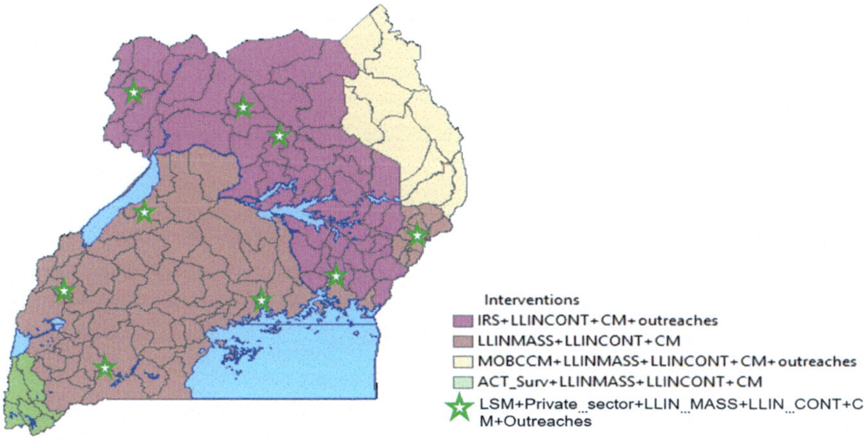

Fig. 9.1 Map of Uganda showing interventions and target areas

entomology, epidemiological surveillance, insecticide resistance monitoring, vector behavior, and bionomics. Uganda has made great progress in reducing malaria transmission over the last 10 years from 42% in 2009 to 9% in 2018. The UMRSP 2014–2020 in line with the WHO (2012) recommended an integrated vector management (IVM) approach in order to improve the impact of vector control interventions in Uganda. Consequently, the abovementioned UMRSP strategic interventions have been expanded under the new IVM strategic approach to reflect IVM components (see Fig. 9.1 and Table 9.2).

9.4.2.7 The National Strategic Plan 2021–2025

The present IVM strategic guideline applies in principle to all VBDs in Uganda but focuses primarily on malaria control. Although the focus of the guideline is human VBDs, there is also significant overlap with livestock diseases, many of which are transmitted by mosquitoes, flies, or ticks. The IVM strategic guideline will be regularly adapted to changes in local eco-epidemiological or socioeconomic conditions. To allow evaluation of the strategic guideline, the NMCP will set operational and impact targets and monitor them separately for each vector control method (WHO 2012).

To accelerate progress toward global malaria targets, the WHO rolled out the High burden, High Impact approach in 10 + 1 countries which include Uganda. The National Strategic Plan 2021–2025 is aligned to the Global Technical Strategy and National Development Plan III with its vision of "A Malaria Free Uganda to enable social economic transformation in line with vision 2040." The strategic focus of the NSP 2021–2025 is to move the country toward elimination by accelerating and sustaining malaria burden reduction in high and moderate transmission areas while reducing transmission intensity in all low transmission areas.

Table 9.2 Summary of the matrix of interventions, mode of delivery, and target populations (also see Fig. 9.2)

Core strategy area	Mode of delivery	Implementation area/target population
Long-lasting insecticide-treated nets (LLINs)	Mass campaign including in refugee settlements	All districts except areas currently implementing IRS and eligible urban cities Household members
	Continuous distribution (Antenatal Care (ANC), EPI, schools, refugee settlements, community groups)	All districts Pregnant women, children under 5 years old, specialized clinics, school-aged children, mobile populations, refugees, and community members
	Private sector—Advocacy	Eligible urban districts Household members
Indoor residual spraying (IRS)	Project led with district engagement Public-private partnership	High-burden districts and boarding institutions, prisons, barracks, health facilities, agricultural and mining settlements, hotels, schools, and higher institutions of learning
Case managements	Public health facilities and private-not-for-profit (PNFPs)	All districts
	Private health facilities	All districts
	Community health workers—iCCM	Hard-to-reach populations, high-burden areas, and refugees
	Community health workers—Outreach services	High-burden populations
	Mobile community health workers including community Intermittent Preventive Treatment for malaria in pregnancy (IPTp)	Nomadic pastoralist populations
Larval source management (LSM)	All types of LSM (larviciding, manipulation, modification, and environmental management) through a community participatory approach for sustainability	All districts (malaria hotspots) Urban cities

Stratification and special populations: The disproportionate and unstable malaria burden and transmission across the country require that a tailored combination of malaria interventions is delivered in varying intensities and approaches based on a specific landscape to accelerate malaria burden reduction. The country has thus been stratified based on epidemiology, entomology, and sociobehavioral characteristics into three main strata, namely, very low-burden areas (<2% prevalence), urban cities, and high-burden (rest of the country) areas. In addition, special population groups including nomadic pastoralists, refugees and internally displaced populations, and hard-to-reach populations will be considered.

9.4.2.8 Targeting Delivery of Interventions for Impact

The NSP will be deployed through a multi-sectoral approach, through decentralization to districts, and in the spirit of leaving no one. Guiding principles will include universal coverage for all populations at risk with proven malaria interventions; equity, equality, and nondiscrimination; human rights and gender sensitivity and availability; accessibility; acceptability; adequacy and quality; and contiguous expansion of interventions.

Universal coverage with malaria prevention interventions and prompt quality malaria diagnosis and treatment will be pursued in all strata through appropriate approaches. LLIN distribution will be conducted in majority of the districts through mass campaign, except in few low prevalence urban cities and districts where IRS is being implemented. To sustain coverage levels, the program will use LLIN continuous distribution channels through antenatal care clinics, expanded immunization clinics, school net distribution, and community groups. Social marketing will be deployed in urban cities (Fig. 9.1).

High-burden districts and boarding institutions will be targeted for indoor residual spraying. IRS will be conducted using both project modes with district engagement to foster ownership and through a public-private partnership. Case management will be sustained nationwide in the public sector, in the private sector, and at the community level. To improve access to quality care, iCCM will target high-burden areas, hard-to-reach populations, and refugees. Community health workers will provide outreach services in high-burden areas where populations are mobile like in Karamoja; these will provide mobile services (Table 9.2).

Various implementation strata have been identified for the delivery of the interventions in order to ensure impact. This was informed by the review of the data on stratification and application of the most effective intervention mixes to the identified population groups. For example, this arrangement informs the activities of the districts, the sub-recipients, and the community-based organizations in the delivery of the services (Fig. 9.2).

The malaria control program has continued to strengthen evidence-based programming through the use of program reviews, evaluations, surveys, and operation research. In 2004, the program launched the malaria research council to coordinate malaria research efforts. New tools like larviciding have been piloted, and development of a gene drive against *Anopheles gambiae* is under way.

9.4.3 Case Study # 3: Integrated Vector Management in Burkina Faso

Burkina Faso continues to bear a heavy brunt of malaria which is endemic throughout the country. Despite all efforts being invested to control the disease, the country is still considered one of those in the world where the numbers of malaria cases and

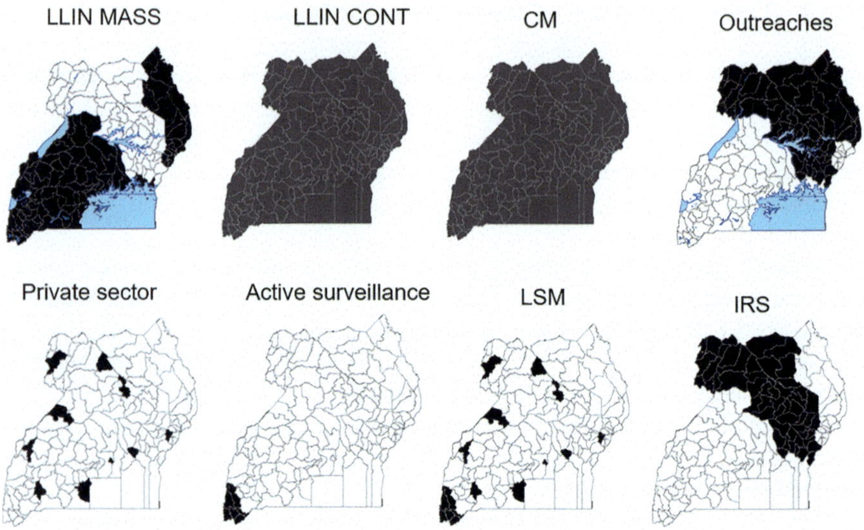

Fig. 9.2 Map of Uganda showing intervention maps designed for districts

deaths are increasing (WHO 2018). According to the *World Malaria Report 2019*, Burkina Faso is among the ten countries with the highest number of malaria cases and deaths. The rate of transmission remains high resulting in almost 12 million malaria cases of which 4000 deaths reported in 2018 (WHO 2018). With this, the country is far from meeting the target of eliminating malaria by 2030, and the objective of increased life expectancy and improved well-being of the populations remains unattainable (Fig. 9.3).

9.4.3.1 National Malaria Control Strategy

The government has developed a malaria control strategy that guides all activities targeting the prevention and control of the disease with the ultimate goal of its elimination by 2030. The strategy is built around five pillars including (1) support for vector control, (2) seasonal malaria chemoprevention for children under 5 years of age, (3) intermittent preventive treatment for pregnant women, (4) case management, and (5) behavior change communication. ITNs (insecticide-treated nets) and IRS that are promoted by the Roll Back Malaria (RBM) initiative led by the World Health Organization (WHO) are the most important vector control tools currently implemented in the country.

Burkina Faso also has a long history for active research—development on malaria vector control, starting from the development of ITNs that are today widely spread. However, with the persistence of the high transmission, it is suggested research should focus on new tools for vector control to complement those being currently

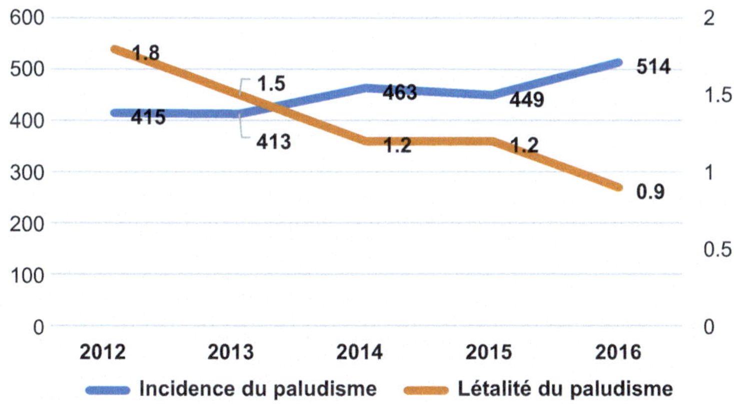

Fig. 9.3 Malaria incidence and case fatality rate (CFR) in Burkina Faso between 2012 and 2016

implemented. Institutions championing national efforts toward malaria elimination include the Ministry of Health and its specialized departments such as the National Malaria Control Program, the National Centre for Malaria Research and Training, and the Ministry of Scientific Research and Innovation through the Health Research Institute (IRSS, *Institut de Recherche en Sciences de la Santé*) under the National Centre of Scientific Research and Technology (*Centre National de la Recherche Scientifique et Technologique, CNRST*).

9.4.3.2 Integrated Vector Management Based on Genetic Engineering and Gene-Drive Approach

Burkina Faso is one of the few African countries selected by Target Malaria to develop and deploy genetically based mosquito control technology as an additional tool for malaria elimination. Target Malaria is a non-for-profit international consortium that aims to tackle malaria at the source, specifically using gene-drive approach to alter the mosquito's ability to reproduce and thus crash malaria mosquito vector populations.

The work in Burkina Faso started in 2012 and is led by the *Institut de Recherche en Sciences de la Santé* (IRSS) which is a government-funded research institute created in 1997 and focused on medical science. The IRSS is active in the fields of public health, biomedical science, traditional medicine, and pharmaceuticals while coordinating research and providing training support for medical students. The work ongoing with the partnership of Target Malaria has three pillar components, namely, (1) the scientific research, (2) the regulatory component, and (3) the stakeholder engagement. Research activities are designed according to the phased development pathway described in the WHO's "Guidance Framework for Testing of Genetically Modified Mosquitoes" and consist of three main phases including (1) lab and small-scale studies using self-limiting sterile constructs (sterile male mosquito strains),

(2) self-limiting fertile constructs (parental male bias mosquito strains), and (3) self-sustaining fertile constructs bearing the actual gene-drive modification.

Decision to move from one phase to another is guided by a rigorous risk-based evaluation and lessons learned from preceding stages. In July 2015, following a series of capacity building activities at both institutional and personnel levels, an initial application was submitted and approved for laboratory studies using male sterile *Anopheles coluzzii* strains. Promising results were obtained for this initial stage, supporting to proceed with the following step for small-scale field experiments using the same *Anopheles coluzzii* male sterile strain [Ac(DsM)2 strain] in which an authorization was granted in August 2018. Research protocols included the release of sterile male mosquitoes marked with fluorescent powder followed by recapture operations executed on daily basis. Those operations were carried out in the village of Bana, western Burkina Faso, 20 km from Bobo-Dioulasso for 3 weeks in July 2019. With this, the IRSS successfully completed the phase 1 of the program and has thus taken the lead, compared to other institutions working with Target Malaria in other countries. Subsequently, an application was prepared and submitted to commence the phase 2 which consists of studies using parental male bias strains of *Anopheles coluzzii*. The application is currently under review by the regulatory body.

The regulatory component is led by the National Biosafety Agency (ANB) which is the national biosafety competent authority operating under the provisions of the biosafety law *Loi n°064-2012 portant régime de sécurité en matière de biotechnologie* (2012). The agency was established in 2006 after the country signed the Cartagena Protocol on Biosafety in 2003. It is currently placed under the Ministry of Scientific Research and Innovation. Biosafety regulators are responsible for evaluating the potential risks that the use of genetically modified mosquitoes may pose to the environment, considering human and animal health. An advisory committee composed of eminent scientists and experts review the applications based on safety criteria adapted from the crop biotechnology risk assessment. While biosafety regulators play a prominent role for the first two phases of the development of the technology, it is expected that regulators in the health sector will take over as the product gets closer to the deployment stage, with the public health benefit valued as the principal criteria for evaluation.

With genetic-based technologies, the concept of integrated vector management goes beyond the use of various strategies and tools but considers effective integration, mobilization, and coordination across different regulatory sectors, especially in the area of biosafety, environment/biodiversity on one hand and health on the other hand. This is where the AUDA-NEPAD focuses its capacity strengthening interventions to building synergies between regulators and scientists in both areas for concerted engagement in moving forward the development of a novel vector control tool with great potential for malaria elimination. Stakeholders in Burkina Faso strive to achieve such a goal.

Following the establishment of the West Africa IVM platform inaugurated in Accra, Ghana, in September 2018, the AUDA-NEPAD has been supporting Burkina Faso on various capacity strengthening needs. Primary outcomes from these interventions include (i) the establishment and operationalization of the Institutional Biosafety Committee (IBC) at the IRSS, Bobo-Dioulasso, and (ii) the establishment of the national integrated vector management platform. The IBC oversees at the institutional level the conduct of the research activities on genetic-based vector control to ensure compliance with regulatory requirements. Creation and operationalization of IBC are considered a prerequisite condition for moving an application to the next step.

The national IVM platform is a translation of the regional one and aims at providing the necessary for constructive cross-sectoral dialogue to facilitate acceptance of the Genetically Based Vector Control (GBVC) strategy. The newly established platform in Burkina Faso brings together the biosafety experts from the national biosafety agency and its scientific advisory committee, experts from the Ministry of Health, members of the institutional biosafety committees from both health and agricultural research institutes, experts from the Ministry of Environment that are focal points for the Convention on Biological Diversity (CBD), and the Nagoya Protocol on Access and Benefit Sharing (ABS) and Communication practitioners.

Stakeholders' engagement is considered a key component for the development of the novel vector control strategy for malaria elimination (Chemonges Wanyama et al. 2021). This is particularly important in Burkina Faso as the country has a very active civil society group culture and has become an essential target to the anti-GMO activists' groups since the early adoption of *Bt* cotton. The IRSS has therefore constituted a strong communication team and developed a stakeholder engagement strategy to ensure that high-level decision-makers and local communities are all carried along as the research progresses on the ground. Interventions include (1) dialogue with high-level government officials for regular update on the progress made by the research team and (2) conversations with rural communities in villages such as Bana surrounding the trial sites to ensure that research activities are clearly understood and accepted. To ensure effective communication with local communities, scientific concepts and terminologies are translated in local languages.

9.4.3.3 Genetically Enhanced Fungus *Metarhizium pinghaense* in Burkina Faso

Several studies have explored the use of insect fungal pathogens to reduce malaria transmission (Bilgo et al. 2018a, b). In Burkina Faso, scientists based at the Institut de la Recherche en Science de la Santé (IRSS) in Bobo-Dioulasso have been working on genetically modified fungus *Metarhizium pinghaense* to improve malarial vector control. *Metarhizium pinghaense* was modified, using *Metarhizium Mcl1 promoter*, to drive the expression of insect-specific neurotoxins. Results from lab studies showed an increased lethality of the GE *Metarhizium* when infesting *An. coluzzii* and *An. gambiae* initially selected resistant to most of the insecticides employed in the country for malaria vector control. Infested mosquitoes showed altered blood feeding behavior, resulting in almost 100% of them unable to transmit malaria parasites within 5 days post-infection. Such level of efficacy was found higher than the threshold set by the WHO for a successful vector control agent (Bilgo et al. 2017).

To optimize pathogen's biocontrol potential, scientists tested additional strategies including fungus that targets the plasmodium at its early development, i.e., the sporozoites when still in the hemolymph. This kills the parasite, and the mosquito insect is no longer able to cause the malaria disease and spread it over the population. Interest was expressed to combine both strategies, one using genetically engineered *Metarhizium pinghaense* as a population suppression method to quickly reduce the mosquito vector populations and the other expressing anti-plasmodial toxins to lower the parasite levels in the mosquito.

Encouraged by the results obtained in the laboratory, scientists conducted semi-field studies in partnership of the US National Institute of Health (NIH) as a project funder, between 2013 and 2017. The aim was to confirm the hypothesis that transgenes have the potential to significantly improve the efficacy of the biocontrol fungus in a malaria disease-endemic setting like Burkina Faso. Conclusions obtained include (Lovett et al. 2019) (1) the tested genetically engineered *Metarhizium* fungus expressed a hypervirulent toxin that kills mosquito insects more rapidly than the wild-type, nonmodified *Metarhizium*; (2) due to the reduced lethal dose of the hypervirulent toxin, the percent of mosquitoes picking up a lethal infection was significantly increased; and (3) genetically modified *Metarhizium* technique employed to infest mosquitoes proved more efficient compared to nonmodified fungus. Consequently, modeling suggests that the use of *Metarhizium pinghaense* genetically modified to produce a hypervirulent toxin could reduce malaria parasite transmission by more than 90%.

The pathway to deploy GE *Metarhizium pinghaense* as a malaria vector control tool has been designed in two phases including (1) development of a final product, i.e., fungus stably modified and (2) conducting a series of field trials starting from the WHO hut trials to small-scale experiments, with increasing size, duration, and complexity, in order to fully assess the efficacy, biology, and environmental risks that genetically engineered pathogens may pose. Transition from one phase to the next is subject to "go/no-go" decision criteria, including efficacy and safety end-points, regulatory approvals, and social acceptance.

9.4.4 Case Study # 4: Integrated Vector Management in Nigeria

9.4.4.1 Introduction

Vector-borne diseases are common in Nigeria (Malaria Consortium 2020a, b). Predominant among them are onchocerciasis, schistosomiasis, malaria, lymphatic filariasis, trypanosomiasis, dracunculiasis, and yellow fever. Malaria, however, tops the list of major public health problem (Ogbonna et al. 2016). The economic cost of treatment of malaria may be as high as 1.30% of economic growth per annum (WHO 2019). Most cases in the country are due to *Plasmodium falciparum*, the life-threatening form of the malaria disease. According to the WHO, malaria accounts for about 110 million clinical cases annually, with 45–60% of all outpatient attendances of all hospital admissions (FMOH 2009). The Roll Back Malaria has played an intervention role with the supply of long-lasting insecticide-treated nets (LLINs)/ vector management with linkages to other crosscutting issues and promotion of the use of insecticide-treated nets (ITNs). Integrated vector management (IVM) was adopted as a strategic approach plan to combat malaria, and these includes vector management approaches, pesticides use and management, and policy, legislative, and regulatory framework.

9.4.4.2 Nigerian Health-Care System

The National Health Policy guideline was adopted in October 1988 by the various arms of government to give appropriate direction to the providers of health care in both public and private sectors. "Federal, State and Local Government shall support in a coordinated manner a three-tier system of health care. Essential features of the system shall be its comprehensive nature, multi-sectoral inputs, community involvement and collaboration with non-governmental providers of health care." This is with a view to achieving the goal for all using the primary health-care approach as its main strategy (FMOH 2014). This is supported by well-mobilized Nigerian community, with adequate collaboration of other related sectors and civil societies, using appropriate locally available technology.

Significant improvements were observed in the health sector. The laudable objectives contained therein were sincerely pursued within the available resources through faithful implementation, close monitoring, and continuous evaluation of the various programs. Vector control unit exists in some states of the federation but is more defined or articulated at the federal level.

9.4.4.3 Malaria Control Strategy

Nigeria is a signatory to the Abuja Declaration to global approach to fight malaria and halve the burden of malaria by 2010 in partnership with the Roll Back Malaria (RBM) that was launched in 1998 which commits the country to deploy all necessary resources and raise funds to achieve the Abuja target. In addition to the RBM efforts, the Support to the National Malaria Programme (SuNMaP) worked in close cooperation with Nigeria's National Malaria Control Elimination Programme (NMEP) in ten states covering around 40% of Nigeria's population which ended in 2016 (Health Partners International 2016).

With the change in leadership at the helm of Nigeria's National Malaria Control Program (NMCP) and significant support for malaria control from the office of the president, the program has received a fresh impetus. The government strategy to accelerate and intensify efforts on malaria control is reflected in the National Malaria Strategic Plan that was developed by the NMCP, in partnership with the RBM Partners, States Ministries of Health and their LGAs, and other stakeholders (FMOH 2011). The collaboration and involvement of all these partners are expected to enable a national scale-up of key preventive and curative interventions. The program aims to reduce the burden of malaria by 50% by 2020, which is expected to reduce all-cause child mortality by 20%.

To achieve this objective, the program envisages a massive scale-up of vector control interventions for all at-risk population and a strong focus on preventive measures, coupled with the introduction of more effective case management for the treatment of outpatient uncomplicated malaria cases, with focus on children under 5 years old (SMC Partnership 2020; Chanda et al. 2017). The Country Strategic Plan builds on a set of core as well as crosscutting interventions such as access to treatment with good diagnosis. To achieve insecticide-treated mosquito net (ITN/LLIN) distribution and coverage of at least 80% by 2020 through integrated vector management, selected areas with suitable epidemiological characteristics were focused by IRS interventions in all eligible households, with prevention during pregnancy, with women to have access to directly observed Intermittent Preventive Treatment (IPT) with sulfadoxine-pyrimethamine (SP) twice during the second and third trimesters of pregnancy.

9.4.4.4 Integrated Vector Management (IVM) in Nigeria

The IVM is broadly defined as a process of evidence-based decision-making procedures aimed to plan, implement, monitor, and evaluate targeted, cost-effective, and sustainable combinations of regulatory and operational vector control measures. The IVM is a process for the management of vector populations to reduce or interrupt transmission of disease. It applies multidisciplinary methodologies and ecosystem approaches with judicious blending of disease control programs. The characteristic features of IVM include methods based on knowledge of factors influencing local vector biology, disease transmission and morbidity, and utilization of a range of interventions, often in combination and synergistically.

Integrated vector management involves collaboration within the health sector and with other public and private sectors that impact on vectors, engagement with local communities and other stakeholders, and public health regulatory and legislative framework. The content of IVM plan includes vector management approaches, pesticide use and management, and policy, legislative, and regulatory framework. The IVM approach may be broadly classified into three categories: chemical [larvicide treatment (LT) (Afrane et al. 2016), insecticide-treated bed nets (ITN), indoor residual spraying (IRS)], biological, and environmental control (Federal Republic of Nigeria 2009).

9.4.4.5 Legislative, Policy, and Regulatory Framework

The long-term sustainability of any economic growth requires that the development interventions be well conceived and that a regulatory framework with enforcement capacity exists (Richardson et al. 2020). The overarching regulatory frameworks guiding this Integrated Vector Management Plan (IVMP) for Malaria Control are the National Health Policy and the National Policy on the Environment (WHO 2013). However, other national and international policies, laws, agreements, and regulations were important in the case of IVMP in Nigeria such as national policies and regulations FEPA decree 58 of 1988 as amended by decree 59 of 1992 and 1999 but complemented by rules and regulations such as FEPA S.1.5 and FEPA S.1.9 dealing with disposal and distribution/use of pesticides NAFDAC decree 15 of 1993, as amended by decree 19 of 1999 which ensures that active components of drugs are up to acceptable standards. The Harmful Waste (Special Criminal Provisions, etc.) decree 42 of 1988. International Treaties International transport and use of pesticides are governed by three major international treaties: the Basel Convention on the Control of Transboundary Movements of Hazardous Wastes and their Disposal (Basel Convention 1989), the Rotterdam Convention on the Prior Informed Consent Procedure for Certain Hazardous Chemicals and Pesticides in International Trade (Rotterdam Convention 2004), and the Stockholm Convention on Persistent Organic Pollutants (POPs) (Stockholm Convention 2001). Nigeria is a signatory to the Stockholm convention on POPs, and malaria control is prioritized in the draft policy framework for development and implementation of integrated vector management in Nigeria prepared by the FMOH in July 2005.

9.4.4.6 Future Prospective

Promotion of inter-sectoral collaboration coupled with massive environmental management activities has yielded more results in many malaria-eliminating countries. Mass mobilization of all stakeholders and national level with accelerated dissemination of information across board in collaboration with malaria control agencies should be strictly upheld. This should be without prejudice to disseminating information and building a stronger health system. Enforcement of environmental protection services are needed with massive environmental sanitation, clearing of irrigation canals, strict sewage disposal and management mechanisms, and drainage of stagnant waters depending on the need of the region involved (Walshe et al. 2013; Kusumawathie et al. 2008). Integrated vector management has proven to be a realistic approach, which has worked for many countries and has set many on the track of malaria elimination and eradication when it is consistently implemented and sustained. Nigeria is a malaria-endemic country where people have incidence of malaria throughout the year with huge economic burden. Embracing the IVM without prejudice could place the country on a fast track to achieving elimination and outright eradication. This should be powered on the platform of people-oriented policies and sound management framework (Ogbonna et al. 2016).

9.5 Innovative and Emerging Technologies for Malaria Vector Control in Africa

Given the lack of a highly effective malaria vaccine and rapid development of insecticide and drug resistance, there is an urgent need for the development and scaling of novel tools to mitigate disease burden worldwide. As an example of innovative and emerging technologies and approaches, four approaches are described below.

9.5.1 Sterile Insect Technique (SIT): An Innovative Method for Mosquito Control

Sterile insect technique (SIT) is one of the most promising genetic tools which can potentially offer innovative, species-specific, environmentally friendly methods for mosquito control. The SIT is the only genetic approach that has succeeded to control vectors on an extensive practical scale with operational effectiveness. The SIT is based on mass rearing, radiation for sterilization of males, and release of a large number of sterile male insects into the target population (Fig. 9.4). When sterile males mate with normal females, the laid eggs will not hatch. If adequate numbers of competitive sterile males are introduced continuously into a natural population, the population will be suppressed or even eliminated (Knipling et al. 1968). Genetic control should be seen not as a magic solution but as a set of powerful new methods that can be integrated with current methods (Alphey 2014).

Benedict and Robinson (2003) reviewed 28 trials of genetic control of mosquitoes. Field trials of SIT against mosquitoes were conducted on *Culex pipiens quinquefasciatus* in Florida, USA, *Anopheles albimanus* in El Salvador, *An. stephensi* in southern India, and *An. arabiensis* in Sudan (Wood 2005).

9.5.2 Feasibility Study of Mosquito SIT in Northern State of Sudan

Northern State, Sudan, was selected for a feasibility study on the use of SIT against the malaria vector *An. arabiensis*. The selection criteria included the existence of a single species of malaria vector, low density of the target population, geographical isolation of the target population, and actual or potential malaria incidence (Malcolm

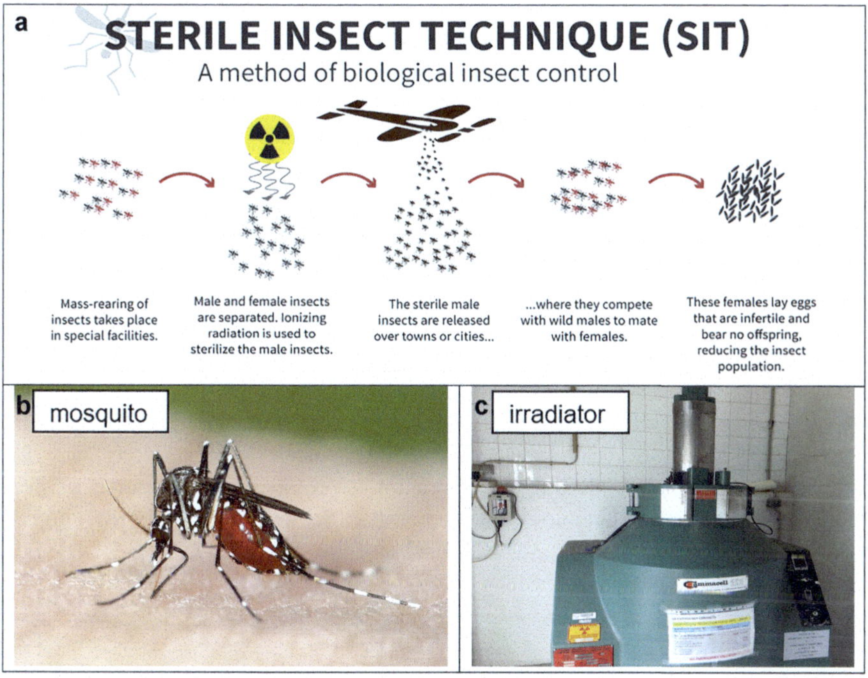

Fig. 9.4 Graphic representation of the (**a**) process of the sterile insect technique (SIT), on (**b**) the mosquito *Aedes aegypti*, (**c**) using irradiator. (*Source*: www.iaea.org)

et al. 2009). *An. arabiensis* was confirmed as the only malaria vector (Dukeen 1986; Azrag et al. 2016). Molecular study using microsatellite and mitochondrial genetic differentiation of *An. arabiensis* confirmed that the population is almost entirely isolated (Azrag et al. 2016). A comprehensive study on spatial and temporal distribution of the malaria mosquito *An. arabiensis* and the influence of environmental factors was carried using geographical information system (Ageep et al. 2014).

The success of SIT depends on genetic uniformity of the target populations. The multi-locus estimate of N_m between *An. arabiensis* populations indicated a high rate of gene flow (Azrag et al. 2016). Microsatellite allelic diversity was reduced among laboratory colonies presumably due to founder effect and the loss of apparently abundant rare alleles (Azrag et al. 2016).

The effect of irradiation on adult biology was evaluated (Helinski et al. 2006). Effects of irradiation and transportation on male fitness and competitiveness were studied (Helinski et al. 2008). Results indicated no major negative effect on male fitness and competitiveness (Hassan et al. 2010; Helinski and Harrington 2013). Ensuring ability of sterile males to mate with wild females is mandatory and essential for SIT. Findings of field studies proved that the sterile males had the ability to find and participate in wild swarms and that the survival and distance traveled by old males were more than those by young males (Ageep et al. 2014; Hassan et al. 2014).

Many innovative equipment prototypes are being developed at the IAEA laboratories. A larva/pupa separator was developed and is being used (Balestrino et al. 2011); automated racks and trays were developed for mosquito mass rearing. The system has the advantages of reducing space of rearing, facilitating larval manipulations, and reducing time and manpower (Balestrino et al. 2012).

With all these developed capacities and the great knowledge gained, successful releases of sterile males were able to reduce the population of *An. arabiensis* in a sector within the pilot site by 98.6% (unpublished data). The results are encouraging for validation by expansion of conduct of releases over the entire pilot site. Before starting any releases in the field site, it was necessary to consider ethical, legal, and social aspects of the approach in Sudan (El-Sayed et al. 2009).

Among all potential methods of genetic control of disease vectors, SIT has the greatest potential of being effective and economically viable when combined with other measures; it can extend the effective life of insecticides by slowing down the evolution of resistance (Curtis and Townson 1998). It is still a long way for population replacement, although research on it provides a rich source of new information on vector physiology and genomics (Seawright et al. 1978; Thomas et al. 2000).

9.5.3 Gene-Drive Technology as a Potential Biological Control Tool for Malaria Vectors in Africa

Gene drive is a subset of gene editing which covers a broad range of practices. Gene editing refers to "the practice of making targeted interventions at the molecular level of DNA or RNA function, deliberately to alter the structural or functional characteristics of biological entities." It can be applied to a variety of organisms, such as plants, animals, or bacteria, to potentially change a wide range of characteristics from the color of flowers to the fertility of a species. Gene editing builds on a long experience in plant and animal breeding but has recently gained more attention due to the advent of new tools, such as CRISPR/Cas9, that have made the precision and relative ease of introducing modifications much greater. It builds on particular characteristics of certain genes that allow them to spread in a population at a higher rate than the usual 50% inheritance observed for the majority of genes.

Gene drive is a technique that promotes the inheritance of a particular gene to increase its prevalence in a population. Gene drives change the way that certain genes (and therefore traits) are inherited—or passed down through generations. Through this process of "biased inheritance," a gene can increase rapidly in frequency in a population over multiple generations. Gene-drive systems are hence "self-sustaining": this is the key differentiating characteristic from other forms of genetic modification, which either are applied only to one generation or are eventually selected out, if disadvantageous, over a few generations.

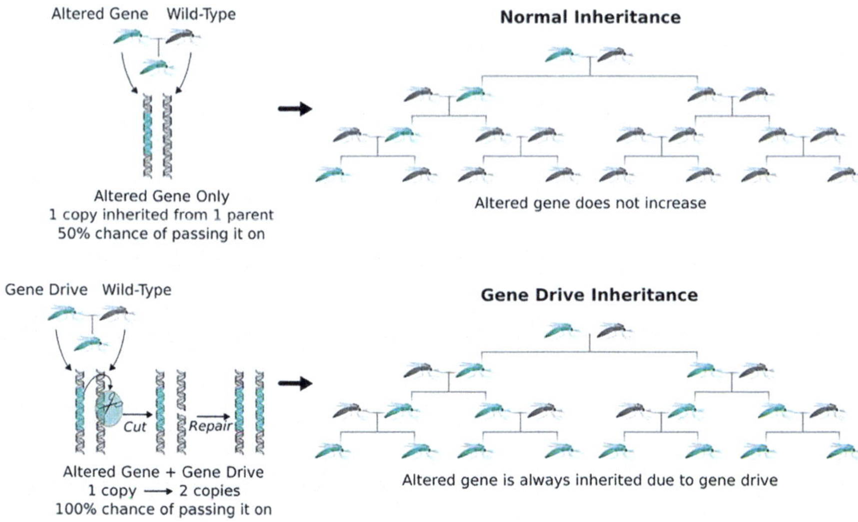

Fig. 9.5 Gene-drive technology. (*Source*: AUDA-NEPAD)

The gene-drive technology offers a great promise to combat vector-borne human diseases such as malaria, Zika, dengue, and chikungunya in Africa and worldwide. Gene drive could be applied either to modify mosquito such that they are not able to transmit malaria parasites or to drastically reduce their size population. Yet while the gene-drive technology offers enormous potential for addressing problems in health sector, for it to be implemented, a sound regulatory system needs to be in place for a country or a region to harness its potential benefits (Fig. 9.5). It is expected that the regulatory system should guide and provide answers to concerns raised by the public, policy makers, regulators, and other stakeholders with an aim to protect public health, welfare, safety, and the environment (African Union/NEPAD 2018; AUDA-NEPAD 2020).

There are several molecular strategies to develop synthetic gene-drive systems; the main alternatives use (1) sequence-specific DNA-cutting enzymes ("endonucleases"), (2) systems of toxins and antidotes, or (3) chromosomal translocations. Of the three, progress has been most rapid with the DNA-cutting enzymes, particularly since the advent of CRISPR/Cas9 technology, and this is also where future progress is likely to be most rapid. CRISPR/Cas9 is a particular architecture for sequence-specific endonucleases.

The potential of genetic technologies for use in humans, animals, and plants in solving global problems such as environmental pressures, hunger, malnutrition, and disease does not respect borders. R&D should move concurrently with creation of an enabling policy and regulatory environment for deployment. Policies to guide potential use of new technologies such as CRISPR/Cas9 and gene drives in disease vector control, preventing the spread of diseases such as Zika and malaria, can be a direct precursor to economic development for developing countries. The regulatory approval process for use of gene drive for malaria control establishes phased mandatory requirements for the conduct of research at laboratory, containment, and confinement as well as during release. The process lays out procedures for soliciting and involving members of a community during research, communication of the project's goals and methods, and public education throughout the development and deployment periods.

9.5.4 Wolbachia-*Based Biological Control of Malaria Vectors*

Wolbachia is a genus of gram-negative endosymbiotic proteobacteria that are vertically transmitted and commonly found in nematodes and arthropods (Werren et al. 2008). Several strains of *Wolbachia* are able to manipulate host reproduction by a mechanism known as cytoplasmic incompatibility (CI) (Lovett et al. 1990; O'Neill and Karr 1990), which allows *Wolbachia* to reach high prevalence in natural populations (Turelli and Hoffmann 1991, 1995) (Fig. 9.6).

Dr. Zhiyong Xi, research scientist, and his team based at Michigan State University have developed a novel natural mosquito biocontrol strategy referred to as malaria mosquito population replacement or "MMPR." In MMPR, natural disease-susceptible mosquito populations are replaced with malaria-resistant mosquito populations by using the endosymbiotic bacterium *Wolbachia* strategy. *Wolbachia* resides naturally in 28% of mosquito species, although not in *A. stephensi*. In 2012,

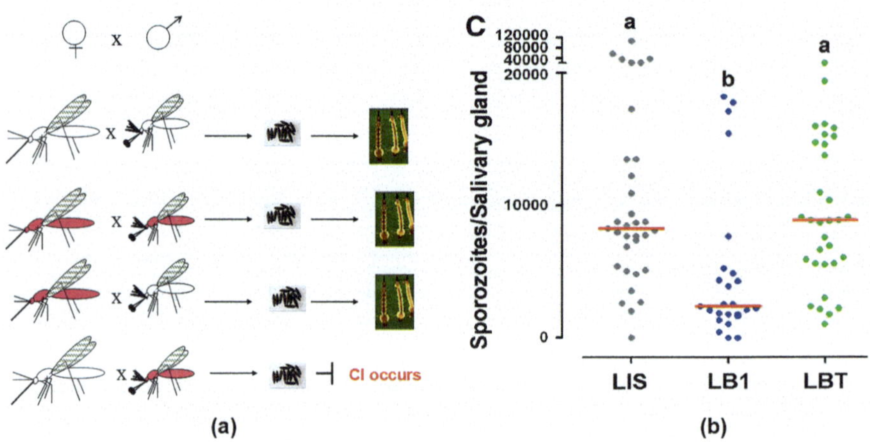

Fig. 9.6 *Wolbachia* technology, (**a**) *Wolbachia* induces cytoplasmic incompatibility (embryo death) in *Anopheles stephensi* when wild-type (white) female mates with infected (red) male, resulting in invasion of *Wolbachia* into wild population, and (**b**) *Wolbachia* induces resistance to malaria parasites in *Anopheles stephensi* (LB1, *Wolbachia* infected; LIS and LBT, uninfected). (*Source*: Dr. Zhiyong Xi, Michigan State University, USA)

MSU researchers successfully introduced *Wolbachia* into *A. stephensi*, and the bacterium has been maintained in the laboratory mosquito population to the present.

Subsequent laboratory research and controlled field trials in China have demonstrated that there are two significant scientific outcomes to MMPR: (1) *Wolbachia* alters the reproduction of its hosts such that it is rapidly assimilated into the general mosquito population (Fig. 9.6a); and (2) introduction of *Wolbachia* can make *A. stephensi* inhospitable to the malaria parasites (Fig. 9.6b) (Xi et al. 2005; Bian et al. 2013). The research has also demonstrated that this bacterium can spread into a wild-type laboratory population, such that after seven reproductive cycles (generations), 100% of the *A. stephensi* population will carry this bacterium with the malaria transmission potential being significantly diminished.

9.5.5 Genetically Engineered (GM) Mosquitoes

The Oxitec company has developed world's leading insect-pest biological control system to safely and sustainably control insects that transmit diseases (https://www.oxitec.com/en/our technology). The technology developed by the Oxitec company for insect control contains a self-limiting gene. When this gene is passed onto their offspring, the offspring do not survive to adulthood which results in the reduction in the target insect pest population. In this method, the males are released, because it is the females that are directly responsible for spreading diseases or producing larvae. This method can be applied to all kinds of insect pests including mosquitoes that serve as vector-borne human diseases.

9.6 Way Forward

Sustainable management of malaria vectors will require sound public health policies and a strong commitment of national and local governments and regional bodies with active engagement and participation of public. The cooperation and collaboration among various ministries and departments and other relevant organizations (health, environment, agriculture, etc.) are critical for the successful planning and implementation of malaria vector control programs. It also implies a robust resource availability globally which in present-day research is inevitably indispensable for a fair exchange of databases without duplication work (vide Appendix).

Experiences in Africa and globally suggest that no single approach or tactic will solve the long standing problem of malaria vectors. Therefore, an interdisciplinary and integrated vector management (IVM) program will be critical for sustainable management of malaria vectors. Although the IVM approach is emphasized, historically, the implementation of integrated approaches to vector control has been a slow and complicated process.

Additionally, an area-wide and landscape ecological approaches must be considered and integrated in the design and implementation of IVM programs. These programs should take into consideration the sociocultural aspects of local communities as well as the emerging threats of climate change on geographic and seasonal distributions of vector populations. Community mobilization and behavioral change mechanisms are significant for the success of all malaria prevention activities and programs. This can occur in the form of public health communication, i.e., the use of information, education, communication and outreach materials, media, and community-based activities. Sometimes, using people with respect and influence in the community and educating them on the benefits and correct use of malaria prevention tools helps communities to understand and communicate better about the malaria vectors and disease (Tizifa et al. 2018).

The innovative and emerging new technologies such as sterile insect technique (SIT), gene drive, and others will provide additional innovative options for IVM programs. However, like any new technology, they will need to be carefully evaluated and regulated for the safe deployment in the local environments. As the IVM programs evolve, the institutional and human capacities will need to be continually enhanced to support and sustain these programs.

For global information, it is significant to have ready in hand the resources from all across the world.

Acknowledgments The authors would like to sincerely appreciate and acknowledge the contributions of several senior officials, scientists, and experts for their invaluable contribution for this chapter through four country case studies. The case study of Sudan was contributed by Prof. Badria Babiker El-Sayed from the Tropical Medicine Research Institute of the National Centre for Research, Khartoum, Sudan; Prof. Badria also contributed the section on sterile insect technique; the case study of Uganda was contributed by Dr. Agapitus Kato Babumba from the Uganda Virus Research Institute; the case study of Burkina Faso was contributed by Dr. Moussa Savadogo from the African Union Development Agency (AUDA-NEPAD), Dr. Adama Gansane from Centre National de Recherche et de Formation sur le Paludisme, Burkina Faso, Dr. Moussa Guelbeogo from Centre National de Recherche et de Formation sur le Paludisme, Burkina Faso, Dr. Diabate Abdoulaye from Institut de Recherche en Sciences de la Sante, Burkina Faso, and Dr. Barro Nicolas from Agence nationale de biosécurité, Burkina Faso; and the case study of Nigeria was contributed by Dr. Olalekan Akinbo from AUDA-NEPAD, Burkina Faso, and Dr. Silas Obukosia from AUDA-NEPAD, Kenya.

Appendix: Global Resources on Vector-Borne Diseases

1. World Health Organization (WHO), Geneva
   http://www.who.int/news-room/fact-sheets/detail/vector-borne-diseases
2. World Mosquito Program, Monash University, Australia
   https://www.worldmosquitoprogram.org/
3. International Center for Insect Physiology and Ecology (ICIPE), Nairobi, Kenya
   http://www.icipe.org/research/human-health/malaria-programme
4. The Infectious Diseases Research Collaboration (IDRC), Uganda

http://idrc-uganda.org/

5. Building out Vector-borne Diseases in Sub-Saharan Africa
   www.bovanetwork.org
6. National Institute of Malaria Research (NIMR), Dwarka, New Delhi, India
   http://www.nimr.org.in/
7. Center for Disease Control (CDC)—Centers for Disease Control and Prevention, USA
   http://www.cdc.gov/necezid/index.html
8. National Institute of Allergy and Infectious Diseases (NIAID), part of the National Institutes of Health (NIH), USA
   https://www.nih.gov/about-nih/what-we-do/nih-almanac/national-institute-allergy-infectious-diseases-niaid
9. Johns Hopkins Malaria Research Institute, USA
   https://hub.jhu.edu/2017/04/24/malaria-research-institue-awarded-10-million/
10. European Virtual Institute for Malaria Research
    https://en.wikipedia.org/wiki/European_Virtual_Institute-awarded-10-million/
11. Malaria-Bill and Melinda Gates Foundation, Seattle, USA
    https://www.gatesfoundation.org/What-We-Do/Global-health/Malaria
12. National Center for Engineering and Zoonotic Infectious Diseases (NCEZID), Atlanta, Georgia, USA
    https://www.cdc.gov/ncezid/dvbd/index.html
13. The Global Fund to Fight AIDS, Tuberculosis and Malaria, Australia
    https://www.theglobalfund.org/en/
14. Burnet Institute—Medical Research, Melbourne, Australia
    https://www.burnet.edu.au/
15. German Centre for Infection Research (Deutsches Zentrum für Infektionsforschung, DZIF, Germany)
    http://www.dzif.de/en/research/malaria/
16. Tata Trusts, Mumbai, India
    https://journosdiary.com/2018/02/09/tata-trusts-funding-tb-malaria/

References

Afrane YA, Mweresa NG, Wanjala CL et al (2016) Evaluation of long-lasting microbial larvicide for malaria vector control in Kenya. Malar J 15:577. https://doi.org/10.1186/s12936-016-1626-6

African Union Development Agency/NEPAD (2018) Gene-drives: for malaria control and elimination in Africa. High level African Union Panel on Emerging Technologies (APET) for Africa report. NEPAD, Midrand

African Union Development Agency/NEPAD (2020) Position paper on integrated vector management: strengthening AU members' regulatory capacities for responsible research towards elimination of malaria in Africa. AUDA-NEPAD, Midrand

Ageep TB, Damiens D, Alsharif B, Ahmed A, Salih EHO, Ahmed FTA, Diabaté A, Lees RS, Gilles JRL, El-Sayed BB (2014) Participation of irradiated Anopheles arabiensis males in swarms following field release in Sudan. Malar J 13(1):484. https://doi.org/10.1186/1475-2875-13-484

ALMA (2016) African leaders malaria alliance. African heads of state adopt roadmap to eliminate malaria in Africa by 2030. Available at: http://alma2030.org/content/african-heads-state-adopt-roadmap-eliminate-malaria-africa-2030

Alphey L (2014) Genetic control of mosquitoes. Annu Rev Entomol 59:502–524

Azrag RS, Ibrahim K, Malcolm C, El Rayah E, El-Sayed B (2016) Laboratory rearing of Anopheles arabiensis: impact on genetic variability and implications for Sterile Insect Technique (SIT) based mosquito control in northern Sudan. Malar J 15:–432. https://doi.org/10.1186/s12936-016-1484-2

Balestrino F, Gilles JRL, Soliban SM, Nirschl A, Benedict QE, Benedict MQ (2011) Mosquito mass rearing technology: A cold-water vortex device for continuous unattended separation of Anopheles arabiensis pupae from larvae. J Am Mosq Control Assoc 27(3):227–235

Balestrino F, Benedict MQ, Gilles JRL (2012) A new larval tray and rack system for improved mosquito mass rearing. J Med Entomol 49(3):595–605

Basel Convention (1989) The Basel Convention on the control of transboundary movements of hazardous wastes and their disposal. Available at https://www.bing.com/search?q=Basel+Convention+on+the+Control+of+Transboundary+Movements+of+Hazardous+Wastes+and+their+Disposal&form=. Accessed 1 Feb 2021

Beier JC, Keating J, Githure JI, Macdonald MB, Impoinvil DE, Novak RJ (2008) Integrated vector management for malaria control. Malar J 7(Suppl 1):S4. https://doi.org/10.1186/1475-2875-7-S1-S4. http://www.malariajournal.com/content/7/S1/S4

Benedict MQ, Robinson AS (2003) The first releases of transgenic mosquitoes : an argument for the sterile insect technique. Trends Parasitol 19(8):349–355

Benelli G, Beier JC (2017) Current vector control challenges in the fight against malaria. Acta Trop 174:91–96. https://doi.org/10.1016/j.actatropica.2017.06.028. Epub 2017 Jul 3

Bian G, Joshi D, Dong Y, Lu P, Zhou G, Pan X, Xu Y, Dimopoulos G, Xi Z (2013) *Wolbachia* invades *Anopheles stephensi* populations and induces refractoriness to *Plasmodium* infection. Science 340(6133):748–751

Bilgo E, Lovett B, Fang W, Bende N, King GF, Diabate A, St Leger RJ (2017) Improved efficacy of an arthropod toxin expressing fungus against insecticide-resistant malaria-vector mosquitoes. Sci Rep 7(1):3433. https://doi.org/10.1038/s41598-017-03399-0

Bilgo E, Lovett B, Bayili K, Millogo AS, Saré I, Dabiré RK, Sanon A, St Leger RJ, Diabate A (2018a) Transgenic Metarhizium pingshaense synergistically ameliorates pyrethroid-resistance in wild-caught, malaria-vector mosquitoes. PLoS One 13(9):e0203529. https://doi.org/10.1371/journal.pone.0203529. eCollection 2018

Bilgo E, Lovett B, St. Leger RJ, Dabiré RK, Sanon A, Diabaté A (2018b) Native entomopathogenic *Metarhizium* spp. from Burkina Faso and their virulence against the malaria vector *Anopheles coluzzii* and off-target insects. Parasit Vectors 11:209. https://doi.org/10.1186/s13071-018-2796-6

Brooke B, Koekemoer L, Kruger P, Urbach J, Misiani E, Coetzee M (2013) Malaria vector control in South Africa. S Afr Med J 10. https://doi.org/10.7196/SAMJ.7447. www.bovanetwork.org. Accessed 2020. Building out vector-borne diseases in Africa

Carter R, Karunaweera ND (2020) The role of improved housing and living environments in malaria control and elimination. Malar J 19:385. https://doi.org/10.1186/s12936-020-03450-y

Chanda E, Ameneshewa B, Bagayoko M, Govere JM, Macdonald MB (2017) Harnessing integrated vector management for enhanced disease prevention. Trends Parasitol 33(1):30–41. https://doi.org/10.1016/j.pt.2016.09.006

Chemonges Wanyama E, Dicko B, Pare Toe L, Coulibaly BM, Barry N, Bayala Traore K, Diabate A, Drabo M, Kayondo KJ, Kekele S, Kodio S, Dinyiri AK, Ronny LR, Magala M, Ihibna MW, Mukwaya S, Namukwaya A, Robinson B, Samoura H, Sanogo K, Thizy D, Traoré F (2021) Co-developing a common glossary with stakeholders for engagement on new genetic approaches for malaria control in a local African setting. Malar J 20:53. https://doi.org/10.1186/s12936-020-03577-y

Curtis CF, Townson H (1998) Malaria: existing methods of vector control and molecular entomology. Br Med Bull 54:311–325

Duchet C, Allan R, Pierre C (2013) Vector control: some new paradigms and approaches, vol 23, pp 705–753. https://doi.org/10.5772/56258

Dukeen MYH (1986) Ecology of the malaria vector Anopheles arabiensis Patton (Diptera: Culicidae) by the Nile in northern Sudan. Bull Entomol Res 76:451–467

El-Sayed BB, Malcolm CA, Babiker A, Malik EM, El Tayeb MAH, Saeed NS, Nugud AH, Knols BGJ (2009) Development of the sterile insect technique for African malaria vectors: ethical, legal and social aspects of the approach in Sudan. Malar J 8(Suppl 2):S3

Federal Ministry of Health (FMOH) Nigeria (2011) Pharmaceutical country profile Nigeria. Available: https://www.who.int/hna/country/nga.pdf

FMOH (2009) Integrated vector management plan for the Malaria control booster project. http://documents1.worldbank.org/curated/en/361241468098984323/pdf/E21060v20P11501NAL1MCBP1IVMP1REPORT.pdf

FMOH (2014) National malaria elimination programme reviews progress in Nigeria. https://www.afro.who.int/news/national-malaria-elimination-programme-reviews-progress-nigeria

Furnival-Adams J, Olanga EA, Napier M, Garner P (2020) House modifications for preventing malaria. Cochrane Database Syst Rev (10):CD013398. https://doi.org/10.1002/14651858.CD013398.pub2

Hassan M'a M, El Motasim W, Ahmed RT, El-Sayed BB (2010) Prolonged colonization, irradiation and transportation do not impede mating and competitiveness of male anopheles arabiensis mosquitoes under semi field conditions in northern Sudan. Malar World 1:1

Hassan MM, Zain HM, Basheer MA, Elhaj H-EF, El-Sayed BB (2014) Swarming and mating behavior of male Anopheles arabiensis Patton (Diptera: Culicidae) in an area of the Sterile Insect Technique Project in Dongola, northern Sudan. Acta Trop 132:S64–S69

Health Partners International (2016) Support to National Malaria elimination programme, Nigeria. http://healthpartners-int.co.uk/our-projects/support-to-the-national-malaria-programme-nigeria. Accessed 1 Feb 2021

Helinski MEH, Harrington LC (2013) Considerations for male fitness in successful genetic vector control programs. In: Ecology of parasite-vector interactions. Wageningen Academic Publishers, Wageningen, pp 221–244

Helinski MEH, El-Sayed BB, Knols BGJ (2006) The Sterile Insect Technique: can established technology beat malaria? Trop Med 66(1):13–20

Helinski MEH, Hassan M'a M, El-Motasim WM, Malcolm CA, Knols BGJ, El-Sayed B (2008) Towards a sterile insect technique field release of *Anopheles arabiensis* mosquitoes in Sudan: irradiation, transportation, and field cage experimentation. Malar J 7:65

Killeen GF, Tatarsky A, Diabate A, Chaccour CJ, Marshall JM, Okumu FO, Brunner S, Newby G, Williams YA, Malone D, Tusting LS, Gosling RD (2017) Developing an expanded vector control toolbox for malaria elimination. BMJ Glob Health 2:e000211. https://doi.org/10.1136/bmjgh-2016-000211

Knipling EF, Laven H, Craig GB, Pal R, Kitzmiller JB, Smith CN et al (1968) Genetic control of insects of public health importance. Bull Org Mond Sante Bull Wld Hlth Org 38:421–438

Kusumawathie PH, Wickremasinghe AR, Karunaweera ND, Wijeyaratne MJ (2008) Larvivorous potential of the guppy, poecilia reticulate, the anopheline mosquito control in riverbed pools below the Kotmale dam, Sri Lanka. Asia Pac J Public Health 20:56–63

Lobo NF, Achee NL, Greico J, Collins FH (2018) Modern vector control. Cold Spring Harb Perspect Med 8(1):a025643. https://doi.org/10.1101/cshperspect.a025643

Lovett B, Breeuwer JA, Werren JH (1990) Microorganisms associated with chromosome destruction and reproductive isolation between two insect species. Nature 346:558–560

Lovett B, Bilgo E, Millogo SA, Ouattarra AK, Sare I, Gnambani EJ, Dabire RK, Diabate A, St Leger RJ (2019) Transgenic Metarhizium rapidly kills mosquitoes in a malaria-endemic region of Burkina Faso. Science 364(6443):894–897. https://doi.org/10.1126/science.aaw8737

Mafra-Neto A, Dekker T (2019) Novel odor-based strategies for integrated management of vectors of disease. Curr Opin Insect Sci 34:105–111. https://doi.org/10.1016/j.cois.2019.05.007

Malaria Consortium (2020a) Seasonal malaria chemoprevention program 2019–2020 Malaria Consortium's annual report on the use of philanthropic funding for seasonal malaria chemoprevention. https://www.malariaconsortium.org/mediadownloads/1339/Seasonal%20malaria%20chemoprevention%20program%20Philanthropy%20Report%202019-2020. Accessed 1 Feb 20

Malaria Consortium (2020b) Seasonal malaria chemoprevention program 2019–2020. https://www.malariaconsortium.org/mediadownloads/1339/Seasonal%20malaria%20chemoprevention%20program%20Philanthropy%20Report%202019-2020

Malcolm CA, El-Sayed B, Babiker A, Girod R, Fontenille D, Knols BG (2009) Field site selection: getting it right first time around. Malar J 8(Suppl 2):S9

MOH (2014) Uganda malaria reduction strategic plan

Mutero CM, Schlodder D, Kabatereine N, Kramer R (2012) Integrated vector management for malaria control in Uganda: knowledge, perceptions and policy development. Malar J 11:21. http://www.malariajournal.com/content/11/1/21

O'Neill SL, Karr TL (1990) Bidirectional incompatibility between conspecific populations of Drosophila simulans. Nature 348:178–180

Ogbonna BO, Ejim CE, Ele GN, Ohiaeri IG, Ezenekwe LN, Nwako CN, Ilonzo NO (2016) Beyond drugs in eliminating malaria scourge in Nigeria; scaling-up vector control strategies. J Adv Med Pharm Sci 9(3):1–8

Raghavendra K, Barik TK, Niranjan Reddy BP, Sharma P, Dash AP (2011) Malaria vector control: from past to future. Parasitol Res 108:757–779. https://doi.org/10.1007/s00436-010-2232-0

Richardson S, Ibinaiye T, Nikau J, Oresanya O, Marasciulo M, Roca-Feltrer A, Rassi C, Adesoro O (2020) COVID-19 knowledge, beliefs, prevention behaviours and misinformation in the context of an adapted seasonal malaria chemoprevention campaign in six northern Nigeria. Trop Med Health 48:101. https://doi.org/10.1186/s41182-020-00288-7

Rotterdam Convention (2004) Rotterdam convention on the prior informed consent procedure for certain hazardous chemicals and pesticides in international trade. https://www.bing.com/search?q=ConsentProcedure+for+Certain+Hazardous+Chemicals+and+Pesticides+in+International+Trade&form=. Accessed 1 Jan 2021

Seawright JA, Keiser PE, Dame DA, Lofgren CS (1978) Genetic method for preferential elimination of females of *Anopheles albimanus*. Science 300:1303–1304

SMC Partnership (2020) Effectiveness of seasonal malaria chemoprevention at scale in west and central Africa: an observational study. Lancet 396:1829–1840. https://www.malariaconsortium.org/media-downloads/1405/Effectiveness%20of%20seasonal%20malaria%20chemoprevention%20at%20scale%20in%20west%20and%20central%20Africa:%20An%20observational%20study

Sougoufara S, Ottih EC, Tripet F (2020) The need for new vector control approaches targeting outdoor biting anopheline malaria vector communities. Parasite Vectors 13:295. https://doi.org/10.1186/s13071-020-04170-7

Stockholm Convention (2001) Overview of the Stockholm convention on persistent organic pollutants was adopted by the conference of plenipotentiaries on 22 May 2001 in Stockholm. http://www.pops.int/TheConvention/Overview/tabid/3351/Default.aspx. Accessed 1 Feb 2021

Thomas DD, Donnelly CA, Wood RJ, Alphey LS (2000) Insect population control using a dominant, repressed lethal genetic system. Science 287:2474–2476

Tizifa TA, Kabaghe AN, McCann RS, van den Berg H, Van Vugt M, Phiri KS (2018) Prevention efforts for malaria. Curr Trop Med Rep 5(1):41–50. https://doi.org/10.1007/s40475-018-0133-y

Turelli M, Hoffmann AA (1991) Rapid spread of an inherited incompatibility factor in California Drosophila. Nature 353:440–442

Turelli M, Hoffmann AA (1995) Cytoplasmic incompatibility in Drosophila simulans: dynamics and parameter estimates from natural populations. Genetics 140:1319–1338

Walshe DP, Garner P, Abdel-Hameed AAA, Pyke GH, Burkot T (2013) Larvivorous fish for preventing malaria transmission. Cochrane Database Syst Rev 12:CD008090

Werren JH, Baldo L, Clark ME (2008) Wolbachia: master manipulators of invertebrate biology. Nat Rev Microbiol 6:741–751

WHO (2012) Handbook for integrated vector management. Available at: http://apps.who.int/iris/bitstream/handle/10665/44768/9789241502801_eng.pdf;jsessionid=AEBA0AD5AC73FA89620518571327864E?sequence=1

WHO (2013) Guidance note on capacity building in malaria entomology and vector control. World Health Organization, Geneva

WHO (2015a) Global technical strategy for Malaria (2016–2030)

WHO (2015b) WHO's assessment of malaria epidemic in Uganda: findings and recommendations: WHO, NMCP and MOH report, Oct 2015

WHO (2016a) World malaria report 2015. World Health organization (WHO), Geneva

WHO (2016b) A toolkit for integrated vector management in sub-Saharan Africa. World Health Organization (WHO), Geneva

WHO (2018) World malaria report 2018. World Health Organization, Geneva

WHO (2019) World malaria report 2019. World Health Organization, Geneva

WHO (2020a) About Nigeria. https://www.afro.who.int/countries/nigeria. Accessed 1 Feb 2021

WHO (2020b) World malaria report 2020. World Health Organization (WHO), Geneva

Wood RJ (2005) Genetic control of vectors. In: Marquardt WC (ed) Biology of disease vectors. Elsevier Academic Press, Amsterdam, pp 661–669

Xi Z, Khoo CC, Dobson SL (2005) *Wolbachia* establishment and invasion in an *Aedes aegypti* laboratory population. Science 310(5746):326–328

Yaw AA, Mweresa NG, Wanjala CL, Gilbreath TM III, Zhou G, Lee M-C, Githeko AK, Yan G (2016) Evaluation of long-lasting microbial larvicide for malaria vector control in Kenya. Malar J 15:577. https://doi.org/10.1186/s12936-016-1626-6

Chapter 10
Engineering RNA Interference-Based Dengue Virus Resistance in the Mosquito Vector *Aedes aegypti*: The Current Status and Future Directions

S. D. Denipitiyage, Y. I. N. S. Gunawardene, Zurita Federico, and R. S. Dassanayake

Abstract Dengue is an acute, febrile disease caused by the dengue viruses (DENV) comprising four serotypes and transmitted by the mosquito vector *Ae. aegypti*. DENV are single-stranded, positive-sense RNA viruses of the family *Flaviviridae*. Dengue is declared as a current significant challenge in the Southeast Asia, imposing growing burden on infected populations. To date, dengue control has mostly relied on vector control strategies which have largely become ineffective. There is, therefore, an urgent need for novel vector control strategies. Development of genetically modified mosquito vectors to manipulate disease-vectoring populations has gathered increased interest in recent time. RNAi-mediated viral resistance contributes to the suppression of viruses, including DENV in the mosquito vector *Ae. aegypti*. With recent advances in the field of molecular biology, we and other scientists are continuing to engineer genes that confer virus resistance to reduce transmission rates of DENV and introducing these genes into the mosquito genome. Even though scientists successfully generated mosquito refractory to DENV2–4, no mosquito refractory to all four serotypes has been developed to date. This limitation can be overcome by systematic analysis of the molecular mechanisms of RNAi in the mosquito vector *Ae. aegypti*. An enhanced understanding of RNAi function in the mosquito vector *Ae. aegypti* will facilitate the application of RNAi to control the transmission of the dengue disease in the future. Here, based on current understanding of the RNAi, we discuss the mechanisms of RNAi in the mosquito vector *Ae.*

S. D. Denipitiyage · R. S. Dassanayake (✉)
Department of Chemistry, Faculty of Science, University of Colombo, Colombo, Sri Lanka
e-mail: sdenipitiyage@chem.cmb.ac.lk; rsdassanayake@chem.cmb.ac.lk

Y. I. N. S. Gunawardene
Molecular Medicine Unit, Faculty of Medicine, University of Kelaniya, Ragama, Sri Lanka
e-mail: nilminisg@kln.ac.lk

Z. Federico
Department of Genetics, University of Granada, Granada, Spain
e-mail: f.zurita@ugr.es

B. K. Tyagi (ed.), *Genetically Modified and other Innovative Vector Control Technologies*, https://doi.org/10.1007/978-981-16-2964-8_10

aegypti. We also provide guidelines for optimal design of RNAi experiments in *Ae. aegypti* with the possible risks associated with them along with proposed solutions.

Keywords Dengue · Dengue virus · *Ae. aegypti* · RNA interference

10.1 Introduction

Mosquitoes are competent vectors of transmitting a large number of medically important arthropod-borne viral pathogens (arboviruses) to humans and animals. Dengue viruses cause the most common arboviral infection in humans, dengue, which is an acute, febrile disease causing symptoms ranging from high fever and flu-like symptoms to the deadly dengue hemorrhagic fever (DHF) and dengue shock syndrome (DSS) (Westaway et al. 1985).

DENV, consisting of four antigenically distinct serotypes (DENV1–4), are enveloped, single-stranded, positive-sense RNA viruses of the family *Flaviviridae*. The RNA genome of DENV is approximately 11 kb and encodes a precursor polyprotein containing three structural proteins (capsid [C], precursor membrane [prM], and envelope [E]) and seven nonstructural (NS) proteins (NS1, NS2A, NS2B, NS3, NS4A, NS4B, and NS5). The open reading frame is flanked by 5′ and 3′ untranslated regions (5′ and 3′ UTR). The genomic RNA is capped at the 5′ end but not polyadenylated at the 3′ end (Fig. 10.1). The polyprotein precursor is co- and posttranslationally processed into individual proteins by cellular and viral proteases (Chambers et al. 1990). DENV is primarily transmitted between humans by the urban-adapted mosquito vector *Ae. aegypti*, together with *Aedes albopictus* as a secondary vector.

The global occurrence of dengue has grown rapidly in recent years. Prior to 1970, only nine countries had reported dengue epidemics. However, at present, more than 100 countries are experiencing dengue epidemics, causing a significant challenge to Southeast Asia (https://www.who.int/en/news-room/fact-sheets/detail/dengue-and-severe-dengue). Today, nearly 390 million infections are occurring annually with an estimated 3.6 billion people living in areas at risk for epidemic transmission, draining annually an estimated $40 billion for health-care spending and lost productivity in affected countries (Gjenero-Margan et al. 2011).

Currently a licensed tetravalent vaccine to treat dengue infection is not available (Capeding et al. 2014; Villar et al. 2015). To date, dengue control has mostly relied on vector control strategies such as the use of insecticides, biological control agents,

Fig. 10.1 Dengue viral genome. (*Source*: This work)

conventional sterile insect technique, and *Wolbachia*-based approaches to suppress mosquito populations. However, current approaches have largely failed to reduce disease transmission in almost all dengue-endemic countries. Furthermore, vaccination mainly reduces urban transmission. Possible enzootic circulation of virus carries the risk of mutation accumulation and spillover infections (Wolfe et al. 2001; Sun et al. 2006). Therefore, novel vector control strategies are essential to prevent virus transmission between mosquitoes and hosts.

While most arboviruses induce significant morbidity and/or mortality in vertebrate hosts, mosquito vectors remain infected without getting diseased (Sun et al. 2006). Therefore, the question is what could be the nature of arbovirus manipulation in mosquito cells. RNAi has been shown to be a key mosquito innate defense against DENV and other arboviruses, resulting in infections in mosquito vectors nonpathogenic (Campbell et al. 2008; Sánchez-Vargas et al. 2009; McFarlane et al. 2014). RNAi is an evolutionary-conserved, sequence-specific, posttranscriptional gene-silencing process triggered by double-stranded RNA (dsRNA) molecules. RNAi plays an important role in gene regulation and innate defense against invading viruses by promoting the degradation of mRNA. Therefore, suppression of DENV replication can be boosted by enhancing the native RNAi machinery of the host through engineering mosquito vectors to express viral-specific dsRNA (Mathur et al. 2010; Franz et al. 2011). Understanding these innate defenses of mosquito vectors against arboviruses has shed light on developing RNAi-based DENV-resistant transgenic mosquitoes.

However, efficiency of RNAi experiments varies among genes and insect species, upon mode of gene delivery, and between different laboratories (Terenius et al. 2011). Therefore, the objective of this article is to describe the principles of designing RNAi experiments in mosquito vector *Ae. aegypti* to control the transmission of DENV as a single protocol cannot be applied for every insect. The article is primarily divided into three sections. In the first section, current understanding of the mechanisms of RNAi in *Ae. aegypti* is reviewed. This information offers a guide to the most suitable strategy for the *Ae. aegypti* and strengthens the timely needed future innovation in RNAi technology. In the second section, the design of RNAi experiments is addressed to identify successful experimental designs, effective methods for RNAi delivery, and informative indices of RNAi efficacy. In the third section, the application of RNAi for the control of dengue disease and the unique opportunities and challenges associated with these applications are discussed.

10.2 Mechanisms of RNAi

The three pathways of RNAi (microRNA, miRNA; small interfering RNA, siRNA; Piwi-interacting RNA, piRNA) in which small ~20–30 nucleotide noncoding RNAs and their associated proteins are involved in RNAi-mediated gene regulation. Through endogenous miRNA processing pathway, miRNA precursors are processed by two RNase III-type endonucleases, Drosha and Dicer, in two stages. In the

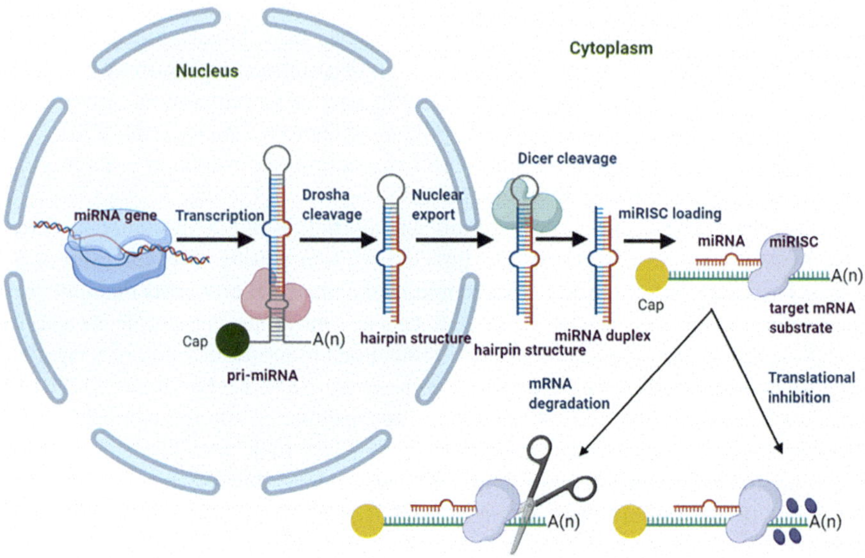

Fig. 10.2 miRNA pathway. (*Source*: This work)

nucleus, primary miRNA (pri-miRNA) is cleaved into ~70-nt-long hairpins by Drosha, followed by exporting into the cytoplasm by Exportin 5. These hairpins are cleaved into ~22-nt miRNA duplexes by Dicer in the cytoplasm. miRNA duplexes are loaded into Ago-1 or Ago-2 proteins in miRNA-induced silencing complexes (miRISCs), based on their structural differences, where the "passenger" strand is cleaved and/or dissociated and the "guide" strand is retained (Förstemann et al. 2007; Ghildiyal et al. 2010; Yang et al. 2013). By recognizing the complementary sequence of the target RNA, miRISCs execute silencing through RNA degradation, translational inhibition, or both (Wilczynska and Bushell 2015; Li and Rana 2014), thereby executing posttranscriptional gene silencing (Fig. 10.2).

The replication strategy used by positive-sense RNA viruses such as DENV helps to explain the role of RNAi in antiviral response. When the adult female ingests an infectious blood meal, DENV enter the mosquito vector, following the infection of the midgut epithelial cells. The exogenous siRNA pathway, another pathway of RNAi, is initiated when long, virus-derived dsRNA is recognized and cleaved by Dicer 2 into 21-bp-long siRNAs as dsRNA replicative intermediates (RI) are generated inside infected cells (Westaway et al. 1997). These duplex siRNAs are loaded onto the RNA-induced silencing complex (RISC), degrading the passenger strand and using the guide strand for targeted degradation of single-stranded viral RNA that has a complementary sequence (cf. Sanchez-Vargas et al. 2004) (Fig. 10.3). It has been proposed that sequestration of flavivirus replication complexes (RC) in double membrane-enclosed vesicles in mosquito cells restricts access of Dicer 2 to dsRNA replicative intermediates (Welsch et al. 2009). The intricate mechanism adopted by flaviviruses to hide the RI suggests that this adaptation may have evolved to limit

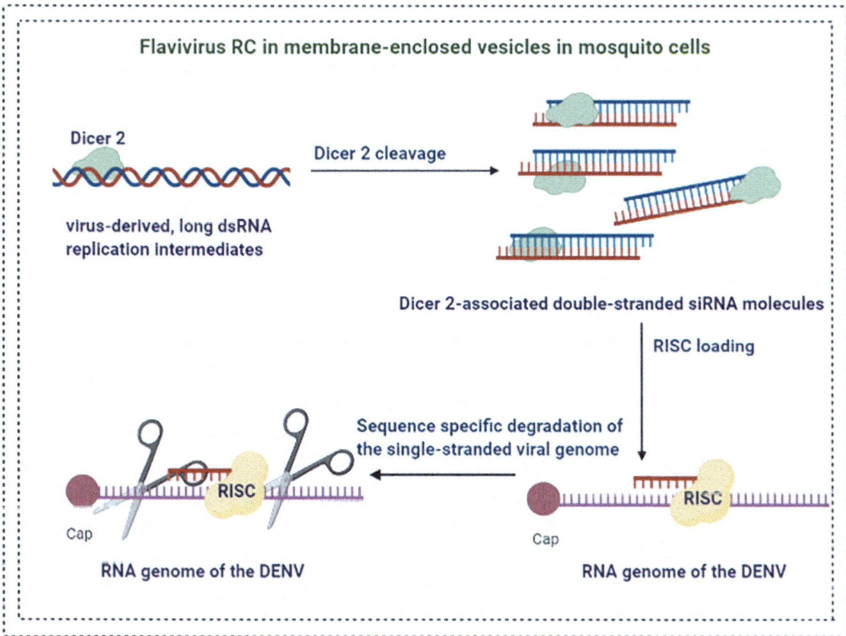

Fig. 10.3 siRNA pathway. (*Source*: This work)

exposure of the RC to host defense mechanism, RNAi. Therefore, the host cell and flaviviruses have coevolved defense-counter defense mechanisms. Following viral replication in the epithelial cells, DENV disseminate to the salivary glands. Virus is subsequently entered into the salivary gland ducts, after further replication in the salivary glands. Transmission occurs through the infected saliva in a subsequent bite. The time between initial infection of the midgut and transmission of virus in saliva is termed the extrinsic incubation period (EIP). The EIP for DENV in *Ae. aegypti* is about 7–14 days (Scott and Burrage 1984; Chandler et al. 1998).

10.3 Design of RNAi Experiments

As described above, RNAi application and efficacy remain variable depending on several factors despite the greatest potential of RNAi in dengue vector control. Therefore, RNAi experiments should be designed to minimize potential challenges by optimizing experimental factors such as the design of the RNAi molecule and the mode of delivery.

10.3.1 The RNAi Molecule

Experimental evaluation of RNAi as an antiviral tool against DENV has evaluated in both in vitro and in vivo systems, with the objectives of assessing efficiency of target sequences to inhibit the DENV replication in cultured cell lines and blocking disease transmission in the mosquito vector, respectively. As the generation of transgenic mosquitoes that are refractory to DENV has already shown promise in controlling the dengue disease, here we summarize designing of RNAi experiments to be evaluated in the mosquito vector.

An anti-DENV transgene is mainly comprised of effector molecules, specific promoters to express effector molecules, and a transposon. The success of an RNAi experiment to boost the native, antiviral response of mosquito vectors depends on the specific effector molecules [in the form of long inverted repeat RNA (IR RNA), miRNA, or siRNA] used to target the single-stranded RNA genome of the DENV. Franz and others (Franz et al. 2006) have used siRNA derived from IR RNA corresponding to 578-bp fragment of the prM region of DENV2 genome generating a larger number of siRNAs enhancing the efficiency of antiviral effect. Alternatively, polycistronic cluster of miRNA/siRNA embedded in a naturally occurring miRNA molecule has used to avoid nontarget effects associated with long IR RNA (Yen et al. 2018; Ramyasoma et al. 2020). Two approaches are broadly in use to design siRNA/miRNA sequences to target the individual genomes of each of the four DENV strains, algorithms such as Dharmacon (Dharmacon siRNA Design Center) (Reynolds et al. 2004; Birmingham et al. 2007) to improve functionality and specificity of small RNA and designing small RNA targeted against a sequence that is highly conserved across the DENV virome (conserved small RNA) to avoid mutant escapes.

Ae. aegypti mosquitoes harboring this synthetic anti-DENV transgene should be essentially expressed in the midgut tissue of female mosquitoes. Expression of the anti-DENV transgene in the midgut of female mosquitoes soon after ingestion of viremic blood ensures the generation of small RNAs when the virus is in its most vulnerable state at the onset of replication and before the formation of infection foci inside the mosquito to reduce DENV infection, dissemination, and transmission rates. The *Ae. aegypti* carboxypeptidase A (AeCPA) gene promoter has been often used to elicit blood meal-inducible, midgut-specific expression of the effector molecules (Moreira et al. 2000). However, the ubiquitous promoter *Aedes polyubiquitin* (AePUB) has also been used, despite the fact that it could increase nonspecific effects imposing fitness cost to the host due to continuous expression of the transgene.

Transposons such as the *piggyBac* element and *mariner* have used to deliver anti-DENV transgenes to mosquito vectors (Fraser et al. 1996; Coates et al. 1998). The use of a specific transposon is determined by its characteristic movement and behavior in a genome. Associated helper plasmid is used to express transposase in trans, facilitating the movement of the transposon carrying the transgene into the host genome. The screening of transformants is facilitated by incorporating

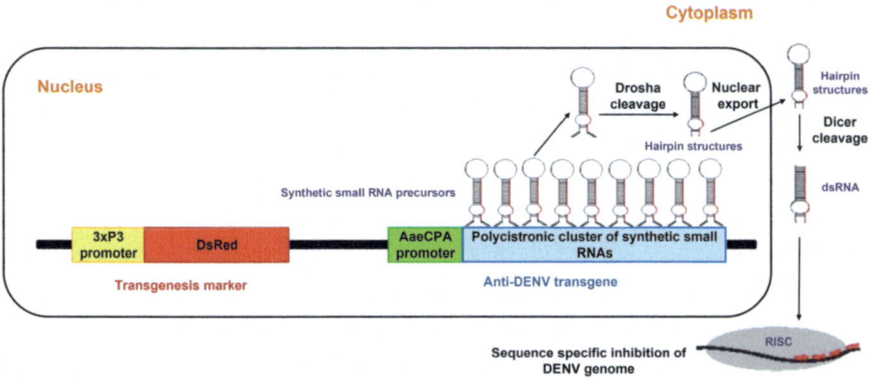

Fig. 10.4 Design and processing mechanism of the anti-DENV transgene. (*Source*: This work)

fluorescent marker genes that are under the expression of various promoters (Berghammer et al. 1999). Figure 10.4 and Tables 10.1 and 10.2 summarize both the in vitro and in vivo RNAi experiments conducted against DENV to date.

10.3.2 RNAi Delivery, Dosage, and Evaluation of Experiments

Microinjection as a method of gene transfer has played an important role in research as a direct method for introducing DNA into either cytoplasm or nucleus where injected DNA is subjected to the subsequent integration to the genome facilitating germline transformation. To date, germline transformation has efficiently demonstrated in several mosquito species including the dengue vector mosquito *Ae. aegypti* (Coates et al. 1998; Lobo et al. 2002). The requisite dose of RNAi molecules is approximately 200–500 ng/μL and varies depending on the size of the gene construct. Evaluation of DENV inhibition will be done through qPCR, miRNA expression analysis, and mosquito experimental infections followed by virus detection in salivary glands.

10.4 Application of RNAi for the Control of Mosquito Vector *Ae. aegypti*

10.4.1 RNAi and the Control of Mosquito Vector Ae. aegypti

Recent advances in in vivo and in vitro RNAi research on the mosquito vector *Ae. aegypti* and numerous cell lines have already shown potential ability to reduce viral

Table 10.1 Summary of RNAi-based, heritable DENV suppression evaluated in *Ae. aegypti* [mode of gene delivery, microinjection of preblastoderm stage embryos; evaluated system, in vivo (*Ae. aegypti*)]

Reference	Type of the effector molecule	Targeted region of the DENV genome	Sequence (5′–3′)	Basis of the target sequence selection	Promoter	Type of the transposon	Strains used	Efficiency
Franz et al. (2006)	dsRNA derived from IR RNA/siRNA	prM	578-bp fragment derived from the prM region of DENV2 Jamaica 1409 genome	N/A	AeCPA	*Mariner MosI*	Mosquito vector, Higgs' white eye	High level of resistance to DENV2 but not to DENV4
							DENV, DENV2 (Jamaica 1409), and DENV4 (Philippines H241)	
Yen et al. (2018)	Polycistronic cluster of hairpin RNA/miRNA	NS2B	TCTCATTGTTCCATCATCATCA	Sequence conservation	AeCPA, AePUb	*Mariner MosI*	Mosquito vector, *Ae. aegypti* Orlando strain	Transmission efficiency of DENV3 is <10%
		NS3	CCTGTGTGTTCAGATTTTGTTG					
		NS3	AATATGACCAGCCTCCTCTTCC				DENV, DENV3 (Cambodia)	
		NS5	CATTTATCATGGAGGAGGCTGA					

Ramyasoma et al. (2020)	Polycistronic cluster of hairpin RNA/siRNA	5′ UTR	AGACCAGAGATCCTGCTGTCT	Sequence conservation	AeCPA	*PiggyBac*	Mosquito vector, Sri Lankan wild *Ae. aegypti* strain	Resist the transmission of DENV2 and DENV4
		3′ UTR	TTAGAGAGCAGATCTCTGATG					
		NS5	AACTCTGGAGCAAATGCAAAG				DENV1, DENV2, DENV3, and DENV4	
		C	AACCGTCTATCAATATGCTGA					

Source: This work

Table 10.2 Summary of RNAi-based, nonheritable DENV suppression evaluated in in vitro and small animal models

Reference	Type of the effector molecule	Targeted region of the DENV genome	Sequence (5′–3′)	Basis of the target sequence selection	Promoter	Type of the vector	Evaluated system	Mode of gene delivery	Strains used for dengue challenging	Efficiency
Adelman et al. (2002a)	dsRNA derived from IR RNA/siRNA	prM	290-bp fragment derived from the prM region of DENV2 New Guinea C genome	N/A	hr5 enhancer and immediate-early 1 (IE-1) promoter region of *Autographa californica* nuclear polyhedrosis virus	Modified pIE1–3 plasmid (Novagen, Madison, Wis.) (Huynh and Zieler 1999)	In vitro C6/36 cells, *Ae. albopictus* (ATCC no. CRL-1660)	Transfection	DENV2 (Jamaica 1409)	DENV2 resistance in cell lines.
Stein et al. (2011)	siRNA	E	GGGCAAUGGUUGUGGGCUA	Dharmacon algorithm (Dharmacon siRNA Design Center) (Reynolds et al. 2004; Birmingham et al. 2007)	N/A	N/A	In vitro Huh7 cells (a human hepatoma-derived cell line) In vivo AG129 mice	Reverse transfection, retro-orbital intravenous administration of siRNA	For in vivo experiments, DENV1 (west-Pac, DVU88535), DENV2 (New Guinea C, AF038403), DENV3 (H87, M93130), and DENV4 (H241, AY947539) For in vivo experiments, DENV2 (strain S221)	By more than 98% against all four serotypes compared to control siRNA in the cell line
		NS5	GGAUGGAGCUUGAGAGAAA							
		NS3	CGGGAAAGACGAAGAGAUA							
		NS5	CCAAAGAGGUAGUGGACAA							
		NS5	GAGGAAUGCUUGUGAGAAA							
		NS5	GGAUGGAGCCUUAGAGAAA							
		NS5	CCAAAGAGGUAGUGGACAA							
		NS5	GGAUGGAGCUUAAGAGAAA							
		5′ UTR	AGUUGUUAGUCUACGUGGAC	Sequence conservation						
		5′ UTR	AUUAGAGAGCAGAUCUCUG							
		C	UGCUGAAACGCGAGAGAAA							
		3′ UTR	GGUUAGAGGAGACCCCUCC							
		3′ UTR	GGACUAGAGGUUAGAGGAG							
		3′ UTR	AACAGCAUAUUGACGCUGG							
		3′ UTR	CCAGAGAUCCUGCUGUCUC							

Korrapati et al. (2012)	shRNA/ siRNA	SLA	TTGTTAGTCTACGTGGACCGA	Sequence conservation	Human U6 promoter (http://www.addgene.org/10879/)	pLKO.1 TRC vector, replication-defective human adenovirus type 5 vector	In vitro Monkey kidney Vero cell line (American Type Culture Collection, Virginia, USA)	Transfection, virus infection	DENV1 (Nauru Island, U88535), DENV2 (New Guinea C, AF038403), DENV3 (H87, M93130), and DENV4 (Dominica, M14931)	Inhibition of all four DENV serotypes
		5′ UAR	TTAGACAGCAGATCTCTGATG							
		Nr	GACTACCGGTTAGAGCAGACC							
		Nr	AAGGACTAGAGGTTACAGGAGA							
		3′ CS	AAACACCATATTGACGCTGGG							
		3′ UAR	AGACCAGAGATCCTGCTGTCT							
		3′ SL	TCCAGGCACAGAACGCCAGAA							
		3′ SL	GGAATGGTGCTCTTGAATCAA							
Villegas et al. (2018)	siRNA	NS43	CTAGTGGCTCATTATGCCATAATA	Sequence conservation	N/A	N/A	In vitro Vero cells, C6/36 cells	Transfection	DENV1 (Hawaii), DENV2 (Yucatán 17438), DENV3 (H87), and DENV4 (H241)	Efficiently reduced the amount of the secreted NS1 antigen and viral yield (more than 90%) in both Vero and C6/36 cells in DENV1–4
		NS5	GAGCCTGAGAGAAACTGCATGCCTA							
		NS5	TTATAGGGAATGAGGATTA							
		NS5	AGAGAAAGTGGACACGAGA							

Source: This work

infection, dissemination, and transmission rates of DENV. There are several lines of evidence that prove RNAi is an important antiviral defense strategy in combating DENV. Knock-down experiments targeting RNAi components such as Dicer 2, R2D2, and Ago2 in *Ae. aegypti* showed increased viral loads or decreased EIPs in mosquitoes (Sánchez-Vargas et al. 2009). Furthermore, virus replication is suppressible in cultured mosquito cell lines expressing long dsRNA molecules designed to target the viral genome (Adelman et al. 2002b). RNAi-based, virus-resistant mosquitoes were developed expressing long dsRNAs targeting DENV2 under the control of a blood meal-inducible gene promoter, conferring a strong serotype-specific, virus-resistance phenotype (Travanty et al. 2004; Franz et al. 2006; Mathur et al. 2010). Moreover, the first miRNA-based genetically engineered mosquito, refractory to DENV3 and chikungunya simultaneously, was developed recently (Yen et al. 2018). Another mosquito refractory to DENV2 and DENV4 was developed using second-generation short hairpin RNA (shRNA) (Ramyasoma et al. 2020). To date, however, no anti-DENV genes refractory to all four serotypes in any mosquito have been developed. Therefore, greatest success can be expected with RNAi experiments to control dengue, as it induces a more robust antiviral RNAi pathway in *Ae. aegypti*.

10.4.2 The Evolutionary Consistency of RNAi-Based Management of Mosquito Population

RNAi acts in a sequence-specific manner against DENV. However, DENV has a RNA-dependent RNA polymerase (encoded by NS5) that has an intrinsically high error rate (~1×10^{-4}, corresponding to approximately one mutation per 11 kb DENV genome), similar to other RNA viruses (Lauring and Andino 2010). This genetic variation that exists within DENV serotypes could be challenging to the application of RNAi for dengue control as less RNAi effectiveness could result, depending on the degree of mismatch present between the small RNA molecule and the target region of the viral genome, leading to the evolution of resistance generating mutant escapes.

Simultaneously, integration of transgenes containing effector molecules to the mosquito genome and off-target effect might be caused by the small RNAs that would be expected to result in little or more fitness cost to the mosquito host. Therefore, long-term benefits and efficiency of RNAi-based mosquito vector control will require more research studies to delay the evolution of resistance and to minimize selective pressures.

10.4.3 RNAi Risks and Regulation

The examples described so far demonstrate the ability of RNAi approach to combat the major health burden of DENV, along with major concerns about the stability of RNAi strategies.

To avoid mutation escapes, target sequence selection should be done using sequence data available in various bioinformatics resources considering the sequence coverage and targeted regions to ensure effective viral suppression and evolutionary stability. Each of highly conserved regions across DENV serotypes could be targeted by a single or several synthetic small RNAs when designing an RNAi experiment.

Despite high efficiencies of engineered genes in reducing virus transmission rates, introduction of these genes into the mosquito genome typically reduces evolutionary fitness and may be eliminated by natural selection. Therefore, a mechanism to disseminate these genes into wild populations at epidemiologically significant rates is necessary. A potential solution is development of a gene drive system that can force inheritance in a super-Mendelian way, enabling it to increase itself and any linked "cargo" genes, in frequency with each generation even without adding fitness advantages to the host (Burt 2014; Champer et al. 2016). Such a method could be used to spread desirable cargo genes, such as virus resistance, rapidly through wild populations, altering vector populations to be disease refractory (Sinkins and Gould 2006). In addition, to verify if any off-target effect might be caused by the small RNAs, all the sequences of each antiviral small RNA should be examined for the off-target effect prediction. Moreover, estimation of the environmental and ecological risks associated with RNAi technologies is still in development, and potential toxicity to nontarget organisms and environmental fate should be systemically evaluated. Semi-field and field level experiments beyond laboratory experiments are essential to promote the responsible and sustainable use of the technology, making them candidates for effective control of dengue in population replacement strategies.

Acknowledgments We acknowledge the Accelerating Higher Education Expansion and Development (AHEAD) operation for financial support to carry out research studies on high tech approaches to control dengue by genetic engineering mosquito vector *Ae. aegypti.*

References

Adelman ZN, Jasinskiene N, James AA (2002a) Development and applications of transgenesis in the yellow fever mosquito, *Aedes aegypti*. Mol Biochem Parasitol 121(1):1–10

Adelman ZN, Sanchez-Vargas I, Travanty EA, Carlson JO, Beaty BJ, Blair CD, Olson KE (2002b) RNA silencing of dengue virus type 2 replication in transformed C6/36 mosquito cells transcribing an inverted-repeat RNA derived from the virus genome. J Virol 76(24):12925–12933

Berghammer AJ, Klingler M, Wimmer EA (1999) A universal marker for transgenic insects. Nature 402(6760):370–371

Birmingham A, Anderson E, Sullivan K, Reynolds A, Boese Q, Leake D, Karpilow J, Khvorova A (2007) A protocol for designing siRNAs with high functionality and specificity. Nat Protoc 2 (9):2068

Burt A (2014) Heritable strategies for controlling insect vectors of disease. Philos Trans R Soc B Biol Sci 369(1645):20130432

Campbell CL, Keene KM, Brackney DE, Olson KE, Blair CD, Wilusz J, Foy BD (2008) *Aedes aegypti* uses RNA interference in defense against Sindbis virus infection. BMC Microbiol 8 (1):47

Capeding MR, Tran NH, Hadinegoro SRS, Ismail HIHM, Chotpitayasunondh T, Chua MN, Luong CQ, Rusmil K, Wirawan DN, Nallusamy R, Pitisuttithum P (2014) Clinical efficacy and safety of a novel tetravalent dengue vaccine in healthy children in Asia: a phase 3, randomised, observer-masked, placebo-controlled trial. Lancet 384(9951):1358–1365

Chambers TJ, Hahn CS, Galler R, Rice CM (1990) Flavivirus genome organization, expression, and replication. Annu Rev Microbiol 44(1):649–688

Champer J, Buchman A, Akbari OS (2016) Cheating evolution: engineering gene drives to manipulate the fate of wild populations. Nat Rev Genet 17(3):146

Chandler LJ, Blair CD, Beaty BJ (1998) La Crosse virus infection of *Aedes triseriatus* (Diptera: Culicidae) ovaries before dissemination of virus from the midgut. J Med Entomol 35 (4):567–572

Coates CJ, Jasinskiene N, Miyashiro L, James AA (1998) Mariner transposition and transformation of the yellow fever mosquito, *Aedes aegypti*. Proc Natl Acad Sci 95(7):3748–3751

Förstemann K, Horwich MD, Wee L, Tomari Y, Zamore PD (2007) *Drosophila* microRNAs are sorted into functionally distinct argonaute complexes after production by dicer-1. Cell 130 (2):287–297

Franz AW, Sanchez-Vargas I, Adelman ZN, Blair CD, Beaty BJ, James AA, Olson KE (2006) Engineering RNA interference-based resistance to dengue virus type 2 in genetically modified *Aedes aegypti*. Proc Natl Acad Sci 103(11):4198–4203

Franz AW, Jasinskiene N, Sanchez-Vargas I, Isaacs AT, Smith MR, Khoo CC, Heersink MS, James AA, Olson KE (2011) Comparison of transgene expression in *Aedes aegypti* generated by mariner Mos1 transposition and ΦC31 site-directed recombination. Insect Mol Biol 20 (5):587–598

Fraser MJ, Clszczon T, Elick T, Bauser C (1996) Precise excision of TTAA-specific lepidopteran transposons piggyBac (IFP2) and tagalong (TFP3) from the baculovirus genome in cell lines from two species of Lepidoptera. Insect Mol Biol 5(2):141–151

Ghildiyal M, Xu J, Seitz H, Weng Z, Zamore PD (2010) Sorting of *Drosophila* small silencing RNAs partitions microRNA* strands into the RNA interference pathway. RNA 16(1):43–56

Gjenero-Margan I, Aleraj B, Krajcar D, Lesnikar V, Klobučar A, Pem-Novosel I, Kurečić-Filipović S, Komparak S, Martić R, Đuričić S, Betica-Radić L (2011) Autochthonous dengue fever in Croatia, August–September 2010. Eurosurveillance 16(9):19805

Huynh CQ, Zieler H (1999) Construction of modular and versatile plasmid vectors for the high-level expression of single or multiple genes in insects and insect cell lines. J Mol Biol 288 (1):13–20

Korrapati AB, Swaminathan G, Singh A, Khanna N, Swaminathan S (2012) Adenovirus delivered short hairpin RNA targeting a conserved site in the 5′ non-translated region inhibits all four serotypes of dengue viruses. PLoS Negl Trop Dis 6(7):e1735

Lauring AS, Andino R (2010) Quasispecies theory and the behavior of RNA viruses. PLoS Pathog 6(7):1001005

Li Z, Rana TM (2014) Therapeutic targeting of microRNAs: current status and future challenges. Nat Rev Drug Discov 13(8):622–638

Lobo NF, Hua-Van A, Li X, Nolen BM, Fraser MJ Jr (2002) Germ line transformation of the yellow fever mosquito, *Aedes aegypti*, mediated by transpositional insertion of a piggyBac vector. Insect Mol Biol 11(2):133–139

Mathur G, Sanchez-Vargas I, Alvarez D, Olson KE, Marinotti O, James AA (2010) Transgene-mediated suppression of dengue viruses in the salivary glands of the yellow fever mosquito, *Aedes aegypti*. Insect Mol Biol 19(6):753–763

McFarlane M, Arias-Goeta C, Martin E, O'Hara Z, Lulla A, Mousson L, Rainey SM, Misbah S, Schnettler E, Donald CL, Merits A (2014) Characterization of *Aedes aegypti* innate-immune pathways that limit chikungunya virus replication. PLoS Negl Trop Dis 8(7):2994

Moreira LA, Edwards MJ, Adhami F, Jasinskiene N, James AA, Jacobs-Lorena M (2000) Robust gut-specific gene expression in transgenic *Aedes aegypti* mosquitoes. Proc Natl Acad Sci 97 (20):10895–10898

Ramyasoma HPBKD, Dassanayake RS, Hapugoda M, Capurro ML, Silva Gunawardene YIN (2020) Multiple dengue virus serotypes resistant transgenic *Aedes aegypti* fitness evaluated under laboratory conditions. RNA Biol 17:918–929

Reynolds A, Leake D, Boese Q, Scaringe S, Marshall WS, Khvorova A (2004) Rational siRNA design for RNA interference. Nat Biotechnol 22(3):326–330

Sanchez-Vargas I, Travanty EA, Keene KM, Franz AW, Beaty BJ, Blair CD, Olson KE (2004) RNA interference, arthropod-borne viruses, and mosquitoes. Virus Res 102(1):65–74

Sánchez-Vargas I, Scott JC, Poole-Smith BK, Franz AW, Barbosa-Solomieu V, Wilusz J, Olson KE, Blair CD (2009) Dengue virus type 2 infections of *Aedes aegypti* are modulated by the mosquito's RNA interference pathway. PLoS Pathog 5(2):1000299

Scott TW, Burrage TG (1984) Rapid infection of salivary glands in *Culiseta melanura* with eastern equine encephalitis virus: an electron microscopic study. Am J Trop Med Hyg 33(5):961–964

Sinkins SP, Gould F (2006) Gene drive systems for insect disease vectors. Nat Rev Genet 7(6):427

Stein DA, Perry ST, Buck MD, Oehmen CS, Fischer MA, Poore E, Smith JL, Lancaster AM, Hirsch AJ, Slifka MK, Nelson JA (2011) Inhibition of dengue virus infections in cell cultures and in AG129 mice by a small interfering RNA targeting a highly conserved sequence. J Virol 85 (19):10154–10166

Sun W, Nisalak A, Gettayacamin M, Eckels KH, Putnak JR, Vaughn DW, Innis BL, Thomas SJ, Endy TP (2006) Protection of Rhesus monkeys against dengue virus challenge after tetravalent live attenuated dengue virus vaccination. J Infect Dis 193(12):1658–1665

Terenius O, Papanicolaou A, Garbutt JS, Eleftherianos I, Huvenne H, Kanginakudru S, Albrechtsen M, An C, Aymeric JL, Barthel A, Bebas P (2011) RNA interference in Lepidoptera: an overview of successful and unsuccessful studies and implications for experimental design. J Insect Physiol 57(2):231–245

Travanty EA, Adelman ZN, Franz AW, Keene KM, Beaty BJ, Blair CD, James AA, Olson KE (2004) Using RNA interference to develop dengue virus resistance in genetically modified *Aedes aegypti*. Insect Biochem Mol Biol 34(7):607–613

Villar L, Dayan GH, Arredondo-García JL, Rivera DM, Cunha R, Deseda C, Reynales H, Costa MS, Morales-Ramírez JO, Carrasquilla G, Rey LC (2015) Efficacy of a tetravalent dengue vaccine in children in Latin America. N Engl J Med 372(2):113–123

Villegas PM, Ortega E, Villa-Tanaca L, Barrón BL, Torres-Flores J (2018) Inhibition of dengue virus infection by small interfering RNAs that target highly conserved sequences in the NS4B or NS5 coding regions. Arch Virol 163(5):1331–1335

Welsch S, Miller S, Romero-Brey I, Merz A, Bleck CK, Walther P, Fuller SD, Antony C, Krijnse-Locker J, Bartenschlager R (2009) Composition and three-dimensional architecture of the dengue virus replication and assembly sites. Cell Host Microbe 5(4):365–375

Westaway EG, Brinton MA, Gaidamovich SY, Horzinek MC, Igarashi A, Kääriäinen L, Lvov DK, Porterfield JS, Russell PK, Trent DW (1985) Flaviviridae. Intervirology 24(4):183–192

Westaway EG, Mackenzie JM, Kenney MT, Jones MK, Khromykh AA (1997) Ultrastructure of Kunjin virus-infected cells: colocalization of NS1 and NS3 with double-stranded RNA, and of NS2B with NS3, in virus-induced membrane structures. J Virol 71(9):6650–6661

Wilczynska A, Bushell M (2015) The complexity of miRNA-mediated repression. Cell Death Differ 22(1):22

Wolfe ND, Kilbourn AM, Karesh WB, Rahman HA, Bosi EJ, Cropp BC, Andau M, Spielman A, Gubler DJ (2001) Sylvatic transmission of arboviruses among Bornean orangutans. Am J Trop Med Hyg 64(5):310–316

Yang JS, Smibert P, Westholm JO, Jee D, Maurin T, Lai EC (2013) Intertwined pathways for Argonaute-mediated microRNA biogenesis in *Drosophila*. Nucleic Acids Res 42(3):1987–2002

Yen PS, James A, Li JC, Chen CH, Failloux AB (2018) Synthetic miRNAs induce dual arboviral-resistance phenotypes in the vector mosquito *Aedes aegypti*. Commun Biol 1(1):11

Chapter 11
Wolbachia: Biological Control Strategy Against Arboviral Diseases

Ipsita Mohanty, Animesha Rath, and Rupenangshu Kumar Hazra

Abstract Arboviral diseases like dengue, chikungunya, and Zika are among the major causes of mortality and morbidity in human population. The limited control methods together with lack of antiviral therapies and effective vaccines have paved way for new approaches. One such approach to reduce the ever alarming conflagration of vector-borne diseases is based on biological strategy that reduces or blocks pathogen transmission in the vector. In this context, *Wolbachia*, an endosymbiont in mosquitoes, is explored as a novel and ecofriendly control strategy. *Wolbachia* seems to confer resistance to diverse RNA viruses protecting lives from virus-induced mortality. This review envisages the deployment of *Wolbachia* technology in controlling several arboviral diseases.

Keywords *Wolbachia* · *Aedes* · Arboviral diseases

11.1 Introduction

The last few decades have experienced significant advancements in development of therapeutics and vaccines against arboviral diseases which seemed to be fruitless so far. Therefore, efficient vector control strategies still acted as the primary intervention tool. The arboviruses that inhabit mosquitoes mainly belonging to genera *Culex* and *Aedes* are classified under three families: *Flaviviridae*, *Togaviridae*, and *Bunyaviridae*. Most of the control strategies have been implemented on *Aedes aegypti* and *Ae. albopictus* that are responsible for transmitting dengue virus (DENV), chikungunya virus (CHIKV), yellow fever virus (YFV), etc. (Tabachnick

I. Mohanty · R. K. Hazra (✉)
ICMR-Regional Medical Research Centre, Bhubaneswar, Odisha, India

A. Rath
ICMR-Regional Medical Research Centre, Bhubaneswar, Odisha, India

KIIT School of Biotechnology, Kalinga Institute of Information Technology, Bhubaneswar, Odisha, India

B. K. Tyagi (ed.), *Genetically Modified and other Innovative Vector Control Technologies*, https://doi.org/10.1007/978-981-16-2964-8_11

1991; Reiter et al. 2006; Benedict et al. 2007). *Culex* species are also targeted for prevention of Japanese encephalitis (JEV) and West Nile virus (WNV) (Solomon and Vaughn 2002). With the expansion of the vectors and the arboviruses, innovative biological control strategies are inevitable. The concept of elimination of disease vectors is quite impractical that focuses on the reduction in the number of competent vectors. The previous vector control approaches included chemical control, biological control, and environmental management aiming at reduction in the size of the vectors. Chemical control had negative effects both on non-target arthropod and target species in the long run invariably leading, in the latter case, to insecticide resistance. Biological control relies on predatory or parasitic organisms that target vectors. The source reduction in the environment has an impact on reducing vector population size. A brief insight on the biology of the vectors and pharmacological mechanisms of the insecticides together with the behavior of predatory or parasitic organisms is essential for reduction of immature and adult vectors.

The deployment of integrated vector management (IVM) for control of arboviruses has aimed at the study of vector biology that has reduced the burden during interepidemic periods (WHO 2004; Achee et al. 2015). The progress by IVM is reversed with the growth of multiple resistance mechanisms (Wilke and Marrelli 2015; Alout et al. 2017). In this regard, there has been an emerging interest on alternatives to chemical strategies, i.e., the use of biological control employing *Wolbachia*. This review purely focuses on implementation of *Wolbachia* technology for mosquito vectors of public health importance. *Wolbachia* is a promising arsenal against mosquitoes as they can influence both mosquito reproduction and pathogen loads. Here we have provided an insight on *Wolbachia*, its prevalence, diversity, and cytoplasmic incompatibility (CI) induction in major mosquito vector, *Ae. albopictus*.

11.2 *Wolbachia*

Wolbachia are gram-negative, obligate, aerobic, intracellular (endosymbiotic), maternally transmitted bacterium with reduced genomes. They can only multiply within the eukaryotic host. They show a parasitic symbiotic association with the mosquito hosts.

11.3 Systematic Classification and Nomenclature

Wolbachia belongs to the family Anaplasmataceae (phylum Proteobacteria: order Rickettsiales under α-subdivision), along with the members of the genera *Anaplasma*, *Ehrlichia*, *Cowdria*, and *Neorickettsia* (Dumler et al. 2001) (Table 11.1). The type species for the *Wolbachia* genus is *Wolbachia pipientis* first described in the mosquito *Culex pipiens* (Hertig and Wolbach 1924) and is

Table 11.1 Classification of *Wolbachia*

Classificatory level	Hierarchical nomenclature
Domain	Bacteria
Kingdom	Eubacteria
Phylum	Proteobacteria
Class	Alphaproteobacteria
Subclass	Rickettsidae
Order	Rickettsiales
Family	Anaplasmataceae
Genus	*Wolbachia*

Source: This work

distinct from its other such as, for example, *W. persica*, *W. popcorn*, *W. postica*, and *W. trichogrammae* (Dumler et al. 2001).

The general format "*w*host" (viz., *w*Pip, a strain of *Wolbachia* isolated from *Culex pipiens*) is followed while naming strains (Zhou et al. 1998) with modifications to make a distinction between numerous strains in the same host species.

The geographic locality has been used to distinguish strain from diverse populations of *Drosophila simulans*, i.e., *w*Cof (Coffs Harbour), *w*Ha (Hawaii), and *w*Ri (Riverside) (Zhou et al. 1998).

11.4 Occurrence and Distribution

Wolbachia exists as an endosymbiont in majority of hosts, thereby maintaining a neutral relationship with its host. They chiefly reside the reproductive tissues like ovaries and testes of insect hosts (Fig. 11.1).

These are harbored in mature eggs but not in mature sperms. The bacterium concentrates itself at posterior pole of mature oocyte (the future site of germline) that helps in its maternal transmission (Kose and Karr 1995). *Wolbachia* is also found in nervous system and digestive and metabolic tissues such as gut, salivary glands, fat body, muscles, wings, hemocytes, and malpighian tissues of arthropods (Dobson et al. 1999; Min and Benzer 1997, Faria and Sucena 2013; Hughes et al. 2011). The nurse cells of ovaries where multiplication occurs have highest number of *W. pipientis* (Louis and Nigro 1989, Zchori-Fein et al. 1992). Nurse cells empty its content through cytoplasmic bridges to the developing egg where *W. pipientis* associates with the microtubules. This association induces cytoplasmic incompatibility and parthenogenesis (Kose and Karr 1995).

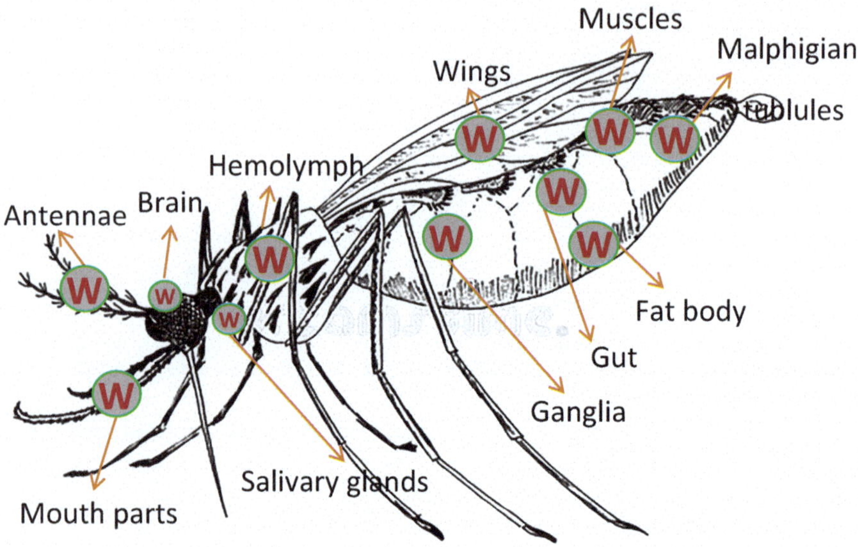

Fig. 11.1 Somatic (tissue) distribution of *Wolbachia* (grey) in mosquito. (*Source*: This work)

11.5 Microscopic/Morphological/Ultrastructure of *Wolbachia*

The general characteristics of rickettsiae are reflected in *Wolbachia*. They are a dimorphic gram-negative bacterium that ranges from small (rod-shaped: 0.5–1.3μm in length; coccoid forms: 0.25–1μm) to large (1–1.8μm in diameter) forms (Hertig 1936). Pleomorphic forms appear with an increase in host age (Wright 1979). *W. pipientis* is enclosed in a vacuole layered by three membranes, i.e., bacterial plasma membrane being the innermost layer surrounded by bacterial cell wall and outermost layer of host origin (Louis and Nigro 1989). The multiplication of *Wolbachia* cells is through binary fission within the vacuoles (Wright et al. 1978). Multiple membranes commonly surrounding intracellular bacteria are considered to take part in host's control over the prokaryote (Wright 1979). The microscopic structure of *Wolbachia* is depicted in Fig. 11.2.

11.6 Molecular Identification of *Wolbachia*

The major breakthrough to study *Wolbachia* lies with polymerase chain reaction (PCR) and DNA sequencing techniques. PCR primers specific to 12S, 16S, or 23S rDNAs and *wsp* (*Wolbachia* surface protein) and *fts*Z (bacterial cell division protein) and *gro*ELI (bacterial heat shock protein) genes are used to detect the bacteria in host tissues. MLST (multilocus sequence typing) uses *fts*Z, *gat*B, *cox*A, *hcp*A, and *fbp*A

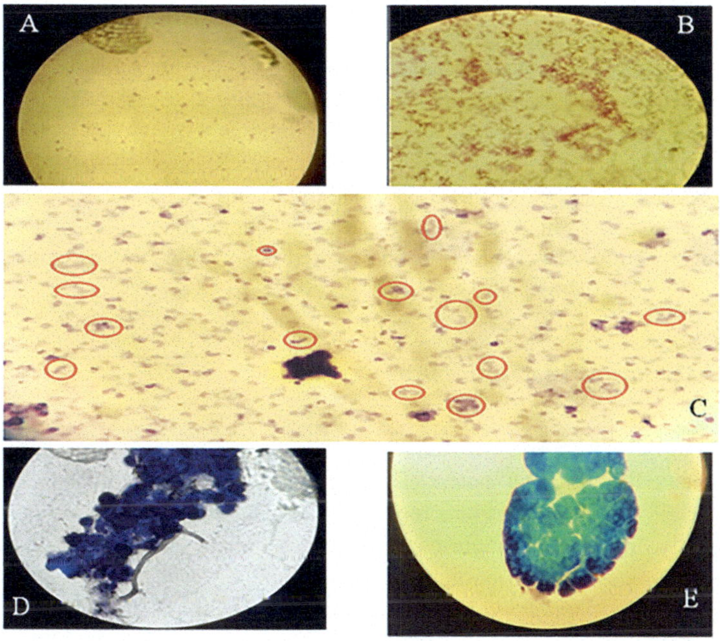

Fig. 11.2 (**a**) Giemsa-stained smear of squashed ovary of *Ae. aegypti* without *Wolbachia*; (**b**) Giemsa-stained smear of squashed ovary of *Ae. albopictus* with *Wolbachia* colonies; (**c**) Giemsa-stained smears showing pleomorphic forms (red rounds) of *Wolbachia* in *Cx. quinquefasciatus*; (**d**) Giemsa-stained ovary of *Ae. aegypti* without *Wolbachia*; and (**e**) purple-stained *Wolbachia* on the follicle cells of ovary in *Ae. albopictus*. (*Source*: Author's own paper, viz., Mohanty et al. 2018; permitted)

genes (Baldo et al. 2006) for typing and relating any new strains to strains already recorded in *Wolbachia* MLST databases (http://pubmlst.org/wolbachia/).

11.7 Phylogeny and Diversity

16S rRNA gene of *W. pipientis* forms a monophyletic clade within the class *Alphaproteobacteria* (Scola et al. 2015). The genus *Wolbachia* being greatly diverse is categorized into 16 "supergroups" (A to Q) with an exception/inclusion of G basing on nucleotide sequence data. All *Wolbachia* spp. are considered to represent a single species *W. pipientis* (Lo et al. 2007; Ilinsky and Kosterin 2017). The supergroups A and B are commonly found among arthropods (Werren et al. 2008; Glowska et al. 2015). Supergroups C and D are limited only to filarial nematodes (Sironi et al. 1995; Glowska et al. 2015). Supergroup F is found in both nematodes and major arthropod orders (Campbell et al. 1992; Covacin and Barker 2007). The supergroups E and H are less common with E occurrence in springtails (Collembola)

(Timmermans et al. 2004) and H in one genus of termites (Isoptera) (Bordenstein and Rosengaus 2005). Supergroup G is detected from Australian spiders (Araneae) (Rowley et al. 2004), with an argument that these strains are not a distinct lineage but recombinants between supergroups A and B (Baldo and Werren 2007). Distinct *Wolbachia* strains have been detected in fleas (Siphonaptera) (Gorham et al. 2003), the filarial nematode (Spirurida) (Covacin and Barker 2007), and the pseudoscorpion (Pseudoscorpionida) (Zeh et al. 2005), without being assigned any supergroup. Supergroup K has been recently detected in a spider mite species (Ros et al. 2009). Supergroups M and N have been detected from aphids (Wang et al. 2014).

11.8 Attraction to *Wolbachia*

The upsurge in interest for *Wolbachia* is due to its widespread occurrence, reproductive manipulations, and potential applications in pests and disease control (Bourtzis 2008). Briefly, enormous research in *Wolbachia* is because of the following: (1) firstly, invasive distribution and effects upon hosts are implications for key evolutionary processes contributing to rapid speciation (Prout 1994; Dobson 2003); (2) secondly, since *Wolbachia* modify early development and mitotic processes in their hosts (Stouthamer et al. 1993; Hoffmann and Turelli 1997), they might be used to study these basic processes; and (3) thirdly, *Wolbachia* can be used as a vector for spreading desirable genetic modifications in insect populations (Hilgenboecker et al. 2008).

11.9 *Wolbachia* Genome

11.9.1 Chromosomal Genome

Wolbachia is an endosymbiont with minute size (~0.5–1 μm) and small genomes (1.08–1.7 Mb) which comes within the range of the Rickettsiales (0.8–2.1 Mb) lacking the typical minimal genome content and high stability unlike other obligate endosymbionts, *Buchnera*, *Candidatus*, and *Blochmannia* (Tamas et al. 2002). *Wolbachia pipientis* have a small circular chromosome containing ~1–1.5 million base pair (bp) of DNA (deoxyribonucleic acid) that is composed of ~65% of A (adenine) and T (thiamine) bps (eol.org/pages/97655/details).

Wolbachia genomes in insects have a high % of repetitive elements such as insertion sequences, group II introns, duplicated segments of prophages, and multigene families such as the ankyrin (ANK) repeat genes (Kent and Bordenstein 2010; Leclercq et al. 2011). These may assist *Wolbachia* to hijack sexual reproduction in arthropods. ANK with a tandem motif of ~33 amino acids help in protein-protein interactions (Mosavi et al. 2004). Proteins with ANK domains are rare in bacteria and archaea but very commonly found in viruses and eukaryotes (Walker et al. 2007;

Al-Khodor et al. 2010) that mediate protein-protein interactions of host and pathogen in *Anaplasma* spp. (Ijdo et al. 2007) and *Ehrlichia* spp. (Rikihisa and Lin 2010). ANK proteins help in transcriptional regulation of transcription, signal transduction, intracellular trafficking, cytoskeleton interactions, development, sex differentiation, and call cycle and can also play role as toxins (Bork 1993; Al-Khodor et al. 2010). Huge architectural variation presence of transmembrane domains in orthologue ANK proteins from closely related *Wolbachia* strains and number of ANK repeats induce different phenotypes in their hosts (Iturbe-Ormaetxe et al. 2005).

Recent studies suggest that host-pathogen interactions and evolution of infections are mainly controlled by ANK proteins (new family of bacterial type IV effectors) (Pan et al. 2008). For transferring DNA and/or proteins to eukaryotic cells, a T4SS (type IV secretion system) transport mechanism is used by numerous bacteria. Certain pathogenic intracellular bacteria secrete ANK proteins into the host cytoplasm that interact with host factors and thereby modulate diverse functions (Lin et al. 2007; Rikihisa and Lin 2010). The *Wolbachia* genome also possesses transcriptionally active T4SS gene components.

11.9.2 Phage WO Genome

Wolbachia genome data analysis revealed that prophage genes play a key role affecting its ecology in terms of density within the host (Bordenstein et al. 2006). Some phage elements in *Wolbachia* are actively transcribed and expressed through a lytic cycle (Landmann et al. 2009). Bacteriophages have a major role in integrating and spreading ANK genes within *Wolbachia*. This implies that to study reproductive manipulations, both are promising candidates (Walker et al. 2007). CI in insects is known to be caused by the WO *Wolbachia* bacteriophage that infects 90% of insect *Wolbachia* strains (Gavotte et al. 2004). *Wolbachia*'s pathogenicity is linked with a series of WO-encoded genes, like the virulence factor *Vrl*C.

Advances in genomics can be best explored for study of the biology of *Wolbachia*. At present, four complete genome sequences of *Wolbachia* are available, while several others are in progress (Salzberg et al. 2005, 2009; Klasson et al. 2008).

Up to 4% of the total number of genes in the insect *Wolbachia* strains are comprised by the ANK genes, viz., *w*Ri, *w*Mel, and *w*Pip (Klasson et al. 2008). The microarray-based comparative genome hybridization analysis showed that ankyrin-repeat encoding genes, certain predicted genes, and prophages make up the *Wolbachia* genomes (Ishmael et al. 2009).

11.10 Reproductive Effects on Host

Wolbachia can stimulate numerous reproductive alterations in their hosts, such as cytoplasmic incompatibility (CI), parthenogenesis induction (PI), feminization, and male killing (MK) (Fig. 11.3) that impart selective advantage to the endosymbiont (Rousset et al. 1992; Turelli 1994) which enhances their vertical transmission facilitating its spread and maintenance in hosts. ~16% of insect species that comprises of the main insect orders (Werren 1997), isopods and mites (Johanowicz and Hoy 1995), and terrestrial crustaceans and nematodes (Sironi et al. 1995) harbors *Wolbachia*. The fertility of nematodes and arthropods is influenced by *Wolbachia* (Bandi et al. 1998). These phenotypes enhance the frequency of infected females in the host population. Such parasitic effects on hosts are usually known as "reproductive parasitism" (Bandi et al. 2001).

11.10.1 Cytoplasmic Incompatibility

CI is the developmental arrest of offspring that results from sperm-egg incompatibility (Breeuwer and Werren 1990; Clancy and Hoffmann 1996) (Figs. 11.3 and 11.4). The unidirectional and bidirectional forms result in embryonic mortality (Bourtzis et al. 1998). Complete compatibility occurs in reciprocal crosses (infected

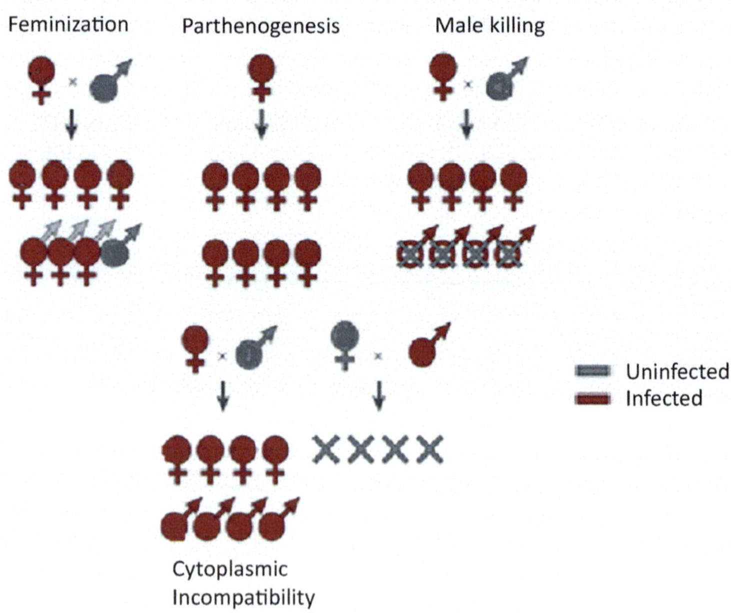

Fig. 11.3 *Wolbachia*-induced reproductive phenotypes. (*Source*: This work)

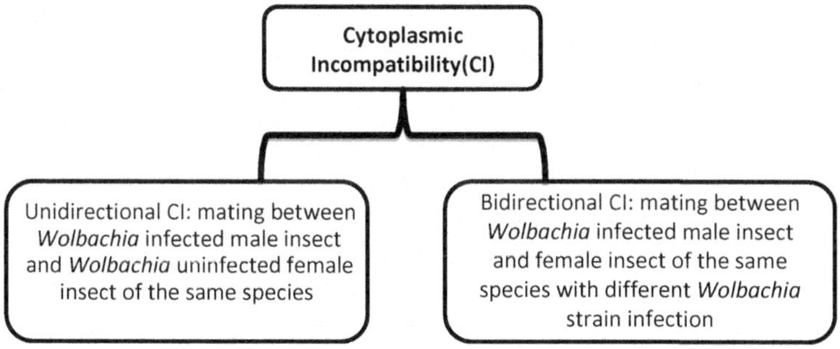

Fig. 11.4 Schematic representation of forms of CI. (*Source*: This work)

female × uninfected male) and crosses between individuals infected with same strain.

CI mechanism at molecular level is unclear; however, presently it is explained using terminologies, "modification," and "rescue." *Wolbachia*-induced "modification" of sperm during spermatogenesis and a "rescue" of this modification in embryos infected with the same *Wolbachia* type. In absence of appropriate *Wolbachia* in developing embryo, with sperm modification, embryonic development is disrupted (Werren 1997).

11.10.2 Parthenogenesis Induction (PI)

In thelytokous PI, diploid females instead of haploid males are produced from unfertilized eggs (Stouthamer et al. 1993). PI bacteria prevents arrhenotokous development (unfertilized haploid eggs develop to males) and allows infected females to produce infected female offspring (through maternal transmission) without mating (Weeks and Breeuwer 2001). Thelytokous parthenogenesis by *Wolbachia* is less frequent than CI.

11.10.3 Feminization

In feminization, infected males are converted into functional phenotypic females (Rousset et al. 1992). Phenotypic males fertilized these feminized insects to generate progeny. This conversion facilitates *Wolbachia* transmission through mothers (Rigaud 1999). In isopods, *Wolbachia* prevents development of the androgenic gland and hence production of the androgenic hormone within genetic males (Azzouna et al. 2004). *Wolbachia* infection may abolish phenotypic males leading to extinction of both the host population and the symbiont.

11.10.4 Male Killing (MK)

In MK, infected male embryos are selectively killed (Rousset et al. 1992), thereby increasing the production of females. MK by *Wolbachia* occurs mainly during embryogenesis resulting in more food availability to the surviving female progeny. In absence of *Wolbachia*, genetic females die during larval development, whereas in the presence of *Wolbachia*, genetic males become feminized and die during larval development. Thus, MK seems to occur through lethal feminization (Kageyama and Traut 2004).

11.11 *Wolbachia* and Mosquitoes

Wolbachia is widespread in various invertebrate hosts because of its reproductive manipulations and horizontal transmission (Charlat et al. 2003; Saridaki and Bourtzis 2010). *Wolbachia* infects ~65% of the insect species (Glowska et al. 2015). Several mosquito species ~28% from different genera are known to carry different *Wolbachia* strains.

11.12 Native and Non-native *Wolbachia* Infections

The study on CI (Yen and Barr 1971) renewed further research on *Wolbachia*. The endo cellular bacterium *Wolbachia* was reported in ovaries and eggs of *Aedes scutellaris* group (Yen 1975). The arthropod strains of *Wolbachia* was observed in some *Culex* species, while filarial strains were also observed in *Aedes* species (Ricci et al. 2002). *Wolbachia* infection from field-collected mosquito species ranged between 7.1% and 55.5% within a given geographical location (Kittayapong et al. 2000; Osei-Poku et al. 2012). *Wolbachia* is present in 39.5% of the 147 Culicinae species screened but is absent in *Ae. aegypti*. *Ae. albopictus* is naturally infected with *Wolbachia* (Hertig 1936; Ravikumar et al. 2010). Till date, none of the anophelines except *An. gambiae* and *An. coluzzii* screened have natural *Wolbachia* (Yen 1972; Osei-Poku et al. 2012; Shaw et al. 2016; Baldini et al. 2014). It is reported ~100% prevalence of *Wolbachia* is found in field populations of *Ae. albopictus* prevalent in most part of the globe (Kittayapong et al. 2002; Armbruster et al. 2003: Behbahani 2012).

The inbuilt potential of *Wolbachia* is hampered owing to its absence in dominant vector species like *Ae. aegypti*. The adaptability of this bacterium to new intracellular environments has been exploited to transfect it via embryonic microinjection to vector species like *Ae. aegypti* and *Ae. albopictus* (Takken and Koenraadt 2013; Bourtzis et al. 2014). The life span of *Ae. aegypti* was reduced through the introduction of life-shortening strain *w*MelPop-CLA from *D. melanogaster*. The *w*MelPop derivative has been also used to block dengue and chikungunya transmitted by *Ae. aegypti*, while *w*Mel *Wolbachia* strain *w*Mel_Br is successful against Zika (Dutra et al. 2016; Moreira et al. 2009).

A long-term association of host and native *Wolbachia* provides an increased fitness to the host (Turelli and Hoffmann 1991; Turelli 1994; McGraw et al. 2002; Weeks et al. 2007). These native associations likely represent a stable relationship. The native *w*AlbA and *w*AlbB strains of *Ae. albopictus* and the native *w*Pol of *Ae. polynesiensis* had null effect on CHIKV and DENV, respectively (Mousson et al. 2012; Lu et al. 2012). Further, it was also observed that *w*AlbB resulted in reduced transmission of DENV by affecting the infection in salivary gland of *Ae. albopictus* (Mousson et al. 2012). However, the transfection of *Wolbachia* into *Ae. aegypti* is known to reduce its vector competence (Sinkins 2004). *Wolbachia* transfection into a host *Ae. aegypti* provided either less or a stronger pathogen protection than its native host. *w*AlbB strain is unable to block DENV in its natural host *Ae. albopictus* as against its non-native host *Ae. polynesiensis* and *Ae. aegypti* (Bian et al. 2010; Lu et al. 2012). The *w*Pip strain also blocks DENV infection in *Ae. aegypti* with a significant cost to its fitness (Fraser et al. 2017). Immune priming also causes pathogen-blocking effect in its non-native host *Ae. aegypti* than native host *Ae. albopictus* (Van den Hurk et al. 2012; Chrostek et al. 2014) with *Wolbachia* titers higher in non-native hosts (McGraw et al. 2002; Chrostek et al. 2014) than native ones (Lu et al. 2012).

11.13 Arbovirus and *Aedes* Mosquitoes

Arboviruses (**ar**thropod-**bo**rne **viruses**) are viral infections transmitted to vertebrates from insects. The arthropod feeds on viremic (infectious) blood meal from an infected vertebrate and proliferates it for further (horizontal) transmission to another vertebrate host. This arbovirus is then transferred from its arthropod host to its offspring through transovarian/vertical transmission. Several *Aedes* species are known to transmit arboviruses causing massive outbreaks. Subgenus *Stegomyia* is of prime medical importance followed by genera *Finalaya*, *Aedimorphs*, and *Diceromyia*. *Aedes aegypti*, *Ae. albopictus*, *Ae. vittatus*, *Ae. scutellaris*, *Ae. pseudoscutellaris*, *Ae. polynesiensis*, *Ae. bromeliae*, and *Ae. africanus* are the main vectors of the subgenus *Stegomyia*.

The global disease burden is due to rapid geographical spread of mosquito-borne diseases (WHO 2014). *Aedes aegypti* and *Ae. albopictus* are well-known for their vectorial role in the spread of dengue, Zika, chikungunya, and yellow fever (Weaver 2014; Simmons et al. 2012; Leparc-Goffart et al. 2014). The World Health Organization (WHO) reported that malaria and dengue contribute up to 17.0% of the global burden of infectious diseases (WHO 2014). Recent studies have shown a decline in malaria and upsurge in dengue and Zika incidences (Benelli and Mehlhorn 2016). The number of dengue cases is on hike with ~more than half of the world's population under its risk (Bhatt et al. 2013). Dengue and chikungunya were previously confined to Africa and Asia but now has spread to the Caribbean, South America, and Europe. The role of *Ae. albopictus* in transmission of ZIKV is also evident from the presence of the ZIKV from the field samples in South America (Musso and Gubler 2016).

11.14 Global Distribution of *Aedes*

The global expansion of *Ae. aegypti* and *Ae. albopictus* led to an upsurge of dengue and chikungunya cases (Kraemer et al. 2015). Africa marked the origin of *Ae. aegypti*, with its ancestral form named *Ae. aegypti formosus* (zoophilic tree-hole mosquito) (Brown et al. 2011). The discrete geographical niche genetically distinguishes the domestic form of *Ae. aegypti*. The vector *Ae. aegypti* was hypothesized to be introduced to the New World from Africa following its spread globally to tropics and subtropics on account of harsh conditions and onset of slave trade (Brown et al. 2014). Asia marked the origin of *Ae. albopictus*, with its expansion to Indian and Pacific Oceans (Delatte et al. 2009), Europe, Brazil, and the United States (Carvalho et al. 2014). Both the species have now invaded most Asian cities and large part of Americas (Lambrechts et al. 2011). *Aedes aegypti* rest indoors and feed on humans in daytime (Scott and Takken 2012), whereas *Ae. albopictus* being exophagic opportunistically bites humans and animals (Paupy et al. 2009). However, both are anthropophilic in nature (Ponlawat and Harrington 2005; Delatte et al. 2009). The global distribution of these species is dependent on climatic factors, with a special emphasis to temperature (Brady et al. 2014) and precipitation (Benedict et al. 2007; Campbell et al. 2015).

Aedes vittatus is prevalent throughout tropical Asia, Africa, and the Mediterranean region of Europe (Melero-Alcíbar 2006). It is also found in Africa and Europe (Diallo et al. 2014). It is a voracious human biter and incriminated as a vector of YFV in Africa (Service 1974). Several other viruses, viz., dengue, chikungunya, and Zika, have been isolated from the mosquito demonstrating its potential to replicate and transmit these viruses experimentally.

11.15 Modes of Arboviral Transmission

Arboviral maintenance and amplification cycles involve horizontal, vertical, and venereal transmissions that are interconnected among them.

11.16 Vector Competence of *Aedes*

After taking an infectious viremic blood meal, the virus is ingested in the blood meal, thereby infecting and replicating in the midgut epithelial cells. The virus successfully escapes from the midgut to infect salivary glands and ovaries followed by its release to the salivary ducts for either oral transmission to vertebrates or transovarian transmission to offspring (Hardy et al. 1983). Both the salivary gland infection and escape barriers determine the virus replication and dissemination into mosquito saliva for transmission to a new vertebrate host (Black et al. 2002; Richards et al. 2012). The chance that a vector contacts a susceptible vertebrate host depends on the population, composition, and behavior of both mosquito and host (de Araújo Lobo et al. 2014; Hardy et al. 1983). Also the vectorial capacity is determined by longevity, blood feeding rate, population density, and vector competence.

11.17 Arboviral Replication in an Arthropod Cell

Wolbachia host combinations exhibit antiviral activities/pathogen blocking against enveloped (*Togaviridae*, *Flaviviridae*) and non-enveloped (*Dicistroviridae*, *Nodaviridae*, *Reoviridae*) RNA viruses from diverse families (Teixeira et al. 2008; Dutra et al. 2016). *Wolbachia* interference with virion attachment and entry has no evidence till date. *Wolbachia* presence perturbs cellular lipid levels (Molloy et al. 2016; Geoghegan et al. 2017) affecting receptor binding and attachment, virion internalization, and replication (Kielian and Helenius 1984; Belov 2014). The preexisting *Wolbachia* in a host cell could exhibit an alteration in the vesicular trafficking compromising delivery of endosomal viruses (Geoghegan et al. 2017).

Since *Wolbachia* lacks essential amino acid biosynthesis gene, they are known to forage amino acids from host cell. *Wolbachia* infection induces cellular stress conditions (Wu et al. 2004; Geoghegan et al. 2017) that are known to limit viral translation and replication. However, four basic steps are needed for RNA virus replication in an arthropod cell: (a) virus attachment, entry, and intracellular localization; (b) viral genome replication; (c) hijack of the host translational mechanism; and (d) genome packaging and exit from the cell (Fig. 11.5).

11.18 *Wolbachia*'s Role in Modification of Host Intracellular Environment

Wolbachia is known to depend on host features to ensure its replication, transmission, and modification of the host (Fig. 11.6). *Wolbachia* infects not only developing oocytes but also other tissues and cell types of the host (Dobson et al. 1999; Pietri et al. 2016). Thus, *Wolbachia* has a drastic effect on host physiology including

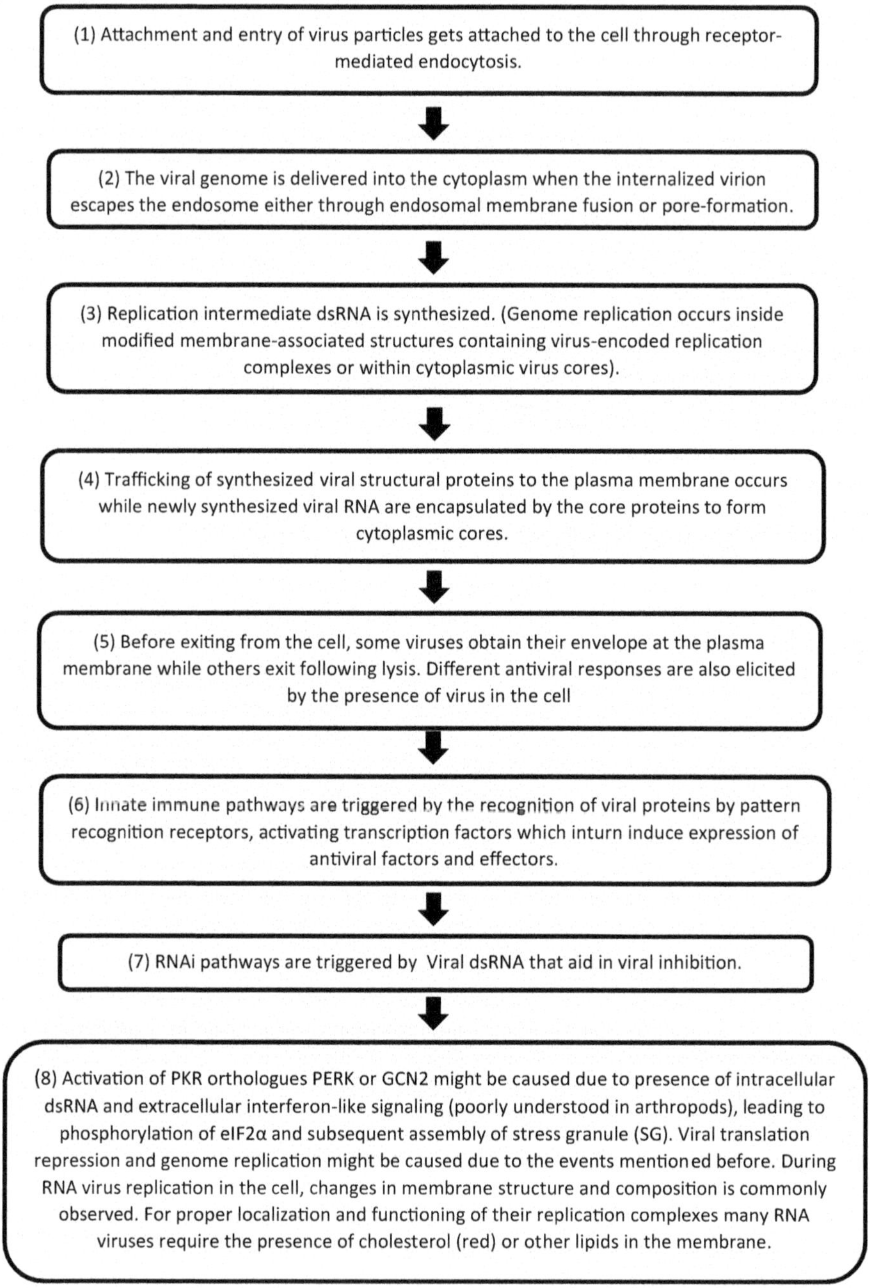

Fig. 11.5 Flow chart showing the RNA replication in an arthropod cell. (*Source*: This work)

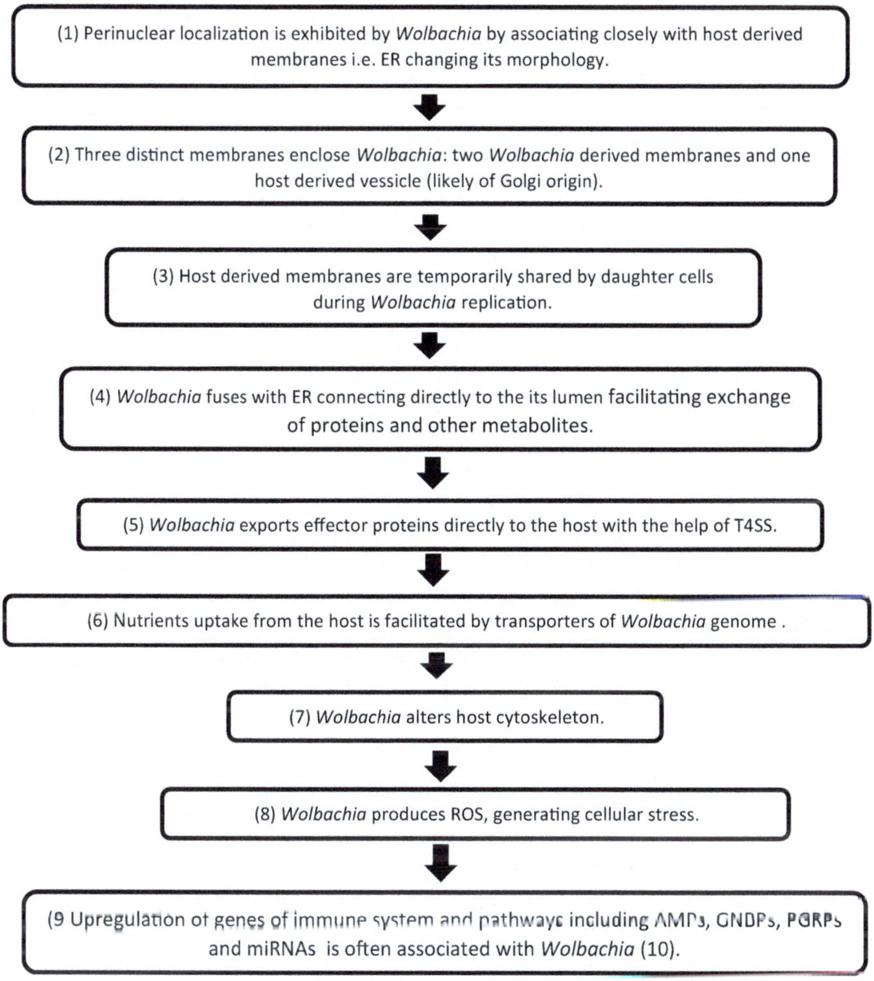

Fig. 11.6 Flowchart showing *Wolbachia* modifying the intracellular environment of the host. (*Source*: This work)

effects on gene expression (Xi et al. 2008; Kremer et al. 2009, 2012), macromolecule availability (Molloy et al. 2016), fecundity (Stouthamer and Luck 1993, Vasquez et al. 2011), behavior (Moreau et al. 2001), and speciation (Jaenike et al. 2006).

Wolbachia affects host physiology by directly targeting host processes and indirectly as a consequence of its nativity in the host cell. It depends upon the host for the nutrients as it is an obligate intracellular endosymbiont with a reduced genome (Wu et al. 2004). *Wolbachia* also benefits the host in the form of metabolic provisioning (Brownlie et al. 2009; Gerth and Bleidorn 2016). All these account to the modification of host intracellular environment that favors its own reproduction and transmission. This changed environment is less favorable for incoming viruses.

11.19 Mechanism of *Wolbachia*-Mediated Virus Inhibition in Arthropods

The host intracellular environment is modified by *Wolbachia* in such a way that it becomes refractory to RNA virus infection (Rainey et al. 2016; Bhattacharya et al. 2017). Among the pathogen-blocking phenotypes, *Wolbachia*-induced stress is the elementary constituent of cellular antiviral response (Brennan et al. 2012; Wong et al. 2015) (Fig. 11.7).

11.19.1 Induction of Cellular Stress

The association of native *Wolbachia* in Aa23 *Ae. albopictus* cells produces several host antioxidant proteins unbalancing redox homeostasis in the cell. An increase in superoxide dismutase (SOD) suggested a cytoplasmic oxidative stress (Brennan et al. 2008). This cytoplasmic ROS (reactive oxygen species) houses the bacterium, and *Wolbachia* density is sensitive to ROS (Brennan et al. 2012). An unusual high ROS levels can lead to the disruption of biological macromolecules at physiological levels. ROS activates signaling pathways like extracellular signal-regulated kinase (ERK) pathway (Müller et al. 1997; Thannickal and Fanburg 2000) known to protect against RNA viruses.

11.19.2 Cholesterol and Lipid Imbalance

The viral entry, replication, and exit depend on cellular cholesterol (Kielian and Helenius 1984; Carro and Damonte 2013). The competition for cholesterol is hypothesized to occur with *Wolbachia* virus coinfections, with victory of *Wolbachia* because of its precedence. Evidences suggest that a hike in amount of esterified cholesterol in *Wolbachia*-infected cells decreases free cholesterol of cell that is required for virus replication (Geoghegan et al. 2017).

11.19.3 Competition for Cellular Resources

Wolbachia (*w*Mel) genome contains numerous predicted amino acid transporters, making amino acid as a primary nutrient source (Wu et al. 2004). Viruses also depend on these amino acid pools, thereby propagating on host translational machinery as confirmed with the success of depletion of amino acids from growth media by CHIKV and DENV (Sasao et al. 1980; Caragata et al. 2014). Further depletion of

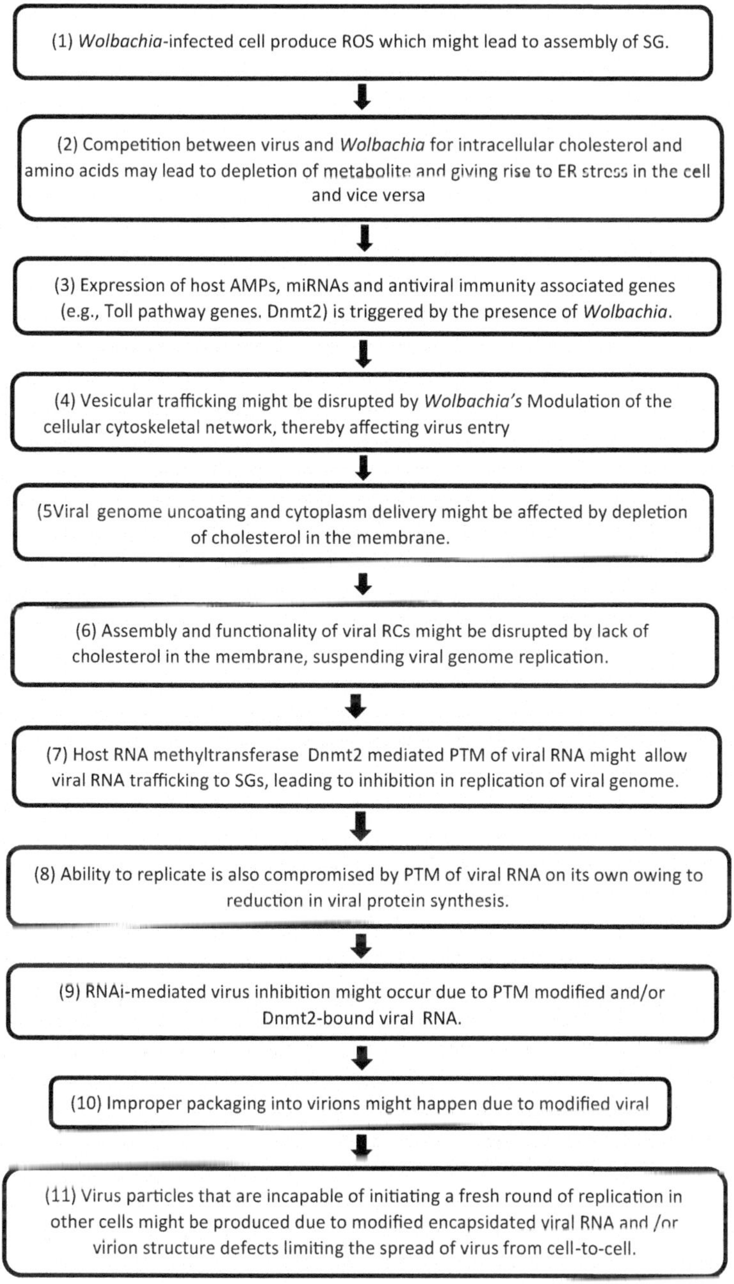

Fig. 11.7 Flow chart explaining the diagrammatic representation of *Wolbachia*-mediated virus inhibition in arthropods. (*Source*: This work)

amino acid pools can lead to translational arrest via eIF2α phosphorylation that might block viral replication (Harding et al. 2003).

11.19.4 RNA Methyltransferase Dnmt2

It is a host-encoded gene known for control of viruses and its *Wolbachia*-mediated viral inhibition (Brennan et al. 2008; Bhattacharya et al. 2017). Upon *w*Mel infection, an enhanced expression of Dnmt2 along with a reduction in virion infectivity was observed (Bhattacharya et al. 2017). AaDnmt2 was downregulated in wMelPop-CLA-infected *Ae. aegypti* mosquitoes via a *Wolbachia*-induced host miRNA which was further correlated with DENV inhibition (Zhang et al. 2013). *Wolbachia* presence leads to induction of oxidative/ER stress where Dnmt2 methylates viral RNA degrading it. Alternatively, the methylated viral RNA can be packaged into virions that upon reinfection into mammalian or arthropod cells cause no productive infectivity restricting virus dissemination within the host.

11.19.5 Host RNA Interference (RNAi) Pathway

Terradas et al. (2017) observed targeted depletion of argonaute-2 (Ago-2) in DENV recovery involving *w*Mel-infected *Ae. aegypti* Aag2 cells. This strengthened the significance of RNAi pathway for DENV inhibition in *Wolbachia*-infected mosquitoes. *Wolbachia*-mediated antiviral effect requires the recognition of viral dsRNA (Weber et al. 2006).

11.19.6 Antimicrobial Peptides and Toll Pathway Genes

Non-native *Wolbachia* strains transfection in mosquitoes results in upregulation of innate immune genes as compared to that of native *Wolbachia* host associations (Rancès et al. 2012). An elevated immune gene expression was a result of reduction in average life span of the *w*MelPop-CLA-infected *Ae. aegypti*. The virulent *w*MelPop-CLA upregulated a higher number of immune genes as compared to *w*Mel transfection in *Ae. aegypti* (Rancès et al. 2012). These immune genes are AMPs such as cecropins, defensin, diptericin, and several Toll pathway genes (PGRP-SA, GNBPB4, GNBPA1).

11.19.7 Wolbachia Density

The strong correlation between *Wolbachia* density and extent of pathogen-blocking phenotype is studied by Lu et al. (2012) and Schultz et al. (2017) (Fig. 11.8).

11.20 Implementation of *Wolbachia*-Based Strategy to Control Arboviral Transmission

Wolbachia has been well explored as a promising tool for controlling arboviral transmissions. The strategy mainly relies on its phenotypic effects on arthropod vectors and resistance to viruses. In 1967, this strategy was first deployed in Burma against filariasis vector with an aim to eliminate the local *Culex quinquefasciatus* population with the release of *Wolbachia*-infected male *Cx. quinquefasciatus* (Laven 1967). The following *Wolbachia*-mediated vector control methods are depicted in Fig. 11.9.

11.20.1 CI-Mediated Population Suppression

It employs incompatible insect technique (IIT) to reduce vector population. This occurs when *Wolbachia*-infected males are released into a population of naturally non-infected females or females that do not harbor the same strain, resulting in embryonic mortality owing to CI. In due course of time, the population of vectors will reduce (O'Connor et al. 2012).

11.20.2 CI-Mediated Population Modification/Pathogen Blocking

Wolbachia-infected females (resistant to arboviruses like dengue, Zika, and chikungunya) and males (either wild type or WolB) are released over a period of 12–16 weeks, producing *Wolbachia*-infected offspring with reduced vector competence (Moreira et al. 2009). Here CI imparts a reproductive advantage to *Wolbachia*-infected females, facilitating their spread and establishment in the population.

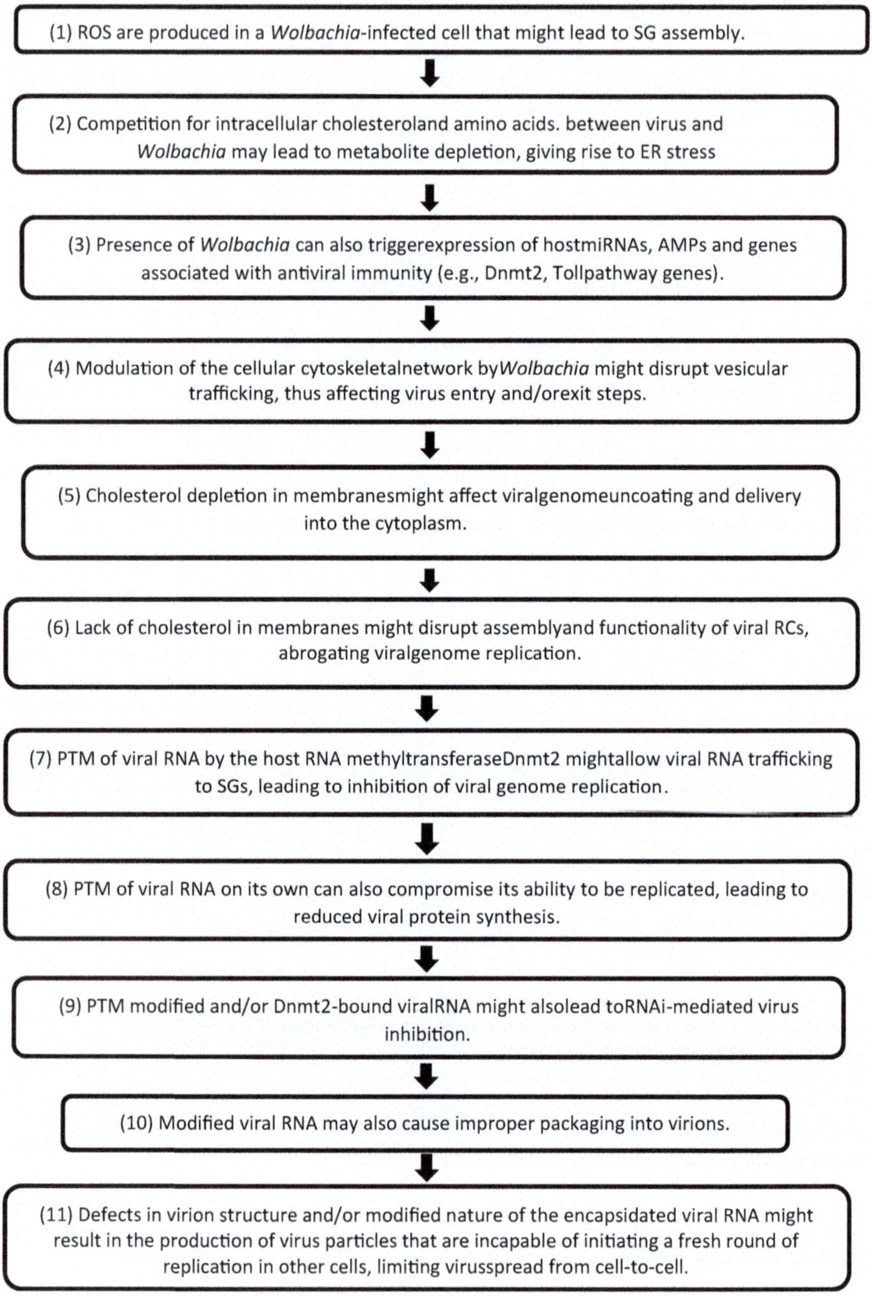

Fig. 11.8 Flow chart explaining the diagrammatic representation *Wolbachia*-mediated virus inhibition in arthropods. (*Source*: This work)

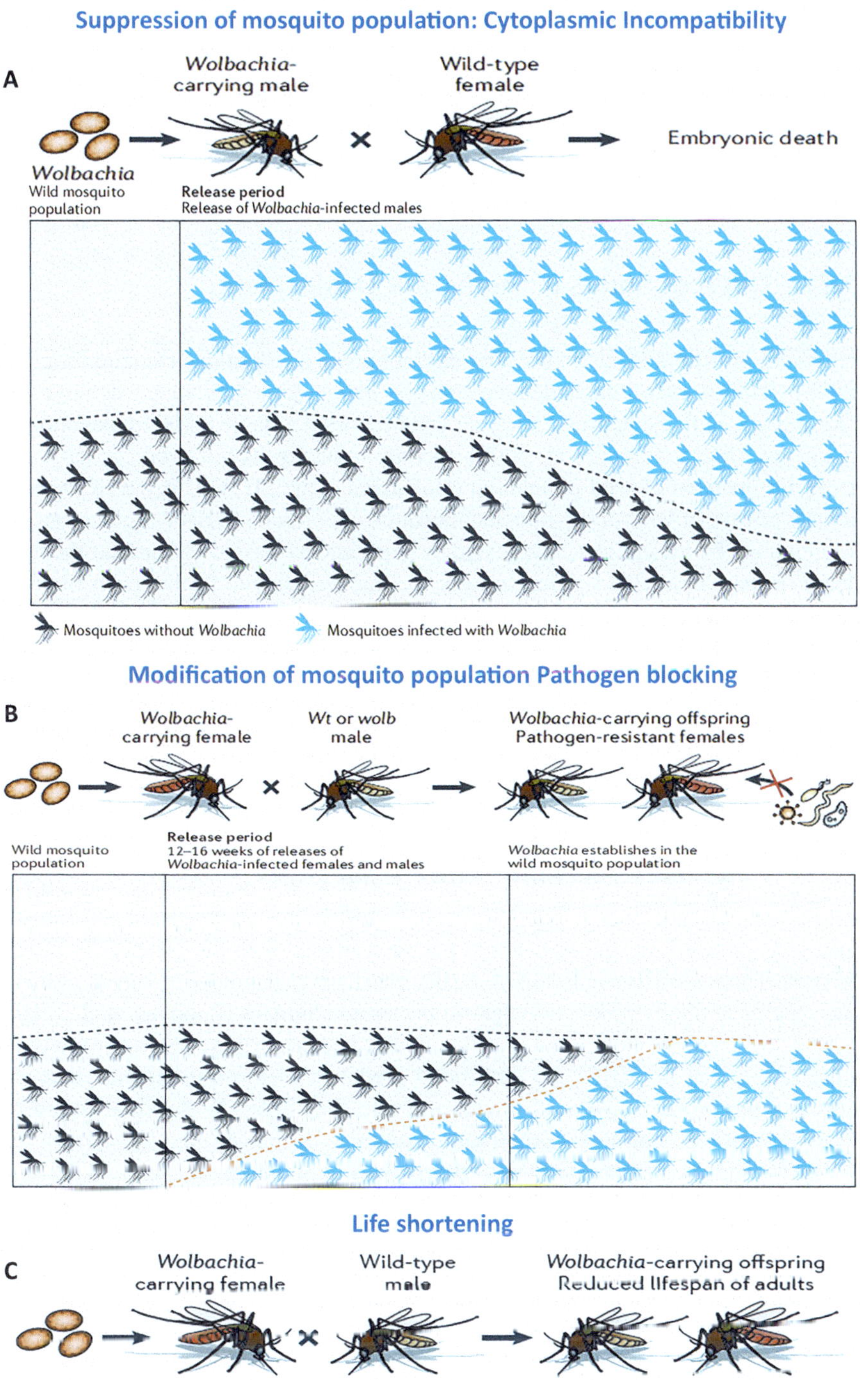

Fig. 11.9 Vector population (**a**) reduction, (**b**) modification, and (**c**) life shortening using *Wolbachia*. (*Source*: McGraw and O'Neill 2013; Flores and O'Neill 2018; permitted)

11.20.3 Wolbachia-*Mediated Life Shortening*

On releasing wMelPop via females, along with pathogen blocking and spread through CI, there will be a reduction in life span of insects as older insects are well-known to transmit disease (Cook et al. 2008).

11.21 Current Status of *Wolbachia* Deployment

The World Mosquito Program (WMP), earlier known as Eliminate Dengue Program (2008) (http://www.eliminatedengue.com), is undertaking *Wolbachia* deployment in *Ae. aegypti* in five countries. The year 2011 reported a stable *w*Mel transfection to *Ae. aegypti* (McMeniman et al. 2009), thereby reducing its vectorial capacity and successful invasion of wild population (Hoffmann et al. 2011). This paved the way for mass release of *Wolbachia*-infected mosquito suppressing DENV replication and dissemination by mosquitoes as revealed by experiments carried 1 year after the field release (Frentiu et al. 2014). This success was succeeded with further trials in other dengue endemic countries like Columbia, Indonesia, Vietnam, and Brazil. *Wolbachia*-infected mosquitoes have also been employed for controlling CHIKV (Moreira et al. 2009), JEV (Jeffries and Walker 2015), and YFV (van den Hurk et al. 2012). Dutra et al. (2016) reported that *Wolbachia*-infected *Ae. aegypti* can help reduce Zika transmission.

11.22 Potential Risks and Future Challenges

Wolbachia-infected mosquitoes should not be confused with the genetically modified ones. There is no transmission of the bacteria from mosquitoes to human as there is no specific antibody production against bacteria in humans (Popovici et al. 2010). There is evidence of horizontal transfer of the bacterium between arthropods in nature (Ahmed et al. 2015). So, arthropod predators of mosquitoes may become infected with *Wolbachia* strains transfected into their prey. With initial round of success, WHO is currently prompting countries to boost the deployment of *Wolbachia*-based control strategies (http://www.who.int/emergencies/zika-virus/articles/mosquito-control/en/).

However before implementation of this strategy, various issues need to be addressed like accounting for geographical specificities, choosing an optimum *Wolbachia* strain, monitoring evolutionary changes, and obtaining community acceptance.

11.23 Conclusion

This review emphasizes the current state of knowledge about the association between mosquitoes and *Wolbachia*. This endocellular symbiotic bacterium, *Wolbachia*, provides an ecofriendly alternative to chemical insecticides in vector control. Genetic manipulation has imparted improvement in existing genetic tools by genetic engineering of entomopathogenic microbes and releasing male mosquitoes with dominant lethal genes. This has led to pave path for more innovative technologies in the near future. With the reemergence and resurgence of arboviruses across geographical borders due to an increase in human and cargo movement, biological control seems to be of utmost importance for reducing vector population size.

References

Achee NL, Gould G, Perkins TA, Reiner RC Jr, Morrison AC, Ritchie SA, Gubler DJ, Teyssou R, Scot TW (2015) A critical assessment of vector control for dengue prevention. PLoS Negl Trop Dis 9(5):e0003655

Ahmed MZ, Araujo-Jnr EV, Welch JJ, Kawahara AY (2015) *Wolbachia* in butterflies and moths: geographic structure in infection frequency. Front Zool 12:16

Al-Khodor S, Price CT, Kalia A, Abu Kwaik Y (2010) Functional diversity of ankyrin repeats in microbial proteins. Trends Microbiol 18:132–139

Alout H, Labbé P, Chandre F, Cohuet A (2017) Malaria vector control still matters despite insecticide resistance. Trends Parasitol 33(8):610–618

Armbruster P, Damsky WE Jr, Giordano R, Birungi J, Munstermann LE, Conn JE (2003) Infection of new- and old-world *Aedes albopictus* (Diptera: Culicidae) by the intracellular parasite *Wolbachia*: implications for host mitochondrial DNA evolution. J Med Entomol 40:356–360

Azzouna A, Greve P, Martin G (2004) Sexual differentiation traits in functional males with female genital apertures (male symbol fga) in the woodlice *Armadillidium vulgare* Latr. (Isopoda, Crustacea). Gen Comp Endocrinol 138:42–49

Baldini F, Segata N, Pompon J, Marcenac P, Shaw WR, Dabiré RK, Diabaté A, Levashina EA, Catteruccia F (2014) Evidence of natural *Wolbachia* infections in field populations of *Anopheles gambiae*. Nat Commun 5:3985

Baldo L, Werren JH (2007) Revisiting *Wolbachia* supergroup typing based on *wsp*: spurious lineages and discordance with MLST. Curr Microbiol 55:81–87

Baldo L, Dunning Hotopp JC, Jolley KA, Bordenstein SR, Biber SA, Choudhury RR, Hayashi C, Maiden MC, Tettelin H, Werren JH (2006) Multilocus sequence typing system for the endosymbiont Wolbachia pipientis. Appl Environ Microbiol 72(11):7098–7110

Bandi C, Anderson TJC, Genchi C, Blaxter ML (1998) Phylogeny of *Wolbachia* in filarial nematodes. Proc R Soc Lond Biol 265:2407–2413

Bandi C, Dunn AM, Hurst GD, Rigaud T (2001) Inherited microorganisms, sex-specific virulence and reproductive parasitism. Trends Parasitol 17:88–94

Behbahani A (2012) *Wolbachia* infection and mitochondrial DNA comparisons among *Culex* mosquitoes in South West Iran. Pak J Biol Sci 15:54–57

Belov GA (2014) Modulation of lipid synthesis and trafficking pathways by picornaviruses. Curr Opin Virol 9:19–23

Benedict MQ, Levine RS, Hawley WA, Lounibos LP (2007) Spread of the tiger: global risk of invasion by the mosquito *Aedes albopictus*. Vector Borne Zoonotic Dis 7:76–85

Benelli G, Mehlhorn H (2016) Declining malaria, rising of dengue and Zika virus: insights for mosquito vector control. Parasitol Res 115:1747–1754

Bhatt S, Gething PW, Brady OJ, Messina JP, Farlow AW, Moyes CL, Drake JM, Brownstein JS, Hoen AG, Sankoh O, Myers MF, George DB, Jaenisch T, Wint GR, Simmons CP, Scott TW, Farrar JJ, Hay SI (2013) The global distribution and burden of dengue. Nature 496:504–507

Bhattacharya T, Newton ILG, Hardy RW (2017) *Wolbachia* elevates host methyltransferase expression to block an RNA virus early during infection. PLoS Pathog 13:e1006427

Bian G, Xu Y, Lu P, Xie Y, Xi Z (2010) The endosymbiotic bacterium *Wolbachia* induces resistance to dengue virus in *Aedes aegypti*. PLoS Pathog 6:e1000833

Black WCT, Bennett KE, Gorrochotegui-Escalante N, Barillas-Mury CV, Fernandez-Salas I, de Lourdes Munoz M, Farfan-Ale JA, Olson KE, Beaty BJ (2002) Flavivirus susceptibility in *Aedes aegypti*. Arch Med Res 33:379–388

Bordenstein S, Rosengaus RB (2005) Discovery of a novel *Wolbachia* supergroup in Isoptera. Curr Microbiol 51:393–398

Bordenstein SR, Marshall ML, Fry AJ, Kim U, Wernegreen JJ (2006) The tripartite associations between bacteriophage, *Wolbachia*, and arthropods. PLoS Pathog 2:e43

Bork P (1993) Hundreds of ankyrin-like repeats in functionally diverse proteins-mobile modules that cross phyla horizontally. Proteins 17:363–374

Bourtzis K (2008) *Wolbachia*-based technologies for insect pest population control. Adv Exp Med Biol 627:104–113

Bourtzis K, Dobson SL, Braig HR, O'Neill SL (1998) Rescuing *Wolbachia* has been over looked. Nature 391:852–853

Bourtzis K, Dobson SL, Xi Z, Rasgon JL, Calvitti M, Moreira LA, Bossin HC, Moretti R, Baton LA, Hughes GL, Mavingui P, Gilles JR (2014) Harnessing mosquito-*Wolbachia* symbiosis for vector and disease control. Acta Trop 132(1):S150–S163

Brady OJ, Golding N, Pigott DM, Kraemer MU, Messina JP, Reiner RC Jr, Scott TW, Smith DL, Gething PW, Hay SI (2014) Global temperature constraints on *Aedes aegypti* and *Ae. albopictus* persistence and competence for dengue virus transmission. Parasit Vectors 7:338

Breeuwer JAJ, Werren JH (1990) Microorganisms associated with chromosome destruction and reproductive isolation between two insect species. Nature 346:558–560

Brennan LJ, Keddie BA, Braig HR, Harris HL (2008) The endosymbiont *Wolbachia pipientis* induces the expression of host antioxidant proteins in an *Aedes albopictus* cell line. PLoS One 3:e2083

Brennan L, Haukedal J, Earle J, Keddie B, Harris H (2012) Disruption of redox homeostasis leads to oxidative DNA damage in spermatocytes of *Wolbachia*-infected *Drosophila simulans*. Insect Mol Biol 21:510–520

Brown JE, McBride CS, Johnson P, Ritchie S, Paupy C, Bossin H, Lutomiah J, Fernandez-Salas I, Ponlawat A, Cornel AJ, Black WC IV, Gorrochotegui-Escalante N, Urdaneta-Marquez L, Sylla M, Slotman M, Murray KO, Walker C, Powell JR (2011) Worldwide patterns of genetic differentiation imply multiple 'domestications' of *Aedes aegypti*, a major vector of human diseases. Proc Biol Sci 278:2446–2454

Brown JE, Evans BR, Zheng W, Obas V, Barrera-Martinez L, Egizi A, Zhao H, Caccone A, Powell JR (2014) Human impacts have shaped historical and recent evolution in *Aedes aegypti*, the dengue and yellow fever mosquito. Evolution 68:514–525

Brownlie JC, Cass BN, Riegler M, Witsenburg JJ, Iturbe-Ormaetxe I, McGraw EA, O'Neill SL (2009) Evidence for metabolic provisioning by a common invertebrate endosymbiont, *Wolbachia pipientis*, during periods of nutritional stress. PLoS Pathog 5:e1000368

Campbell BC, Bragg TS, Turner CE (1992) Phylogeny of symbiotic bacteria of four weevil species (Coleoptera: Curculionidae) based on analysis of 16S ribosomal DNA. Insect Biochem Mol Biol 22:415–421

Campbell LP, Luther C, Moo-Llanes D, Ramsey JM, Danis-Lozano R, Peterson AT (2015) Climate change influences on global vector distributions for dengue and chikungunya viruses. Philos Trans R Soc Lond B Biol Sci 370:1665

Caragata EP, Rancès E, O'Neill SL, McGraw EA (2014) Competition for amino acids between *Wolbachia* and the mosquito host, *Aedes aegypti*. Microbiol Ecol 67:205–218

Carro AC, Damonte EB (2013) Requirement of cholesterol in the viral envelope for dengue virus infection. Virus Res 174:78–87

Carvalho RG, Lourenço-de-Oliveira R, Braga IA (2014) Updating the geographical distribution and frequency of *Aedes albopictus* in Brazil with remarks regarding its range in the Americas. Mem Inst Oswaldo Cruz 109:787–796

Charlat S, Le Chat L, Mercot H (2003) Characterization of non-cytoplasmic incompatibility inducing *Wolbachia* in two continental African populations of *Drosophila simulans*. Heredity 90:49–55

Chrostek E, Marialva MS, Yamada R, O'Neill SL, Teixeira L (2014) High anti-viral protection without immune upregulation after interspecies *Wolbachia* transfer. PLoS One 9:e99025

Clancy DJ, Hoffmann AA (1996) Cytoplasmic incompatibility in *Drosophila simulans*: evolving complexity. Trends Ecol Evol 11:145–146

Cook PE, McMeniman CJ, O'Neill SL (2008) Modifying insect population age structure to control vector-borne disease. Adv Exp Med Biol 627:126–140

Covacin C, Barker SC (2007) Supergroup F *Wolbachia* bacteria parasitise lice (Insecta: Phthiraptera). Parasitol Res 100:479–485

de Araújo Lobo JM, Christofferson RC, Mores CN (2014) Investigations of Koutango virus infectivity and dissemination dynamics in *Aedes aegypti* mosquitoes. Environ Health Insights 8:9–13

Delatte AH, Gimonneau G, Triboire A, Fontenille D, Delatte H (2009) Influence of temperature on immature development, survival, longevity, fecundity, and gonotrophic cycles of *Aedes albopictus*, vector of chikungunya and dengue in the Indian Ocean. J Med Entomol 46:33–41

Diallo D, Sall AA, Diagne CT, Faye O, Faye O, Ba Y, Hanley KA, Buenemann M, Weaver SC, Diallo M (2014) Zika virus emergence in mosquitoes in southeastern Senegal, 2011. PLoS One 9(10):e109442

Dobson SL (2003) Reversing *Wolbachia*-based population replacement. Trends Parasitol 19:128–133

Dobson SL, Bourtzis K, Braig HR, Jones BF, Zhou W, Rousset F, O'Neill SL (1999) *Wolbachia* infections are distributed throughout insect somatic and germ line tissues. Insect Biochem Mol Biol 29:153–160

Dumler SJ, Barbet AF, Bekker CP, Dasch GA, Palmer GH, Ray SC, Rikihisa Y, Rurangirwa FR (2001) Reorganization of genera in the families Rickettsiaceae and Anaplasmataceae in the order Rickettsiales: unification of some species of *Ehrlichia* with *Anaplasma*, *Cowdria* with *Ehrlichia* and *Ehrlichia* with *Neorickettsia*—descriptions of six new species combinations and designation of Ehrlichia equi and "HGE agent" as subjective synonyms of *Ehrlichia phagocytophila*. Int J Syst Evol Microbiol 51:2145–2165

Dutra HLC, Rocha MN, Dias FBS, Mansur SB, Caragata EP, Moreira LA (2016) *Wolbachia* blocks currently circulating zika virus isolates in Brazilian *Aedes aegypti* mosquitoes. Cell Host Microbe 19:771–774

Faria VG, Sucena E (2013) *Wolbachia* in the malpighian tubules: evolutionary dead end or adaptation? J Exp Zool B Mol Dev Evol 320:195–199

Flores HF, O'Neill SL (2018) Controlling vector-borne diseases by releasing modified mosquitoes. Nat Rev Microbiol 16:508–518

Fraser JE, De Bruyne JT, Iturbe-Ormaetxe I, Stepnell J, Burns RL, Flores HA, Neil SL (2017) Novel Wolbachia-transinfected Aedes aegypti mosquitoes possess diverse fitness and vector competence phenotypes. PLoS Pathog 13(12):e1006751

Frentiu FD, Zakir T, Walker T, Popovici J, Pyke AT, van den Hurk A, McGraw EA, O'Neill SL (2014) Limited dengue virus replication in field collected *Aedes aegypti* mosquitoes infected with *Wolbachia*. PLoS Negl Trop Dis 8:e2688

Gavotte L, Vavre F, Henri H, Ravallec M, Stouthamer R, Bouletreau M (2004) Diversity, distribution and specificity of WO phage infection in *Wolbachia* of four insect species. Insect Mol Biol 13:147–153

Geoghegan V, Stainton K, Rainey SM, Ant TH, Dowle AA, Larson T, Hester S, Charles PD, Thomas B, Sinkins SP (2017) Perturbed cholesterol and vesicular trafficking associated with dengue blocking in *Wolbachia*-infected *Aedes aegypti* cells. Nat Commun 8:526

Gerth M, Bleidorn C (2016) Comparative genomics provides a timeframe for *Wolbachia* evolution and exposes a recent biotin synthesis operon transfer. Nat Microbiol 2:16241

Glowska E, Dragun-Damian A, Dabert M, Gerth M (2015) New *Wolbachia* supergroups detected in quill mites (Acari: Syringophilidae). Infect Genet Evol 30:140–146

Gorham CH, Fang QQ, Durden LA (2003) *Wolbachia* endosymbionts in fleas (Siphonaptera). J Parasitol 89:283–289

Harding HP, Zhang Y, Zeng H, Novoa I, Lu PD, Calfon M, Sadri N, Yun C, Popko B, Paules R (2003) An integrated stress response regulates amino acid metabolism and resistance to oxidative stress. Mol Cell 11:619–633

Hardy JL, Houk EJ, Kramer LD, Reeves WC (1983) Intrinsic factors affecting vector competence of mosquitoes for arboviruses. Annu Rev Entomol 28:229–262

Hertig M (1936) The rickettsia, *Wolbachia pipientis* (Gen. Et SP.N.) and associated inclusions of the mosquito *Culex pipiens*. Parasitology 28:453–486

Hertig M, Wolbach SB (1924) Studies on rickettsia-like microorganisms in insects. J Med Res 44:329–374

Hilgenboecker K, Hammerstein P, Schlattmann P, Telschow A, Werren JH (2008) How many species are infected with *Wolbachia*? A statistical analysis of current data. FEMS Microbiol Lett 281:215–220

Hoffmann AA, Turelli M (1997) Cytoplasmic incompatibility in insects. In: O'Neill SL, Hoffmann AA, Werren JH (eds) Influential passengers: inherited microorganisms and arthropod reproduction. Oxford University Press, Oxford, pp 42–80

Hoffmann AA, Montgomery BL, Popovici J, Iturbe-Ormaetxe I, Johnson PH, Muzzi F, Greenfield M, Durkan M, Leong YS, Dong Y, Cook H, Axford J, Callahan AG, Kenny N, Omodei C, McGraw EA, Ryan PA, Ritchie SA, Turelli M, O'Neill SL (2011) Successful establishment of *Wolbachia* in *Aedes* populations to suppress dengue transmission. Nature 476:454–457

Hughes GL, Ren X, Ramirez JL, Sakamoto JM, Bailey JA, Jedlicka AE, Rasgon JL (2011) *Wolbachia* infections in *Anopheles gambiae* cells: transcriptomic characterization of a novel host-symbiont interaction. PLoS Pathog 7:e1001296

Ijdo JW, Carlson ACAC, Kennedy EL (2007) *Anaplasma phagocytophilum* AnkA is tyrosine-phosphorylated at EPIYA motifs and recruits SHP1 during early infection. Cell Microbiol 9:1284–1296

Ilinsky Y, Kosterin OE (2017) Molecular diversity of *Wolbachia* in Lepidoptera: prevalent allelic content and high recombination of MLST genes. Mol Phylogenet Evol 109:164–179

Ishmael N, Dunning Hotopp JC, Ioannidis P, Biber S, Sakamoto J, Siozios S, Nene V, Werren J, Bourtzis K, Bordenstein SR, Tettelin H (2009) Extensive genomic diversity of closely related *Wolbachia* strains. Microbiology 155:2211–2222

Iturbe-Ormaetxe I, Burke GR, Riegler M, O'Neill SL (2005) Distribution, expression, and motif variability of ankyrin domain genes in *Wolbachia pipientis*. J Bacteriol 187:5136–5145

Jaenike J, Dyer KA, Cornish C, Minhas MS (2006) Asymmetrical reinforcement and *Wolbachia* infection in *Drosophila*. PLoS Biol 4:e325

Jeffries CL, Walker T (2015) The potential use of *Wolbachia*-based mosquito biocontrol strategies for Japanese encephalitis. PLoS Negl Trop Dis 9:e0003576

Johanowicz DL, Hoy MA (1995) Molecular evidence for A-*Wolbachia* endosymbiont in the predatory mite *Metaseiulus occidentalis*. J Cell Biochem 21A:198

Kageyama D, Traut W (2004) Opposite sex-specific effects of *Wolbachia* and interference with the sex determination of its host *Ostrinia scapulalis*. Proc R Soc Lond Biol 271:251–258

Kent BN, Bordenstein SR (2010) Phage WO of *Wolbachia*: lambda of the endosymbiont world. Trends Microbiol 18:173–181

Kielian MC, Helenius A (1984) Role of cholesterol in fusion of Semliki forest virus with membranes. J Virol 52:281–283

Kittayapong P, Baisley KJ, Baimai V, O'Neill SL (2000) Distribution and diversity of *Wolbachia* infections in Southeast Asian mosquitoes (Diptera: Culicidae). J Med Entomol 37:340–345

Kittayapong P, Baimai V, O'Neill SL (2002) Field prevalence of *Wolbachia* in the mosquito vector *Aedes albopictus*. Am J Trop Med Hyg 66:108–111

Klasson L, Walker T, Sebaihia M, Sanders MJ, Quail MA, Lord A, Sanders S, Earl J, O'Neill SL, Thomson N, Sinkins SP, Parkhill J (2008) Genome evolution of *Wolbachia* strain wPip from the *Culex pipiens* group. Mol Biol Evol 25:1877–1887

Kose H, Karr TL (1995) Organization of *Wolbachia pipientis* in the *Drosophila* fertilized egg and embryo revealed by an anti-*Wolbachia* monoclonal antibody. Mech Dev 51:275–288

Kraemer MUG, Sinka ME, Duda KA, Mylne AQN, Shearer FM, Barker CM, Moore CG, Carvalho RG, Coelho GE, Van Borte W, Hendrickx G, Schaffner F, Elyazar IRF, Teng HJ, Brady OJ, Messina JP, Pigott DM, Scott TW, Smith DL, Wint GRW, Golding N, Hay SI (2015) The global distribution of the arbovirus vectors *Aedes aegypti* and *Ae. albopictus*. eLife 4:e08347

Kremer N, Voronin D, Charif D, Mavingui P, Mollereau B, Vavre F (2009) *Wolbachia* interferes with ferritin expression and iron metabolism in insects. PLoS Pathog 5:e1000630

Kremer N, Charif D, Henri H, Gavory F, Wincker P, Mavingui P, Vavre F (2012) Influence of Wolbachia on host gene expression in an obligatory symbiosis. BMC Microbiol 12(Suppl 1):S7

Lambrechts L, Paaijmans KP, Fansiri T, Carrington LB, Kramer LD (2011) Impact of daily temperature fluctuations on dengue virus transmission by *Aedes aegypti*. PNAS USA 108:7460–7465

Landmann F, Orsi GA, Loppin B, Sullivan W (2009) *Wolbachia*-mediated cytoplasmic incompatibility is associated with impaired histone deposition in the male pronucleus. PLoS Pathog 5: e1000343

Laven H (1967) Eradication of *Culex pipiens fatigans* through cytoplasmic incompatibility. Nature 216:383–384

Leclercq S, Giraud I, Cordaux R (2011) Remarkable abundance and evolution of mobile group II introns in *Wolbachia* bacterial endosymbionts. Mol Biol Evol 28:685–697

Leparc-Goffart I, Nougairede A, Cassadou S, Prat C, de Lamballerie X (2014) Chikungunya in the Americas. Lancet 383(9916):514

Lin MQ, den Dulk-Ras A, Hooykaas PJJ, Rikihisa Y (2007) *Anaplasma phagocytophilum* AnkA secreted by type IV secretion system is tyrosine phosphorylated by Abl-1 to facilitate infection. Cell Microbiol 9:2644–2657

Lo N, Paraskevopoulos C, Bourtzis K (2007) Taxonomic status of the intracellular bacterium *Wolbachia pipientis*. Int J Syst Evol Microbiol 57:645–657

Louis C, Nigro L (1989) Ultrastructural evidence of *Wolbachia* Rickettsiales in *Drosophila simulans* and their relationships with unidirectional cross incompatibility. J Invert Pathol 54:39–44

Lu P, Bian G, Pan X, Xi Z (2012) *Wolbachia* induces density-dependent inhibition to dengue virus in mosquito cells. PLoS Negl Trop Dis 6(7):e1754

McGraw EA, O'Neill SL (2013) Beyond insecticides: new thinking on an ancient problem. Nat Rev Microbiol 11:181–193

McGraw E, Merritt D, Droller J, O'Neill S (2002) *Wolbachia* density and virulence attenuation after transfer into a novel host. PNAS USA 99:2918–2923

McMeniman CJ, Lane RV, Cass BN, Fong AW, Sidhu M, Wang YF, O'Neill SL (2009) Stable introduction of a life-shortening *Wolbachia* infection into the mosquito *Aedes aegypti*. Science 323:141–144

Melero Alcíbar R (2006) The pupae of Spanish Culicinae II: *Aedes vittatus* Bigot, 1861 (Diptera: Culicidae). Eur Mosq Bull 21:19–22

Min KT, Benzer S (1997) *Wolbachia*, normally a symbiont of *Drosophila*, can be virulent, causing degeneration and early death. PNAS USA 94:10792–10796

Mohanty I, Rath A, Swain S, Pradhan N, Hazra RK (2018) *Wolbachia* population in vectors and non vectors: a sustainable approach towards dengue control. Curr Microbiol 76:133–143

Molloy JC, Sommer U, Viant MR, Sinkins SP (2016) *Wolbachia* modulates lipid metabolism in *Aedes albopictus* mosquito cells. Appl Environ Microbiol 82:3109–3120

Moreau J, Bertin A, Caubet Y, Rigaud T (2001) Sexual selection in an isopod with *Wolbachia*-induced sex reversal: males prefer real females. J Evol Biol 14:388–394

Moreira LA, Iturbe-Ormaetxe I, Jeffery JA, Lu G, Pyke AT, Hedges LM, Rocha BC, Hall-Mendelin S, Day A, Riegler M (2009) A *Wolbachia* symbiont in *Aedes aegypti* limits infection with dengue, chikungunya, and plasmodium. Cell 139:1268–1278

Mosavi LK, Cammett TJ, Desrosiers DC, Peng ZY (2004) The ankyrin repeat as molecular architecture for protein recognition. Protein Sci 13:1435–1448

Mousson L, Zouache K, Arias-Goeta C, Raquin V, Mavingui P, Failloux AB (2012) The native *Wolbachia* symbionts limit transmission of dengue virus in *Aedes albopictus*. PLoS Negl Trop Dis 6:e1989

Müller JM, Cahill MA, Rupec RA, Baeuerle PA, Nordheim A (1997) Antioxidants as well as oxidants activate c-fos via ras-dependent activation of extracellular-signal-regulated kinase 2 and ELK-1. Fed Eur Biochem Soc (FEBS) J 244:45–52

Musso D, Gubler DJ (2016) Zika Virus. Clin Microbiol Rev 29:487–524

O'Connor L, Plichart C, Sang AC, Brelsfoard CL, Bossin HC, Dobson SL (2012) Open release of male mosquitoes infected with a *Wolbachia* biopesticide: field performance and infection containment. PLoS Negl Trop Dis 6:e1797

Osei-Poku J, Han C, Mbogo CM, Jiggins FM (2012) Identification of *Wolbachia* strains in mosquito disease vectors. PLoS One 7:e49922

Pan X, Luhrmann A, Satoh A, Laskowski-Arce MA, Roy CR (2008) Ankyrin repeat proteins comprise a diverse family of bacterial type IV effectors. Science 320:1651–1654

Paupy C, Delatte H, Bagny L, Corbel V, Fontenille D (2009) *Aedes albopictus*, an arbovirus vector: from the darkness to the light. Microb Infect 11:1177–1185

Pietri JE, DeBruh H, Sullivan W (2016) The rich somatic life of *Wolbachia*. Microbiol Open 5:923–936

Ponlawat A, Harrington LC (2005) Blood feeding patterns of *Aedes aegypti* and *Aedes albopictus* in Thailand. J Med Entomol 42:844–849

Popovici J, Moreira LA, Poinsignon A, Iturbe-Ormaetxe I, McNaughton D, O'Neill SL (2010) Assessing key safety concerns of a *Wolbachia*-based strategy to control dengue transmission by *Aedes* mosquitoes. Mem do Inst Oswaldo Cruz 105:957–964

Prout T (1994) Some evolutionary possibilities for a microbe that causes incompatibility in its host. Evolution 48:909–911

Rainey SM, Martinez J, McFarlane M, Juneja P, Sarkies P, Lulla A, Schnettler E, Varjak M, Merits A, Miska EA, Jiggins FM, Kohl A (2016) *Wolbachia* blocks viral genome replication early in infection without a transcriptional response by the endosymbiont or host small RNA pathways. PLoS Pathog 12:e1005536

Rancès E, Ye YH, Woolfit M, McGraw EA, O'Neill SL (2012) The relative importance of innate immune priming in *Wolbachia*-mediated dengue interference. PLoS Pathog 8:e1002548

Ravikumar H, Ramachandraswamy N, Sampathkumar S, Prakash BM, Huchesh HC, Uday J, Puttaraju HP (2010) A preliminary survey for *Wolbachia* and bacteriophage WO infections in Indian mosquitoes (Diptera: Culicidae). Trop Biomed 27:384–393

Reiter P, Fontenille D, Paupy C (2006) *Aedes albopictus* as an epidemic vector of chikungunya virus: another emerging problem? Lancet Infect Dis 6:463–464

Ricci I, Cancrini G, Gabrielli S, D'Amelio S, Favia G (2002) Searching for *Wolbachia* (Rickettsiales: Rickettsiaceae) in mosquitoes (Diptera: Culicidae): large polymerase chain reaction survey and new identifications. J Med Entomol 39(4):562–567

Richards SL, Anderson SL, Lord CC, Smartt CT, Tabachnick WJ (2012) Relationships between infection, dissemination, and transmission of West Nile virus RNA in Culex pipiens quinquefasciatus (Diptera: Culicidae). J Med Entomol 49:132–142

Rigaud T (1999) Further *Wolbachia* endosymbiont diversity: a tree hiding in the forest. Trends Ecol Evol 14:212–213

Rikihisa Y, Lin M (2010) *Anaplasma phagocytophilum* and *Ehrlichia chaffeensis* type I IV secretion and Ank proteins. Curr Opin Microbiol 13:59–66

Ros VID, Fleming VM, Feil EJ, Breeuwer JAJ (2009) How diverse is the genus *Wolbachia*? Multiple-gene sequencing reveals a putatively new *Wolbachia* supergroup recovered from spider mites (Acari: Tetranychidae). Appl Environ Microbiol 75:1036–1043

Rousset F, Bouchon D, Pintureau B, Juchault P, Solignac M (1992) *Wolbachia* endosymbionts responsible for various alterations of sexuality in arthropods. Proc R Soc Lond Biol 250:91–98

Rowley SM, Raven RJ, McGraw EA (2004) *Wolbachia pipientis* in Australian spiders. Curr Microbiol 49:208–214

Salzberg SL, Hotopp JCD, Delcher AL, Pop M, Smith DR, Eisen MB, Nelson WC (2005) Serendipitous discovery of *Wolbachia* genomes in multiple *Drosophila* species. Genome Biol 6:R23

Salzberg SL, Puiu D, Sommer DD, Nene V, Lee NH (2009) Genome sequence of the *Wolbachia* endosymbiont of *Culex quinquefasciatus* JHB. J Bacteriol 191:1725–1725

Saridaki A, Bourtzis K (2010) *Wolbachia*: more than just a bug in insects genitals. Curr Opin Microbiol 13:67–72

Sasao F, Igarashi A, Fukai K (1980) Amino acid requirements for the growth of *Aedes albopictus* clone C6/36 cells and for the production of dengue and chikungunya viruses in the infected cells. Microbiol Immunol 24:915–924

Schultz MJ, Isern S, Michael SF, Corley RB, Connor J, Frydman HM (2017) Variable inhibition of zika virus replication by different *Wolbachia* strains in mosquito cell cultures. J Virol 91: e00339–e00317

Scola BL, Bandi C, Raoult D (2015) Wolbachia. In: Bergey's manual of systematics of archaea and bacteria. John Wiley & Sons Ltd., Chichester, pp 1–12

Scott TW, Takken W (2012) Feeding strategies of anthropophilic mosquitoes result in increased risk of pathogen transmission. Trends Parasitol 28:114–121

Service MW (1974) Survey of the relative prevalence of potential yellow fever vectors in Northwest Nigeria. Bull World Health Org 50:487–494

Shaw WR, Marcenac P, Childs LM, Buckee CO, Baldini F, Sawadogo SP, Dabiré RK, Diabaté A, Catteruccia F (2016) *Wolbachia* infections in natural *Anopheles* populations affect egg laying and negatively correlate with *Plasmodium* development. Nat Commun 7:11772

Simmons CP, Farrar JJ, Chau NVV, Wills B (2012) Dengue. N Engl J Med 366:1423–1432

Sinkins SP (2004) *Wolbachia* and cytoplasmic incompatibility in mosquitoes. Insect Biochem Mol Biol 34:723–729

Sironi M, Bandi C, Sacchi L, Di Sacco B, Damiani G, Genchi C (1995) Molecular evidence for a close relative of the arthropod endosymbiont *Wolbachia* in a filarial worm. Mol Biochem Parasitol 74:223–227

Solomon T, Vaughn DW (2002) Pathogenesis and clinical features of Japanese encephalitis and West Nile virus infections. In: Japanese encephalitis and West Nile viruses. Springer, Berlin

Stouthamer R, Luck R (1993) Influence of microbe-associated parthenogenesis on the fecundity of *Trichogramma deion* and *T. pretiosum*. Entomol Exp Appl 67:183–192

Stouthamer R, Breeuwer JAJ, Luck RF, Werren JH (1993) Molecular identification of microorganisms associated with parthenogenesis. Nature 361:66–68

Tabachnick WJ (1991) Evolutionary genetics and arthropod-borne disease: the yellow fever mosquito. Am Entomol 37:14–26

Takken W, Koenraadt CJ (2013) Ecology of parasite-vector interactions. Wageningen Academic Publishers, Wageningen

Tamas I, Klasson L, Canback B, Naslund AK, Eriksson AS, Wernegreen JJ, Sandström JP, Moran NA, Andersson SG (2002) 50 million years of genomic stasis in endosymbiotic bacteria. Science 296:2376–2379

Teixeira L, Ferreira Á, Ashburner M (2008) The bacterial symbiont *Wolbachia* induces resistance to RNA viral infections in *Drosophila melanogaster*. PLoS Biol 6:e1000002

Terradas G, Joubert DA, McGraw EA (2017) The RNAi pathway plays a small part in *Wolbachia*-mediated blocking of dengue virus in mosquito cells. Sci Rep 7:43847

Thannickal VJ, Fanburg BL (2000) Reactive oxygen species in cell signaling. Am J Physiol Lung Cell Mol Physiol 279:L1005–L1028

Timmermans MJTN, Marien J, Roelofs D, van Straalen NM, Ellers J (2004) Evidence for multiple origins of *Wolbachia* infection in springtails. Pedobiologia 48:469–475

Turelli M (1994) Evolution of incompatibility-inducing microbes and their hosts. Evolution 48:1500–1513

Turelli M, Hoffmann AA (1991) Rapid spread of an inherited incompatibility factor in California *Drosophila*. Nature 353:440–442

Van den Hurk AF, Hall-Mendelin S, Pyke AT, Frentiu FD, McElroy K, Day A, Higgs S, O'Neill SL (2012) Impact of *Wolbachia* on infection with chikungunya and yellow fever viruses in the mosquito vector *Aedes aegypti*. PLoS Negl Trop Dis 6:e1892

Vasquez CJ, Stouthamer R, Jeong G, Morse JG (2011) Discovery of a CI-inducing Wolbachia and its associated fitness costs in the biological control agent *Aphytis melinus* DeBach (Hymenoptera: Aphelinidae). Biol Control 58:192–198

Walker T, Klasson L, Sebaihia M, Sanders MJ, Thomson NR, Parkhill J, Sinkins SP (2007) Ankyrin repeat domain-encoding genes in the *w*Pip strain of *Wolbachia* from the *Culex pipiens* group. BMC Biol 5:39

Wang Z, Su X-M, Wen J, Jiang L-Y, Qiao G-X (2014) Widespread infection and diverse infection patterns of *Wolbachia* in Chinese aphids. Insect Sci 21:313–325

Weaver SC (2014) Arrival of chikungunya virus in the new world: prospects for spread and impact on public health. PLoS Negl Trop Dis 8:e2921

Weber F, Wagner V, Rasmussen SB, Hartmann R, Paludan SR (2006) Double-stranded RNA is produced by positive-strand RNA viruses and DNA viruses but not in detectable amounts by negative-strand RNA viruses. J Virol 80:5059–5064

Weeks AR, Breeuwer JA (2001) *Wolbachia*-induced parthenogenesis in a genus of phytophagous mites. Proc R Soc Lond Biol 268:2245–2251

Weeks AR, Turelli M, Harcombe WR, Reynolds KT, Hoffmann AA (2007) From parasite to mutualist: rapid evolution of *Wolbachia* in natural populations of *Drosophila*. PLoS Biol 5:e114

Werren JH (1997) Biology of *Wolbachia*. Annu Rev Entomol 42:587–609

Werren JH, Baldo L, Clark ME (2008) *Wolbachia*: master manipulators of invertebrate biology. Nat Rev Microbiol 6:741–751

Wilke ABB, Marrelli MT (2015) Paratransgenesis: a promising new strategy for mosquito vector control. Parasit Vectors 8:1

Wolbachia MLST databases (2006). http://pubmlst.org/wolbachia/. Accessed 16 Aug 2018

Wong ZS, Brownlie JC, Johnson KN (2015) Oxidative stress correlates with *Wolbachia*-mediated antiviral protection in *Wolbachia*-Drosophila associations. Appl Environ Microbiol 81:3001–3005

World Health Organization (2004) Global strategic framework for integrated vector management, WHO/CDS/CPE/PVC/2004.10. WHO, Geneva

World Health Organization (2014) A global brief on vector-borne diseases. World Health Organization, Geneva. Available from: http:// www.who.int/campaigns/world-healthday/2014/global-brief/en/. Accessed 10 May 2015

World Mosquito Program (2008). http://www.eliminatedengue.com. Accessed 22 Oct 2018

Wright JD (1979) The etiology and biology of cytoplasmic incompatibility in *Aedes scutellaris* group. PhD thesis, University of California, Los Angeles, p 199

Wright JD, Sjostrand FS, Portaro JK, Barr AR (1978) The ultrastructure of the rickettsia-like microorganism *Wolbachia pipientis* and associated virus like bodies in the mosquito *Culex pipiens*. J Ultrastruct Res 63:79–85

Wu M, Sun LV, Vamathevan J, Riegler M, Deboy R, Brownlie JC, McGraw EA, Martin W, Esser C, Ahmadinejad N, Wiegand C, Madupu R, Beanan MJ, Brinkac LM, Daugherty SC, Durkin AO, Kolonay JF, Nelson WC, Mohamoud Y, Lee P, Berry K, Young MB, Utterback T, Weidman J, Nierman WC, Paulsen IA, Nelson KE, Tettelin H, O'Neill SL, Eisen JA (2004) Phylogenomics of the reproductive parasite *Wolbachia pipientis* *w*Mel: a streamlined genome overrun by mobile genetic elements. PLoS Biol 2:327–341

Xi Z, Gavotte L, Xie Y, Dobson SL (2008) Genome-wide analysis of the interaction between the endosymbiotic bacterium *Wolbachia* and its *Drosophila* host. BMC Genomics 9:1

Yen JH (1972) The microorganismal basis of cytoplasmic incompatibility in the *Culex pipiens* complex. PhD thesis, University of California, Los Angeles

Yen JH (1975) Transovarial transmission of Rickettsia-like microorganisms in mosquitoes. Ann N Y Acad Sci 266:152–161

Yen JH, Barr AR (1971) The etiological agent of cytoplasmic incompatibility in *Culex pipiens*. J Invert Pathol 22:242–250

Zchori-Fein E, Roush RT, Hunter MS (1992) Male production induced by antibiotic treatment in *Encarsia formosa* (Hymenoptera: Aphelinidae), an asexual species. Cell Mol Life Sci 48:102–105

Zeh DW, Zeh JA, Bonilla MM (2005) *Wolbachia*, sex ratio bias and apparent male killing in the harlequin beetle riding pseudoscorpion. Heredity 95:41–49

Zhang G, Hussain M, O'Neill SL, Asgari S (2013) *Wolbachia* uses a host micro RNA to regulate transcripts of a methyltransferase, contributing to dengue virus inhibition in *Aedes aegypti*. PNAS USA 110:10276–10281

Zhou W, Rousset F, O'Neill SL (1998) Phylogeny and PCR-based classification of *Wolbachia* strains using *wsp* gene sequences. Proc Biol Sci 265:509–515

Chapter 12
Wolbachia Endosymbiont and Mosquito Vectors, with Emphasis on Lymphatic Filariasis Elimination

I. P. Sunish

Abstract *Wolbachia* are maternally inherited intracellular bacteria, known to alter early development and mitotic processes in their hosts. They are frequently observed as a reproductive parasite, capable of inducing feminization, parthenogenesis, male killing, or cytoplasmic incompatibility. A total of 18 clades of *Wolbachia* have been reported, almost exclusively in arthropods. *Wolbachia*-based strategies have been proposed for the control of disease vectors. *Wolbachia*-based population suppression and transmission blocking can work in species not commonly infected with *Wolbachia* in the wild. However, efficient maintenance and spread of *Wolbachia* infection into field populations is crucial to the success of this strategy. Property of cytoplasmic incompatibility (CI) can be used to reduce the density of mosquito field populations through inundative releases of incompatible males in order to sterilize females. In semi-field condition at La Reunion, the LR[wPip(Is)] males of *Culex quinquefasciatus* successfully competed with field males in mating with field females. Depletion of *Wolbachia* endobacteria by antibiotic therapy prevents larval moulting and kills adult filarial worms. This strategy could act as an adjunct to vector control and is being exploited for the elimination of lymphatic filariasis.

Keywords *Wolbachia* · Bacterium · Cytoplasmic incompatibility · Vector control Lymphatic filariasis

12.1 Introduction

Wolbachia is an endosymbiont and is distributed widely in different organisms worldwide. It is reported to affect the organisms in its evolutionary life process. It was implicated to have significant role in the process of speciation, especially among different insect groups (Breeuwer and Werren 1990; Laven 1959, 1967a). *Wolbachia* are reported to modify the development process during mitosis in the

I. P. Sunish (✉)
ICMR-Regional Medical Research Centre, Port Blair, Andaman & Nicobar Islands, India

B. K. Tyagi (ed.), *Genetically Modified and other Innovative Vector Control Technologies*, https://doi.org/10.1007/978-981-16-2964-8_12

host organisms (Glover et al. 1990; Lassy and Karr 1996; Reed and Werren 1995; Stouthamer and Kazmer 1994). This endosymbiont has a worldwide interest and in few instances has been utilized as a biological control agent to enhance the potential effect on the target organisms (Stouthamer 1993). It has also been employed to disseminate useful genetic information into the insect groups (Beard et al. 1993; Curtis and Adak 1974). Over the past few decades, this endosymbiont has been studied in-depth for their action on population biology of insects and evolution.

12.2 *Wolbachia* Endosymbiont

Wolbachia endosymbionts (Fig. 12.1), a rickettsia-like microorganism, was discovered in *Culex pipiens* mosquito tissues and also in other insects (Hertig and Wolbach 1924). This microorganism is known to cause cytoplasmic incompatibility, which are maternally inherited. They are also reported to be responsible for inducing sterility in some insect groups.

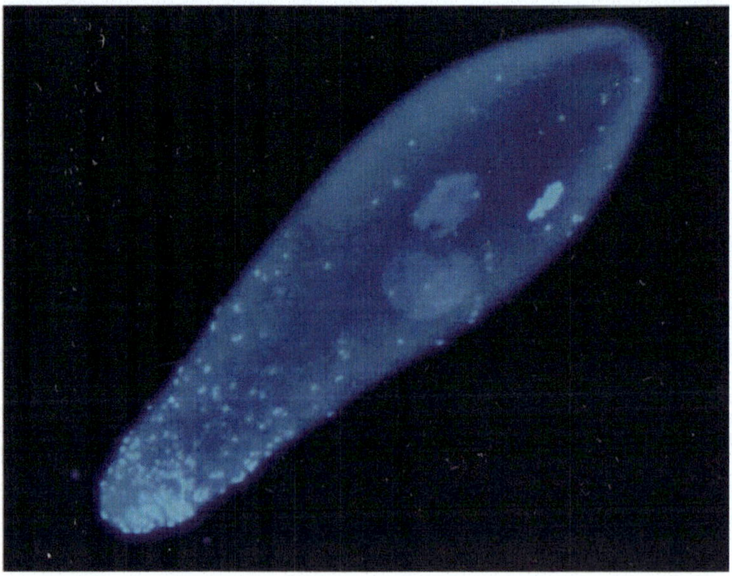

Fig. 12.1 *Wolbachia* observed as bright spots in the egg of an insect. (*Source*: Jonathan Knight 2001; permission being sought)

12.3 Taxonomical Status of *Wolbachia*

Surveys on the distribution and taxonomic status of different forms of endosymbiont are now possible due to advanced molecular assay methods (Hoffmann et al. 1994; Werren et al. 1995; Werren and Jaenike 1995). Molecular studies by using 16S rDNA and *ftsZ* facilitated in various surveys. The endosymbiont was identified initially in arthropods (supergroups A and B). The supergroups, C and D, were found to be associated with filarial nematodes (Bandi et al. 1998; Kozek 1977). There are presently 18 clades of endosymbionts that have been identified (supergroups A to R), mainly in arthropods (Lefoulon et al. 2016; Glowska et al. 2015; Wang et al. 2016). Among the various clades, C, D and J are found in Onchocercidae only, while arthropods and nematodes were found with the supergroup F (Fig. 12.2). At genomic level, the separation between various *Wolbachia* supergroups belonging to arthropods and filariae are strong. This shows the establishment of symbiotic relationship of *Wolbachia* with a host species.

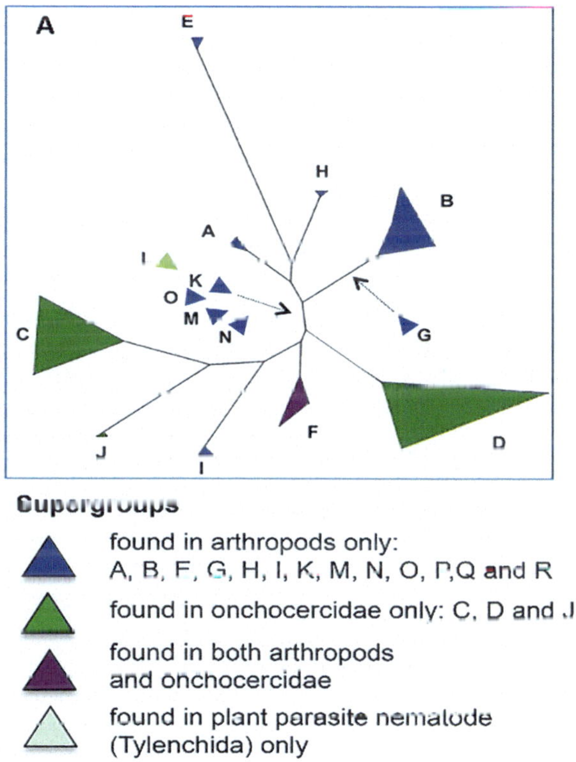

Fig. 12.2 Phylogenic tree (unrooted) of *Wolbachia* supergroups. (*Source*: Landmann 2019; permission being sought)

12.4 Role of *Wolbachia* in Altering Insect Population

Artificial association of *Wolbachia* in mosquitoes needs to be developed, in order to implement suppression or replacement strategies of insect population. It has been opined that presence of an insecticide resistance gene in mosquitoes may control the density of endosymbiont to a lesser extent (Berticat et al. 2002). This has been reported in *Culex pipiens* mosquito, where insecticide-resistant gene significantly modified the density of *Wolbachia*. The resistance-conferring genes present in those strains were found to be infected more compared to those of susceptible strains with similar genetic background.

The wide dissemination of *Wolbachia* in arthropods can be associated with the ability of the endobacteria to influence the reproduction of organisms, in which it resides. In most arthropods, *Wolbachia* is generally found as a reproductive parasite, which is responsible for causing feminization and cytoplasmic incompatibility (O'Neill et al. 1995; Werren 1997; Werren et al. 2008).

12.5 Feminization Effect of *Wolbachia*

Females of few insect groups, viz. Hemiptera, Isopoda and Lepidoptera, can result in genetic male individuals. *Wolbachia*, which induces feminization, has long been reported to be present in isopod crustaceans (Legrand et al. 1987), and woodlouse, *Armadillidium vulgare* (a terrestrial isopod), is the best studied (Rigaud et al. 1991). The endobacterium suppresses the androgen gland and converts males to females, which are reproductively competent. The intersex individuals are also formed. Feminizing characteristics of *Wolbachia* is interesting, as the presence of these endobacterium can cause potential genetic conflict with respect to sex determination. This can cause evolution in sex-determining system (Juchault et al. 1994; Rigaud and Juchault 1993), which appeared to have happened in *Armadillidium vulgare*. In this isopod, female heterogamete (ZZ for males and ZW for females) is the standard mechanism of sex determination. Due to the feminizing *Wolbachia* and other factors present in variable frequencies in the isopods, the male determining chromosomes (W) are lost. Thus the sex-determining mechanism gets disrupted. As a secondary effect, in some population, there is an enhanced storage of suppressing genes responsible for male determination (Rigaud and Juchault 1993). In the terrestrial isopod, *Armadillidium vulgare*, three *Wolbachia* strains (*wVulM*, *wVulC*, *wVulP*), specific for feminization and are tissue-specific, were studied in-depth by Dittmer et al. (2014). In this study it was observed that each strain of *Wolbachia* had specific distribution pattern in tissues and the strain; wVulM had lower titres in almost all the tissues. The pattern of distribution was similar in various host genetic system and could be a determining factor for their evolutionary relationship between *Wolbachia* and the terrestrial isopod being studied. Females infected by *Wolbachia* endobacterium had higher load of bacteria in several tissues, depending on the strain being infected.

12.6 Effect of Endosymbiont on Parthenogenesis

Parthenogenesis enables females to produce individuals of that sex, which was observed in few groups: Acari, Hymenoptera and Thysanoptera. It was observed that antibiotic treatment of these can revert this process, as observed in the strains of *Trichogramma* wasps where parthenogenetic females produced males (Stouthamer et al. 1990). The mechanism of parthenogenesis was later observed to be due to the presence of a bacteria, *Wolbachia*, which is cytoplasmically associated (Stouthamer et al. 1993; Stouthamer and Werren 1993). These bacteria which induces parthenogenesis are observed in both A and B clades of *Wolbachia*. There are evidences that this characteristic has evolved in this bacteria independently, over many generations. Parthenogenesis by *Wolbachia* has evolved by a simple biochemical mechanism. These *Wolbachia* strains which are responsible for inducing parthenogenesis are being studied extensively for their role in biological control of pests and vectors (Aeschilimann 1990; Stouthamer 1993; Zchori-Fein et al. 1995). In classical biological control programmes, parasitoid wasps are being used. Several advantages have been identified by Stouthamer et al. (1993): (a) There was higher rate of population increase for parthenogens and also the rate of stinging. (b) As the problem of finding mate is not a concern, these parthenogens are better colonizers and can establish sufficiently well at low population densities. (c) The mass rearing programmes with parthenogens are cost-effective, and hence the production is not wasted on males. But the effective number of females is less formed and has a large fertility cost, which is disadvantageous. Due to this, bacterial populations which have low fertility costs on host need to be investigated (Girin and Bouletreau 1995).

12.7 Male-Killing Characteristics of *Wolbachia*

In few arthropod groups, viz. Coleoptera, Lepidoptera, Diptera and Pseudoscorpiones, the endobacteria *Wolbachia* is able to kill the males in order to facilitate the survival of infected females. In the LIN strain of *Culex pipiens* mosquito, the females infected with the bacteria had no fitness cost, which was studied using the cohort life table analysis. The bacteria can cause a decline in male life span to almost 30%. *Culex pipiens LIN* strains which are infected with *Wolbachia* is active in a sex-specific manner. As enhanced mortality in older males is not a significant selective force, the bacterial strains that cause a decline in the survival of males can evolve, under laboratory conditions, where mosquitoes mate at young age (Rasgon 2012).

12.8 Cytoplasmic Incompatibility

In diploid species, due to cytoplasmic incompatibility, caused by the endobacteria, death of the zygote happens, as the reproduction will not be compatible between sperm and the egg (Yen and Barr 1971). Production of male occurs in haplodiploid species (Breeuwer and Werren 1990; Ryan and Saul 1968; Saul 1961). *Wolbachia* is reported to be transmitted through eggs, but not through sperms (Binnington and Hoffmann 1989; Bressac and Rousset 1992). There are two forms of cytoplasmic incompatibility, viz. unidirectional and bidirectional. In the former (unidirectional incompatibility), the egg of a female, uninfected by *Wolbachia*, is not fertilized by the sperm from an infected male. However, the reverse cross is compatible, i.e. sperm from an uninfected male fertilizing an egg from the infected female. The females which are infected with *Wolbachia* produce viable embryo with both the infected (with the same strain as in females) and uninfected males. In the bidirectional incompatibility, different strains of endobacteria (which are not mutually compatible) are harboured in male and female organisms (Mercot et al. 1995; O'Neill and Karr 1990; Perrot-Minnot et al. 1966). There are two components in incompatibility, viz. modification and rescue of the bacteria, in the sperm and the egg, respectively. The developing sperm gets modified by the *Wolbachia* bacteria, which is rescued in the egg during fertilization. The modification of the sperm could be through chromatin-binding proteins. During the rescue process by the egg, the *Wolbachia* strain should be similar. In the presence of different strain of bacteria, rescue does not occur but results in incompatibility between the sperm and the egg (Fig. 12.3). This model is equally applicable to both unidirectional and bidirectional incompatibility, where in the former, sperm of a *Wolbachia*-infected male is not

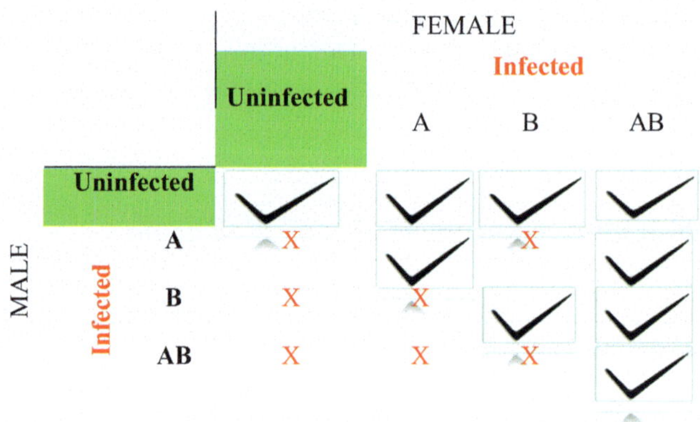

Fig. 12.3 Types of cytoplasmic incompatibility. The notations A and B are infections with *Wolbachia*, which occur in the cells either singly (A, B) or together (AB). Mating with similar notations (A X A) is part both unidirectional and bidirectional CI. Crosses indicate incompatibility, and tick marks represent compatibility. A complex pattern of incompatibility results when strains with different *Wolbachia* status are intercrossed. (*Source*: Hoffmann 2005, attached)

rescued in the uninfected female. While in the latter, different modification and rescue processes are employed by different *Wolbachia* strains. In *Culex pipiens* mosquito, collected from different geographical regions, the incompatibility relationship was explored and reported by various workers (Laven 1951, 1959, 1967b; Barr 1980; French 1978; Subbarao et al. 1974; Wright and Wang 1980; Zchori-Fein et al. 1995). The new function of *rescue* can be explained by various determinants of *Wolbachia* resulting in different types of cytoplasmic incompatibility. These types of incompatibility evolved gradually in *Cx. pipiens* mosquitoes. Here, in addition to the compatibility of self, there is co-existence of other rescue functions in one host. In order to evolve the modification function, co-infection of hosts with mutant *Wolbachia* strains should be present, which was opined by many workers. This makes the cytoplasmic incompatibility a dynamic process (Duron et al. 2012). Both types of incompatibilities (unidirectional and bidirectional) occur in a complex pattern. But the role of *Wolbachia* in all the cross incompatibilities in *Cx. pipiens* needs to be accessed thoroughly with caution (Curtis 1992; McClelland 1967). It was observed in few cases that the inheritance pattern is cytoplasmic (Subbarao et al. 1974; Yen and Barr 1971). The assessment of genetic effects on the bacterial strain distribution is very much essential. These incompatibilities (both unidirectional and bidirectional) were observed among the *Aedes* mosquito species (McClelland 1967; Trpis et al. 1981).

12.9 Distribution of *Wolbachia* in Different Mosquito Vectors

Wolbachia endosymbiont is distributed in a wide array of arthropods, including mosquito vectors of disease pathogens. Adult *Culex quinquefasciatus* mosquitoes ($n = 750$), the primary vector of lymphatic filarial parasites, were analysed for infection with *Wolbachia* in Tamil Nadu state, India (Sunish et al. 2011). The prevalence range of endobacterium in adult males and females were 88–96% and 84–100% respectively. There was 99% similarity in the *wsp* gene sequence of *Wolbachia* in *Cx. quinquefasciatus* with that of the endobacterium in *Culex pipiens* collected from various geographic areas. *Wolbachia* isolates were found to be associated with *Culex pipiens* and also in *Niphotettix virescens* (order: Hemiptera) and *Cnaphalocrocis medinalis* (order: Lepidoptera), based on molecular assay with *wsp* gene. In another study, out of 20 wild collected mosquito species, 8 were infected with the endobacterium, in the surveys carried out at Karnataka and Pondicherry, India. Of these, two species had singe infection with A group, while four species had infection with B group, and two species had both A and B group *Wolbachia* (double infection). All the *Wolbachia*-infected mosquitoes were also found to have *phage WO* (Ravikumar et al. 2010).

There is a wide spread infection of *Wolbachia* in mosquitoes, which is revealed by molecular assay for *Wolbachia* in *Ae. polynesiensis* (vector of subperiodic form of *W. bancrofti* in French Polynesia). All the samples analysed by sequencing by means of *wsp* and *ftsZ* genes demonstrated *A* type of *Wolbachia*. The investigators opined that in the study area, polymorphism of *Wolbachia* infection was absent. This property, as a strategy for the control of vectors, encouraged cytoplasmic incompatibility (Plichart and Legran 2005).

There was 51% ($n = 29$) infection with *Wolbachia* among wild-caught mosquitoes of Taiwan (Tsai et al. 2004). Among these, three, eight and four mosquito species were found to be infected with *Wolbachia*: A, B and AB (dual infection), respectively. These infections in different mosquito species were further divided into subgroups. But there were few isolates which did not belong to any subgroups already known. This shows that there are unidentified subgroups that still exist. Even though ovary tissues are more susceptible to *Wolbachia* infection, these bacteria were also found to be distributed in other tissues of the host organism. This was revealed while studying the tissue tropism in the mosquito species, viz. *Ae. albopictus* and *Ar. subalbatus*.

Wolbachia infection was 87.3% in the 260 field collected *Culex pipiens* mosquitoes in Iran (Karami et al. 2016). In the female and male adult mosquitoes, the bacterial infection ranged from 61.5% to 100% and 80% to 100%, respectively. The Iranian *Culex pipiens* was found with *B* supergroup of *Wolbachia* (molecular analysis of *wsp* gene fragment), which showed close similarity with the members of the same species reported earlier.

Of 330 mosquitoes screened, almost 87 mosquitoes (which is 26%), belonging to 4 species, were found positive for *Wolbachia* in Sri Lanka. The 330 mosquitoes analysed belonged to 22 species in 7 genera (Nugapola et al. 2017). The four species which were positive for bacteria include, viz. *Aedes albopictus*, *Culex quinquefasciatus*, *Armigeres subalbatus* and *Mansonia uniformis*. *Aedes aegypti*, the primary vector of dengue virus; the vectors of Japanese encephalitis virus, viz. *Cx. gelidus and Cx. tritaeniorhynchus*; and the anopheline mosquitoes, were found negative for the bacterium. High infection rates were reported in *Ae. albopictus* (which is the secondary vector of dengue virus) and *Culex quinquefasciatus* (vector of nocturnal periodic form of *Wuchereria bancrofti*). When analysed, *Ae. albopictus* was found with both the supergroups *A* and *B*. The sequences of nine *Wolbachia* products were deposited in the GenBank and was analysed by comparing other data available. By carrying out molecular assay with strain-specific primers, two major branches were observed and thus confirmed the identity.

In Lahore, Pakistan, the endobacteria was characterized by molecular methods (polymerase chain reaction) from 145 pairs of ovaries in *Culex quinquefasciatus*, using 3 target genes, viz. *wsp*, 16S *rRNA* and *ftsZ* (Sarwar et al. 2018). Of the 145 specimens, there was 82% infection rate with *Wolbachia* endobacteria. These endosymbionts showed cent percent similarity with the B supergroup *Wolbachia* and the *wPip* strains.

It was observed that mosquito vectors transmitting dengue virus and malaria-causing pathogens, which are infected with the endobacteria, have reduced vectorial capacity for transmission of these pathogens. Reverse was also observed in *Ae. fluviatilis* mosquito, where mosquitoes infected with *Wolbachia* (native strain of *wFlu*) showed an enhanced level of infection with *Plasmodium gallinaceum* (Caragata et al. 2017).

The density of *Wolbachia* in the ovaries and other tissues (which are not gonadal) was not related in the organisms belonging to same family. This suggests that different mechanisms are involved in determining the *Wolbachia* densities in each tissue. It was observed that cytoplasmic genotype determines the density of the endobacteria in ovaries, while both cytoplasmic and nuclear genotypes, along with the epistatic interactions, will determine the density of *Wolbachia* in non-gonadal tissues. In the tissues present in *Cx. quinquefasciatus*, there was 23% variance in the quantitative trait loci.

12.10 Controversy on the Evidence of *Wolbachia* in Anophelines and *Ae. aegypti*

The endobacteria *Wolbachia* has long been thought of not present in the natural population of anopheline mosquitoes (Hughes et al. 2014). However, in Africa, three malaria vectors belonging to *An. gambiae* complex, viz. *An. coluzzii*, *An. gambiae* and *An. arabiensis*, were found to harbour *Wolbachia* in the wild (Baldini et al. 2014, 2018; Gomes et al. 2017; Shaw et al. 2016). Similarly, in the Central African region of Gabon, 16 anopheline species (including 5 major malaria parasite transmitting anophelines) out of 25 were found with the endobacterium, in the natural population (Ayala et al. 2019). The presence of *Wolbachia* in several species of anopheline mosquitoes provides an opportunity to select particular phenotype which can suppress the transmission of malaria parasite (*Plasmodium*) and also the reproductive potential of anophelines. This will facilitate in reducing the burden of malaria in African region. However, Chrostek and Gerth (2019) later reported that, although *Wolbachia* were detected in *An. gambiae* complex at increasing frequencies, the unusual properties of these *Wolbachia* sequences render them insufficient to diagnose natural infections in *An. gambiae*. The protocol employed for the amplification of *wSpec* was sensitive to detect the endobacteria in filarial worm, which reside in the gut of *An. coustani* mosquito, and hence the method was suggested as a diagnostic test for determining the presence of *Wolbachia* and their interaction with the anopheline host (Ayala et al. 2019). In order to demonstrate the symbiotic relationship between the endobacteria and *An. gambiae*, characterization of *Wolbachia* sequences is required. The *Wolbachia* infection in anopheline is crucial in determining the vector potential of anophelines for transmitting malaria parasites.

Similar to the observation in anophelines, the dengue vector *Ae. aegypti* has also long been believed not to harbour *Wolbachia* in the natural population (Kittayapong et al. 2000; Ricci et al. 2002; Rasgon and Scott 2004; Hilgenboecker et al. 2008; Wiwatanaratanabutr 2013; Nugapola et al. 2017). A global survey, from a total of 27 countries, also reported the absence of endobacterium in *Ae. aegypti* (Gloria-Soria et al. 2018). However, several studies contradicted and demonstrated the presence of *Wolbachia* in this species of mosquito, for which the first observation was made from Malaysia in the immature samples. Since the size of the samples was less ($n = 16$), the findings were not considered (Teo et al. 2017). Low count of this endobacterium was reported from the USA and Thailand, in *Ae. aegypti* midgut, while screening for bacterial communities (Coon et al. 2016; Thongsripong et al. 2018). However, based on the molecular markers, viz. *wsp*, 16S rRNA and *ftsZ*, the *Wolbachia* was detected in the natural population of *Ae. aegypti* from India (Balaji et al. 2019). Further, endobacteria from this mosquito species were also found using the markers, viz. *16S rDNA*, *gatB* and *ftsZ*, from the specimens collected in the USA (Kulkarni et al. 2019). In these two reports, *Wolbachia* was also detected in different developmental stages of *Ae. aegypti*, based on cytological and molecular assay. In the Philippines, the endobacteria was demonstrated in 11.90% ($n = 80$) of the total mosquitoes of *Ae. aegypti* screened, by using two molecular markers (*wsp* and 16S rDNA), specific for *Wolbachia* (Carvajal et al. 2019). The supergroups A and B *Wolbachia* were found in these mosquitoes. Lately, Ross et al. (2020) opined that the molecular evidence of the presence of *Wolbachia* put forth by above workers is insufficient to confirm an active infection of *Wolbachia* in *Ae. aegypti* population, naturally (Chrostek and Gerth 2019), and suggested for a detailed in-depth studies. For confirming *Wolbachia* infection in this mosquito species, the researchers suggested minimum three requirements, which include transmission by maternal means, localization of the bacteria intracellularly and removal of the endobacteria.

12.11 *Wolbachia* to Control Mosquito Vectors

For the control of mosquito vectors, responsible for the transmission of pathogens, several strategies using *Wolbachia* have been proposed. For the mosquito to transmit any disease pathogens, the life span of the host (mosquito) should be at least the time duration of the pathogens to multiply and develop in the host and later reach the salivary gland of the mosquito, from where the pathogen is delivered to the vertebrates/humans. This period is termed as extrinsic incubation period (*Eip*), which is 10–14 days for dengue virus (Salazar et al. 2007). Thus, transmission of any pathogens is highly dependent on mosquito age. Female mosquitoes older than *Eip* are capable of transmitting the pathogens to the susceptible hosts to cause disease and are considered as epidemiologically significant. Therefore, the control strategies to reduce the burden of disease should consider methods to reduce the vector (mosquito) life span (Brownstein et al. 2003; Cook et al. 2008; Rasgon et al. 2003; Sinkins and O'Neill 2000). *Wolbachia*-based strategy involves the introduction of *wMelpop* strain into the wild mosquito population (*Ae. aegypti*), which will shorten the life span of the vector mosquito. This strategy will remove the older females of the vector, which are the ones responsible for transmitting disease-causing pathogens. Maintenance of the spread of *Wolbachia* endobacteria efficiently in nature is important in the control strategies.

By the release of males, which are not compatible, the females of the species get sterilized due to cytoplasmic incompatibility, and thus the mosquito population in the field declines. This strategy is termed as incompatible insect technique or IIT. For the control of *Culex quinquefasciatus* in the field, a candidate line, viz. LR[wPip (Is)], was constructed in southwestern Indian Ocean Islands (Atyame et al. 2015). In semi-field conditions, incompatible males were assessed to determine their mating competitiveness. Two experiments were carried out. The La Reunion wild females from 19 different localities (collected as larvae), when crossed with incompatible strain of males (LR[wPip(Is)]), resulted in cent percent mortality of the embryo. Thus the sterilizing capacity of LR[wPip(Is)] strain was confirmed. In the second set of experiment, mating competitiveness was assessed in the presence of wild females and males from La Reunion, in a semi-field condition. The LR[wPip(Is)] strain of males was allowed to cross with wild females in the presence of wild males. Crossing was carried out in different ratios of LR[wPip(Is)] males and wild males. The results showed that the LR[wPip(Is)] males were competent to mate with wild females, as compared to wild males.

The *wMel* strain of *Wolbachia* in the mosquito vectors was reported to be highly efficient in reducing the transmission of disease-causing pathogens, especially arboviruses, viz. dengue and Zika viruses. The genes for the synthesis of lipopoly saccharides are not likely to be present in the genome of *Wolbachia*. These endobacteria, in their membrane, incorporate cholesterol (Wu et al. 2004; Lin and Rikihisa 2003). In the presence of *Wolbachia*, this metabolite is esterified in the cells of the mosquito, and thus the cholesterol gets accumulated in the membrane. This cholesterol homeostasis prevents accessing it for the dengue virus (Geoghegan et al. 2017). In small-scale field trials, the use of *Wolbachia* endobacteria, which are insect-specific, tested successfully for their biological control potential, in inhibiting the transmission of disease pathogens. This strategy has been replicated in large-scale control programmes in Townsville, Australia. Presently, this programme has been widely used in ten other countries to control dengue transmission, as a part of "Eliminate Dengue Program". The usage of this strategy has been increasing under the "World Mosquito Program" (O'Neill 2018)

12.12 Distribution of *Wolbachia* in Parasites of Lymphatic Filariasis (LF)

The *Wolbachia* endobacteria have been detected in most of the filarial nematodes, in the order Spirurida, belonging to Onchocercidae family. These nematodes include *Litomosoides sigmodontis*, *Wuchereria bancrofti*, *Brugia malayi*, *Dirofilaria immitis* and *Onchocerca volvulus* (Bandi et al. 1999; Taylor et al. 1999; Casiraghi et al. 2000). However, members of the superfamily Dracunculoidea (*Dracunculus medinensis*) in the order Spirurida were found to be free of *Wolbachia* (Foster et al. 2014). Similarly, no evidence for *Wolbachia* symbiosis was observed in *Loa*

loa (McGarry et al. 2003). Nematodes and *Wolbachia* have a symbiotic relationship between them, and the bacteria facilitate in the development, fertility and viability of the worm. The endobacteria are reported to be transmitted from maternal source to microfilaria. Almost cent percent occurrence of endobacteria has been found in different developmental stages of the nematode (Bandi et al. 2001). Microfilaria collected from different geographical regions of South India was found to harbour the obligate *Wolbachia* endosymbiont (Hoti et al. 2003). Similarly, in West Bengal, Gayen et al. (2010) found the *Wolbachia* in the microfilariae of *W. bancrofti*.

12.13 Elimination of LF Through Depletion of *Wolbachia* in Adult Filarial Worm

In order to interrupt the transmission of lymphatic filarial (LF) parasite and to reduce morbidity to LF infection, the World Health Organization initiated the Global Programme to Eliminate Lymphatic Filariasis (GPELF) in 2000. Mass administration of diethylcarbamazine (DEC) citrate and albendazole (ALB), annually, is the strategy followed. Due to this strategy, prevalence of disease declined drastically and showed a significant impact (Ramaiah and Ottesen 2014). The WHO has acknowledged almost nine countries for achieving the elimination target. Almost 44.3% of the LF burden is contributed by India, on a global scale. In India, all 256 districts (belonging to 21 states), which are endemic for LF, are under the national programme for the elimination of LF, since its inception in 2004 (WHO 2012). This includes 610 million population at risk for LF infection. In India, two parasites of LF are prevalent, viz. *Wuchereria bancrofti* and *Brugia malayi*. The latter parasite mainly occurs in six states.

Wolbachia present in the nematode worm facilitate the worm to escape from the immune mechanism of the host and from the degranulating eosinophils, by attracting neutrophils. The endobacteria induce a response through *Toll-like receptors 2 and 4* and help the worm to evade the immune system (Brattig et al. 2004; Darby et al. 2012). *Wolbachia* endobacteria which are mutualistic are employed as target for various drugs to combat filariasis, in which *ivermectin* sterilizes the adult worm and can be used for antihelmintic treatment (Slatko et al. 2014). It has been reported that the antibiotic *doxycycline* depletes *Wolbachia* and also prevents the moulting of larvae, in addition to their role in killing the adult worm (Foster et al. 2013; Taylor et al. 2005; Slatko et al. 2010; Turner et al. 2017). Thus, the host immune response was found to be correlated with the loss of *Wolbachia*, resulting in the reduced life span of the worm (Hansen et al. 2011). More than 90% of *Wolbachia* endobacteria in the worm tissues was observed to be eliminated by treatment with doxycycline for a period of 4–6 weeks (Mand et al. 2008). After 18–24 months of treatment with doxycycline, embryogenesis was prevented and also caused adult worm sterility. However, contraindication due to doxycycline treatment for children and pregnant women prevents the use of this drug for public health. Another antibiotic, rifampicin drug, at higher doses administered for 7 days resulted in more than 90% elimination of *Wuchereria bancrofti*, and *O. ochengi* required 14 days of treatment (Aljayyoussi et al. 2017). Researchers opined that short-course treatment of 1–2 weeks period with *rifampicin* can be used for targeting LF parasite, which was also found to be safe to humans. This was concluded after pharmacodynamics modelling study by mouse-human bridging analysis.

12.14 Conclusion

The age structure of natural population of mosquito vectors can be modified by *Wolbachia* endobacteria and thus cause population replacement of mosquitoes. This could reduce the transmission of disease pathogens. The natural population of *Cx. pipiens* was reduced successfully by the release of males, infected with *Wolbachia*. The endobacteria, *Wolbachia*, can also be employed to drive the required genes into the mosquito vector population and thus result in population replacement of insects targeted. In 1967, *Cx. pipiens fatigans* was successfully controlled from Okpo, Myanmar, through the strategy of bidirectional cytoplasmic incompatibility, in which the males infected with *Wolbachia* produced sterile offspring when mated with females with different strain of *Wolbachia* (Laven 1967a). In recent trials at Lexington, Kentucky, and California and New York, USA, population of *Ae. albopictus*, which is the vector of dengue, Zika, and West Nile viruses, was suppressed by this strategy. Depleting *Wolbachia* in adult filarial worms could be an effective strategy, to achieve the goal of LF elimination much faster.

References

Aeschilimann JP (1990) Simultaneous occurrence of thelytoky and bisexuality in Hymenopteran species, and its implications for the biological control of pests. Entomophaga 35:3–5

Aljayyoussi G, Tyrer HE, Ford L et al (2017) Short-course, high-dose rifampicin achieves *Wolbachia* depletion predictive of curative outcomes in preclinical models of lymphatic filariasis and Onchocerciasis. Sci Rep 7(1):210

Atyame CM, Cattel J, Lebon C et al (2015) *Wolbachia*-based population control strategy targeting *Culex quinquefasciatus* mosquitoes proves efficient under semi-field conditions. PLoS One 10 (3):e0119288

Ayala D, Akone-Ella O, Rahola N, Kengne P et al (2019) Natural *Wolbachia* infections are common in the major malaria vectors in Central Africa. Evol Appl 12:1583–1594

Balaji S, Jayachandran S, Prabagaran SR (2019) Evidence for the natural occurrence of *Wolbachia* in *Aedes aegypti* mosquitoes. FEMS Microbiol Lett 366:fnz055

Baldini F, Segata N, Pompon J et al (2014) Evidence of natural *Wolbachia* infections in field populations of *Anopheles gambiae*. Nat Commun 5:3985. https://doi.org/10.1038/ncomms4985

Baldini F, Rouge J, Kreppel K et al (2018) First report of natural *Wolbachia* infection in the malaria mosquito *Anopheles arabiensis* in Tanzania. Parasit Vectors 11(1):635. https://doi.org/10.1186/s13071-018-3249-y

Bandi C, Anderson TJC, Genchi C, Blaxter ML (1998) Phylogeny of *Wolbachia* in filarial nematodes. Proc Biol Sci 265:2407–2413. https://doi.org/10.1098/rspb.1998.0591

Bandi C, McCall JW, Genchi C, Corona S, Venco L, Sachhi L (1999) Effects of tetracycline on the filarial worms *Brugia pahangi* and *Dirofilaria immitis* and their bacterial endosymbionts *Wolbachia*. Int J Parasitol 29:357–364

Bandi C, Trees AJ, Brattig NW (2001) Wolbachia in filarial nematodes: evolutionary aspects and implications for the pathogenesis and treatment of filarial diseases. Vet Parasitol 98 (1–3):215–238

Barr AR (1980) Cytoplasmic incompatibility in natural populations of a mosquito, *Culex pipiens* L. Nature 283:71–72

Beard CB, O'Neill SL, Tesh RB, Richards FF, Aksoy S (1993) Modification of arthropod vector competence via symbiotic bacteria. Parasitol Today 9:179–183

Berticat C, Rousset F, Raymond M, Berthomieu A, Weill M (2002) High Wolbachia density in insecticide-resistant mosquitoes. Proc Biol Sci 269(1498):1413–1416

Binnington KL, Hoffmann AA (1989) *Wolbachia*-like organisms and cytoplasmic incompatibility in *Drosophila simulans*. J Invertebr Pathol 54:344–352

Brattig NW, Bazzocchi C, Kirschning CJ, Reiling N et al (2004) The major surface protein of *Wolbachia* endosymbionts in filarial nematodes elicits immune responses through TLR2 and TLR4. J Immunol 173:437–445. https://doi.org/10.4049/jimmunol.173.1.437.124

Breeuwer JAJ, Werren JH (1990) Microorganisms associated with chromosome destruction and reproductive isolation between two insect species. Nature 346:558–560

Bressac C, Rousset F (1992) The reproductive incompatibility system in *D. simulans*: DAPI-staining analysis of the *Wolbachia* symbionts in sperm cysts. J Invertebr Pathol 61:226–230

Brownstein JS, Hett E, O'Neill SL (2003) The potential of virulent *Wolbachia* to modulate disease transmission by insects. J Invertebr Pathol 84:24–29

Caragata EP, Pais FS, Baton LA, Silva JB, Sorgine MH, Moreira LA (2017) The transcriptome of the mosquito *Aedes fluviatilis* (Diptera: Culicidae), and transcriptional changes associated with its native *Wolbachia* infection. BMC Genomics 18(1):6

Carvajal TM, Hashimoto K, Harnandika RK, Amalin DM, Watanabe K (2019) Detection of *Wolbachia* in field-collected *Aedes aegypti* mosquitoes in metropolitan Manila, Philippines. Parasite Vectors 12:361

Casiraghi M, Anderson TJC, Bandi C, Bazzocchi C, Genchi C (2000) A phylogenetic analysis of filarial nematodes: comparison with the phylogeny of *Wolbachia* endosymbionts. Parasitology 122:93–103

Chrostek E, Gerth M (2019) Is *Anopheles gambiae* a natural host of *Wolbachia*? mBio 10:e00784-19

Cook PE, McMeniman CJ, O'Neill SL (2008) Modifying insect population age structure to control vector-borne disease. Adv Exp Med Biol 627:126–140

Coon KL, Brown MR, Strand MR (2016) Mosquitoes host communities of bacteria that are essential for development but vary greatly between local habitats. Mol Ecol 25:5806–5826

Curtis CF (1992) Selfish genes in mosquitoes. Nature 355:511–515

Curtis CF, Adak T (1974) Population replacement in *Culex fatigans* by means of cytoplasmic incompatibility. Bull World Health Org 51:249–255

Darby AC, Armstrong SD, Bah GS et al (2012) Analysis of gene expression from the *Wolbachia* genome of a filarial nematode supports both metabolic and defensive roles within the symbiosis. Genome Res 22:2467–2477. https://doi.org/10.1101/gr.138420.112

Dittmer J, Beltran-Bech S, Lesobre J, Raimond M, Johnson M, Bouchon D (2014) Host tissues as microhabitats for *Wolbachia* and quantitative insights into the bacterial community in terrestrial isopods. Mol Ecol 23(10):2619–2635

Duron O, Bernard J, Atyame CM, Dumas E, Weill M (2012) Rapid evolution of *Wolbachia* incompatibility types. Proc Biol Sci 279(1746):4473–4480

Foster JM, Hoerauf A, Slatko BE, Taylor MJ (2013) The *Wolbachia* bacterial endosymbionts of filarial nematodes. In: Kennedy MW, Harnett W (eds) Parasitic nematodes: molecular biology, biochemistry and immunology, 2nd edn. CABI, Wallingford, pp 308–336

Foster JM, Landmann F, Ford L, Johnston KL, Elsasser SC, Schulte-Hostedde AI et al (2014) Absence of *Wolbachia* endobacteria in the human parasitic nematode *Dracunculus medinensis* and two related *Dracunculus* species infecting wildlife. Parasit Vectors 7:140

French WL (1978) Genetic and phenogenetic studies on the dynamic nature of the cytoplasmic inheritance system in *Culex pipiens*. Genetics 88:447–455

Gayen P, Maitra S, Datta S, Sinha Babu SP (2010) Evidence for *Wolbachia* symbiosis in microfilariae of *Wuchereria bancrofti* from West Bengal, India. J Biosci 35(1):73–77

Geoghegan V, Stainton K, Rainey SM et al (2017) Perturbed cholesterol and vesicular trafficking associated with dengue blocking in *Wolbachia* infected *Aedes aegypti* cells. Nat Commun 8:526. https://doi.org/10.1038/s41467-017-00610-8

Girin C, Bouletreau M (1995) Microorganism-associated variation in host infestation efficiency in a parasitoid wasp, *Trichogramma boutatachae* (Hymenoptera, Trichogrammatidae). Experientia 51:398–401

Gloria-Soria A, Chiodo TG, Powell JR (2018) Lack of evidence for natural *Wolbachia* infections in *Aedes aegypti* (Diptera: Culicidae). J Med Entomol 55:1354–1356

Glover DM, Raff J, Karr TL, O'Neil SL, Lin H, Wolfner MF (1990) Parasites in *Drosophila* embryos. Nature 348:117

Glowska E, Dragun-Damian A, Dabert M, Gerth M (2015) New *Wolbachia* supergroups detected in quill mites (Acari: Syringophilidae). Infect Genet Evol 30:140–146. https://doi.org/10.1016/j.meegid.2014.12.019.10

Gomes FM, Hixson BL, Tyner MDW et al (2017) Effect of naturally occurring *Wolbachia* in *Anopheles gambiae* s.l. mosquitoes from Mali on *Plasmodium falciparum* malaria transmission. PNAS 114(47):12566–12571. https://doi.org/10.1073/pnas.17161.81114

Hansen RDE, Trees AJ, Bah GS et al (2011) A worm's best friend: recruitment of neutrophils by *Wolbachia* confounds eosinophil degranulation against the filarial nematode *Onchocerca ochengi*. Proc Biol Sci 278:2293–2302. https://doi.org/10.1098/rspb.2010.2367

Hertig M, Wolbach SB (1924) Studies on Rickettsia-like micro-organisms in insects. J Med Res 44:329–374

Hilgenboecker K, Hammerstein P, Schlattmann P, Telschow A, Werren JH (2008) How many species are infected with *Wolbachia*?—a statistical analysis of current data. FEMS Microbiol Lett 281:215–220

Hoffmann AA (2005) Incompatible mosquitoes. Nature 436(1):189

Hoffmann AA, Clancy DJ, Merton E (1994) Cytoplasmic incompatibility in Australian populations of *Drosophila melanogaster*. Genetics 136:993–999

Hoti SL, Sridhar A, Das PK (2003) Presence of *Wolbachia* endosymbionts in microfilariae of *Wuchereria bancrofti* (Spirurida: Onchocercidae) from different geographical regions in India. Mem Inst Oswaldo Cruz 98:1017–1019

Hughes GL, Dodson BL, Johnson RM et al (2014) Native microbiome impedes vertical transmission of *Wolbachia* in Anopheles mosquitoes. Proc Natl Acad Sci 111(34):12498–12503

Juchault P, Frelon M, Bouchon D, Rigaud T (1994) New evidence for feminizing bacteria in terrestrial isopods: evolutionary implications. C R Acad Sci Paris Life Sci 317:225–230

Karami M, Moosa-Kazemi SH, Oshaghi MA et al (2016) *Wolbachia* endobacteria in natural populations of *Culex pipiens* of Iran and its phylogenetic congruence. J Arthropod Borne Dis 10(3):347–363

Kittayapong P, Baisley KJ, Baimai V, O'Neill SL (2000) Distribution and diversity of *Wolbachia* infections in southeast Asian mosquitoes (Diptera: Culicidae). J Med Entomol 37:340–345

Knight J (2001) Meet the Herod bug. Nature 412:12–14

Kozek WJ (1977) Transovarially-transmitted intracellular microorganisms in adult and larval stages of *Brugia malayi*. J Parasitol 63:992–1000. https://doi.org/10.2307/3279832

Kulkarni A, Yu W, Jiang J et al (2019) *Wolbachia pipientis* occurs in *Aedes aegypti* populations in New Mexico and Florida, USA. Ecol Evol 9:6148–6156

Landmann F (2019) The Wolbachia endosymbionts. Microbiol Spectrum 7(2):BAI-0018-2019. https://doi.org/10.1128/microbiolspec.BAI-0018-2019

Lassy CW, Karr TL (1996) Cytological analysis of fertilization and early embryonic development in incompatible crosses of *Drosophila simulans*. Mech Dev 57:47–58

Laven H (1951) Crossing experiments with *Culex* strains. Evolution 5:370–375

Laven H (1959) Speciation by cytoplasmic isolation in the *Culex pipiens* complex. Cold Spring Harb Symp Quant Biol 24:166–173

Laven H (1967a) Eradication of *Culex pipiens fatigans* through cytoplasmic incompatibility. Nature 216:383–384

Laven H (1967b) Speciation and evolution in *Culex pipiens*. In: Wright JW, Pal R (eds) Genetics of insect vectors of disease. Elsevier, Holland, pp 251–275

Lefoulon E, Bain O, Makepeace BL et al (2016) Breakdown of coevolution between symbiotic bacteria *Wolbachia* and their filarial hosts. Peer J 4:e1840. https://doi.org/10.7717/peerj.1840.9

Legrand JJ, Legrand-Hamelin E, Juchault P (1987) Sex determination in Crustacea. Biol Rev 62:439–470

Lin M, Rikihisa Y (2003) *Ehrlichia chaffeensis* and *Anaplasma phagocytophilum* lack genes for lipid A biosynthesis and incorporate cholesterol for their survival. Infect Immun 71:5324–5331. https://doi.org/10.1128/IAI.71.9.5324-5331.2003

Mand S, Büttner DW, Hoerauf A (2008) Images in clinical tropical medicine Bancroftian Filariasis—absence of *Wolbachia* after doxycycline treatment. Am J Trop Med Hyg 78(6):854–855

McClelland GAH (1967) Speciation and evolution in *Aedes*. In: Wright JW, Pai R (eds) Genetics of insect vectors of diseases. Elsevier, Amsterdam

McGarry HF, Pfarr K, Egerton G et al (2003) Evidence against *Wolbachia* symbiosis in *Loa loa*. Filaria J 2:1–7

Mercot H, Llorente B, Jacques M, Atlan A, Montchamp-Moreau C (1995) Variability within the Seychelles cytoplasmic incompatibility system in *Drosophila simulans*. Genetics 141:1015–1023

Nugapola NWNP, De Silva WAPP, Karunaratne SHPP (2017) Distribution and phylogeny of *Wolbachia* strains in wild mosquito populations in Sri Lanka. Parasit Vectors 10:230

O'Neill SL (2018) The use of *Wolbachia* by the world mosquito program to interrupt transmission of *Aedes aegypti* transmitted viruses. Adv Exp Med Biol 1062:355–360

O'Neill SL, Karr TL (1990) Bidirectional incompatibility between conspecific populations of *Drosophila simulans*. Nature 348:178–180

O'Neill SL, Pettigrew MM, Andreadis TG, Tesh RB (1995) An in-vitro system for culturing *Wolbachia* symbionts of arthropods. J Cell Biochem 21A:210

Perrot-Minnot MJ, Guo L, Werren JH (1966) Double and single *Wolbachia* infections in *Nasonia vitripennis*. Genetics 143:961–972

Plichart C, Legran A-M (2005) Detection and characterization of *Wolbachia* infections in *Wuchereria bancrofti* (Spirurida: Onchocercidae) var. Pacifica and *Aedes (Stegomyia) polynesiensis* (Diptera: Culicidae). Am J Trop Med Hyg 73(2):354–358

Ramaiah K, Ottesen EA (2014) Progress and impact of 13 years of the global programme to eliminate lymphatic filariasis on reducing the burden of filarial disease. PLoS Negl Trop Dis 8: e3319

Rasgon JL (2012) *Wolbachia* induces male-specific mortality in the mosquito *Culex pipiens* (LIN strain). PLoS One 7(3):e30381

Rasgon JL, Scott TW (2004) Impact of population age structure on *Wolbachia* transgene driver efficacy: ecologically complex factors and release of genetically modified mosquitoes. Insect Biochem Mol Biol 34:707–713

Rasgon JL, Styer LM, Scott TW (2003) *Wolbachia*-induced mortality as a mechanism to modulate pathogen transmission by vector arthropods. J Med Entomol 40:125–132

Ravikumar H, Ramachandraswamy N, Sampathkumar S et al (2010) A preliminary survey for *Wolbachia* and bacteriophage WO infections in Indian mosquitoes (Diptera: Culicidae). Trop Biomed 27(3):384–393

Reed KM, Werren JH (1995) Induction of paternal genome loss by the paternal sex-ratio chromosome and cytoplasmic incompatibility bacteria (*Wolbachia*): a comparative study of early embryonic events. Mol Reprod Dev 40:408–418

Ricci I, Cancrini G, Gabrielli S, D'amelio S, Favia G (2002) Searching for *Wolbachia* (Rickettsiales: Rickettsiaceae) in mosquitoes (Diptera: Culicidae): large polymerase chain reaction survey and new identifications. J Med Entomol 39:562–567

Rigaud T, Juchault P (1993) Conflict between feminizing sex ratio distorters and an autosomal masculinizing gene in the terrestrial isopod *Armadillidium vulgare* Latr. Genetics 133:247–252

Rigaud T, Souty-Grosset C, Raimond R, Mocquard JP, Juchault P (1991) Feminizing endocytobiosis in the terrestrial crustacean *Armadillidium vulgare* Latr. (Isopoda): recent acquisitions. Endocytobiosis Cell Res 7:259–273

Ross PA, Callahan AG, Yang Q et al (2020) An elusive endosymbiont: does *Wolbachia* occur naturally in *Aedes aegypti*? Ecol Evol 10:1581–1591

Ryan SL, Saul GB (1968) Post fertilization effect of incompatibility factors in *Mormoniella*. Mol Gen Genet 103:24–36

Salazar M, Richardson JH, Sanchez-Vargas I, Olson KE, Beaty BJ (2007) Dengue virus type 2: replication and tropisms in orally infected *Aedes aegypti* mosquitoes. BMC Microbiol 7:9

Sarwar MS, Jahan N, Shahbaz F (2018) Molecular detection and characterization of *Wolbachia* pipientis from *Culex quinquefasciatus* collected from Lahore, Pakistan. Am J Trop Med Hyg 98 (1):154–161

Saul GB (1961) An analysis of non-reciprocal cross incompatibility in *Mormoniella vitripennis* (Walker). Z Vererbungs 92:28–33

Shaw WR, Marcenac P, Childs LM et al (2016) *Wolbachia* infections in natural *Anopheles* populations affect egg laying and negatively correlate with Plasmodium development. Nat Commun 7:11772

Sinkins SP, O'Neill SL (2000) *Wolbachia* as a vehicle to modify insect populations. CRC press, Boca Raton, FL

Slatko BE, Taylor MJ, Foster JM (2010) The *Wolbachia* endosymbiont as an anti-filarial nematode target. Symbiosis 51:55–65

Slatko BE, Luck AN, Dobson SL, Foster JM (2014) *Wolbachia* endosymbionts and human disease control. Mol Biochem Parasitol 195:88–95

Stouthamer R (1993) The use of sexual versus asexual wasps in biological control. Entomophaga 38:3–6

Stouthamer R, Kazmer DJ (1994) Cytogenetics of microbe associated parthenogenesis and its consequences for gene flow in *Trichogramma* wasps. Heredity 73:317–327

Stouthamer R, Werren JH (1993) Microbes associated with parthenogenesis in wasps of the genus *Trichogramma*. J Invertebr Pathol 61:6–9

Stouthamer R, Luck RF, Hamilton WD (1990) Antibiotics cause parthenogenetic *Trichogramma* (Hymenoptera/Trichogrammatidae) to revert to sex. Proc Natl Acad Sci U S A 87:2424–2427

Stouthamer R, Breeuwer JAJ, Luck RF, Werren JH (1993) Molecular identification of microorganisms associated with parthenogenesis. Nature 361:66–68

Subbarao SK, Krishnamurthy BS, Curtis CF, Adak T, Chandrahas RK (1974) Segregation of cytoplasmic incompatibility properties in *Culex pipiens fatigans*. Genetics 87:381–390

Sunish IP, Rajendran R, Paramasivan R, Dhananjeyan KJ, Tyagi BK (2011) *Wolbachia* endobacteria in a natural population of *Culex quinquefasciatus* from filariasis endemic villages of South India and its phylogenetic implication. Trop Biomed 28(3):569–576

Taylor MJ, Bilo K, Cross HF, Archer JP, Underwood AP (1999) 16S rDNA phylogeny and ultrastructural characterization of *Wolbachia* intracellular bacteria of the filarial nematodes *Brugia malayi, B. pahangi* and *Wuchereria bancrofti*. Exp Parasitol 91:356–361

Taylor MJ, Makunde WH, McGarry HF, Turner JD, Mand S, Hoerauf A (2005) Macrofilaricidal activity after doxycycline treatment of *Wuchereria bancrofti*: a double-blind, randomised placebo-controlled trial. Lancet 365:2116–2121

Teo CH, Lim PK, Voon K, Mak JW (2017) Detection of dengue viruses and *Wolbachia* in *Aedes aegypti* and *Aedes albopictus* larvae from four urban localities in Kuala Lumpur, Malaysia. Trop Biomed 34:583–597

Thongsripong P, Chandler JA, Green AB et al (2018) Mosquito vector-associated microbiota: Metabarcoding bacteria and eukaryotic symbionts across habitat types in Thailand endemic for dengue and other arthropod-borne diseases. Ecol Evol 8:1352–1368

Trpis M, Perrone JB, Reissig M, Parker KL (1981) Control of cytoplasmic incompatibility in the *Aedes scutellaris* complex. J Hered 72:313–317

Tsai KH, Lien JC, Huang CG, Wu WJ, Chen WJ (2004) Molecular (sub) grouping of endosymbiont *Wolbachia* infection among mosquitoes of Taiwan. J Med Entomol 41(4):677–683

Turner JD, Sharma R, Al Jayoussi G et al (2017) Albendazole and antibiotics synergize to deliver short-course anti-*Wolbachia* curative treatments in preclinical models of filariasis. Proc Natl Acad Sci U S A 114:E9712–E9721. https://doi.org/10.1073/pnas.1710845114

Wang GH, Jia LY, Xiao JH, Huang DW (2016) Discovery of a new *Wolbachia* supergroup in cave spider species and the lateral transfer of phage WO among distant hosts. Infect Genet Evol 41:1–7. https://doi.org/10.1016/j.meegid.2016.03.015

Werren JH (1997) Biology of *Wolbachia*. Annu Rev Entomol 42:587–609

Werren JH, Jaenike J (1995) *Wolbachia* and cytoplasmic incompatibility in mycophagous *Drosophila* and their relatives. Heredity 75:320–326

Werren JH, Guo L, Windsor DW (1995) Distribution of *Wolbachia* in neotropical arthropods. Proc R Soc Lond B 262:147–204

Werren JH, Baldo L, Clark ME (2008) Wolbachia: master manipulators of invertebrate biology. Nat Rev Microbiol 6:741–775

Wiwatanaratanabutr I (2013) Geographic distribution of *Wolbachia* infections in mosquitoes from Thailand. J Invertebr Pathol 114:337–340

World Health Organization (2012) Global programme to eliminate lymphatic Filariasis: progress report on mass drug administration, 2011. Wkly Epidemiol Rec 37(87):345–356

Wright JD, Wang BT (1980) Observations on *Wolbachiae* in mosquitoes. J Invertebr Pathol 35:200–208

Wu M, Sun LV, Vamathevan J, Riegler M et al (2004) Phylogenomics of the reproductive parasite *Wolbachia pipientis wMel*: a streamlined genome over run by mobile genetic elements. PLoS Biol 2:e69. https://doi.org/10.1371/journal.pbio.0020069

Yen JH, Barr AR (1971) New hypothesis of the cause of cytoplasmic incompatibility in *Culex pipiens*. Nature 232:657–658

Zchori-Fein E, Faktor O, Zeidan M, Gottlieb Y, Czosnek H, Rosen D (1995) Parthenogenesis-inducing microorganisms in *Aphytis* (Hymenoptera: Aphelinidae). Insect Mol Biol 4:173–178

Chapter 13
Laboratory Biosafety in Handling Genetically Modified Mosquitoes

Jhansi Charles

Abstract One of the novel approaches in controlling vector-borne diseases is to release genetically modified mosquitoes in nature. Trial studies are done in different phases by the researches, both in the laboratory and in the fields. Before a GM mosquito is validated to be ready for field release, the same has to rigorously go through several phase studies, and Phase I being the laboratory is the most significant to set the future of the GM mosquito for future investigations. Though the risk of handling GM mosquitoes in laboratory is low, nevertheless there is a prescribed list of DOs and DON'Ts, and the laboratory workers are needed to strictly follow the SOPs or basic principles of biosafety like handling administrative controls, using biosafety equipment, wearing personal protective equipment, etc. The laboratory also should have a proper design as per the risk assessment. Accordingly, the biosafety laboratories (BSL) are classified into four types: type 1, 2, 3 and 4. The risks are assessed as per the factors in the host, vector and donor sequences and the environmental factors and their activities in such environments. As GMMs are of low risk, BSL 1 and 2 are enough for their manipulation, but occasionally BSL 3 may be needed. There may be chances of spillage on the working surfaces during manipulation of the genes which can be remedied by the spill management protocols. Biological wastes may be generated in all areas of manipulation. These can be properly treated by either chemical disinfection or autoclaving and disposed of by incineration. These wastes should be segregated in colour-coded bags before disposal. There may be some risks while transporting GMMs to distant places. They should be packed securely in triple-layer pack and sent for disposal following IATA and other road rules. For each process of the manipulation of gene, a separate standard operating procedure (SOP) should be maintained which has to be updated whenever any change in the procedure is made.

Keywords Biosafety · Containment · Safety practices · Risk assessment

J. Charles (✉)
Department of Microbiology, Velammal Medical College Hospital and Research Institute, Madurai, Tamil Nadu, India

B. K. Tyagi (ed.), *Genetically Modified and other Innovative Vector Control Technologies*, https://doi.org/10.1007/978-981-16-2964-8_13

13.1 Introduction

When DNA is manipulated and transferred into an organism, it should be done in a confined laboratory with skilled and trained scientists. They should be oriented towards all positive and negative outputs and the risks involved. In GMMs, the genes are modified in such a way that the genes are either reproduced or naturally recombined. GMMs are produced by (a) recombinant technology using host/vector and (b) preparing a heritable material outside the mosquito and introducing it into the mosquito directly by micro injection, macro injection or micro encapsulation. These sterile genetically modified mosquitoes are released in the environment. Such novel approach to mosquito management is tried in many countries either as a field trial or open release of GM mosquitoes. Scientists have shown that the inserted genes do not create any products that are known to act as toxins or to cause allergic reactions in humans. Studies on animals feeding on such genetically modified mosquitoes show that there is no difference in the development, life span, size or survival rates of these animals. Scientists also have proved that there is no adverse effect in the ecosystem where these GMMs are released. As manipulation of genes is done in different steps in various procedures, containment laboratories are needed with proper biosafety levels.

13.2 Biosafety

Biosafety are the principles, procedures and policies which are to be followed by the people, scientists and workers to ensure their personal and surrounding safety. It refers to safety practices that are required to prevent unexpected contact with pathogens and toxins or their unintended release into the surrounding. Collective efforts in laboratories dealing with GM mosquitoes are the underlying principles to ensure biological safety for a clean and safe environment. Biosafety is a group activity to provide a clean and protective environment. The rules and regulations to be adopted in biosafety practices and the various bodies monitoring these activities should be known to people, researchers and workers at the grassroots level itself.

13.3 Regulatory Bodies for Biosafety

In 2003, it was felt that there should be an international agreement on the biosafety needs in genetic engineering research and development activities. Accordingly, "The Cartagena Protocol on Biosafety" (CPB) was developed by 167 parties including 165 UN countries, Niue, and the European Union. The main objectives are as follows:

1. Making the living modified organisms to move safely within boundaries by setting up appropriate procedures.
2. Regularising the principles and methodologies for risk assessment by setting up mechanisms for message sharing through Biosafety Clearing-House (BCH).

For any work related to research on GE and GMMs, prior approval from the appropriate regulatory authorities of the country is mandatory. To follow, all steps needed to reduce biosafety issues by all staff working in that area are very much essential. At the institutional level, the Institutional Biosafety Committee (IBSC) is the primary regulatory body which ensures the presence of all basic biosafety equipment functioning with optimal biosafety levels.

13.4 Containment

The important concept of biosafety is containment which focuses on the various activities adopted to save the laboratory and its environment from infectious agents. The aim of containment is to reduce or remove harmful agents from the workers, working surfaces and working environments. The four important components of containment are as follows:

1. Maintaining administrative controls
2. Following standard safe work practices
3. Using safety equipment including personal protective equipment
4. Proper facility designing

Containments may be either primary or secondary. *Primary containment* is the first safe method preventing direct contact with infectious material as well as protecting the workers and their immediate working surfaces from contact with the effect of harmful disease agents. Primary containment includes using proper collection and storage materials, doing proper processing and using safety equipment like safety cabinets. Secondary containment is the safety to the surrounding places external to the laboratory from contact with the harmful agents by giving proper facility designing and good laboratory practices (Benedict et al. 2003, 2018).

13.4.1 Maintaining Administrative Controls

These controls are not to remove the health hazards but to reduce the exposure to the hazardous agents. They control the various activities done to reduce the contact. The important activities to be managed are the time management of workers, written work policies to be followed by workers, written standard operating policies in various areas of biosafety like training protocols, management of housekeeping, maintenance of biosafety equipment and practices of individual hygiene.

13.4.2 Following Standard Safe Work Practices (Good Laboratory Practices)

Good laboratory practice (GLP) or standard safe work practices focus on minimising the common laboratory accidents produced by human errors, improper working practices and improper usage of equipment. It has a list of policies about the sterile techniques in the usage of laboratory equipment. This will prevent contamination in the procedures, workers and working environment. Adopting safe practices will prevent the formation and spread of aerosols also. Some of the good laboratory practices are as follows:

- Preventing pipetting through mouth.
- Avoiding eating, drinking and smoking inside the laboratory premises.
- Wearing hand and body protectives like gloves and gowns while performing steps in the experiments involving close contact with harmful materials.
- Proper disposal of biological wastes after decontamination with chemicals or treating them with mechanical devices.
- Disinfecting the work areas at least once daily and after each spill of harmful infectious materials.

13.4.3 Spills Management

This is common happening in the laboratory functions, and every effort should be made to tackle such an emergency, in the following manner:

- If any spill is found on the working surfaces, remove them with tissues or paper towels and discard them in colour-coded bag as per CDC guidelines.
- Disinfect the cleaned working surface with 70% ethanol or any other prescribed disinfectant.
- Wash the hands with soap and water and wipe dry or clean the hands with hand sanitisers.

13.4.4 Decontamination of Reused Items

There are always certain items in the biosafety laboratory which are reused, but they cannot be done so just that way since decontamination of all such products or material is essential, in the following manner.

- All reused items should be either sterilised by autoclaving or by high-level disinfection.

- All GMM biological wastes should be sterilised by disinfection/autoclaving or by both first and then disposed of by incineration.

13.5 Using Safety Equipment/Personal Protective Equipment

Everyone who does experiments on manipulation of mosquitoes should follow the following guidelines:

- All doors and windows should be closed tightly when manipulation is in progress.
- Laboratory workers and scientists should wear a full-sleeve laboratory coat.
- Storing food and drinks in the storage area of laboratory reagents should be avoided.
- Female workers should not wear hand ornaments while working. Nails should be as short as possible.
- Ensure centrifuge tubes are closed and caps are dry before keeping in the centrifuge so that spread of aerosol is minimised.
- Hand hygiene should be strictly adopted before and after each step of experiment.

Instead of pipetting through mouth, a mechanical pipette with a pipetting bulb is advisable.

13.5.1 Biosafety Cabinets

In mosquito manipulation, biosafety cabinets play a major role in safeguarding the workers and working environment from the harmful agents if released during the various laboratory processes. The BSC gets rid of aerolised harmful materials through HEPA filters (high-efficiency particulate air filters). The influx of air from the front of the unit is pulled into the front grate of the BSC, passed through a HEPA filter, get filtered and then enter the BSC working surface. Outflux of air also passes through a HEPA filter. Aerosols generated in the work area of the BSC are remained within the BSC which are removed while passing through the filters. Biological safety cabinets (BSCs) are categorised by classes into class I, class II and class III. The class II cabinets are classified into A1, A2, B1 and B2, depending on the amount of air vented out of the laboratory. Class I cabinets give personal protection only. Class II cabinets give both personal and product protection. Class II A2 BSCs are the most common one which recirculates 70% filtered air and exhausts 30% into the room if they are non-ducted (Fig. 13.1). If ducted, the exhaust air is expelled outside the room. In GMM handling, class I or class II cabinets are used. They give a clean work environment for research activities. Class III cabinets are glove-box units which give personal protection and optional product protection.

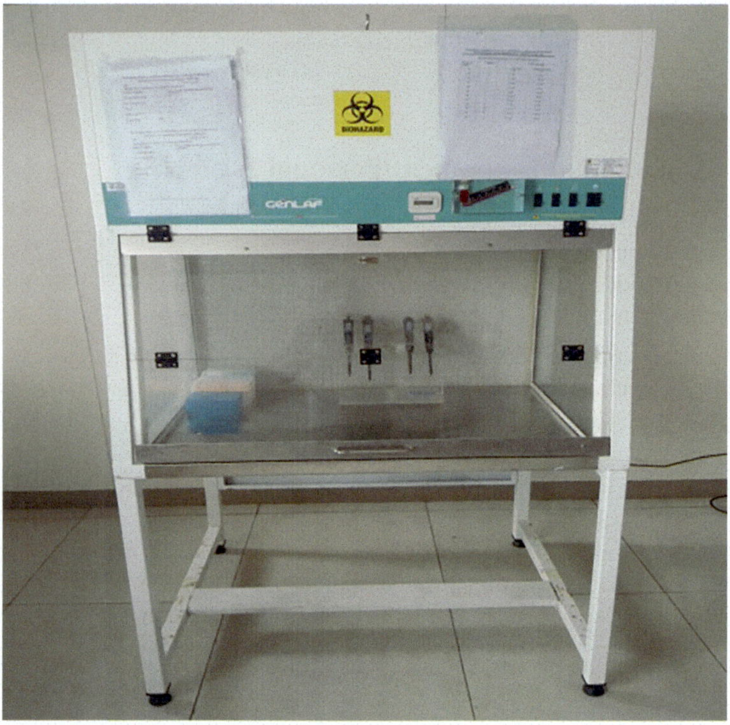

Fig. 13.1 Class II A2 cabinet. (*Source*: This work)

13.5.2 Innovations in the Existing Lab Equipment

The biosafety innovation has changed the design of some laboratory equipment like pipettes, needles, sharps, etc. to improve safety and to reduce risks.

(a) *Pipettes*: Using pipettes with bulbs or mechanical pipettes or pipettes with aerosol barrier pipette tips protects the workers and scientists from risks coming out of mouth pipetting.
(b) *Needles and other sharps*: Many labs use retracting needles, clearly labelled containers and puncture proof containers to minimise harmful effects of GMM if any.

13.5.3 Personal Protective Equipment

This is another very important component of biosafety equipment in the laboratory. Personal protective equipment includes gloves, masks, lab coats and other wearable equipment (such as safety glasses and respirators) that protect laboratory workers from infectious agents and toxins in the laboratory. For level 1 and 2 activities, gloves are not needed. It should be worn if a hazard or prick or cutting injuries are suspected.

13.6 Facility Design

The gene manipulation laboratory should be designed in such a way that the infectious agents or their toxins if liberated should not be released outside. The presence of double doors or two self-closing or interlocked doors, sealed windows and wall surfaces are prerequisites for such lab buildings. Air filtration system, waste management system and filtered ventilation systems are other prerequisites. To prevent access to level 1 and 2 laboratories, these laboratories should not be located near the entrance or exit of floors or buildings or in areas where many people frequent the area.

13.7 Risk Assessment in GMM Laboratories

Risk assessment in GMM lab is very essential because it is not known whether GMMs are either hazardous or safe. It is always advisable to do risk assessment for each case separately. Risk assessment is normally done in Phase I laboratory studies with a cautious approach. The risk assessment can be done as follows:

1. Identify the preliminary risk level by studying the characters of the host, vector and donor sequences.
2. Study the characters of the environment which are likely to be exposed to the GMMs with their type of activities.
3. Based on the above two results, a risk class is determined.

13.8 Risk Management in GMM Labs

When mosquito manipulation is done in GMM laboratories, utmost care should be given to the laboratory staff including workers and scientists. The management should be in such way that live mosquitoes are contained properly and dead mosquitoes and waste materials are disposed of properly. Lab should ensure that the mosquito colonies and feed sources are free from human pathogens. If the laboratory management is optimum, the lab staff will not carry any mosquito-transmissible diseases or will get any mosquito bite from the wandering mosquitoes outside the cages. One of the most important protocols in the lab management is the usage of personal protective equipment by all lab staff.

13.9 Biosafety Levels in GMM Labs

There are four levels of biosafety laboratories: BSL 1, 2, 3 and 4. For handling GMMs, BSL 1 and BSL 2 are commonly used. BSL 3 is occasionally used.

13.9.1 Biosafety Level 1 (BSL 1) Laboratory

- BSL 1 labs are used to study harmful agents not known to cause disease in healthy adults.
- They follow basic safety procedures, called standard safety work practices.
- They need no special equipment or design features.
- They have easily cleaned surfaces which can withstand the basic chemicals used in the laboratory.

13.9.2 Biosafety Level 2 (BSL 2) Laboratory

- BSL 2 laboratories are used to study moderate-risk infectious agents or toxins that pose a risk if accidentally inhaled, swallowed or exposed to the skin.
- Laboratory design includes handwashing sinks, eye washing stations in case of accidents and doors that close automatically and lock.
- BSL 2 labs must also have equipment that can decontaminate and dispose laboratory wastes, like an incinerator, an autoclave, etc.

13.9.3 Biosafety Level 3 (BSL 3) Laboratory

- BSL 3 laboratories are used to study infectious agents or toxins that may be transmitted through the air and cause potentially lethal infection through inhalation.
- Researchers perform all experiments in biosafety cabinets that use carefully controlled air flow or sealed enclosures to prevent infection.
- BSL 3 laboratories are designed in such a way that they can be decontaminated easily.

These laboratories use controlled, or "directional", air flow so that the air flows from non-laboratory areas into laboratory areas. This provides an additional safety measure.

13.10 Shipping of Genetically Modified Strains

GMMs are neither considered dangerous substances nor infectious materials. They can be transported as per the rules and guidelines formed by the competent authority of their own country. If the GMMOs are intended for transport, shipping can be done under the Label "UN 3245", and the shipping name for UN 3245 is "GENETICALLY MODIFIED MICROORGANISMS" or "GENETICALLY MODIFIED ORGANISMS".

13.10.1 Packaging

Genetically modified vectors are packaged in accordance with ADR PI 904 (Road) or IATA PI 913. Genetically modified vectors must be packed according to the procedures, i.e. triple-layer packaging system. There are no limits on the quantity transported by road, but by air, the following limits apply: on both passenger and cargo aircraft, no limit per package, but the maximum quantity in the primary receptacle must not exceed 100 ml or 100 g.

13.10.2 Transport

If transport of genetically manipulated mosquitoes is planned to a destination outside country, then mandatorily one must first get the details of packing, labelling and handling from the recipient country. International Air Transport Association (IATA) is the global organisation authorised for providing guidelines for transporting biological reagents. If the transport is through air or ship, all these guidelines should be strictly followed. If the transport is through roads, road regulations of that particular country have to be followed. To prevent complications, requirements from exporting countries like permits, documents, conditions of materials to be transported, etc. have to be prepared well ahead before shipping.

13.11 Waste Management

Gene manipulation laboratories release solid, liquid and general biological wastes which have to be treated and disposed of immediately in order to maintain containment of organism and to prevent cross-contamination among strains. The common method of treating biological wastes in such laboratories is by disinfection/autoclaving or by both. Therefore, it is appropriate to have an autoclave in the designated containment area.

Fig. 13.2 Various types of colour-coded dustbins. (*Source*: This work)

The following are the types of biological wastes generated in the gene manipulation lab.

1. *Solid*: Filter papers used for sieving eggs, towelling used for biological spills, dead mosquitoes of all life stages, leftover diet (sugar and larval diet), consumables or minor equipment such as air vent filters or sink filters.
2. *Liquid*: Culturing water (sieved to remove any life forms) and wash water from trays, small quantities of blood from feeding equipment and water from handwashing during work in insectaries.
3. *General*: Routine office wastes like paper and linen.

All solid and general wastes are first segregated in colour-coded bags (Fig. 13.2). The various dustbins with colour-coded bags are shown in (Fig. 13.2). The colours used for segregation are black, blue, red and white as per present CDC guidelines. Red-coloured bags are commonly prescribed for segregating biological infective wastes from GMM labs. After segregation, the infective wastes are sent for treatment. The mode of treatment for GMM wastes is either by autoclaving or by disinfection. Occasionally, the wastes are first disinfected and then autoclaved. While using disinfectants, see for their potency, see whether they are active against the particular species of organism you handle and ensure that you comply with all guidelines given by the manufacturer of the disinfectant. While autoclaving, ensure that the temperature, pressure and time contact are strictly followed so that all infectious materials are removed and the waste can be sent for disposal without fear. The common methods of disposal of solid wastes after treatment are incineration or land fill. The liquid wastes are treated with disinfectants as per the protocol given in the manufacturer's guidelines and drained in the common sewage system. While handling chemical disinfectants, the time of exposure and concentration of the disinfectant should be noted so that they do not harm the environment in which they are used. Training on waste management should be given to all personnel involved in GMM labs including maintenance and service providers.

13.12 Operational Measures in GMM Laboratories

All GMM labs should have the following operational guidelines and documents:

1. Laboratory manual
2. Laboratory procedures
3. Study protocols
4. Standard operating procedures for all procedures done in the laboratory
5. Infection control manual
6. Biosafety manual

Standard operating procedures are prepared in such a way that they are specific to the activities of a particular area in the laboratory to ensure their quality control and quality assurance processes and governmental regulations.

13.13 Conclusion

Working with GMMs may provoke risks which are to be managed immediately. The important concept of biosafety in the lab is containment which includes administrative controls, safety practices, protective equipment and proper designing of facilities. During gene manipulation, enclosure of infectious agents has to be done by biosafety cabinets 1, 2 and rarely 3. Strict adherence to standard microbiological practices is essential in all biosafety levels. These include attention to hand hygiene, prohibition of eating or drinking in the laboratory, use of mechanical rather than mouth-pipetting devices, appropriate engineering and practice controls. Planning for management of spills or other laboratory contamination events is essential at all levels of risk. The first principle in the prevention of such risks is to assess the risks occurring in different areas of the laboratory by studying the characters of the host, vector and environment. The risk management in these laboratories include four levels of biosafety. Biological waste management is another mainstay in the biosafety of GMM.

References

Benedict MQ, Tabachnick WJ, Higgs S (2003) Arthropod containment guidelines. Vector Borne Zoonotic Dis 3:1–98

Benedict M, Burt A, Cappuro M, De Barro P et al (2018) Recommendations for laboratory containment and management of gene drive systems in arthropods. Vector Borne Zoonotic Dis 18:2–13

Chapter 14
Arbovirus Detection in Vectors

David T. Williams, Prasad Paradkar, and Stephan Karl

Abstract Detection of arthropod-borne (arbo)viruses is a fundamental element of mosquito surveillance programmes. Moreover, recent advances in modifying mosquitoes for managing arbovirus vector populations rely on sensitive arbovirus detection methods, which are applied at various stages of development, evaluation and production of modified mosquitoes. An increasingly wide range of mosquito trapping, sampling and testing options are available. Although virus culture will remain important for isolating viruses for research and reference purposes, the widespread use and application of real-time reverse transcription polymerase chain reaction (RT-PCR) offers rapid and cost-effective workflows for detecting arbovirus nucleic acid. Advances in next-generation sequencing (NGS) techniques and bioinformatics approaches have also enabled increasingly rapid, accurate and inexpensive arbovirus genome sequencing that can be employed following virus detection using conventional methods or used independently as a stand-alone platform. Unbiased NGS is also a powerful tool for arbovirus discovery and metagenomics. Continued advances in arbovirus detection methods and approaches are expected to provide ever more sophisticated tools for controlling and responding to the threat of pathogenic arboviruses.

Keywords Mosquito traps · Arbovirus detection · Laboratory techniques · Next-generation sequencing

D. T. Williams (✉) · P. Paradkar
CSIRO, Australian Centre for Disease Preparedness, Geelong, VIC, Australia
e-mail: d.williams@csiro.au; prasad.paradkar@csiro.au

S. Karl
Australian Institute of Tropical Health and Medicine, James Cook University, Smithfield, QLD, Australia

Vector-Borne Diseases Unit, PNG Institute of Medical Research, Madang, Madang Province, Papua New Guinea
e-mail: stephan.karl@jcu.edu.au

B. K. Tyagi (ed.), *Genetically Modified and other Innovative Vector Control Technologies*, https://doi.org/10.1007/978-981-16-2964-8_14

14.1 Introduction

As advances are made in the development and application of modified mosquito vectors of human pathogens, the ability to accurately detect target pathogens is critical to assess the effectiveness of genetic or biological modifications to inform field trials and their longer-term implementation. Ensuring laboratory-reared mosquito colonies are pathogen free is important to ensure the integrity of research results. Similarly, arbovirus testing of initiating colonies for modified mosquito production and subsequent screening of derived mass-reared colonies intended for field release is an important element of quality control and biosafety. Detection of vector-borne pathogens also underpins mosquito surveillance programmes for public health purposes: estimation of infection rates allows an assessment of arbovirus transmission risk, while early detection facilitates focussed control efforts. This chapter will focus on mosquito trapping methods and laboratory diagnostic techniques used to detect the major arthropod-borne (arbo)viruses of global public health concern, namely, the flaviviruses dengue virus (DENV), Zika virus (ZIKV), Japanese encephalitis virus (JEV) and West Nile virus (WNV) and the alphavirus chikungunya virus (CHIKV). Each of these viruses has demonstrated an alarming propensity to spread and emerge into new geographic areas and cause disease outbreaks in human and animal populations.

DENV is found throughout the tropics with transmission occurring in over 100 countries and approximately four billion people at risk (Brady et al. 2012). An estimated 394 million infections occurred in 2010, with the majority (~70%) occurring in Asia (Bhatt et al. 2013). First described in Uganda, ZIKV has now spread to over 80 countries in Africa, Asia, the Pacific and the Americas. ZIKV has more recently emerged as an important cause of neurological disease (e.g. congenital Zika and Guillain-Barre syndrome) following major outbreaks in Latin America (Gulland 2016), which were preceded by large outbreaks in the Pacific (Duffy et al. 2009; Musso et al. 2018). JEV is the most important cause of human viral encephalitis in Southeast Asia. Approximately 68,000 human cases occur annually in JE-endemic areas, with high rates of associated morbidity (~30–50%) and mortality (20–30%) (Campbell et al. 2011). WNV is one of the most widely distributed flaviviruses and has been found in Africa, the Middle East, Europe, Asia, Australasia and North America. It has caused large and unexpected outbreaks involving high incidences of neurological infection in humans in Europe and the United States (Hayes and Gubler 2006; Sambri et al. 2013), as well as outbreaks of equine neurological disease and avian mortality events (Castillo-Olivares and Wood 2004; Frost et al. 2012). The alphavirus CHIKV is a major cause of debilitating arthralgia and has been the cause of explosive epidemics. CHIKV has a similar distribution to DENV and can now be found across the tropics. In the past 15 years, CHIKV has spread to the Indian Ocean, the Caribbean and Central and South America; autochthonous transmission has also occurred in Europe and the United States (reviewed in Weaver and Forrester 2015).

These viruses share common mosquito vectors: *Aedes aegypti* and *Ae. albopictus* are major vectors for DENV, ZIKV and CHIKV, while *Culex* species mosquitoes are the main vectors for JEV and WNV. Despite this, major vectors for each virus can vary between regions. For example, *Culex tritaeniorhynchus* is the main vector for JEV in much of Southeast Asia; however, at the edge of its southeastern range in Australasia, the main vector is *Cx. annulirostris* (Ritchie et al. 1997). Hence, understanding the breadth of arbovirus vectors through surveillance is important for focussing control efforts.

Traditional methods of vector surveillance comprise of mosquitoes being collected, identified, pooled by species or other taxonomic grouping, and sent to the laboratory where they are tested for virus infection status using one or more techniques (see below). This approach may be applied for early detection of arboviruses of medical or veterinary importance, determination of virus infection rates, identification of vectors and vector abundance. Depending on resources, surveillance activities may be undertaken year-round in areas endemic for arboviruses or on a seasonal basis in epidemic regions. In areas with low-level mosquito infections or when trapping is undertaken early in the transmission season, efforts should be directed towards performing targeted surveillance at 'hotspots' where a high likelihood of arbovirus presence is suspected; as vector populations increase later in the season, the number of sampling sites should be expanded for broader monitoring.

14.2 Mosquito Traps for Arbovirus Surveillance

A large variety of mosquito traps have been developed for commercial and research and surveillance purposes. Mosquito traps use attractants that are based on olfactory, visual and sometimes temperature and sound signals that mosquitoes are able to sense with their antennae, compound eyes or palpi (Takken and Kline 1989; Daniel 2006).

Commercial traps mostly intended, e.g. for private usage, are usually designed to attract and kill as many mosquitoes and other nuisance insects as possible. They are available in many shape, size and price categories (Mosquito World n.d.). High end models combine different attractants, mostly UV light, CO_2 gas either produced by burning propane or emanated from cylinders, other olfactory baits (e.g. octenol chips) and heat. Many traps use a fan or otherwise generated suction to aspirate mosquitoes once they have come close to the trap (Ritchie et al. 2007).

Traps used in research studies and for vector surveillance are often intended to target mosquito species more selectively and tend to be designed to allow for less destructive sampling, so that the collected mosquitoes can be studied further. In addition, collections involving very many traps, or in studies in remote locations, ease of use and extended operating duration without maintenance are desirable trap features. As such, traps used in research or for surveillance often use more specific cues or attractants believed to lure specific mosquito species or sub-populations (e.g. gravid, male etc.).

Widely used Biogents (BG) Sentinel Traps, for example, rarely use a light source but attract females of many *Aedes* and *Culex* mosquito species simply by colour contrast of a white lid with a black intake. Once mosquitoes have come close to the intake opening, suction from a fan aspirates them into a catch bag. These traps can further be baited with CO_2 or specific lures to attract other mosquito species as well (Maciel-de-Freitas et al. 2006; Williams et al. 2006).

CDC miniature light traps, widely used for research and surveillance purposes, use incandescent white, coloured LED or UV light as a main attractant. They, too, are equipped with an intake fan and can be set up with CO_2 cylinders (Mathenge et al. 2004).

Some traps developed for research and surveillance purposes, such as Passive Box Traps, focus on collecting nucleic acid samples deposited by mosquitoes, e.g. for arbovirus detection. In Passive Box Traps, nucleic acids are deposited by the mosquitoes while sitting on a (honey-baited) filter paper card. The card can then be removed and subjected to molecular analyses (Ritchie et al. 2013).

Gravid or oviposition traps are designed to trap especially gravid female mosquitoes by resembling a container ovipositioning site. Gravid mosquitoes are older and have blood-fed and are thus more likely to carry arboviruses (Liew et al. 2019; Alemayehu et al. 2018). These traps are usually black containers filled with, e.g. hay-infused water and a geometry allowing for entry, but not exit of the mosquitoes.

Promising new trap developments include, for example, the Male Aedes Sound Trap (MAST), which uses a small speaker assembly to imitate the wingbeat frequency of female mosquitoes, thereby attracting males in the vicinity. These traps can operate for weeks without maintenance, making them an attractive alternative for research and surveillance in remote locations (Staunton et al. 2020, 2021).

Trapping efficiency using any of the aforementioned methods is low, in particular for some anophelines, such as *An. farauti*. Alternatives for adult mosquito collection include the widely used human landing catches (Mathenge et al. 2004), barrier screen (or vertical barrier) methods (Keven et al. 2019), animal- or human-baited double net or tent traps (Govella et al. 2009; Tangena et al. 2015), sweep net collections and resting collections using, for example, manually operated aspirators, such as Prokopacks (Vazquez-Prokopec et al. 2009). In particular human landing catches may be viewed as ethically controversial, as humans are exposed directly to potentially infectious mosquito bites. However, it is believed that normally, exposure in human landing catch collections is equal or less than that normally experienced by the collectors (Gimnig et al. 2013). In addition, studies are usually expected to provide malaria prophylaxis and treatment free of charge to collectors.

14.3 Mosquito Sampling

Processing pooled samples of mosquitoes comes with inherent limitations. Testing individual caught mosquitoes offers a very precise means of determining infection rate. However, mosquito populations often have very low carriage rates, and, to increase the probability of detection, large numbers of mosquitoes are usually required. Under such circumstances, individual mosquito testing may be impractical. Gu and Novak (2004) showed that detection of low levels of mosquito infections requires large samples (>1600 individuals) for a high probability of detection. The authors recommended an intensified sampling strategy at sites where potential vector mosquitoes are abundant, or in areas with a history of arbovirus circulation. Such an approach is cost-effective and increases the probability of detection, which is advantageous if the primary objective of the surveillance activity is early detection.

Estimation of mosquito infection rates is important for determining risk of transmission. In this regard, it has been shown that in situations where mosquito infection rates are high (e.g. peak season) and there may be more than one infected mosquito per pool, the use of variable size pools provides more accurate estimates of infection rates than constant size pooling (Gu and Novak (2004).

The size of mosquito pools for testing can vary in size, but typically pool sizes of 25–100 mosquitoes have been used (e.g. Johansen et al. 2000; Kauffman et al. 2003). However, arboviruses can be detected in larger pools of mosquitoes. Using real-time RT-PCR, a single WNV-infected mosquito could be reliably detected in a pool of 500, and virus isolation and commercial antigen capture assays could detect virus in pools of 200 mosquitoes (Sutherland and Nasci 2007). Tang et al. (2020) showed the detection of CHIKV, WNV, ZIKV and Usutu virus could be achieved in pools comprising up to 1600 mosquitoes using real-time RT-PCR. Molecular testing of pools of 100–1000 mosquitoes is now routinely undertaken by some surveillance programmes for early detection of arboviruses.

Following collection, mosquitoes are typically sorted (e.g. male/female, blood-fed), speciated and pooled, ideally on cold tables to minimise degradation. To preserve samples, freezing whole traps soon after collection or pools following sorting is recommended. Nucleic acid preservative solutions can also be used for molecular testing applications. Mosquito pools are typically homogenised using sterile glass beads or ball bearings in PBS or virus transport medium solutions. Clarified homogenates are then used as inoculum for selected culture systems or for direct testing for antigen or viral RNA.

14.4 Conventional Methods of Arbovirus Detection

Effective vector surveillance requires rapid and accurate methods to identify trapped insect samples and screen them for pathogenic arboviruses. Detection and identification of arboviruses also facilitate research into patterns of virus activity and movement, by enabling genetic analysis of geographically and temporally distinct isolates and strains. Laboratory testing of mosquitoes involves direct detection of virus from samples or following virus isolation.

14.4.1 Virus Isolation

Historically, arbovirus isolation involved intracerebral inoculation of suckling mice, guinea pigs or hamsters or inoculation of embryonated chicken eggs via the chorio-allantoic or allantoic membranes or the yolk sac (Beaty et al. 1995). Furthermore, the use of animals for scientific purposes normally requires institutional animal ethics approval to ensure animal welfare standards are followed and adhered to. With the establishment of cell lines, virus isolation in cell culture became the gold standard for arbovirus detection from pools of mosquitoes. Several mosquito cell lines have been established that are susceptible to arboviruses (Walker et al. 2014). Notably the C6/36 clone from *Ae. albopictus* is susceptible to a wide range of arboviruses (Singh 1967), in part due to having a dysfunctional innate antiviral RNA interference response (Brackney et al. 2010). A range of mammalian cell lines have also been used to isolate arboviruses, including African green monkey kidney (Vero), rhesus monkey kidney (LLC-MK2), baby hamster kidney (BHK), rabbit kidney (RK-13) and pig kidney (PS) cells. The PS cell clone PSEK has also been used widely for isolation of arboviruses; however, this should be used with caution since it is known to harbour contaminating pestivirus. Alphaviruses and flaviviruses do not normally cause cytopathic effect (CPE) in mosquito cells such as C6/36, and therefore further culture in vertebrate cells in which CPE occurs and/or detection of viral antigen by immunoassay or viral RNA by RT-PCR is required to confirm virus isolation.

Arboviruses can also be isolated using direct inoculation of laboratory colonies of susceptible mosquitoes or mosquito larvae (Rosen and Gubler 1974; Gajanana et al. 1995; Alera et al. 2015). Susceptible mosquito species such as *Ae. albopictus* or *Toxorhynchites splendens* are typically inoculated intracerebrally or intrathoracically with homogenate samples from pools of trapped mosquitoes. For biosafety reasons, male *Aedes* mosquitoes are used as they do not ingest blood; neither male nor female *Toxorhynchites* species ingest blood. Since arbovirus infection does not overtly affect the inoculated mosquitoes, as for propagation in mosquito cell lines, virus detection is undertaken by immunodetection or RT-PCR. A simple method to test inoculated mosquitoes is to perform an indirect immunofluorescence test using a virus-specific antibody on head smears (Gajanana et al. 1995).

Virus isolation techniques are costly, laborious and time-consuming and require specialised laboratories with highly trained personnel. Virus isolation also relies on the presence of infectious virus, which can be degraded in mishandled samples or when cold chain from the point of collection to the laboratory cannot be maintained. Despite these challenges, the ability to isolate and cultivate arbovirus is essential to provide reference isolates for biological, antigenic and pathogenic characterisation, for generating diagnostic reagents and for diagnostic test and vaccine development.

Virus whole genome sequencing is also more efficient and reliable using high-titred virus cultures. Although culture methods remain important in mosquito surveillance activities, molecular methods are increasingly being used for rapid detection of arboviruses in mosquito collections.

14.4.2 *Antigen Detection by Immunodetection*

Immunodetection methods are commonly employed for detection and identification of arboviruses isolated from or using mosquitoes. These techniques employ species- or group-specific monoclonal antibodies and include antigen capture ELISA to detect arbovirus particles in field-collected mosquito pools, including commercially available tests (Burkhalter et al. 2006; Gajanana et al. 1995; Konishi and Takahashi 1985; Sutherland and Nasci 2007; Kumari et al. 2011), fixed cell enzyme immunoassay following virus propagation (Broom et al. 1998; Johansen et al. 2000; Zhang et al. 1984) and direct or indirect immunofluorescence assay of infected mosquito impression smears from bioassays (Alera et al. 2015; Gajanana et al. 1995). These methods are relatively rapid and inexpensive to perform and, in the case of the ELISA method, can be adapted for high-throughput testing. However, as mentioned above, for virus isolation in cell culture or mosquito bioassay, the preceding steps are laborious and time-consuming.

14.4.3 *Polymerase Chain Reaction (PCR)*

Molecular testing by reverse transcription polymerase chain reaction (RT PCR) has become the most commonly employed tool in diagnostic and research laboratories for detecting arbovirus genetic material in mosquito specimens or following culture. Numerous assays have been reported, including both conventional and real-time tests; specific examples are shown in Table 14.1 for DENV, ZIKV, JEV, WNV and CHIKV.

Many of these assays were designed for detection of a particular virus lineage (s) or genotype(s). Therefore, test selection should consider the lineage/genotype of regionally circulating arboviruses. For instance, the WNV real-time PCR reported by Pyke et al. (2004) was developed for the specific detection of WNV-KUNV (lineage 1b), found only in the Australasian region, where other WNV lineages are exotic and alternative assays are required for their detection. For surveillance studies of several target viruses, multiplex or generic assays can be used. These assays are useful to support syndromic surveillance studies targeting viruses that cause similar disease (e.g. febrile or neurological). Flavivirus or alphavirus generic assays can also be employed; however, these can be less sensitive, particularly in the conventional formats. Generic assays—when coupled with sequencing—can be useful for

Table 14.1 RT-PCR assays used to detect dengue, Zika, Japanese encephalitis, West Nile and chikungunya viruses in mosquitoes

Virus	Format	Genome target(s)	References
Dengue virus 1 to 4	Conventional	E gene	Balingit et al. (2020)
		Capsid/prM genes	Johnson et al. (2005)
	Real time	NS5 (DENV-1) E gene (DENV-2) prM/M gene (DENV-3) prM/M-E genes (DENV-4)	Balingit et al. (2020), Johnson et al. (2005)
		NS5	Hue et al. (2011)
Zika	Conventional	E gene	Faye et al. (2008)
		NS5 gene	Balm et al. (2012)
	Real time	NS5	Faye et al. (2013)
		M-E genes E gene	Lanciotti et al. (2008)
		NS1 gene E genes	Pyke et al. (2014)
Japanese encephalitis	Real time	NS5–3′UTR	Pyke et al. (2004)
West Nile	Conventional	E gene	Johnson et al. (2001)
	Real time	E gene 3′UTR	Lanciotti et al. (2000)
		NS5-3′UTR	Pyke et al. (2004)
Chikungunya	Conventional	nsp1 gene	Hasebe et al. (2002)
	Real time	E1 gene	van den Hurk et al. (2010)
		nsp1 gene nsp4 gene	Lanciotti et al. (2007)
Flavivirus generic	Conventional	NS5-3′UTR	Pierre et al. (1994)
		E gene	Gaunt and Gould (2005)
		NS5 gene	Scaramozzino et al. (2001)
	Real time	NS5 gene	Moureau et al. (2007)
		NS5	Patel et al. (2013)
Alphavirus generic	Conventional	nsP1 gene	Pfeffer et al. (1997)
		nsP4 gene	Sanchez-Seco et al. (2001)
		nsP4 gene	Grywna et al. (2010)
	Real time	nsP4 gene	Giry et al. (2017)
Multiplex	Real time	5′-UTR (DENV) nsp1 gene (CHIKV) E gene (ZIKV)	Santiago et al. (2018)
		3′UTR (DENV) NS5 gene (CHIKV) E gene (ZIKV)	Mansuy et al. (2018)
		NS5 (YFV, JEV, WNV, SLEV, DENV-1 to -4)	Chao et al. (2007)
		NS2A (WNV, JEV)	Barros et al. (2013)

inexpensive and rapid identification of virus isolates from mosquito samples when other available tests have failed.

Other formats of molecular testing have also been reported for arbovirus laboratory diagnosis. RT-LAMP assays have been published for the detection of JEV (Liu et al. 2012; Parida et al. 2006), WNV (Parida et al. 2004), ZIKV (Silva et al. 2019), DENV (Lopez-Jimena et al. 2018) and CHIKV (Parida et al. 2007). Multiplex RT-LAMP assays have also been reported for detecting combinations of DENV, ZIKV, CHIKV, JEV or WNV (Li et al. 2011; Yaren et al. 2017). LAMP offers comparable performance to real-time RT-PCR, in a simple and convenient assay format without the need for sophisticated equipment or highly trained personnel. At the other end of the technology spectrum, RT-PCR-based microsphere array assays have been reported for the multiplex detection of medically important flaviviruses and alphaviruses from mosquitoes (Foord et al. 2014; Glushakova et al. 2019). Although these assays require specialised equipment and trained staff to run the assays, they can offer high-throughput multiplexed testing for arbovirus surveillance activities.

Selection of a molecular test should also consider prior validation using infected mosquitoes. In-house test verification is recommended to ensure the assay is fit for purpose in the laboratory where it will be applied. This may be undertaken using mosquito samples that are known to be infected from prior testing or using mosquito pools spiked with a known quantity of the target virus. This is to ensure that variation in laboratory conditions and equipment, extraction methods and reaction chemistries do not affect the sensitivity or performance of the selected assay.

14.5 Next-Generation Sequencing Methods for Arbovirus Surveillance

Conventional diagnostic test methods typically target certain pathogens based on likelihood and risk; however, this approach fails to detect novel or unexpected arboviruses when present in mosquito samples. Next-generation sequencing (NGS) describes a DNA or RNA sequencing technology which has revolutionised genomic research (Behjati and Tarpey 2013) and has been applied to vector-borne disease surveillance for the identification of both known and previously unknown arboviruses. Although several NGS platforms are available, they all perform sequencing of millions of small fragments of nucleotides in parallel. Appropriate bioinformatics are then used to remove host sequences and assemble target sequence fragments either as de novo synthesis or mapped to known genomes. Total RNA sequencing enables non-targeted, high-throughput detection and characterisation of viruses in a sample, such as mosquitoes.

Combined with metabarcoding, NGS can allow the rapid identification of large numbers of mosquitoes with simultaneous screening for pathogens (Batovska et al.

2018). Based on the sensitivity of the technique, this method also can quantify the number of mosquitoes in a trap. In Australia, viral metagenomics has been used for the identification of multiple arboviruses, including novel rhabdoviruses, bunyaviruses (Quan et al. 2011; Coffey et al. 2014; Briese et al. 2016), ephemeroviruses (Blasdell et al. 2014) and mesoniviruses (Warrilow et al. 2014) from field collected mosquitoes.

By combining unbiased sequencing, rapid data analysis and comprehensive reference databases, metagenomics can be applied for hypothesis-free, universal pathogen detection, providing a promising approach to improved surveillance of arboviruses. Several studies have used NGS approaches to detect viruses in individual mosquitoes or pools of mosquitoes using various technologies. There are currently several NGS platforms available for use, including Illumina (Chandler et al. 2015), Ion Torrent (Hall-Mendelin et al. 2013) and Oxford Nanopore (Batovska et al. 2017), each with its own benefits and limitations. The laboratory workflow is also determined by the technology used, and this can limit feasibility of usage. It is important that the surveillance laboratory should clearly define the intended use, range of pathogens and reporting workflow as this will determine choice of data analysis. Since bioinformatics forms a key aspect of the workflow, it is important to validate the bioinformatics pipeline along with laboratory techniques. The bioinformatics analyses chosen to process the NGS reads (e.g., Flygare et al. 2016; Andrusch et al. 2018; Oulas et al. 2015; Naccache et al. 2014) can also affect sensitivity and specificity. A common method used to detect viruses in a sample is by mapping reads back to viral reference sequences. However, when dealing with short reads, this can lead to false-positive results if a virus is present with partial sequence homology to a virus of interest. Recent advances in bioinformatics can overcome this by de novo assembly of reads to produce longer contiguous sequences (contigs) (Schlaberg et al. 2017).

There are numerous benefits of using NGS for surveillance of mosquito-borne pathogens. Advances in the NGS technology in recent years have allowed for detection of all viruses in mosquito samples in a cost-effective and unbiased manner. This methodology can be used to detect both known and unknown viruses and bacteria. Due to the untargeted approach, the method allows for accurate detection of mosquito species as well and can serve as an early warning system for invasive mosquitoes (Batovska et al. 2018). With appropriate protocols, it can be very sensitive, detecting low quantities of pathogen nucleic acid. Multiplexing with barcoding can help in higher throughput of sample processing (Batovska et al. 2018). NGS data analysis with sufficient read depth and coverage can also inform about mutation and variants circulating or evolving in the environment. This not only can help in performing molecular epidemiology (Ko et al. 2020; Bialosuknia et al. 2019; Maan et al. 2015; Johnson et al. 2012) and identifying source and sinks of pathogens in mosquito populations but also can help in formulating appropriate strategy against vector and pathogen spread.

The major drawback of use of NGS in surveillance is the infrastructure cost of setting up the instruments. The high costs, long sequencing times and slow, unwieldy data analysis tools have made it impractical for wider use of these methods. With advances in instrumentation and sequencing chemistries, the costs for sequencing are dropping and the amount of data being generated is increasing; however, the current bottleneck is the bioinformatics pipeline for large data analysis and appropriate interpretation.

Appropriate sample preparation determines the analytical sensitivity and specificity of the assay, which is important in assessing the transmission risk and temporal changes in virus abundance. Samples can be a single mosquito, pool of mosquitoes or honey-baited FTA cards (Birnberg et al. 2020). The quality of the generated data depends on the stability of RNA in the sample; hence appropriate storage conditions of samples, such as in RNAlater, are necessary. Sensitivity of detection can be increased artificially by enriching for arbovirus using size filtration (Sadeghi et al. 2018), PEG precipitation or sequence-independent amplification (Xiao et al. 2018). While this does increase the number of viral sequences, enrichment can also introduce bias (Conceição-Neto et al. 2015). An alternate way to increase the number of viral sequences is by depleting the mosquito RNA, generally by targeting highly abundant ribosomal RNA (rRNA). A variety of rRNA depletion kits are available; however, these are not specific to mosquitoes, and so custom probes based on mosquito rRNA sequences need to be generated (Fauver et al. 2019).

There are also bioinformatics tools and approaches to increase the specificity of pathogen sequence detection, such as performing de novo assembly, where short reads are assembled into longer contigs, and then comparing these contigs to a database containing viral reference sequences. This approach can improve specificity because longer fragments are taxonomically classified with greater accuracy (McHardy et al. 2007).

A recent study (Batovska et al. 2019) in a lab setting showed that NGS was highly specific in identification of Ross River virus or Umatilla virus in mosquito pools spiked with these viruses, recovering whole genome information and detecting 19 other viruses. However, the method was not as sensitive as RT-qPCR or RT-ddPCR.

NGS generates several million to billion short-read sequences of the DNA and RNA isolated from a sample. In contrast to traditional Sanger sequencing, with read lengths of 500–900 base pairs (bp), short reads of NGS range in size from 75 to 300 bp depending on the application and sequencing chemistry. Newer NGS technologies such as those from PacBio (Rhoads and Au 2015), Nanopore and 10× Genomics (Singh et al. 2019) enable longer read sequences of more than 10 kb.

Until recently, NGS was restricted to the laboratory due to the size of the sequencers available. MinION by Oxford Nanopore Technologies provides a powerful tool for in-field surveillance, allowing non-targeted (unbiased) detection of viruses in a sample within a few hours. The MinION is a relatively low-cost, handheld sequencer. Although it's sequencing accuracy is considerably lower than Illumina MiSeq and HiSeq, with an error rate of approximately 5–10% (Tyson et al. 2018) for the recent MinION R9 chemistry with 2D (double-strand) reads, compared to <0.1% for the Illumina sequencers (Houldcroft et al. 2017), MinION makes it up by producing long reads (up to 233 kb) in real time (Jansen et al. 2017) and achieving >99% accuracy post-data analysis (Wang et al. 2015). It has recently been demonstrated that the MinION can be used for metagenomic arbovirus detection from infected mosquitoes (Batovska et al. 2017), so it could be used during arbovirus outbreaks. Even with its limitations, it is expected that MinION will play a significant role in making sequencing available in real time, helping appropriate public health response.

Xenosurveillance is a novel surveillance technique that leverages and extends mosquito surveillance activities to the detection of non-vectored pathogens using PCR- or NGS-based methods. With mosquitoes feeding on a variety of vertebrate hosts, they also have potential to act as samplers for circulating viruses present in host blood. This can offer an alternative to direct sampling of hosts (such as sentinel animals), with mosquitoes acting as 'syringes'. Other than acting as surveillance for arboviruses (Yang et al. 2015), it can also serve as surveillance for non-vector-borne pathogens (Grubaugh et al. 2015).

These emerging technologies have also translated into trap development as well with development of the next generation of mosquito traps. Other than previously mentioned use of NGS on honey-baited FTA cards to detect circulating viruses in mosquito traps, technology companies such as Microsoft are developing mosquito traps ('Microsoft Premonition' n.d.) which have in situ NGS technology. The idea behind this is using drone technology to position these traps at select locations and using baits to attract female mosquitoes, which are photographed followed by whole genome sequencing to generate pathogen profiles. An alternative to this which has also been proposed includes in situ technology to detect specific arboviruses using lateral flow technology. Although these traps are not used widely, they are currently being trialled in various locations for feasibility studies.

14.6 Conclusion

In recent years there have been numerous advances in the methods and approaches used for arbovirus detection. Together with tried and tested conventional methods, there is an array of options that can be applied to different situations and conditions, as well as budgets and resources. While most of these methods have been developed for and applied to arbovirus surveillance programs, they also have a place in the laboratories of researchers working on modified mosquitoes and the companies and factories that rear the millions of modified mosquitoes required for field release. In the future, molecular diagnostic methods such as real-time RT-PCR will likely remain a workhorse platform for arbovirus laboratories; however, it is expected that advancements in NGS technologies will continue, providing increasingly inexpensive, accessible and sensitive platforms that can be applied to arbovirus surveillance and mosquito testing in the field and laboratory.

References

Alemayehu D, Reyes T, Haas-Stapleton EJ (2018) Field evaluation of a redesigned oviposition trap to monitor gravid invasive *Aedes* mosquitoes in a suburban environment. J Am Mosq Control Assoc 34(1):67–69

Alera MT, Hermann L, Tac-An IA, Klungthong C, Rutvisuttinunt W, Manasatienkij W, Villa D, Thaisomboonsuk B, Velasco JM, Chinnawirotpisan P, Lago CB, Roque VG Jr, Macareo LR, Srikiatkhachorn A, Fernandez S, Yoon IK (2015) Zika virus infection, Philippines, 2012. Emerg Infect Dis 21:722–724

Andrusch A, Dabrowski PW, Klenner J, Tausch SH, Kohl C, Osman AA, Renard BY, Nitsche A (2018) PAIPline: pathogen identification in metagenomic and clinical next generation sequencing samples. Bioinformatics 34(17):i715–i721

Balingit JC, Carvajal TM, Saito-Obata M, Gamboa M, Nicolasora AD, Sy AK, Oshitani H, Watanabe K (2020) Surveillance of dengue virus in individual Aedes aegypti mosquitoes collected concurrently with suspected human cases in Tarlac City, Philippines. Parasit Vectors 13(1):594

Balm MN, Lee CK, Lee HK, Chiu L, Koay ES, Tang JW (2012) A diagnostic polymerase chain reaction assay for Zika virus. J Med Virol 84(9):1501–1505

Barros SC, Ramos F, Zé-Zé L, Alves MJ, Fagulha T, Duarte M, Henriques M, Luís T, Fevereiro M (2013) Simultaneous detection of West Nile and Japanese encephalitis virus RNA by duplex TaqMan RT-PCR. J Virol Methods 193(2):554–557

Batovska J, Lynch SE, Rodoni BC, Sawbridge TI, Cogan NO (2017) Metagenomic arbovirus detection using MinION nanopore sequencing. J Virol Methods 249:79–84

Batovska J, Lynch SE, Cogan NOI, Brown K, Darbro JM, Kho EA, Blacket MJ (2018) Effective mosquito and arbovirus surveillance using metabarcoding. Mol Ecol Resour 18(1):32–40

Batovska J, Mee PT, Lynch SE, Sawbridge TI, Rodoni BC (2019) Sensitivity and specificity of metatranscriptomics as an arbovirus surveillance tool. Sci Rep 9(1):19398

Beaty BJ, Calisher CH, Shope RE (1995) Arboviruses. In: Lennette EH, Lennette DA, Lennette ET (eds) Diagnostic procedures for viral, rickettsial and chlamydial infections, 7th edn. American Public Health Association, Washington, DC, pp 189–212

Behjati S, Tarpey PS (2013) What is next generation sequencing? Arch Dis Child Educ Pract Ed 98 (6):236–238

Bhatt S, Gething PW, Brady OJ, Messina JP, Farlow AW, Moyes CL, Drake JM, Brownstein JS, Hoen AG, Sankoh O, Myers MF, George DB, Jaenisch T, Wint GR, Simmons CP, Scott TW, Farrar JJ, Hay SI (2013) The global distribution and burden of dengue. Nature 496:504–507

Bialosuknia SM, Tan Y, Zink SD, Koetzner CA, Maffei JG, Halpin RA, Muller E, Novatny M, Shilts M, Fedorova NB, Amedeo P, Das SR, Pickett B, Kramer LD, Ciota AT (2019) Evolutionary dynamics and molecular epidemiology of West Nile virus in New York state: 1999–2015. Virus Evol 5(2):vez020

Birnberg L, Temmam S, Aranda C, Correa-Fiz F, Talavera S, Bigot T, Eloit M, Busquets N (2020) Viromics on honey-baited FTA cards as a new tool for the detection of circulating viruses in mosquitoes. Viruses 12(3):274

Blasdell KR, Widen SG, Diviney SM, Firth C, Wood TG, Guzman H, Holmes EC, Tesh RB, Vasilakis N, Walker PJ (2014) Koolpinyah and Yata viruses: two newly recognised ephemeroviruses from tropical regions of Australia and Africa. Vet Microbiol 174 (3–4):547–553

Brackney DE, Scott JC, Sagawa F, Woodward JE, Miller NA, Schilkey FD, Mudge J, Wilusz J, Olson KE, Blair CD, Ebel GD (2010) C6/36 *Aedes albopictus* cells have a dysfunctional antiviral RNA interference response. PLoS Negl Trop Dis 4(10):e856

Brady OJ, Gething PW, Bhatt S, Messina JP, Brownstein JS, Hoen AG, Moyes CL, Farlow AW, Scott TW, Hay SI (2012) Refining the global spatial limits of dengue virus transmission by evidence based consensus. PLoS Negl Trop Dis 6(8):e1760

Briese T, Williams DT, Kapoor V, Diviney SM, Certoma A, Wang J, Johansen CA, Chowdhary R, Mackenzie JS, Lipkin WI (2016) Analysis of arbovirus isolates from Australia identifies novel bunyaviruses including a Mapputta group virus from Western Australia that links Gan Gan and Maprik viruses. PLoS One 11(10):e0164868

Broom AK, Hall RA, Johansen CA, Oliveira N, Howard MA, Lindsay MD, Kay BH, Mackenzie JS (1998) Identification of Australian arboviruses in inoculated cell cultures using monoclonal antibodies in ELISA. Pathology 30(3):286–288

Burkhalter KL, Lindsay R, Anderson R, Dibernardo A, Fong W, Nasci RS (2006) Evaluation of commercial assays for detecting West Nile virus antigen. J Am Mosq Control Assoc 22 (1):64–69

Campbell GL, Hills SL, Fischer M, Jacobson JA, Hoke CH, Hombach JM, Marfin AA, Solomon T, Tsai TF, Tsu VD, Ginsburg AS (2011) Estimated global incidence of Japanese encephalitis: a systematic review. Bull World Health Organ 89(10):766–774, 774A-774E

Castillo-Olivares J, Wood J (2004) West Nile virus infection of horses. Vet Res 35(4):467–483

Chandler JA, Liu RM, Bennett SN (2015) RNA shotgun metagenomic sequencing of Northern California (USA) mosquitoes uncovers viruses, bacteria, and fungi. Front Microbiol 6:185

Chao DY, Davis BS, Chang GJ (2007) Development of multiplex real-time reverse transcriptase PCR assays for detecting eight medically important flaviviruses in mosquitoes. J Clin Microbiol 45(2):584–589

Coffey LL, Page BL, Greninger AL, Herring BL, Russell RC, Doggett SL, Haniotis J, Wang C, Deng X, Delwart EL (2014) Enhanced arbovirus surveillance with deep sequencing: identification of novel rhabdoviruses and bunyaviruses in Australian mosquitoes. Virology 448:146–158

Conceição-Neto N, Zeller M, Lefrère H, De Bruyn P, Beller L, Deboutte W, Yinda CK, Lavigne R, Maes P, Van Ranst M, Heylen E, Matthijnssens J (2015) Modular approach to customise sample preparation procedures for viral metagenomics: a reproducible protocol for virome analysis. Sci Rep 5:16532

Daniel LK (2006) Traps and trapping techniques for adult mosquito control. J Am Mosq Control Assoc 22(3):490–496

Duffy MR, Chen TH, Hancock WT, Powers AM, Kool JL, Lanciotti RS, Pretrick M, Marfel M, Holzbauer S, Dubray C, Guillaumot L, Griggs A, Bel M, Lambert AJ, Laven J, Kosoy O, Panella A, Biggerstaff BJ, Fischer M, Hayes EB (2009) Zika virus outbreak on Yap Island, Federated States of Micronesia. N Engl J Med 360(24):2536–2543

Fauver JR, Akter S, Morales AIO, Black WC IV, Rodriguez AD, Stenglein MD, Ebel GD, Weger-Lucarelli J (2019) A reverse-transcription/RNase H based protocol for depletion of mosquito ribosomal RNA facilitates viral intrahost evolution analysis, transcriptomics and pathogen discovery. Virology 528:181–197

Faye O, Faye O, Dupressoir A, Weidmann M, Ndiaye M, Alpha Sall A (2008) One-step RT-PCR for detection of Zika virus. J Clin Virol 43(1):96–101

Faye O, Faye O, Diallo D, Diallo M, Weidmann M, Sall AA (2013) Quantitative real-time PCR detection of Zika virus and evaluation with field-caught mosquitoes. Virol J 10:311

Flygare S, Simmon K, Miller C, Qiao Y, Kennedy B, Di Sera T, Graf EH, Tardif KD, Kapusta A, Rynearson S, Stockmann C, Queen K, Tong S, Voelkerding KV, Blaschke A, Byington CL, Jain S, Pavia A, Ampofo K, Eilbeck K, Marth G, Yandell M, Schlaberg R (2016) Taxonomer: an interactive metagenomics analysis portal for universal pathogen detection and host mRNA expression profiling. Genome Biol 17(1):111

Foord AJ, Boyd V, White JR, Williams DT, Colling A, Heine HG (2014) Flavivirus detection and differentiation by a microsphere array assay. J Virol Methods 203:65–72

Frost MJ, Zhang J, Edmonds JH, Prow NA, Gu X, Davis R, Hornitzky C, Arzey KE, Finlaison D, Hick P, Read A, Hobson-Peters J, May FJ, Doggett SL, Haniotis J, Russell RC, Hall RA, Khromykh AA, Kirkland PD (2012) Characterization of virulent West Nile virus Kunjin strain, Australia, 2011. Emerg Infect Dis 18(5):792–800

Gajanana A, Rajendran R, Thenmozhi V, Philip Samuel P, Tsai TF, Reuben R (1995) Comparative evaluation of bioassay and ELISA for detection of Japanese encephalitis virus in field collected mosquitoes. Southeast Asian J Trop Med Public Health 26:91–97

Gaunt MW, Gould EA (2005) Rapid subgroup identification of the flaviviruses using degenerate primer E-gene RT-PCR and site specific restriction enzyme analysis. J Virol Methods 128:113–127

Gimnig JE, Walker ED, Otieno P, Kosgei J, Olang G, Ombok M, Williamson J, Marwanga D, Abong'o D, Desai M, Kariuki S, Hamel MJ, Lobo NF, Vulule J, Bayoh MN (2013) Incidence of malaria among mosquito collectors conducting human landing catches in western Kenya. Am J Trop Med Hyg 88(2):301–308

Giry C, Roquebert B, Li-Pat-Yuen G, Gasque P, Jaffar-Bandjee MC (2017) Improved detection of genus-specific alphavirus using a generic TaqMan® assay. BMC Microbiol 17(1):164

Glushakova LG, Alto BW, Kim MS, Hutter D, Bradley A, Bradley KM, Burkett-Cadena ND, Benner SA (2019) Multiplexed kit based on Luminex technology and achievements in synthetic biology discriminates Zika, chikungunya, and dengue viruses in mosquitoes. BMC Infect Dis 19 (1):418

Govella NJ, Chaki PP, Geissbuhler Y, Kannady K, Okumu F, Charlwood JD, Anderson RA, Killeen GF (2009) A new tent trap for sampling exophagic and endophagic members of the *Anopheles gambiae* complex. Malar J 8:157

Grubaugh ND, Sharma S, Krajacich BJ, Fakoli LS III, Bolay FK, Diclaro JW II, Johnson WE, Ebel GD, Foy BD, Brackney DE (2015) Xenosurveillance: a novel mosquito-based approach for examining the human-pathogen landscape. PLoS Negl Trop Dis 9(3):e0003628

Grywna K, Kupfer B, Panning M, Drexler JF, Emmerich P, Drosten C, Kummerer BM (2010) Detection of all species of the genus alphavirus by reverse transcription-PCR with diagnostic sensitivity. J Clin Microbiol 48(9):3386–3387

Gu W, Novak RJ (2004) Short report: Detection probability of arbovirus infection in mosquito populations. Am J Trop Med Hyg 71(5):636–638

Gulland A (2016) Zika virus is a global public health emergency, declares WHO. BMJ 352:i657

Hall-Mendelin S, Allcock R, Kresoje N, van den Hurk AF, Warrilow D (2013) Detection of arboviruses and other micro-organisms in experimentally infected mosquitoes using massively parallel sequencing. PLoS One 8(2):e58026

Hasebe F, Parquet MC, Pandey BD, Mathenge EG, Morita K, Balasubramaniam V, Saat Z, Yusop A, Sinniah M, Natkunam S, Igarashi A (2002) Combined detection and genotyping of Chikungunya virus by a specific reverse transcription-polymerase chain reaction. J Med Virol 67(3):370–374

Hayes EB, Gubler DJ (2006) West Nile virus: epidemiology and clinical features of an emerging epidemic in the United States. Annu Rev Med 57:181–194

Houldcroft CJ, Beale MA, Breuer J (2017) Clinical and biological insights from viral genome sequencing. Nat Rev Microbiol 15(3):183–192

Hue KD, Tuan TV, Thi HT, Bich CT, Anh HH, Wills BA, Simmons CP (2011) Validation of an internally controlled one-step real-time multiplex RT-PCR assay for the detection and quantitation of dengue virus RNA in plasma. J Virol Methods 177(2):168–173

Jansen HJ, Liem M, Jong-Raadsen SA, Dufour S, Weltzien FA, Swinkels W, Koelewijn A, Palstra AP, Pelster B, Spaink HP, Thillart GEVD, Dirks RP, Henkel CV (2017) Rapid *de novo* assembly of the European eel genome from nanopore sequencing reads. Sci Rep 7(1):7213

Johansen CA, van den Hurk AF, Ritchie SA, Zborowski P, Nisbet DJ, Paru R, Bockarie MJ, Macdonald J, Drew AC, Khromykh TI, Mackenzie JS (2000) Isolation of Japanese encephalitis virus from mosquitoes (Diptera: Culicidae) collected in the Western Province of Papua New Guinea, 1997–1998. Am J Trop Med Hyg 62(5):631–638

Johnson DJ, Ostlund EN, Pedersen DD, Schmitt BJ (2001) Detection of North American West Nile virus in animal tissue by a reverse transcription-nested polymerase chain reaction assay. Emerg Infect Dis 7(4):739–741

Johnson BW, Russell BJ, Lanciotti RS (2005) Serotype-specific detection of dengue viruses in a fourplex real-time reverse transcriptase PCR assay. J Clin Microbiol 43(10):4977–4983

Johnson N, Voller K, Phipps LP, Mansfield K, Fooks AR (2012) Rapid molecular detection methods for arboviruses of livestock of importance to northern Europe. J Biomed Biotechnol 2012:719402

Kauffman EB, Jones SA, Dupuis AP II, Ngo KA, Bernard KA, Kramer LD (2003) Virus detection protocols for West Nile virus in vertebrate and mosquito specimens. J Clin Microbiol 41 (8):3661–3667

Keven JB, Katusele M, Vinit R, Koimbu G, Vincent N, Thomsen EK, Karl S, Reimer LJ, Walker ED (2019) Species abundance, composition, and nocturnal activity of female *Anopheles* (Diptera: Culicidae) in malaria-endemic villages of Papua New Guinea: assessment with barrier screen sampling. Malar J 18(1):96

Ko HY, Salem GM, Chang GJ, Chao DY (2020) Application of next-generation sequencing to reveal how evolutionary dynamics of viral population shape dengue epidemiology. Front Microbiol 11:1371

Konishi E, Takahashi J (1985) Detection of chikungunya virus antigen in *Aedes albopictus* mosquitoes by enzyme-linked immunosorbent assay. J Virol Methods 12(3–4):279–285

Kumari R, Kumar K, Chauhan LS (2011) First dengue virus detection in *Aedes albopictus* from Delhi, India: its breeding ecology and role in dengue transmission. Trop Med Int Health 16 (8):949–954

Lanciotti RS, Kerst AJ, Nasci RS, Godsey MS, Mitchell CJ, Savage HM, Komar N, Panella NA, Allen BC, Volpe KE, Davis BS, Roehrig JT (2000) Rapid detection of West Nile virus from human clinical specimens, field-collected mosquitoes, and avian samples by a TaqMan reverse transcriptase-PCR assay. J Clin Microbiol 38(11):4066–4071

Lanciotti RS, Kosoy OL, Laven JJ, Panella AJ, Velez JO, Lambert AJ, Campbell GL (2007) Chikungunya virus in US travelers returning from India, 2006. Emerg Infect Dis 13(5):764–767

Lanciotti RS, Kosoy OL, Laven JJ, Velez JO, Lambert AJ, Johnson AJ, Stanfield SM, Duffy MR (2008) Genetic and serologic properties of Zika virus associated with an epidemic, Yap State, Micronesia, 2007. Emerg Infect Dis 14(8):1232–1239

Li S, Fang M, Zhou B, Ni H, Shen Q, Zhang H, Han Y, Yin J, Chang W, Xu G, Cao G (2011) Simultaneous detection and differentiation of dengue virus serotypes 1–4, Japanese encephalitis virus, and West Nile virus by a combined reverse-transcription loop-mediated isothermal amplification assay. Virol J 8:360

Liew JWK, Selvarajoo S, Tan W, Ahmad Zaki R, Vythilingam I (2019) Gravid oviposition sticky trap and dengue non-structural 1 antigen test for early surveillance of dengue in multi-storey dwellings: study protocol of a cluster randomized controlled trial. Infect Dis Poverty 8(1):71

Liu H, Liu ZJ, Jing J, Ren JQ, Liu YY, Guo HH, Fan M, Lu HJ, Jin NY (2012) Reverse transcription loop-mediated isothermal amplification for rapid detection of Japanese encephalitis virus in swine and mosquitoes. Vector Borne Zoonotic Dis 12(12):1042–1052

Lopez-Jimena B, Bekaert M, Bakheit M, Frischmann S, Patel P, Simon-Loriere E, Lambrechts L, Duong V, Dussart P, Harold G, Fall C, Faye O, Sall AA, Weidmann M (2018) Development and validation of four one-step real-time RT-LAMP assays for specific detection of each dengue virus serotype. PLoS Negl Trop Dis 12(5):e0006381

Maan S, Maan NS, Belaganahalli MN, Rao PP, Singh KP, Hemadri D, Putty K, Kumar A, Batra K, Krishnajyothi Y, Chandel BS, Reddy GH, Nomikou K, Reddy YN, Attoui H, Hegde NR, Mertens PP (2015) Full-genome sequencing as a basis for molecular epidemiology studies of bluetongue virus in India. PLoS One 10(6):e0131257

Maciel-de-Freitas R, Eiras AE, Lourenco-de-Oliveira R (2006) Field evaluation of effectiveness of the BG-Sentinel, a new trap for capturing adult *Aedes aegypti* (Diptera: *Culicidae*). Mem Inst Oswaldo Cruz 101(3):321–325

Mansuy JM, Lhomme S, Cazabat M, Pasquier C, Martin-Blondel G, Izopet J (2018) Detection of Zika, dengue and chikungunya viruses using single-reaction multiplex real-time RT-PCR. Diagn Microbiol Infect Dis 92(4):284–287

Mathenge EM, Omweri GO, Irungu LW, Ndegwa PN, Walczak E, Smith TA, Killeen GF, Knols BG (2004) Comparative field evaluation of the Mbita trap, the Centers for Disease Control light trap, and the human landing catch for sampling of malaria vectors in western Kenya. Am J Trop Med Hyg 70(1):33–37

McHardy AC, Martín HG, Tsirigos A, Hugenholtz P, Rigoutsos I (2007) Accurate phylogenetic classification of variable-length DNA fragments. Nat Methods 4(1):63–72

Microsoft Premonition (n.d.) Microsoft Research (blog). https://www.microsoft.com/en-us/research/project/project-premonition/. Accessed 29 Jan 2021

Mosquito World (n.d.) Mosquito trap reviews 2021. http://www.mosquitoworld.net/mosquito-control/trap-reviews/. Accessed 29 Jan 2021

Moureau G, Temmam S, Gonzalez JP, Charrel RN, Grard G, de Lamballerie X (2007) A real-time RT-PCR method for the universal detection and identification of flaviviruses. Vector Borne Zoonotic Dis 7(4):467–477

Musso D, Bossin H, Mallet HP, Besnard M, Broult J, Baudouin L, Levi JE, Sabino EC, Ghawche F, Lanteri MC, Baud D (2018) Zika virus in French Polynesia 2013–14: anatomy of a completed outbreak. Lancet Infect Dis 18(5):e172–e182

Naccache SN, Federman S, Veeraraghavan N, Zaharia M, Lee D, Samayoa E, Bouquet J, Greninger AL, Luk KC, Enge B, Wadford DA, Messenger SL, Genrich GL, Pellegrino K, Grard G, Leroy E, Schneider BS, Fair JN, Martínez MA, Isa P, Crump JA, DeRisi JL, Sittler T, Hackett J Jr, Miller S, Chiu CY (2014) A cloud-compatible bioinformatics pipeline for ultrarapid pathogen identification from next-generation sequencing of clinical samples. Genome Res 24 (7):1180–1192

Oulas A, Pavloudi C, Polymenakou P, Pavlopoulos GA, Papanikolaou N, Kotoulas G, Arvanitidis C, Iliopoulos I (2015) Metagenomics: tools and insights for analyzing next-generation sequencing data derived from biodiversity studies. Bioinform Biol Insights 9:75–88

Parida M, Posadas G, Inoue S, Hasebe F, Morita K (2004) Real-time reverse transcription loop-mediated isothermal amplification for rapid detection of West Nile virus. J Clin Microbiol 42 (1):257–263

Parida MM, Santhosh SR, Dash PK, Tripathi NK, Saxena P, Ambuj S, Sahni AK, Lakshmana Rao PV, Morita K (2006) Development and evaluation of reverse transcription-loop-mediated isothermal amplification assay for rapid and real-time detection of Japanese encephalitis virus. J Clin Microbiol 44(11):4172–4178

Parida MM, Santhosh SR, Dash PK, Tripathi NK, Lakshmi V, Mamidi N, Shrivastva A, Gupta N, Saxena P, Babu JP, Rao PV, Morita K (2007) Rapid and real-time detection of chikungunya virus by reverse transcription loop-mediated isothermal amplification assay. J Clin Microbiol 45 (2):351–357

Patel P, Landt O, Kaiser M, Faye O, Koppe T, Lass U, Sall AA, Niedrig M (2013) Development of one-step quantitative reverse transcription PCR for the rapid detection of flaviviruses. Virol J 10:58

Pfeffer M, Proebster B, Kinney RM, Kaaden OR (1997) Genus-specific detection of alphaviruses by a semi-nested reverse transcription-polymerase chain reaction. Am J Trop Med Hyg 57 (6):709–718

Pierre V, Drouet MT, Deubel V (1994) Identification of mosquito-borne flavivirus sequences using universal primers and reverse transcription/polymerase chain reaction. Res Virol 145:93–104

Pyke AT, Smith IL, van den Hurk AF, Northill JA, Chuan TF, Westacott AJ, Smith GA (2004) Detection of Australasian Flavivirus encephalitic viruses using rapid fluorogenic TaqMan RT-PCR assays. J Virol Methods 117(2):161–167

Pyke AT, Daly MT, Cameron JN, Moore PR, Taylor CT, Hewitson GR, Humphreys JL, Gair R (2014) Imported zika virus infection from the Cook Islands into Australia, 2014. PLoS Curr 6: ecurrents.outbreaks.4635a54dbffba2156fb2fd76dc49f65e

Quan PL, Williams DT, Johansen CA, Jain K, Petrosov A, Diviney SM, Tashmukhamedova A, Hutchison SK, Tesh RB, Mackenzie JS, Briese T, Lipkin WI (2011) Genetic characterization of K13965, a strain of Oak Vale virus from Western Australia. Virus Res 160(1–2):206–213

Rhoads A, Au KF (2015) PacBio sequencing and its applications. Genomics Proteomics Bioinformatics 13(5):278–289

Ritchie SA, Phillips D, Broom A, Mackenzie J, Poidinger M, van den Hurk A (1997) Isolation of Japanese encephalitis virus from *Culex annulirostris* in Australia. Am J Trop Med Hyg 56 (1):80–84

Ritchie SA, van den Hurk AF, Zborowski P, Kerlin TJ, Banks D, Walker JA, Lee JM, Montgomery BL, Smith GA, Pyke AT, Smith IL (2007) Operational trials of remote mosquito trap systems for Japanese encephalitis virus surveillance in the Torres Strait, Australia. Vector Borne Zoonotic Dis 7(4):497–506

Ritchie SA, Cortis G, Paton C, Townsend M, Shroyer D, Zborowski P, Hall-Mendelin S, Van Den Hurk AF (2013) A simple non-powered passive trap for the collection of mosquitoes for arbovirus surveillance. J Med Entomol 50(1):185–194

Rosen L, Gubler D (1974) The use of mosquitoes to detect and propagate dengue viruses. Am J Trop Med Hyg 23(6):1153–1160

Sadeghi M, Altan E, Deng X, Barker CM, Fang Y, Coffey LL, Delwart E (2018) Virome of >12 thousand *Culex* mosquitoes from throughout California. Virology 523:74–88

Sambri V, Capobianchi M, Charrel R, Fyodorova M, Gaibani P, Gould E, Niedrig M, Papa A, Pierro A, Rossini G, Varani S, Vocale C, Landini MP (2013) West Nile virus in Europe: emergence, epidemiology, diagnosis, treatment, and prevention. Clin Microbiol Infect 19 (8):699–704

Sanchez-Seco MP, Rosario D, Quiroz E, Guzman G, Tenorio A (2001) A generic nested-RT-PCR followed by sequencing for detection and identification of members of the alphavirus genus. J Virol Methods 95(1–2):153–161

Santiago GA, Vázquez J, Courtney S, Matías KY, Andersen LE, Colón C, Butler AE, Roulo R, Bowzard J, Villanueva JM, Muñoz-Jordan JL (2018) Performance of the Trioplex real-time RT-PCR assay for detection of Zika, dengue, and chikungunya viruses. Nat Commun 9(1):1391

Scaramozzino N, Crance J-M, Jouan A, DeBriel DA, Stoll F, Garin D (2001) comparison of universal primer pairs and development of a rapid, highly sensitive heminested reverse transcription-PCR assay for detection of flaviviruses targeted to a conserved region of the NS5 gene sequences. J Clin Microbiol 39(5):1922–1927

Schlaberg R, Chiu CY, Miller S, Procop GW, Weinstock G, Professional Practice Committee and Committee on Laboratory Practices of the American Society for Microbiology; Microbiology Resource Committee of the College of American Pathologists (2017) Validation of metagenomic next-generation sequencing tests for universal pathogen detection. Arch Pathol Lab Med 141(6):776–786

Silva SJRD, Paiva MHS, Guedes DRD, Krokovsky L, Melo FL, Silva MALD, Silva AD, Ayres CFJ, Pena LJ (2019) Development and validation of reverse transcription loop-mediated isothermal amplification (RT-LAMP) for rapid detection of ZIKV in mosquito samples from Brazil. Sci Rep 9(1):4494

Singh KRP (1967) Cell cultures derived from larvae of *Aedes albopictus* (Skuse) and *Aedes aegypti* (*L.*). Curr Sci 36(19):506–508

Singh M, Al-Eryani G, Carswell S, Ferguson JM, Blackburn J, Barton K, Roden D, Luciani F, Giang Phan T, Junankar S, Jackson K, Goodnow CC, Smith MA, Swarbrick A (2019) High-throughput targeted long-read single cell sequencing reveals the clonal and transcriptional landscape of lymphocytes. Nat Commun 10(1):3120

Staunton KM, Rohde BB, Townsend M, Liu J, Desnoyer M, Howell P, Amos B, Crawford J, Snoad N, Ritchie SA (2020) Investigating male *Aedes aegypti* (Diptera: Culicidae) attraction to different oviposition containers using various configurations of the sound gravid *Aedes* trap. J Med Entomol 57(3):957–961

Staunton KM, Crawford JE, Liu J, Townsend M, Han Y, Desnoyer M, Howell P, Xiang W, Burkot TR, Snoad N, Ritchie SA (2021) A low-powered and highly selective trap for male *Aedes* (Diptera: Culicidae) surveillance: the male *Aedes* sound trap. J Med Entomol 58(1):408–415

Sutherland GL, Nasci RS (2007) Detection of West Nile virus in large pools of mosquitoes. J Am Mosq Control Assoc 23(4):389–395

Takken W, Kline DL (1989) Carbon dioxide and 1-octen-3-ol as mosquito attractants. J Am Mosq Control Assoc 5(3):311–316

Tang Z, Yamada H, Kraupa C, Canic S, Busquets N, Talavera S, Jiolle D, Vreysen MJB, Bouyer J, Abd-Alla AMM (2020) High sensitivity of one-step real-time reverse transcription quantitative PCR to detect low virus titers in large mosquito pools. Parasit Vectors 13(1):460

Tangena JA, Thammavong P, Hiscox A, Lindsay SW, Brey PT (2015) The human-baited double net trap: an alternative to human landing catches for collecting outdoor biting mosquitoes in Lao PDR. PLoS One 10(9):e0138735

Tyson JR, O'Neil NJ, Jain M, Olsen HE, Hieter P, Snutch TP (2018) MinION-based long-read sequencing and assembly extends the *Caenorhabditis elegans* reference genome. Genome Res 28(2):266–274

van den Hurk AF, Hall-Mendelin S, Pyke AT, Smith GA, Mackenzie JS (2010) Vector competence of Australian mosquitoes for chikungunya virus. Vector Borne Zoonotic Dis 10(5):489–495

Vazquez-Prokopec GM, Galvin WA, Kelly R, Kitron U (2009) A new, cost-effective, battery-powered aspirator for adult mosquito collections. J Med Entomol 46(6):1256–1259

Walker T, Jeffries CL, Mansfield KL, Johnson N (2014) Mosquito cell lines: history, isolation, availability and application to assess the threat of arboviral transmission in the United Kingdom. Parasit Vectors 7(1):382

Wang J, Moore NE, Deng YM, Eccles DA, Hall RJ (2015) MinION nanopore sequencing of an influenza genome. Front Microbiol 6:766

Warrilow D, Watterson D, Hall RA, Davis SS, Weir R, Kurucz N, Whelan P, Allcock R, Hall-Mendelin S, O'Brien CA, Hobson-Peters J (2014) A new species of mesonivirus from the Northern Territory, Australia. PLoS One 9(3):e91103

Weaver SC, Forrester NL (2015) Chikungunya: evolutionary history and recent epidemic spread. Antivir Res 120:32–39

Williams CR, Long SA, Russell RC, Ritchie SA (2006) Field efficacy of the BG-sentinel compared with CDC backpack aspirators and CO_2-baited EVS traps for collection of adult *Aedes aegypti* in Cairns, Queensland, Australia. J Am Mosq Control Assoc 22(2):296–300

Xiao P, Li C, Zhang Y, Han J, Guo X, Xie L, Tian M, Li Y, Wang M, Liu H, Ren J, Zhou H, Lu H, Jin N (2018) Metagenomic sequencing from mosquitoes in China reveals a variety of insect and human viruses. Front Cell Infect Microbiol 8:364

Yang Y, Garver LS, Bingham KM, Hang J, Jochim RC, Davidson SA, Richardson JH, Jarman RG (2015) Feasibility of using the mosquito blood meal for rapid and efficient human and animal virus surveillance and discovery. Am J Trop Med Hyg 93(6):1377–1382

Yaren O, Alto BW, Gangodkar PV, Ranade SR, Patil KN, Bradley KM, Yang Z, Phadke N, Benner SA (2017) Point of sampling detection of Zika virus within a multiplexed kit capable of detecting dengue and chikungunya. BMC Infect Dis 17(1):293

Zhang YU, Tian ZW, Ge JQ, Wang YI (1984) A simplified method for the identification of arboviruses: virus detection using enzyme immunoassay on cultured cells. J Virol Methods 9:45–51

Chapter 15
Long-Lasting Insecticidal Nets: An Evidence-Based Technology for Malaria Vector Control and Future Perspectives

Vas Dev

Abstract The advent of long-lasting insecticidal nets (popularly known as LLINs) have revolutionized the concept of vector control and are increasingly in demand for wider community acceptance and perceived benefits on account of decreased nuisance due to mosquito bites and relative freedom from malaria. LLINs are currently being promoted worldwide and proven appropriate technology for added advantages of extended residual bio-efficacy and durability that span over 3–4 years (serviceable life of net). These nets employ pyrethroids (a class of insecticide), which have been recommended for use in public health due to their property of negligible mammalian toxicity, rapid knockdown effect, extended longevity of residual effect and affordable costs. LLIN-based intervention has proven efficacious against local disease vectors for containment of malaria transmission in varied ecological settings and is recommended by the World Health Organization for incorporation in health systems prioritizing high-risk population groups. However, population coverage remains miniscule of what is needed for universal coverage to achieve the coveted goal of malaria elimination. The need of hour is greater allocation of resources ensuring equitable access for which innovative strategies are mandated for mass distribution including indigenous production, social marketing, reduced taxes and tariffs and partnership between government, donors and corporate sectors alike, making LLIN a household commodity to combat malaria disease, vectors of malaria and spread of drug-resistant malaria.

Keywords Malaria · *P. falciparum* · LLIN · Vector control · India · Southeast Asia

---

V. Dev (✉)
National Institute of Malaria Research, New Delhi, India

B. K. Tyagi (ed.), *Genetically Modified and other Innovative Vector Control Technologies*, https://doi.org/10.1007/978-981-16-2964-8_15

15.1 Introduction

Malaria continues to inflict high morbidity in tropical countries with large concentration of cases in sub-Saharan Africa contributing 94% of total estimated cases and deaths worldwide (World Malaria Report 2020). In the WHO South-East Asia Region (SEAR), while Sri Lanka and Maldives have been certified to be malaria-free, India (despite COVID-19 pandemic) made laudable progress with a reduction in cases from estimated 20 million cases in 2000 to 5.6 million cases in 2019. Ironically, India alone contributed 86% of the deaths reported in the SEA Region countries. *Plasmodium falciparum* constituting nearly 50% of the reported cases is a fast-evolving multi-drug-resistant parasite which results in extended morbidity and is held solely responsible for deaths in the country (Anon. 2020a). The National Vector Borne Disease Control Programme's malaria control rests mainly on two pillars, viz., (1) disease surveillance and (2) vector control (Anon. 2020b). Disease epidemiology is highly complex governed my multiplicity of mosquito vectors and vast and varied terrains and contextual ecological determinants (Anvikar and Dev 2020). At the same time, vector control relies mostly on insecticide residual spraying (IRS) supplemented by distribution of insecticide-treated nets (ITNs) with pyrethroids (Babu 2020). IRS operations are besieged with several operational constraints including high refusal rates, logistics of transportation and storage, rising costs and emerging insecticide resistance (Yadav et al. 2020). While manually impregnated ITNs were held effective in preventing malaria, their coverage remained off target, and they invariably required re-treatment on a 6-monthly basis (Jana-Kara et al. 1995; Dev and Dash 2008). Instead the advent of long-lasting insecticidal nets (LLINs) obviated the need of re-treatment of net, could stay effective about 3 years with or without repeated washes and proved epidemiologically significant and cost-effective on long-term basis. LLIN has thus revolutionized the concept of vector control with added advantages (Dev 2009). Different varieties of LLINs are currently recommended by WHO (Sharma 2020), some of which, for instance, PermaNet, Olyset net, Interceptor net and Duranet, have been applied in malaria hyper-endemic Indian states of Assam, Odisha, Chhattisgarh, Madhya Pradesh and Uttar Pradesh for containment of disease transmitted by different vector species and assessing much needed community responses (Sharma et al. 2006; Sreehari et al. 2007; Dev et al. 2010a, b; Bhatt et al. 2012; Gunasekaran et al. 2014). Results from these evaluations were made available to the program and policy managers reporting substantial disease transmission reduction. Included in this chapter are discussed research findings from field evaluation of LLINs in malaria endemic blocks highlighting the characteristics and advantages of LLIN-based intervention for data-based control in difficult terrains of Assam State in northeastern India (cf. Box 15.1: Frequently Asked Questions).

Box 15.1 Frequently Asked Questions

What is a long-lasting insecticidal net?

Long-lasting insecticidal net (popularly known as LLIN) is a ready-to-use factory-treated insecticidal mosquito net which does not require re-treatment for 3–4 years (the net serviceable life span). The insecticide (pyrethroid) is either coated around polyester netting fibre (type-1) or incorporated into polyethylene polymer before fibre extrusion (type-2). These nets can be washed multiple times and still retain adequate residual bio-efficacy against target mosquito vector species for extended periods so called LLIN.

What are pyrethroids?

Pyrethroids are class of insecticides, viz. deltamethrin, alpha-cypermethrin and permethrin, which have been recommended for use in public health for having relatively low toxicity to humans, rapid knockdown effect, extended longevity of residual effect and comparative low costs. These compounds are neurotoxic and have multiple modes of action against the mosquito vector/insect pests which involve opening of sodium channels resulting in continuous nerve excitation, paralysis and death of the mosquito. Pyrethroids also have an irritant effect resulting in hyperactivity, feeding inhibition and shorter landing period on human host, all of which help reducing mosquito vector population density and infective bites.

Why do we need long-lasting insecticidal nets?

Despite decades of attempted vector control using insecticide residual spraying (IRS), malaria transmission remained uninterrupted in tropical countries. While conventionally used mosquito nets conferred some degree of personal protection, these nets when treated with insecticide (ITNs), efficacy is increased manifold but needed re-impregnation every 6 months for continued protection. This limitation is overcome by advent of long-lasting insecticidal net obviating the need for re-treatment exercises of the community-owned mosquito nets optimizing much needed community compliance. The ITNs and LLINs look physically alike, but LLINs are the one that ensures long-term protection against malaria transmitting mosquitoes without having had the requirement to re-impregnate with the insecticide of choice.

What are the recommended products of long-lasting insecticidal nets by WHO?

Number of LLIN products have been field-evaluated and qualified the criterion for extended residual bio-efficacy and durability for minimum of 3 years of continuous use and accorded either interim or full approval by the World Health Organization, to name a few, viz. Olyset net® (permethrin incorporated), Interceptor net® (alpha-cypermethrin coated) and PermaNet® (deltamethrin coated) which have been accorded full recommendation.

(continued)

Box 15.1 (continued)

LLINs are made available in different colours, shapes and sizes as per given preferences.

Are these long-lasting insecticidal nets accepted by the communities?

Use of LLIN is a well-tested intervention for disease vector control in many countries with considerable success and held advantageous over indoor residual spraying. Community acceptance is overwhelming and forthcoming as personal guard against malaria and other pest insects, e.g. bedbugs and head lice. LLINs are held operationally feasible, community-based and sustainable intervention for low operational costs and being promoted as key intervention for universal coverage of risk populations.

15.2 LLINs Have Extended Residual Bio-Efficacy

LLINs are factory-treated ready-to-use mosquito nets made either of polyester (type-1) or polyethylene (type-2) that are resistant to multiple washings. Number of branded LLINs has been accorded interim or full recommendations based on WHO specifications for use in control operations (Table 15.1). All LLINs are treated with pyrethroid at doses which are proven safe to humans and efficacious against target disease vectors. In type-1 category, polyester net is resin coated with target dose of pyrethroid, while in type-2, the insecticide is incorporated into polyethylene fibre in process of polymerization. Both categories of LLINs are proven highly potent in preventing mosquito bites by killing action as well as repelling vector mosquitoes, thereby reducing transmission intensities for extended periods (WHO 2005). For data based on Olyset net (permethrin-incorporated LLIN), persistence of the residual bio-efficacy monitored by standard techniques by contact cone-bioassay revealed 100% mortality post 24 h exposure against principal vector mosquito species, *Anopheles minimus*, in Assam for first few months (March–December 2007), but it gradually declined from 98% in January 2008 down to 81% in January 2010 after continuous use in field conditions; yet mortality range was greater than requisite 80% as per WHOPES criteria of retention of residual efficacy for minimum of 3 years period (Fig. 15.1). Similar trends were observed in % knockdown mosquitoes subject to 3 min exposure that decreased significantly from 74% in March 2007 to 53% in January 2010 ($P \leq 0.05$). These observations are suggestive of gradual depletion of available insecticide on net fibre over time in use by wear and tear alone.

These observations were further corroborated by data on ring-net bioassay (a measure to monitor the residual bio-availability of insecticide on net fibre) studies that revealed steady increase in knockdown time for 1st mosquito (5.00–10.20 m), 6th mosquito (8.20–14.00 m) and 11th mosquito (14.00–21.20 m) in the corresponding study period (Fig. 15.2). Based on the univariate *t*-test for change from baseline observations, the increase in knockdown time for three categories of

Table 15.1 List of long-lasting insecticidal nets recommended by WHO[a] (*Source:* Sharma 2020)

S. No.	Product name	Fabric	Product type	Status of WHO recommendation
1.	DawaPlus 2.0	Polyester	Deltamethrin coated	Interim
2.	DawaPlus 3.0	Polyester (side panels); polyethylene (roof)	Combination of deltamethrin coated on polyester and deltamethrin + PBO incorporated into polyethylene	Interim
3.	DawaPlus 4.0	Polyethylene	Deltamethrin + PBO incorporated	Interim
4.	Duranet	Polyethylene	Alpha-cypermethrin incorporated	Full
5.	Interceptor	Polyester	Alpha-cypermethrin coated	Full
6.	Interceptor G2	Polyester	Alpha-cypermethrin and chlorfenapyr coated	Interim
7.	LifeNet	Polypropylene	Deltamethrin incorporated	Interim
8.	Magnet	Polyethylene	Alpha-cypermethrin incorporated	Full
9.	MiraNet	Polyethylene	Alpha-cypermethrin incorporated	Interim
10.	Olyset net	Polyethylene	Permethrin incorporated	Full
11.	Olyset Plus	Polyethylene	Permethrin + PBO incorporated	Interim
12.	Panda Net 2.0	Polyethylene	Deltamethrin incorporated	Interim
13.	PermaNet 2.0	Polyester	Deltamethrin coated	Full
14.	PermaNet 3.0	Polyester (side panels); polyethylene (roof)	Deltamethrin coated on side panels and deltamethrin + PBO incorporated into roof	Interim
15.	Royal Sentry	Polyethylene	Alpha-cypermethrin incorporated	Full
16.	SafeNet	Polyester	Alpha-cypermethrin coated	Full
17.	Veeralin	Polyethylene	Alpha-cypermethrin + PBO incorporated	Interim
18.	Yahe	Polyester	Deltamethrin coated	Interim
19.	Yorkool	Polyester	Deltamethrin coated	Full

[a]Source: WHO Pesticide Evaluation Scheme (WHOPES), updated 29 June 2017

mosquitoes was significant ($P \leq 0.05$), suggesting substantial depletion of bio-availability of insecticide.

15.3 LLINs Are Wash-Resistant

LLINs nets are proven to withstand residual bio-efficacy for up to 20 serial washings (gently with soap and water) tested at fortnightly intervals for having >80% mortality against target vector species. For data based on Duranet® LLIN (alpha-

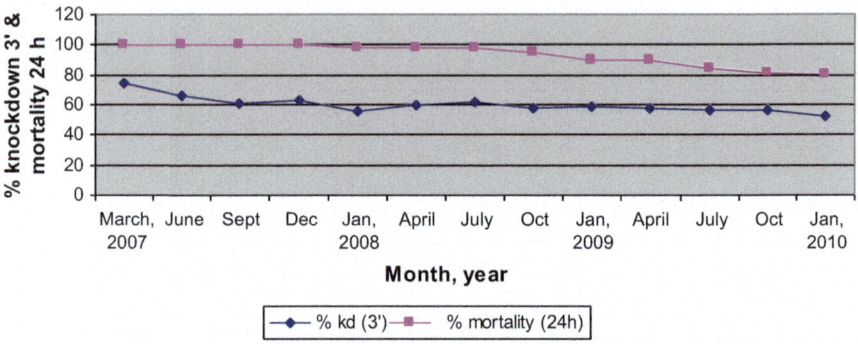

Fig. 15.1 Residual bio-efficacy of Olyset® net in use by households expressed in terms of percent knockdown of *Anopheles minimus* group of mosquito species post 3 min exposure and mortality 24-h recovery period for data based in Assam, India (March 2007–January 2010)

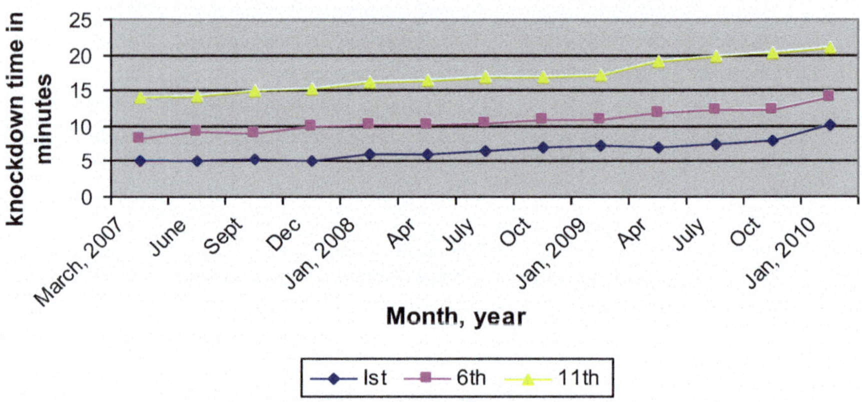

Fig. 15.2 Residual bio-efficacy of net fibre monitored by ring-net bioassay of Olyset net® in use by households against *Anopheles minimus* group of mosquito species in Assam, India, for the extended follow-up study period (March 2007–January 2010)

cypermethrin-incorporated net), mortality was >95% in serially washed and 100% in unwashed nets post 24 h exposure (Table 15.2).

However, there was significant reduction ($P = 0.002$) in % knockdown mosquitoes post 3 min exposure both in washed (58–28%) and unwashed nets (73–40%). Assuming that community-owned nets are washed at quarterly intervals, persistence of residual bio-efficacy up to 20 washes was held adequate lasting estimated serviceable net life of 3–4 years. Similar study results were reported in Olyset as well as Interceptor LLINs on serial washings reaffirming their capacity to retain residual efficacy despite repeated washings (Sharma et al. 2006, 2010; Dev et al. 2010a, b; Bhatt et al. 2012).

Table 15.2 Wash resistance of Duranet® LLIN expressed in terms of percent knockdown of *Anopheles minimus* group of mosquito species post 3 min exposure and mortality 24-h recovery period in laboratory conditions in response to serial washing at periodic intervals[a]

	Serially washed Duranet LLIN		Unwashed Duranet LLIN		Washed untreated net
S. No. washing (fortnightly interval) of net	No. mosquitoes knockdown post 3 min exposure (% knockdown)	No. mosquitoes dead post 24 h recovery period (% mortality)	No. mosquitoes knockdown post 3 min exposure (% knockdown)	No. mosquitoes dead post 24 h recovery period (% mortality)	No. mosquitoes dead post 24 h recovery period (% mortality)
1	35 (58)	60 (100)	22 (73)	30 (100)	1 (5)
2	33 (55)	60 (100)	20 (67)	30 (100)	0 (0)
3	31 (51)	60 (100)	18 (60)	30 (100)	1 (5)
4	29 (48)	60 (100)	15 (50)	30 (100)	0 (0)
5	21 (52)	40 (100)	10 (50)	20 (100)	0 (0)
6	22 (55)	40 (100)	11 (55)	20 (100)	1 (5)
7	21 (52)	40 (100)	11 (55)	20 (100)	0 (0)
8	19 (48)	40 (100)	12 (60)	20 (100)	0 (0)
9	20 (50)	40 (100)	11 (55)	20 (100)	1 (5)
10	21 (52)	40 (100)	10 (50)	20 (100)	2 (10)
11	19 (48)	40 (100)	09 (45)	20 (100)	1 (5)
12	17 (42)	40 (100)	10 (50)	20 (100)	1 (5)
13	15 (37)	40 (100)	10 (50)	20 (100)	0 (0)
14	14 (35)	40 (100)	09 (45)	20 (100)	0 (0)
15	16 (40)	40 (100)	10 (50)	20 (100)	2 (10)
16	13 (33)	40 (100)	09 (45)	20 (100)	1 (5)
17	14 (35)	40 (100)	08 (40)	20 (100)	2 (10)
18	12 (30)	40 (100)	09 (45)	20 (100)	3 (15)
19	11 (28)	39 (98)	08 (40)	20 (100)	3 (15)
20	11 (28)	38 (95)	08 (40)	20 (100)	2 (10)

[a]Based on exposure of ten anopheline mosquitoes per cone-bioassay

15.4 LLINs Are Durable

Given the guidelines for monitoring the durability of long-lasting insecticidal mosquito nets under operational conditions (WHO 2011), cross sectional community surveys in the beneficiary population groups revealed that earlier distributed LLINs 3 years ago were still in possession and were being used for personal protection. However, physical inspection of community-used LLINs (both inclusive of Olyset and Interceptor) revealed that 48.7% (243/499) were torn for having holes inclusive of small (0.5–2 cm), medium (2–10 cm) and large size of (>10 cm) diameter (Dev et al. 2016). However, small-sized holes were of more frequent occurrence with cumulative average of 3.4 (832/243) holes per LLIN. Majority of the holes in torn LLINs, however, were concentrated in lower half of the net with an overall average of 3.3 (806/243) per LLIN. Of total nets inspected, some of the community users (80/243, 33%) had repaired the torn net by stitching which appeared to be the common practice rather than tying knots and applying patches. Of those checked for physical appearance, 55% (274/499) were clean, 29% (146/499) bit dirty and remaining dirty to very dirty. Nevertheless, a substantial number of LLINs were torn after 3 years of continuous use yet conferred protection against mosquito nuisance evidenced by the mere absence of mosquito vector populations and reduced transmission. Similar observations were recorded in other physiographic zones of India providing evidence of durability, ownership and residual efficacy in community-used LLINs (Bhatt et al. 2012; Anuse et al. 2015; Raghavendra et al. 2017).

15.5 LLINs Are Community-Based and Socially Acceptable

Based on the periodic questionnaire-based surveys during different timings of community usage, it was observed that householders were fully aware of benefits of using LLINs as personal protection method, and compliance was 100% of the individuals surveyed (Table 15.3).

In the first round of community surveys conducted during December 2006–January 2007 after few months of usage (initially distributed in October 2006), 36% (81/225) of users complained of skin-related disorders (skin irritation/itching), 8.8% (20/225) of users reported eye irritation, and 0.04% (1/225) complained of headache; but all these were reported to be transitory in nature lasting few days of initial use. However, during follow-up community surveys conducted in March 2007, November 2008 and May 2009, there was no such complaining except 1% (2/200) of users reported skin irritation. All surveys combined, majority 86% (609/706) reported reduction in mosquito bites, and 24.4% (104/425) observed collateral benefits such as decreased nuisance due to body lice, head lice and bedbugs. Most householders 666/706 (94%) recommended the use of LLIN as personal guard against malaria and other pest insects and clearly preferred LLIN over residual spraying (Fig. 15.3).

15.6 LLINs Are Proven Intervention Against Malaria Transmission

Investigations based in Assam revealed that LLINs served as effective intervention between mosquito and human host preventing infective bites compared to untreated (plain net) and no-net control villages, thus disrupting malaria transmission. While mosquito landing rates in LLIN intervention villages were substantially reduced

Table 15.3 Cross-sectional community response surveys among LLIN users for data based on Interceptor net earlier distributed in October 2006 in experimental villages of the Sonapur Primary Health Centre (Dimoria Block), Kamrup District, Assam, India

S. No.	Question	% Users[a] ($N = 225$)	% Users[b] ($N = 200$)	% Users[c] ($N = 281$)	% Users[d] ($N = 200$)
1.	Are you familiar of the benefit of using mosquito net?	100	100	100	100
2.	Are you using any other indigenous method for mosquito control?	No	No	No	No
3.	Do you sleep inside the LLIN?	100	100	100	100
4.	**Did you suffer any of the following?**				
	Skin irritation	12	1	0	0
	Nausea	0	0	0	0
	Vomiting	0	0	0	0
	Itching	24	0	0	0
	Headache	0.4	0	0	0
	Drowsiness	0	0	0	0
	Eye irritation	8.8	0	0	0
	Difficulty in breathing	0	0	0	0
	Any other	0	0	0	0
5.	**Observations/perceptions using LLINs**				
	Reduction in mosquito bites	85.7	74.7	95	–
	Reduction in nuisance due to bedbugs	18.6	02	No response	–
	Reduction in nuisance due to head lice	16.4	9.5	No response	–
	Reduction in nuisance due to body lice	0.8	0	No response	–
6.	**Do you recommend use of LLIN in future?**	100	80	100	–

[a]First survey conducted in December 2006–January 2007
[b]Second round in March 2007
[c]Third survey done in November 2008
[d]Fourth survey done in May 2009

(Dev et al. 2010a, b), extended follow-up investigations revealed that vector populations virtually had disappeared (Fig. 15.4). These research findings were corroborated by data on comparative malaria incidences in experimental villages. There was substantial transmission reduction in LLIN villages almost by 80% opposed to untreated nets intervention, while incidence in no-net villages had increased manifold in the corresponding study periods (Table 15.4).

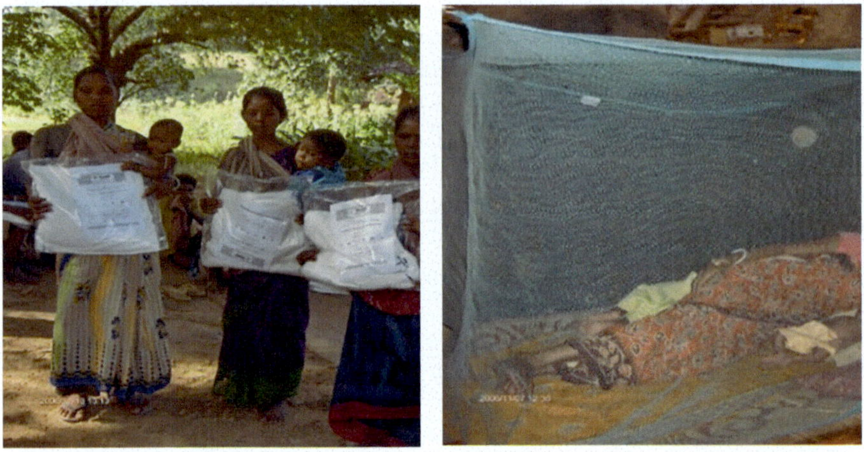

Fig. 15.3 (Left) Distribution of long-lasting insecticidal nets in vulnerable population groups in high-risk areas; (right) mother and child sleeping under LLIN

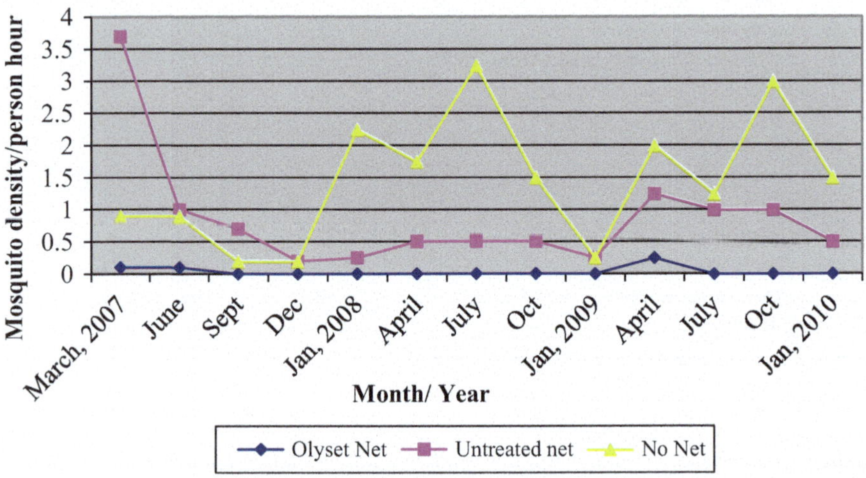

Fig. 15.4 Relative abundance of *Anopheles minimus* per person-hour in study area including Olyset net users, untreated net intervention and without net control villages of the Sonapur Primary Health Centre (Dimoria Block) of Kamrup District of Assam during the follow-up study period March 2007–January 2010

15.7 Conclusions and the Way Forward

In essence, LLINs are proven intervention against malaria in containing disease vector mosquitoes resulting in substantial transmission reduction world over. These nets are widely accepted and increasing popular for perceived benefits of decreased nuisance due to biting insects and relative freedom from malaria. Nevertheless, there

Table 15.4 Data on malaria incidence in experimental villages of the Sonapur Primary Health Centre (Dimoria Block), Kamrup District, Assam, India

Category of intervention villages	Study period[a]	Pop.	No. of malaria cases (age in years) (<5)	(5–15)	(>15)	Total	Cases/ 1000 pop.
Interceptor net[b]	Aug–Sept 2006	2494	16	23	16	55	22
	Oct 06–April 2007		0	04	06	10	4
Plain net	Aug–Sept 2006	2199	24	33	24	81	37
	Oct 06–April 2007		07	47	29	82	37
No net	Aug–Sept 2006	2092	15	12	12	39	19
	Oct 06–April 2007		40	91	102	233	111

[a]Malaria incidence for Aug.–Sept. 2006 is the baseline data
[b]Nets were distributed in October 2006

is scope of research on nets, which are robust, and insecticides that are more potent to meet the emerging threat of pyrethroid resistance (Yadav et al. 2020). In this context, advent of next-generation LLINs, viz. PermaNet 3.0; Olyset Plus combining mix of different technologies incorporating synergists, viz. piperonyl butoxide (PBO); Interceptor G2 nets treated with chlorfenapyr + alpha-cypermethrin (Pennetier et al. 2013; Corbel et al. 2010; N'Guessan et al. 2014); and nets incorporating antimalarials inhibiting development of parasite in mosquito host, seems to be promising in defeating insecticide resistance (Hemingway 2019: Paton et al. 2019). There is body of evidence that LLINs continue to provide protection against malaria irrespective of prevailing insecticide resistance given the wider coverage and community compliance (Tokponnon et al. 2019). Requirements are huge, but distribution remains patchy and far from universal coverage for which greater allocation of resources is need of the hour (Anuse et al. 2015). Yet presently available options of LLINs conferred protection for just about 3 years of usage for waning efficacy and durability warranting fresh supply for which there exists no such provision of net replacement of torn/worn-out nets providing protection uninterrupted. While many countries are certified malaria-free, transmission of malaria is clearly deaccelerating in group of E-2020 countries, but much more can be achieved by universal access of interventions to each one and everyone at any risk (WHO 2017a, b). India has made a huge leap forward in containing transmission and embarked upon malaria elimination in the foreseeable future (Narain and Nath 2018). With the present-day knowledge on disease vectors and transmission dynamics, malaria elimination is well within reach for which rollout of evidence-based interventions are crucial to end malaria (Dhiman et al. 2018; Dev 2020).

Acknowledgements This study was partly funded by the Clarke Mosquito Control Products, Inc., Illinois, USA; M/S Sumitomo Chemical India Ltd.; and BASF India Ltd. for logistic supplies of Duranet, Olyset and Interceptor long-lasting insecticidal nets, respectively. The funding agencies did not participate in the experimental design and influence the study results in any manner. The author also wishes to thank the state health directorate of Assam for coordination, local communities for extending cooperation and research support staffs for technical expertise.

References

Anon (2020a) Malaria situation in India. National Vector Borne Disease Control Programme, Directorate General of Health Services, Ministry of Health and Family Welfare, Government of India, India. https://www.nvbdcp.gov.in/malaria. Accessed 5 Jan 2020

Anon (2020b) Malaria control strategies. National Vector Borne Disease Control Programme, Directorate General of Health Services, Ministry of Health and Family Welfare, Government of India, New Delhi. https://www.nvbdcp.gov.in/malaria. Accessed 5 Jan 2020

Anuse SS, Sahu SS, Subramanian S et al (2015) Usage pattern, physical integrity & insecticidal efficacy of long-lasting insecticidal nets in Odisha state, India. Indian J Med Res 142:71–78

Anvikar A, Dev V (2020) Malaria transmission in India: disease distribution and prevalence of mosquito vectors in physiographic zones. In: Dev V (ed) Vector biology and control, an update for malaria elimination initiative in India. The National Academy of Sciences, New Delhi, pp 117–128

Babu CJ (2020) Malaria vector control in India: present perspectives and continuing challenges. In: Dev V (ed) Vector biology and control, an update for malaria elimination initiative in India. The National Academy of Sciences, India, pp 149–164

Bhatt RM, Sharma SN, Uragayala S et al (2012) Effectiveness and durability of Interceptor long-lasting insecticidal nets in a malaria endemic area of Central India. Malar J 11:189

Corbel V, Chabi J, Dabire RD et al (2010) Field efficacy of a new mosaic long lasting net (PermaNet 3) against pyrethroid resistant malaria vectors: a multi-centre study in Western and Central Africa. Malar J 9:113

Dev V (2009) Long-lasting insecticidal nets for malaria control. Curr Sci 97:469–470

Dev V (2020) Vector biology and control: an update for malaria elimination initiative in India. The National Academy of Sciences, New Delhi, p 272

Dev V, Dash AP (2008) Insecticide-treated nets, the key element for rolling back malaria in North-Eastern India: policy and practice. Open Entomol J 2:14–20

Dev V, Raghavendra K, Barman K et al (2010a) Wash-resistance and field efficacy of Olyset net, a permethrin incorporated long-lasting insecticidal netting against *Anopheles minimus* transmitted malaria in Assam, northeastern India. Vector Borne Zoonotic Dis 10:403–410

Dev V, Raghavendra K, Singh SP et al (2010b) Wash resistance and residual efficacy of long-lasting polyester netting coated with alpha-cypermethrin (Interceptor) against malaria-transmitting mosquitoes in Assam, Northeast India. Trans R Soc Trop Med Hyg 104:273–278

Dev V, Barman K, Khound K (2016) A cross-sectional study for assessing residual bio-efficacy and durability of field distributed long-lasting insecticidal nets in malaria endemic ethnic communities of Assam, Northeast India. J Infect Pub Health 9:298–307

Dhiman S, Singh V, Dev V (2018) Declining transmission of malaria in India: accelerating towards elimination. In: Manguin S, Dev V (eds) Towards malaria elimination – a leap forward. InTech Open, London, pp 257–280

Gunasekaran K, Sahu SS, Vijayakumar T et al (2014) Comparison of efficacy of five types of long-lasting insecticidal nets against *Anopheles fluviatilis*, the primary malaria vector in East-Central India. J Med Entomol 51(4):785–794

Hemingway J (2019) Malaria parasite tackled in mosquitoes. Nature 567:185–186

Jana-Kara BR, Wajihullah WA, Shahi B et al (1995) Deltamethrin impregnated bed nets against *Anopheles minimus* transmitted malaria in Assam, India. J Trop Med Hyg 98:73–83

N'Guessan R, Ngufor C, Kudom AA et al (2014) Mosquito nets treated with a mixture of Chlorfenapyr and Alphacypermethrin control Pyrethroid resistant *Anopheles gambiae* and *Culex quinquefasciatus* mosquitoes in West Africa. PLoS One 9(2):e87710

Narain JP, Nath LM (2018) Eliminating malaria in India by 2027: the countdown begins. Indian J Med Res 148:123–126

Paton DG, Childs LM, Itoe MA et al (2019) Exposing Anopheles mosquitoes to antimalarials blocks Plasmodium parasite transmission. Nature 567:239–243

Pennetier C, Bouraima A, Chandre F et al (2013) Efficacy of Olyset Plus, a new long-lasting insecticidal net incorporating Permethrin and Piperonil Butoxide against multi-resistant malaria vectors. PLoS One 8(10):e75134

Raghavendra K, Chourasia MK, Swain DK et al (2017) Monitoring of long-lasting insecticidal nets (LLINs) coverage versus utilization: a community-based survey in malaria endemic villages of Central India. Malar J 16:467

Sharma SK (2020) Malaria prevention through personal protection: insecticide-treated mosquito nets and allied intervention tools. In: Dev V (ed) Vector biology and control, an update for malaria elimination initiative in India. The National Academy of Sciences, New Delhi, pp 205–218

Sharma SK, Upadhyay AK, Haque MA et al (2006) Wash resistance and bio-efficacy of Olyset Net – a long-lasting insecticide treated net against malaria vectors and nontarget household pests. J Med Entomol 43:884–888

Sharma SK, Upadhyay AK, Haque MA et al (2010) Wash-resistance and field evaluation of alphacypermethrin treated long-lasting insecticidal net (Interceptor) against malaria vectors, *Anopheles culicifacies* and *Anopheles fluviatilis* in tribal area of Orissa, India. Acta Trop 116:24–30

Sreehari U, Mittal PK, Razdan RK et al (2007) Efficacy of PermaNet 2.0 against *Anopheles culicifacies* and *Anopheles stephensi*, malaria vectors in India. J Am Mosq Control Assoc 23 (2):220–223

Tokponnon FT, Sissinto Y, Ogouyémi AH et al (2019) Implications of insecticide resistance for malaria vector control with long-lasting insecticidal nets: evidence from health facility data from Benin. Malar J 18:37

World Health Organization (2005) Guidelines for laboratory and field testing of long-lasting insecticidal mosquito nets. WHO/CDS/WHOPES/GCDPP/2005.11

World Health Organization (2011) Guidelines for monitoring the durability of long-lasting insecticidal mosquito nets under operational conditions. WHO/HTM/NTD/WHOPES, Geneva

World Health Organization (2017a) Regional action plan 2017–2030. Towards 0. Malaria-free South-East Asia Region. Regional Office for South-East Asia, New Delhi. License: CC BY-NC-SA 3.0 IGO

World Health Organization (2017b) Global Malaria Programme. Achieving and maintaining universal coverage with long-lasting insecticidal nets for malaria control. WHO/HTM/GMP/2017.20

World Malaria Report 2020: 20 years of global progress and challenges. Geneva. World Health Organization. Licence: CC BY-NC SA 3.0 IGO Programme

Yadav RS, Raghavendra K, Velamuri PS, Verma V, Uragayala S (2020) Insecticide resistance in malaria vectors and management strategies in India. In: Dev V (ed) Vector biology and control, an update for malaria elimination initiative in India. The National Academy of Sciences, New Delhi, pp 129–148

Chapter 16
Bioefficacy Evaluation of Indigenously Developed Long-Lasting Insecticidal Net (LLIN) 'Defender Net™' for Control of Mosquitoes and Their Diseases in India

Vijay Veer, S. K. Dhiman, K. Chandel, S. N. Tikkar, Bipul Rabha, Diganta Goswami, I. Baruah, P. K. Talukdar, R. Yadav, and B. K. Tyagi

Abstract Indigenous technology for long-lasting insecticidal mosquito net (LLIN) of monofilament of polyethylene was developed by Defence Research and Development Organisation (DRDO), named as Defender Net™, which is effective against mosquitoes in their varied Indian geoclimatic conditions. Defender Net™ contains a synthetic pyrethroid insecticide, deltamethrin at 0.15% ww or 55 a.i./sq.m. and a synergist piperonyl butoxide (PBO) at 0.55% ww and licensed under 9 (3B) of the Insecticides Act in India. Defender Net™ was found effective against disease vectors like *Anopheles stephensi*, *Aedes aegypti* and *Culex quinquefasciatus* in laboratory bioassay (phase I) even after 20 washings. *Cx. quinquefasciatus* is a sturdy mosquito among tested three species and very common in India. Defender Net™ was found effective against malaria vectors and safe to the users in the malaria-endemic Assam state and various culicines and mansonoides in the lymphatic filariasis (LF) as well as Japanese encephalitis (JE)-endemic Tamil Nadu state in a 3-year long investigation (phase II). Defender Net™ can be effectively incorporated within the Integrated Vector Management (IVM) strategy for any mosquito-borne disease management especially in the difficult frontier areas in the country.

Keywords LLIN · Defender Net · Mosquitoes · Vectors · Control · India

---

V. Veer (✉) · S. K. Dhiman · B. Rabha · D. Goswami · I. Baruah · P. K. Talukdar
Defence Research Laboratory, Tezpur, Assam, India

K. Chandel · S. N. Tikkar · R. Yadav
Defence Research & Development Establishment, Gwalior, Madhya Pradesh, India

B. K. Tyagi
SpoRIC, VIT University, Vellore, Tamil Nadu, India

B. K. Tyagi (ed.), *Genetically Modified and other Innovative Vector Control Technologies*, https://doi.org/10.1007/978-981-16-2964-8_16

16.1 Introduction

Mosquitoes are one of the greatest enemies of man. Mosquito-borne diseases like malaria, dengue, chikungunya, filariasis, encephalitis, etc., affecting half of the world population, cause high morbidity and mortality to humans, mainly in the tropical and subtropical countries. Injudicious use of insecticides, especially DDT, to control vectors during past several decades has resulted in development of resistance in mosquitoes, in addition to environment pollution and human health hazards. All these factors led to the rollback of malaria and re-emergence of some of other deadly mosquito-borne diseases like dengue and chikungunya. Thus, a more comprehensive all-inclusive Integrated Vector Management (IVM) strategy was developed so that dependency on the insecticides could be minimized. Vector control is still a mainstay in control of vector-borne diseases particularly where specific antidote is lacking and the vaccine is still a far cry, e.g. dengue, chikungunya, Zika virus, etc. Various methods to control mosquitoes were developed with different level of effectiveness either to suppress or reduce the vector population so that contact between vector and man can be reduced. These methods are biopesticide (based on *Bacillus thuringiensis israelensis* serotype H-14 (bti-14)) and predatory fishes like *Gambusia affinis*, odour-baited traps (OBT), insecticide-based residual spray, mat, vaporizer, repellent cream and lotion, bed net and environmental engineering. Every method/device has its own merits and demerits. Insecticide-impregnated mosquito net has been quite useful in bringing down the malaria cases in the trial areas (Lengler 2004).

Of these vector control methods in vogue, insecticide-treated nets (ITN) proved efficacious in several parts of the world in reducing the mosquito density and also collaterally reducing malaria incidence (Djènontin et al. 2018; Pennetier et al. 2013; WHO 2013; Dev et al. 2016; Dhiman 2019; Bhatt et al. 2012). However, these early ITNs needed periodic re-impregnation, besides erratic impregnation dose and other factors diluted their efficacy in the long run; the impregnated chemicals frequently diluting and the nets needing periodic chemical treatment every 6–12 months which itself warranted trained personnel to handle chemicals safely and visit sites regularly to ensure the ITNs were properly used and treated on appropriate time. To overcome these deficiencies, long-lasting insecticidal nets (LLINs) were developed by incorporating a suitable insecticide in the synthetic yarn. These LLINs are being effectively deployed in many malaria-endemic areas in India and elsewhere, and the World Health Organization (WHO) has recommended deployment of the LLIN as an important component in malaria control. The percentage households owning at least one LLIN in endemic areas of Africa, the severest malaria-affected continent, has increased from 3% in 2000 to >80% in 2019, making it the most widely deployed intervention tool against mosquitoes under malaria control campaign.

Various types of LLINs have been developed with varying degrees of sustainability and performance. Generally, the LLINs have synthetic pyrethroid insecticide (s) either coated on or incorporated into the polyester or polyethylene yarns. In the coating method, a resin is used to bind the insecticide to the surface of the yarn, making it wash-resistant, while in the incorporation method, a suitable pyrethroid is mixed with the raw polyethylene and then extruded into yarns which subsequently evolve the insecticide to surface of the yarn, thereby replenishing the surface concentration.

16.2 History of Insecticidal Mosquito Net

Initially bed nets were treated with existing emulsifiable concentrates of pyrethroids which were diluted with water, and then the bed nets were dipped in the solution followed by air drying which leave an insecticidal deposit on the fibre. Generally known as insecticide-treated nets or ITNs, these bed nets offer about twice the protection as compared to an untreated net. The effectiveness of the ITNs can be further enhanced by inveigling onto fibre with an appropriate repellent bonded with insecticide. The latter can even protect other people in the room outside the net. Early trials with such bed nets were conducted in Gambia, West Africa and India which showed that ITNs having permethrin pyrethroid have more than doubled its efficacy and had a huge impact on malaria incidence (Jambulingam et al. 1989; Mboma et al. 2021). Trials of insecticide-treated nets (ITNs) in the 1980s and 1990s showed that ITNs reduced deaths in young children by an average of 20%. This led to the commercial production of several 'dip' products of pyrethroid insecticides for treating polyester or cotton nets. On the basis of five community-randomised trials, it was concluded that when full coverage is achieved, ITNs reduce all-cause child mortality by an average of 18% (range 14–29%) in sub-Saharan Africa. The general implication of this is that 5.5 lives could be saved per year for every 1000 children under 5 years of age. It was concluded that ITNs reduce clinical episodes of malaria caused by *Plasmodium falciparum* and *P. vivax* infections by 50% on average (range 39–62%), as well as reducing the prevalence of high density parasitemia (Lengler 2004). However, there were some serious drawbacks to ITNs such as this required dilution of chemicals for treatment and dipping of bed nets after every 6–12 months in the solution which required training programmes for personnel to handle chemicals safely and their regular visits to the villages to ensure that bed nets were re-dipped at the appropriate time. While these early products clearly demonstrated the value of adding insecticide to bed nets, the programme LLINs come in with two technologies, one using existing polyester nets to factory for dipping treatment by using a resin to bind the insecticide to the surface of the fibre making it wash-resistant. So bed nets remained efficacious for long period. The other methodology, pioneered by Sumitomo Chemical, was to mix pyrethroid insecticide to the raw (virgin) polyethylene and then extruded it into mono-fibre, and that trapped insecticide subsequently migrated to the fibre surface to give protection from mosquitoes. Thus on use or wash of net, the loss of surface concentration of insecticide is replenished from within the fibre and therefore provided long protection to the users.

The World Health Organization has drafted the guidelines for laboratory and field testing of LLINs to assess their efficacy which is famously called as 'WHO Pesticide Evaluation Scheme' (WHO 2005, 2013). Therefore, for a WHOPES-recommended product, its efficacy and safety are both assured. WHOPES recommendations are delivered in two stages:

1. An interim recommendation is delivered after an accelerated wash resistance and small-scale field testing.
2. Subsequently, this interim recommendation is changed to full recommendation based on large-scale performance of efficacy and durability of the product over a 3-year period under field conditions. At present, only one LLIN (Olyset) has received full recommendation implying a minimum of 3 years of efficacy; all the others have interim recommendation, while field efficacy and durability tests are ongoing (http://www.who.Int/whopes/Long-lasting_insecticidal_nets_ok2.pdf).

Published data as well as observations related to durability of mosquito nets under field conditions are limited. However, surveys carried out in the African nations Senegal by WHO and Tanzania by the Swiss Tropical Institute (STI) (Tami et al. 2004) showed that the majority of polyethylene (Olyset) nets distributed 7 years ago in villages were still hanging over the beds and in reasonably good condition and provided good residual efficacy in laboratory tests against mosquitoes after years of use. A survey carried out in Morogoro Region of Tanzania concluded that the effective life of polyester nets was 2–3 years (Erlanger et al. 2004), confirming observations made by PSI (Population Services International) in the same area. The effectiveness of ITNs often failed due to the cost and logistics of regularly dipping.

16.3 LLIN Status in India

Arthropod vectors, besides their painful bites, transmit several dreadful diseases like malaria, dengue, chikungunya, Japanese encephalitis, filariasis, etc. to human being in the Indian subcontinent. Malaria alone causes several deaths annually in different regions; however, a majority of cases are reported in northeastern region of the country. Indian Armed Forces operating in such difficult terrains and inclement environments are often vulnerable to the painful mosquito bites and the deadly mosquito-borne diseases (Dhiman et al. 2010a, b, 2011). Several methods are in vague to tackle the mosquito menace and disease transmission. One of the most unique and promising tools is the use of mosquito repellents and residual insecticides with low mammalian toxicity, since conventional methods using insecticides are not sufficient to control mosquito populations, besides being hazardous to humans and non-target organisms as well as the environment. Therefore, though the use of cotton bed net has been known for many decades for protection from mosquitoes, yet the method had several limitations and did not provide foolproof protection during sleep hours if not properly used; more than often the users in sleep would make contact with the ambushing mosquitoes on the net walls. Mosquitoes also entered the bed net if the net was not properly tucked below the mattress.

Mosquito nets treated with pyrethroid insecticides have been shown to cause a decline in malaria morbidity and mortality in several trials carried out in different countries (Djènontin et al. 2018; Pennetier et al. 2013; Dev et al. 2016; Tami et al. 2004; Erlanger et al. 2004). Although insecticide-dipped nets killed the mosquitoes which were coming in its contact, periodic re-impregnation, low treatment rates and impact of repeated washing on the efficacy of nets have been a serious drawback in affecting the application of insecticide-treated nets on longer duration, in addition to health hazards to both the handlers of insecticides and the users at sleep particularly children and pregnant women, and high cost and labour associated with recurrent bed net treatment.

To overcome all these problems, a need was felt to develop insecticide-incorporated net which should be effective against mosquitoes for several years and even after many washes. Such LLINs developed by a few companies are now available in the international market. Yet, in such cases of nets, the insecticide with the aid of specially designed binders is incorporated only once in the synthetic yarn at the manufacturing stage in such a way that during net use insecticide migrates to the surface of the yarn in controlled fashion and offers to the net the desired mosquito control capability even after several washes.

In India, as such the LLIN technology is not available to the Indian industry. LLINs are either imported or manufactured by multinational companies through their undisclosed technology and are available at exorbitant price. Moreover, those imported LLINs are never tested against mosquitoes in local environments and under varied Indian geoclimatic conditions where their behaviour changes very frequently. Considering the circumstances, Defence Research and Development Organisation took upon to invent country's first, indigenous LLIN with the primary objective to provide protection to the Indian Armed Forces from vexatious bites of mosquitoes and other blood-sucking arthropods. The LLINs can be an important component of Integrated Vector Management Programme by reducing the load of vector-borne disease transmission in both army and civil populations. Indigenously developed LLIN technology could be able to make the country self-reliant in this area especially new start ups.

16.4 WHO Requirement for LLIN

The following criteria have been recommended by the WHO for LLIN (WHO 2006):

- 20 standard washes
- 3 years of efficacy for polyesters
- 5 years of efficacy for polyethylene

16.4.1 Local Criteria for Ideal LLIN Followed

- Highly toxic to mosquitoes but nontoxic to mammals at the concentration used.
- At least 80% mortality after 24 h of exposure and 95% knockdown in mosquitoes after 60 min of exposure (WHO requirement).
- Effective even after 20 washings.
- Be resistant to the hazards of use, light, rubbing and pressing.
- Available at low cost and with pleasant fragrance.
- Be efficacious for at least 5 years.

16.5 Development of Indigenous LLIN (Defender Net™) and Regulatory Registration in India

The Central Insecticide Board (CIB) & Registration Committee (RC), Directorate of Plant Protection, Quarantine & Storage, Ministry of Agriculture, Government of India has issued conditional registration the DRDO's LLIN (Defender Net™) (Registration No.: CIR-1639/2014 (352))—long-lasting insecticidal net (LLIN) dated the 20th March, 2015, containing a synthetic pyrethroid, deltamethrin 0.15% ww, or 55 a.i./sq.m. and a synergist Piperonyl butoxide (PBO) at 0.55% ww in monofilament of polyethylene under 9 (3B) of the insecticide Act. LLINs were having 100 ± 7 denier thick thread and 56 holes per square inch. This LLIN works on slow release of insecticide on demand basis, and technology has been patented in India.

16.6 Bioefficacy Evaluation of Defender Net™

Indigenously developed LLIN, Defender Net™, was evaluated as per phase I and phase II criteria in malaria-endemic areas in Assam and in lymphatic filariasis and JE-endemic areas of Tamil Nadu in India.

16.6.1 Phase I Evaluation

Phase I evaluation is mainly laboratory evaluation of LLIN (WHO 2006). The main objective of phase I evaluation is to determine the laboratory scale efficacy and wash resistance and to demonstrate the dynamics of insecticide on the net fibre (regeneration time). The primary objective is to establish efficacy after standard washings, so that the results should be consistent among different products and testing laboratories.

16.6.2 *Methodology*

At least four LLINs were required to be taken from two different batches. Altogether 56 pieces (14 pieces for each LLIN; 25 cm × 25 cm dimension) were taken as per WHO recommended pattern (Fig. 16.1) (WHO 2006).

Of these, 8 pieces were used to determine the regeneration time and 28 pieces to evaluate the wash resistance, while the remaining 20 pieces were wrapped in aluminium foil and stored at 4 °C for estimation of active molecules at intermittent time.

16.6.2.1 Bioassays

Five insecticide susceptible, non-blood-fed, 2–5-day-old female mosquitoes were exposed to netting material (25 cm × 25 cm) under standard cones, and efficacy was monitored using WHO standard cone bioassay (WHO 2005, 2006, 2013) (Fig. 16.2). Mosquitoes are exposed for 3 min, after which they are held for 24 h with access to sugar solution. Knockdown is recorded after 60 min, whereas mortality is noted after 24 h post-exposure. At least 50 mosquitoes on each net piece (10 replicates; 4–5 mosquitoes in each replicate) were tested. Mosquitoes exposed to plain nets (not treated with insecticide) were used as controls. Bioassay war carried out at 25 ± 2 °C and 70 ± 10% RH.

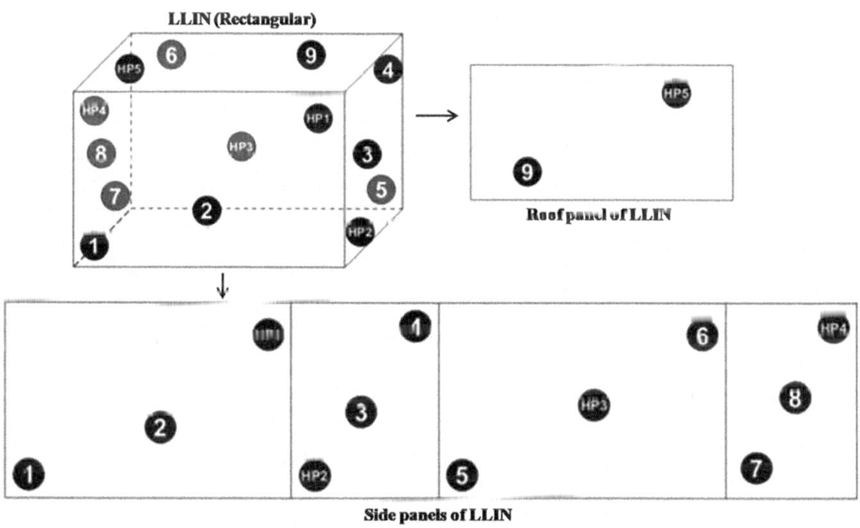

Fig. 16.1 LLIN sampling scheme for phase I evaluation (14 samples from an LLIN including HP1–HP5 for active ingredient assay). (*Source*: WHO 2013; autopermitted with acknowledgement)

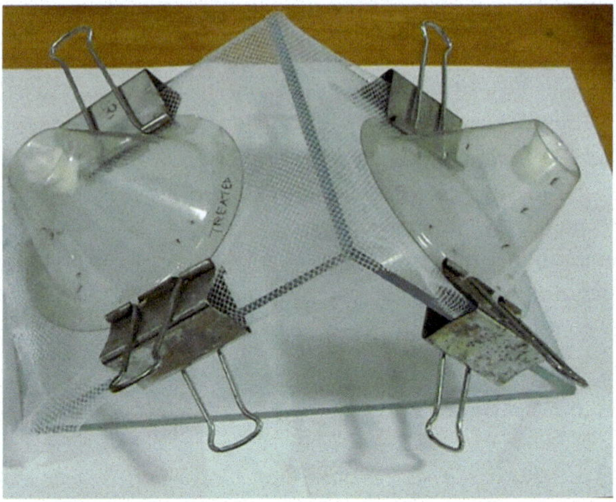

Fig. 16.2 WHO cone bioassay for mosquitoes using LLIN. (*Source*: This work)

16.6.2.2 Determination of Regeneration Time

Regeneration time is determined to estimate the time taken by insecticide to come to the surface after washing (WHO 2005, 2006, 2013). LLINs were washed and dried consecutively for 3 times following the standard WHO method and kept at 30 °C. WHO cone bioassays were carried out at 0, 1, 3 and 5 days post-washing. Efficacy curves of 24 h mortality and 60 min knockdown were drawn, and the time taken (in days) to reach a plateau was considered the regeneration time.

16.6.2.3 Wash Resistance

The resistance of an LLIN to washing was determined through standard bioassays carried out on nets washed at intervals required for regeneration using the standard WHO wash and dried and held at 30 °C (WHO 2006). Bioassays were done after every washing for *Cx. quinquefasciatus* for 30 min. Exposure and *Ae. aegypti* for 5 min. Exposure, and at every 0, 1, 5, 10, 15 and 20 washes for *An. stephensi* for 3 min. Exposure. Each bioassay was done just before the next wash. Net samples (25 cm × 25 cm) were introduced into 2 L beakers containing 1 L deionized water, with 1 g/L soap (pH 9–11) added just before and fully dissolved. Beakers were immediately kept at hotplate water temperature maintained at 30 °C and shaken for 10 min at 155 movements per minute. The samples were then removed and rinsed twice for 10 min in clean, deionized water in the same shaking condition as stated above. Nets were dried at room temperature and stored at 30 °C in the dark between washes.

16.7 Defender Net™ Phase I Evaluation Outcome

16.7.1 Exposure Time Optimization

WHO cone bioassay suggested that after 5 min of exposure, 1 h knockdown and 24 h mortality were 60% in *Cx. quinquefasciatus* mosquitoes; furthermore, both knockdown and mortality were found to be 100% each after 30 min of exposure to Defender Net™ (Table 16.1).

On the other hand, female *Ae. aegypti* recorded 80% knockdown and 90% mortality post 3 min exposure, while 100% knockdown and mortality were found after 5 in of exposure in WHO cone bioassay (Table 16.2).

16.7.2 Determination of Regeneration Time

Regeneration time (days) was determined by estimating the knockdown and mortality in *Ae. aegypti* and *Cx. quinquefasciatus* mosquitoes after different time intervals for one and three washings (Figs. 16.3 and 16.4). For *Ae. aegypti*, after one washing, knockdown was 98.0 ± 2.0 (95% CI 92.5–103.6), whereas mortality was 100.0 ± 0.0, respectively. Similarly, after 5 days, the knockdown (%) and mortality (%) were found to be 100.0 ± 0.0 and 96.0 ± 2.4 (95% CI 89.2–102.8), respectively (Fig. 16.3a). Furthermore, after three consecutive washes, the knockdown was 100.0 ± 0.0, whereas mortality was found to be 98.0 ± 2.0 (95% CI: 92.5–103.6), respectively, after 5 days of regeneration time (Fig. 16.3b).

For *Cx. quinquefasciatus*, after one washing, knockdown was 96.0 ± 2.4 (95% CI 89.2–102.8), whereas mortality was 92.0 ± 3.7 (95% CI 81.6–102.4). However, after 5 days, the knockdown (%) and mortality (%) were found to be 98.0 ± 2.0

Table 16.1 Optimization of exposure time in *Cx. quinquefasciatus* ($N = 50$ for each trial) (*Source*: This work)

Exposure time (min)	Knockdown post 1 h of exposure (%)	Mortality post 24 h (%)
3	50	40
5	60	60
10	70	80
15	90	100
30	100	100

Table 16.2 Optimization of exposure time in *Ae. aegypti* ($N = 50$ for each trial) (*Source*: This work)

Exposure time (min)	Knockdown post 1 h of exposure (%)	Mortality post 24 h (%)
3	80	90
5	100	100

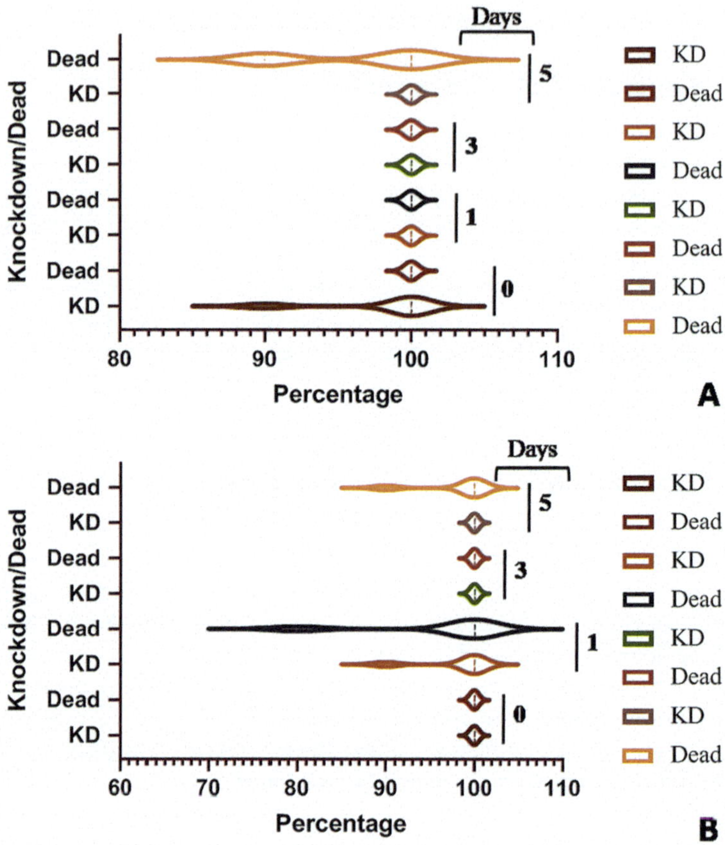

Fig. 16.3 Regeneration time (days) determined against *Ae. aegypti* for one wash (**a**) and three washes (**b**). (*Source*: This work)

(95% CI 92.5–103.6) each (Fig. 16.4a). Similarly, after three consecutive washes, the knockdown and mortality were found to be 94.0 ± 4.0 (95% CI 82.9–105.1) and 98.0 ± 2.0 (95% CI 92.5–103.6), respectively, after 5 days of regeneration time (Fig. 16.4b).

16.7.3 Laboratory Bioassay of Defender Net™ After 20 Washes

Bioefficacy was determined after different washes against *Cx. quinquefasciatus*, *Ae. aegypti* and *An. stephensi* mosquitoes. For all the three tested species, it was found that efficacy was 100% for knockdown and mortality except 80% mortality for *An. stephensi*. For *Cx. quinquefasciatus*, it was found that knockdown declined to 86%

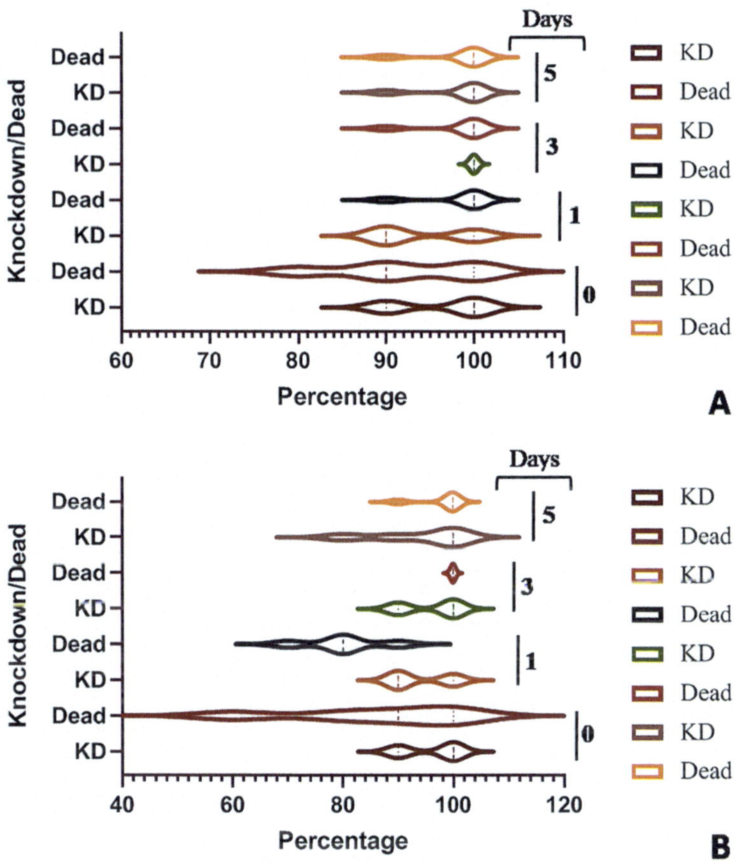

Fig. 16.4 Regeneration time of Defender Net™ determined against *Cx. quinquefasciatus* for one wash (**a**) and three washes (**b**). (*Source*: This work)

while mortality had declined to 92% after 20 standard washes (Fig. 16.5). In case of *Ae. Aegypti*, the knockdown and mortality after 20 washes reduced to 72% and 30%, respectively (Fig. 16.6). In contrast, for *An. stephensi*, the knockdown percentage and mortality were 100% and 80% post 20 WHO standard washes in laboratory (Fig. 16.7).

However, the efficacy was found improved when the LLINs were regenerated for 60 °C for 1 h after 20 washes. It was observed that in case of *Cx. quinquefasciatus*, after regeneration, the knockdown was 100%, while mortality was 98%. The efficacy did not improve significantly in case of *Ae. aegypti* after 1 h regeneration, as the knockdown and mortality stayed at 96% and 44%, respectively.

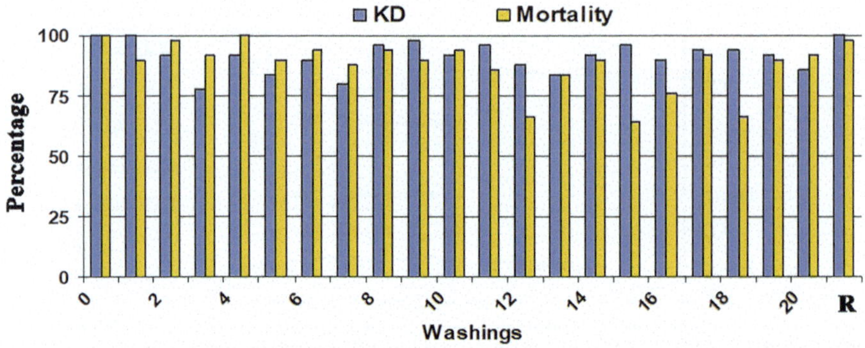

Fig. 16.5 Bioassay of LLIN Defender Net™ to *Cx. quinquefasciatus* after different washes (*KD* knockdown, *M* mortality, *R* regenerated for 1 h at 60 °C). (*Source*: This work)

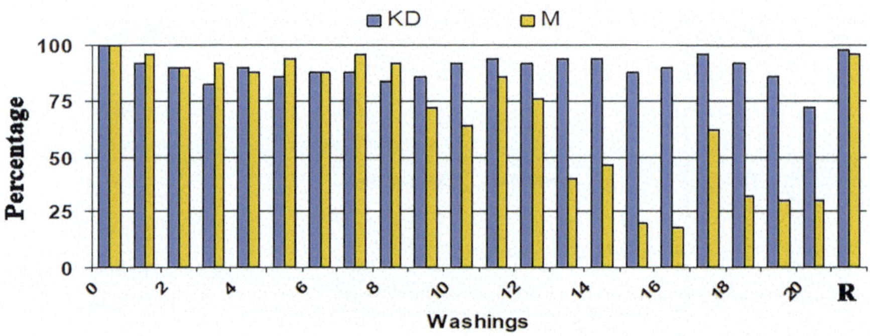

Fig. 16.6 Bioassay of LLIN Defender Net™ to *Ae. aegypti* after different washes (*KD* knockdown, *M* mortality, *R* regenerated for 1 h at 60 °C). (*Source*: This work)

16.8 Phase II Trial

Phase II field evaluation conducted in Missamari area of Assam, among armed forces troops in Assam, and in civil population of Tirukoilur block of Villupuram District, Tamil Nadu.

16.8.1 Field Evaluation in Assam (Anonymous 2015a)

The evaluation of Defender Net™ was carried out in malaria-endemic Assam state (India) with the following major objectives:

1. Biological activity of Defender Net™ in terms of effectiveness in deterring known mosquito vectors in the study area.

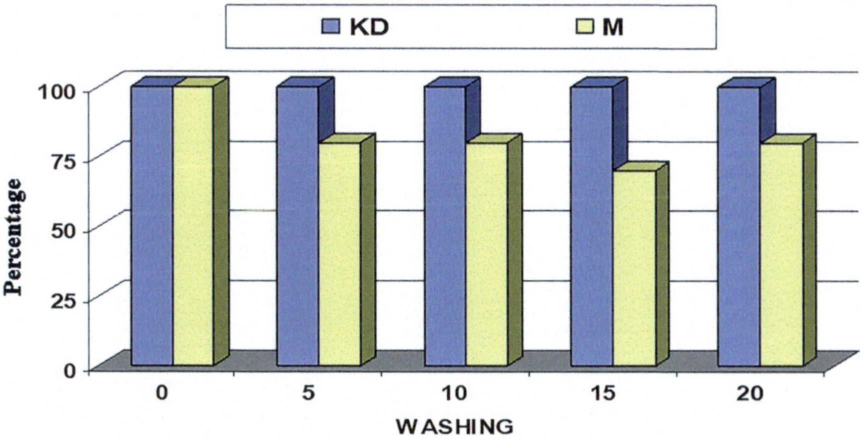

Fig. 16.7 Bioassay of LLIN Defender Net™ to *An. stephensi* after different washes (*KD* knock-down, *M* mortality). (*Source*: This work)

2. Evaluation of biological integrity of the Defender Net™ after various standard washings using the nets practically used in the operational condition.
3. Acceptance level and the safety observation of the users.

16.8.1.1 Study Area

The study was conducted in three villages situated on the north bank of Brahmaputra River in Assam, Northeast India. The villages selected have been reporting malaria cases for many years, and several efficient malaria vectors are reported in high density (Dhiman et al. 2012, 2018; Yadav et al. 2012). The villages surround a military cantonment having a highly susceptible troop population posted there from different non endemic regions of the country (Dhiman et al. 2010a, b, 2012). The study area falls under Balipara Primary Health Centre of Sonitpur district, which has reported a number of *Plasmodium falciparum* cases regularly. A previous study conducted in the area to identify the malaria risk zones in Sonitpur district suggested that the weighted annual parasitic index (API) ranged from 4 to 9, weighted annual blood examination rate (ABER) was 10 and weighted *P. falciparum* percentage (Pf %) ranged from 4 to 9 (Nath et al. 2013). The study area was identified to be at high risk of malaria.

The climate of study area is hot and humid, average temperature ranges from 32 to 35 °C in summer and 15 to 20 °C in winter, and humidity is around 80–90%, while the average annual rainfall lies between 170 and 220 cm. The monsoon period starts from June to September, though the rainfall starts from the early part of April (Nath et al. 2013). Study villages are inhabited by the Assamese, Bodo, Nepali and Adivasi people. The prevailing climatic conditions provide numerous breeding locations which help in vector mosquitoes breeding and proliferation throughout

Fig. 16.8 Different types of mosquito breeding habitats in the study area; (**a**) rice field, (**b**) marshy area, (**c**) water pools during rain, (**d**) house surrounded by rain water

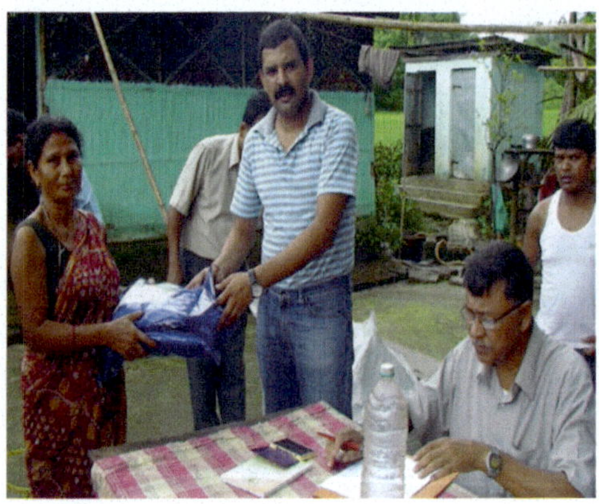

Fig. 16.9 Distribution of LLIN Defender Net™ among the study participants. (*Source*: This work)

the year (Fig. 16.8a–d). The villagers mainly depend on agriculture, while some are engaged in the tea industry (Rabha et al. 2011). The houses for the experiments were selected carefully keeping in mind the WHO design suggestion for the experiment huts. All the selected houses had only one room where at least two people used to sleep daily (Fig. 16.9). The study included three arms of which one each arm was for plain net (control), Defender Net™ (test) and insecticide treated net (ITN) which was conventionally treated using chemical (deltamethrin) equivalent to serve as positive control. Since Assam is highly endemic for malaria, therefore, no-net arm (fourth study arm without any net) was not taken in the present study. During the study, the indoor residual spray (IRS) and use of other insecticides were prohibited in the experimental houses. Informed consents were obtained from the participants, and the study was approved by the ethical review committee.

16.8.1.2 Mosquito Collection

All the adult mosquitoes entering the experimental houses were collected using 6 V battery-operated CDC miniature light traps (John Hock, Gainesville, FL, USA) during 1800–0600 h. Indoor resting collections were made using handheld aspirators and flashlights during 0500–0700 h. The adult mosquitoes were identified using the standard morphological keys.

16.8.1.3 WHO Cone Bioassay of Washed Defender Net™

The biological integrity of Defender Net™ with or without any wash after use in operational conditions was evaluated using WHO cone bioassay against laboratory reared adult *Ae. albopictus* mosquitoes. The results were recorded as corrected mortality and knockdown time (KDT). In addition insecticide susceptibility to 0.05% deltamethrin was also determined in adult *Ae. albopictus* mosquito following standard WHO tube bioassay.

16.9 Results

16.9.1 Pre-distribution Density and Randomization of the Study Villages

The mosquito collection reported different species in the area during March 2013–May 2013. For analysis purpose, the *Anopheles* mosquitoes were presented as anophelines. The known *Culex* vector mosquitoes were presented as culicines, while *Mansonia* were presented as Mansonioides

The density of *An. annularis*, *An. vagus*, *An. culicifacies* and *An. philippinensis*/*nivipes* and all anopheline among the three arms has been depicted in Fig. 16.10a–c. There was variation in the density of *Anopheles* mosquitoes, density of *An. annularis* and *An. vagus* being comparatively higher in each of the study arms. There was no significant difference in the total anopheline density among the three study arms ($p \geq 0.07$; KW ≤ 4.5).

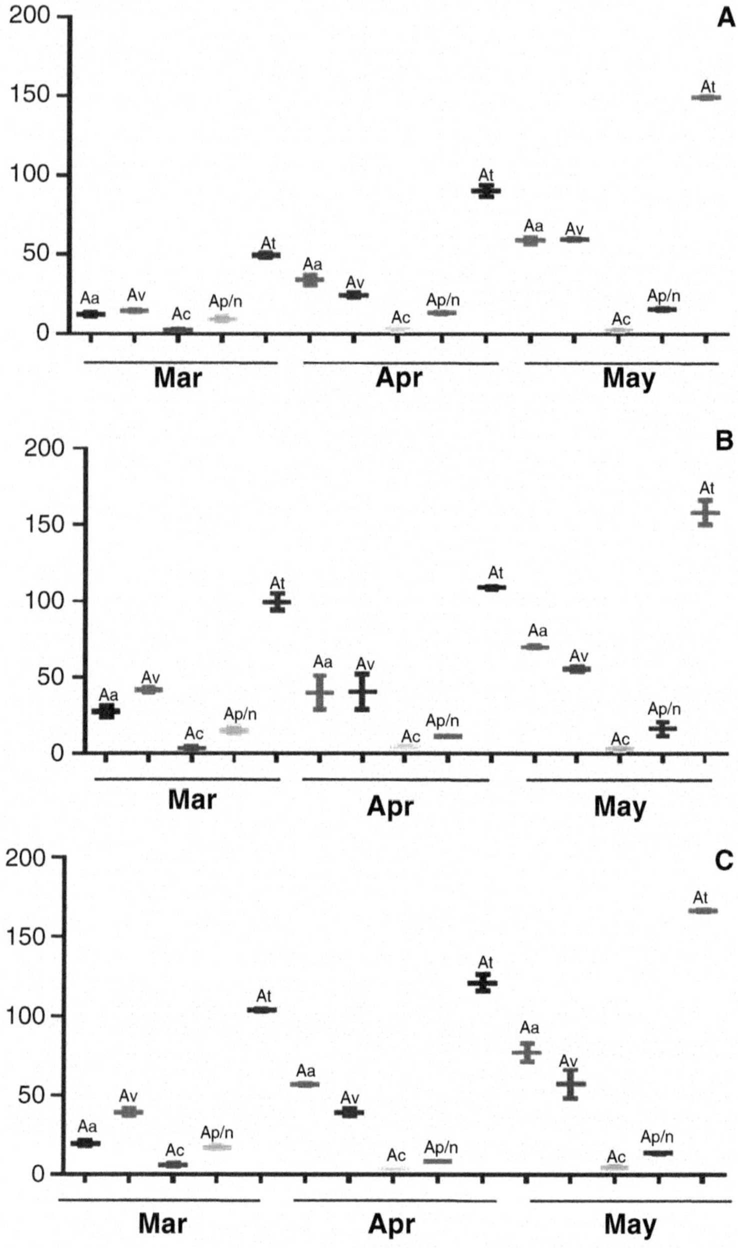

Fig. 16.10 Anophelines density (pre-distribution during Mar–May, 2013) (mean ± SEM) in all the three study arms selected for the evaluation study; Aa (*An. annularis*), Av (*An. vagus*), Ac (*An. culicifacies*), Ap/n (*An. philippinensis*/*nivipes*), At (anopheline total). (*Source*: This work)

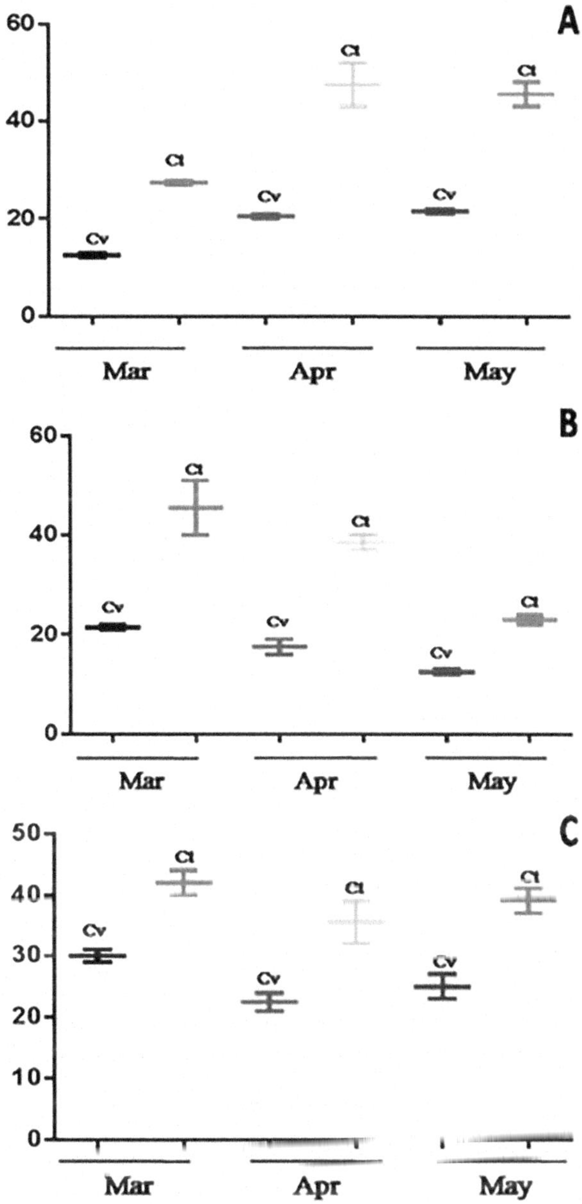

Fig. 16.11 Culicine density (pre-distribution during Mar–May 2013) (mean ± SEM) in all the three study arms (**a–c**) selected for the evaluation study; Cv = *Cx. vishnui* subgroup; Ct = Culicines total. (*Source*: This work)

The density recorded for culicine mosquitoes did not differ statistically among the three arms during the pre-evaluation study period. Among culicines, *Cx. vishnui* was recorded in large number, but the density was statistically similar among the three arms ($p \geq 0.07$; KW ≤ 4.6) (Fig. 16.11). Similarly, the density of known *Mansonia* vectors also did not differ statistically among the study arms in the pre-distribution months ($p \geq 0.06$; KW ≤ 4.7). Since the pre-distribution density of known vector mosquito was similar in all the study villages, the historical malaria records were similar. For the present study, control net, Defender Net™ and ITN were distributed in study arms 1, 2 and 3, respectively.

16.9.2 Post-distribution Density in the Trial Villages

In the post-distribution months (Jun–Nov 2013), there was significant decrease in the density of the vector mosquitoes in the study arm where Defender Net™ was deployed. The anopheline density was reduced to a bare minimum, and the decrease was statistically significant ($p \leq 0.02$; $F \geq 34.1$). Similarly, the density of culicines and Mansonioides was also significantly decreased statistically as compared to the plain net study arm ($p \leq 0.03$; $F \geq 13.4$ for culicines and $p \leq 0.08$; $F \geq 6.4$ for Mansonioides). Furthermore, there was immediate decline in the anopheline and culicines density in the Defender Net™ arm as compared to the ITN arm after 1 month of the start of the study ($p = 0.004$; $t = 81.0$ for anophelines and $p = 0.02$; $t = 15.0$ for culicines). The density of Mansonioides mosquitoes declined in the first month itself of the post-distribution ($p = 0.01$; $t = 51.0$).

16.9.3 Blood-Feeding Inhibition

Number of blood-fed mosquitoes corresponding to anophelines, culicines and Mansonioides were recorded from the Defender Net™ and plain net index houses and analysed to determine the percent protection provided by Defender Net™. The number recorded in each month during the post-intervention period has been shown in Fig. 16.12.

Every month there was significant reduction in the total blood-fed mosquitoes collected from the Defender Net™ houses as compared to the plain net houses during the post-intervention months ($p \leq 0.04$; $t \geq 4.9$). Total blood-fed mosquitoes collected in the Defender Net™ houses and plain net houses during the study have been depicted in Fig. 16.13.

Further, the mosquito blood-feeding inhibition achieved using Defender Net™ in the study places during the current study ranged from 82.1% to 100%. The percent protection recorded during the study has been presented in Fig. 16.14.

16.9.4 Mortality and Knockdown Bioefficacy After Washing

The corrected morality, KDT_{50} and KDT_{10} values have been depicted in Table 16.3. The results suggest that there was no difference in the mortality obtained in

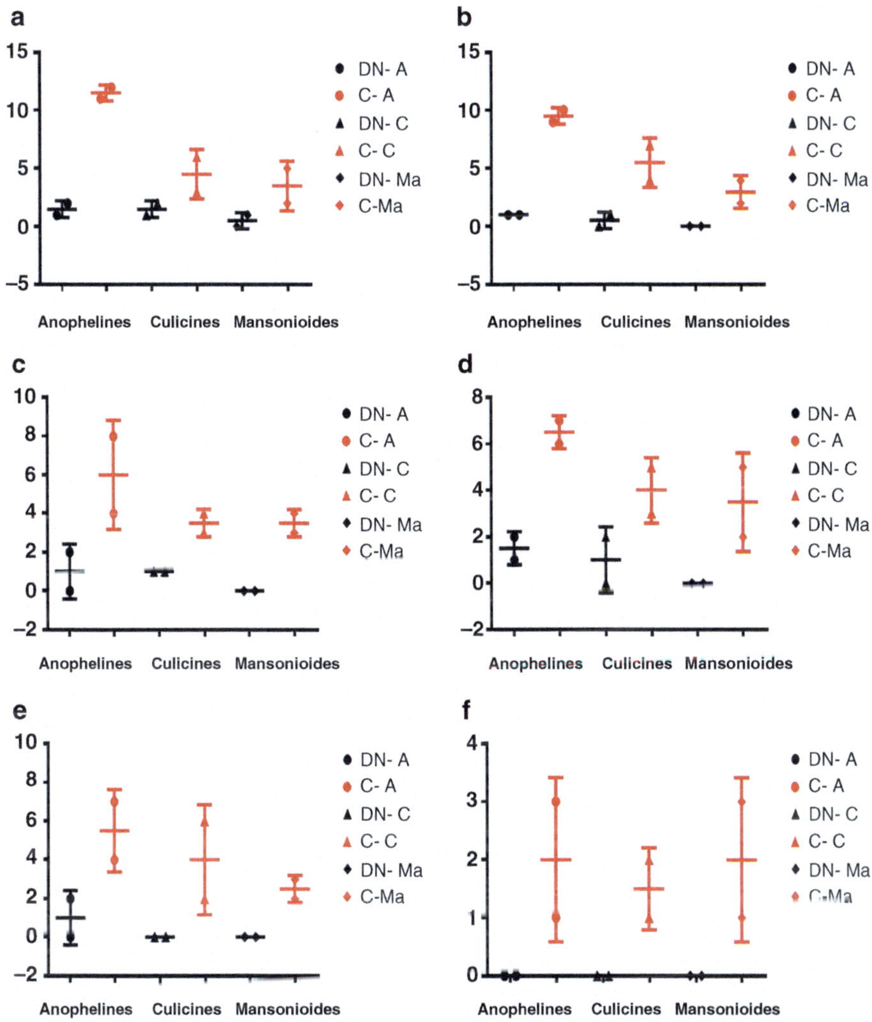

Fig. 16.12 Blood-fed anopheline, culicines and Mansonioides mosquitoes collected during post-intervention months in 2013: (**a**) June, (**b**) July, (**c**) August, (**d**) September, (**e**) October, (**f**) November. DN-A/C/Ma—Defender Net™ anophelines/culicines/mansonioides; CA/C/Ma—Control anophelines/culicines/mansonioides. (*Source*: This work)

unwashed and 5, 10 or 15 times washed Defender Net™ using WHO cone bioassay. The statistical values and coefficient of variance among the replicates have been shown in Table 16.3. The variation in the correction mortality replicates along with standard error mean has been depicted in Fig. 16.10, whereas the variation for KDT_{50} and KDT_{10} has been shown in Fig. 16.15. WHO tube bioassay suggested that in wild collected adult *Ae. albopictus*, the knockdown rate after 1 h (KDR_{1h}) of exposure to 0.05% deltamethrin was 100%, suggesting the tested vector mosquito is

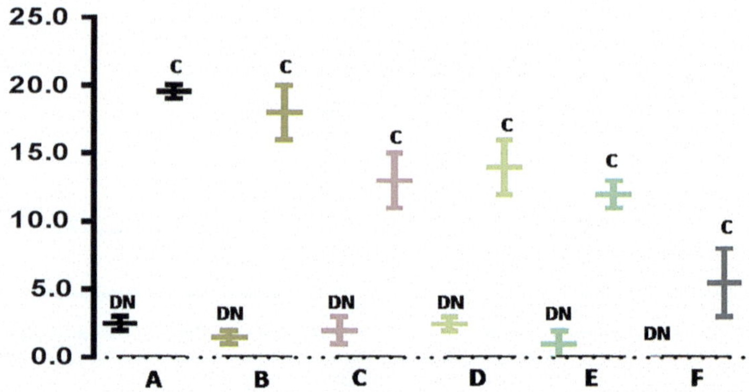

Fig. 16.13 Blood-fed mosquitoes encountered during post-intervention months in 2013: (A) June, (B) July, (C) August, (D) September, (E) October, (F) November. *DN* Defender Net™, *C* control. (*Source*: This work)

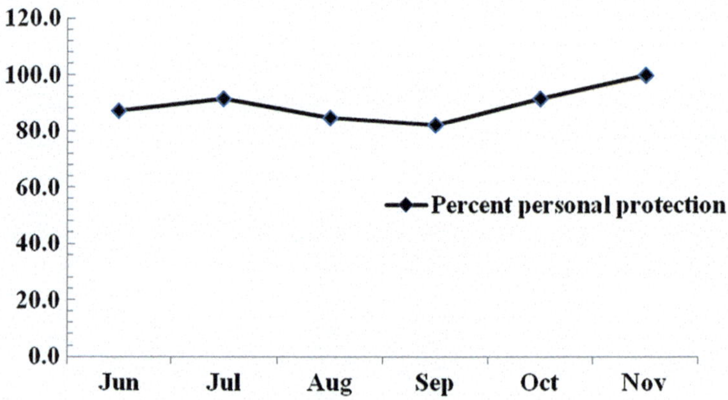

Fig. 16.14 Mosquito blood-feeding inhibition (%) using Defender Net™ during the post-intervention period (Jun–Nov, 2013). (*Source*: This work)

knockdown-sensitive to deltamethrin. Further, 98.8% mortality was recorded after 24 h post-exposure, suggesting that the mosquito was fully susceptible to deltamethrin.

16.9.5 Safety Observations

At pre-distribution day (day 0), there was no sign of illness among the participants to whom Defender Net™ were distributed (Anonymous 2015a). After 7 days of the use, 20% ($n = 1$) still did not know how to use the LLIN. A total of 40% ($n = 2$) still used indigenous methods such as mosquito repellents. All the participants (100%;

Table 16.3 Mortality and knockdown time (KDT) obtained in unwashed and washed (5, 10, 15 times) Defender Net™ (after use in operational conditions) against *Ae. albopictus* (LS) using WHO cone bioassay[a] (*Source*: This work)

Mortality					KDT_{50}					KDT_{10}				
No of wash	CM	CV %	95% CI	ANOVA	No of wash	Time (min)	CV %	95% CI	ANOVA	No of wash	Time (min)	CV %	95% CI	ANOVA
UW	96.7 ± 2.1	5.34	91.3–102.1	$p = 0.3$ $F = 1.3$	UW	13.6 ± 2.9	52.9	6.1–21.2	$p = 0.5$ $F = 0.8$	UW	2.1 ± 0.5	58.6	0.8–3.4	$p = 0.6$ $F = 0.6$
5 W	93.3 ± 3.2	8.75	84.7–101.9		5 W	17.7 ± 3.1	42.4	9.8–25.6		5 W	2.6 ± 0.3	30.8	1.7–3.4	
10 W	98.3 ± 1.7	4.2	94.1–102.6		10 W	19.8 ± 3.3	40.9	11.3–28.3		10 W	2.5 ± 0.5	49	1.2–3.7	
15 W	91.7 ± 3.1	8.2	83.8–99.6		15 W	18.5 ± 2.7	36.3	11.4–36.3		15 W	3.0 ± 0.5	39.4	1.7–4.2	

[a]Ordinary ANOVA performed; *CM* corrected mortality; *CV* coefficient of variance; values expressed in mean ± SEM (six replicates); Brown-Forsythe test has been performed for equality of variance; KDT_{50}/KDT_{10}, knockdown time 50/10 percent

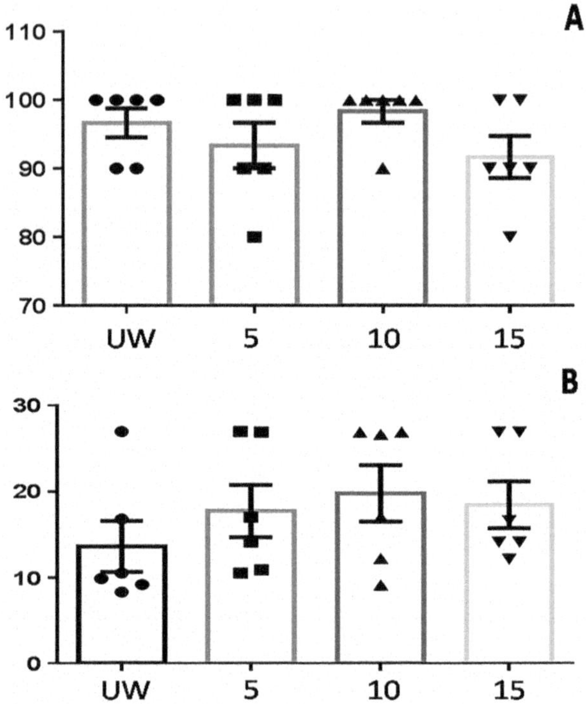

Fig. 16.15 Corrected mortality and KDT_{50} values (mean ± SEM) in *Ae. albopictus* obtained in unwashed, 5 times washed, 10 times washed and 15 times washed Defender Net™ (points represent the value of replicates). *UW* unwashed; 5/15/15–5/10/15 times washed Defender Net™ under operational conditions. (*Source*: This work)

n = 5) slept in the Defender Net™, but a few reported itching (20%; *n* = 1), headache (20%; *n* = 1) and fear of toxicity (20%; *n* = 1). All the participants reported that there had been reduction in the presence of the bed bug and head lice after using the Defender Net™. After 30 days of use, all the participants were aware of the use of Defender Net™, none used indigenous method for mosquito control, and only 20% (*n* = 1) reported mild headache. None reported any fear of poisoning on using the Defender Net™, and the villagers (those who were not distributed Defender Net™) asked for gifting more such nets. Therefore, no adverse effect which could be attributed to the use of Defender Net™ was observed or reported among the participants.

The results indicated that the Defender Net™ was highly effective in the control of not only anopheline vectors but also culicine and Mansonioides vectors. There was a substantial reduction in the vector density immediately after the start of use in the experimental houses. The tested LLIN was wash-resistant, and there was no loss in the biological integrity after 15 standard washes while in use under field conditions. The acceptance of currently tested LLIN was high, and the users reported decreased nuisance due to biting mosquitoes. The Defender Net™ proved safe to the users as no adverse events were observed and/or reported.

16.10 Defender Net™ Evaluation in Tamil Nadu (Anon 2015b)

The LLIN bed nets (Defender Net™) designed and produced by DRDO were provided to the Centre for Research in Medical Entomology (CRME), Madurai (TN, India) for the evaluation of its efficacy on the control of vectors of lymphatic filariasis (LF) and Japanese encephalitis (JE) in endemic areas (Anonymous 2015b). A detailed Field Bioefficacy Evaluation study of the Defender Net™ against vectors of JE and LF was conducted by the CRME in Tirukoilur and Vridhachalam, respectively. The study had following objectives:

1. To assess the efficacy of LLINs on disease vectors.
2. To assess the persistence of the insecticide on the nets by chemical assay.
3. To evaluate the bioefficacy of LLINs in relation to number of washes.
4. To study the impact of LLINs on disease prevalence (LF and JE).
5. To assess the social acceptability and collateral benefits.

16.10.1 Baseline Study

16.10.1.1 Selection of Villages

Studies were conducted in the area which is endemic for either LF or JE and where vector mosquitoes are in high density. In Tirukoilur block of Villupuram District, out of 11 villages surveyed, 3 villages named Veeranampattu, Thathanur and Paradapattu were selected based on the high density of *Culex quinquefasciatus*. In Vridhachalam block of Cuddalore District, the selected villages were G. Kudikadu, Ko. Kollathankurichi and Nanthapadi. The above selected six villages of two study areas were selected based on identical physiotopographic features and vector abundance. Villages in each study area were located at least 10 km apart from each other with a population size of 2000–5000.

16.10.1.2 Evaluation Study Design

From the three study villages in Tirukollur block, one village (Thathanur) was intended for distribution of the treated nets (Defender Net™) and second (Paradapattu) for plain nets (plain net control or no Defender Net™), and in the third village (Veeranampattu), no net was distributed at all (no net control) (Table 16.1). Simultaneously, studies were undertaken during the pre-intervention period to assess the vector density and parasitological data for microfilaraemia prevalence.

Among the three selected villages from Vridhachalam block, G. Kudikadu village was selected for insecticide-treated bed nets, Kollathankurichi village was selected for plain net control (placebo), and Nanthapadi was selected as control village, respectively.

16.10.1.3 Entomological Evaluation

Entomological monitoring for mosquitoes was carried out in the experimental and control villages at monthly intervals in the randomly selected outdoor areas and/or human dwellings during pre- and post-intervention periods, i.e. before and after the distribution of the nets.

For LF Vector

1. *Indoor Resting Collection (IRC)*
   Indoor resting collection of mosquitoes had been carried out by hand-catch method. Female adult *Cx. quinquefasciatus* mosquitoes were collected and dissected to determine the filarial infection under a compound microscope. Infection rate was calculated from mosquitoes which were infected with microfilariae and/or other filarial larval stages. Infectivity rate was calculated based on the number of mosquitoes with third-stage larvae (L3; infective stage). Transmission indices like vector density, infection and infectivity rates and transmission intensity index (TII) were calculated for evaluation. The JE vector mosquitoes were also collected outdoors during dusk hours from respective three villages and pooled for virus detection/isolation.
2. *Parasitological Survey*
   Survey was carried out in 10% of the population in each village by simple random sampling method, to determine filarial infection in human. Basic demographic information including population, sex, details of households and filarial case details were obtained from local health department. Blood samples were collected in households selected at random. Twenty microliters (20μl) of finger-prick blood was collected from all age groups during 21.00–00.00 h in all the three villages to determine the *mf* prevalence for LF.

For JE Vectors

1. *Dusk Collection*
   Mosquito collection was carried out in two parameters such as dusk collection and indoor resting collection in the three index villages, viz. G. Kudikadu (treated), Ko. Kollathankurichi (placebo) and Nandhapadi (control). The PMH density of *Cx. tritaeniorhynchus* was compared before and after distribution of the Defender Net™.
2. *Indoor Resting Collection*

Indoor resting collection of mosquitoes had been carried out by hand-catch method in all the study villages.

3. *Vector Infection*

JE vector mosquitoes collected during the survey were pooled and sent to the CRME laboratory for JEV detection/isolation.

16.10.2 Results of Entomological Monitoring for the LF and JE Vectors During Pre-intervention Period (Anon 2015b)

16.10.2.1 LF Vector

In the month of January 2013, a total of 224 adult female *Cx. quinquefasciatus* mosquitoes were collected from the three arms. The overall per man hour density of female *Cx. quinquefasciatus* was 35.8, 0.75 and 3.00 in Thathanur, Paradapattu and Veeranampattu villages, respectively. In the month of February 2013, overall 12 adult female *Culex quinquefasciatus* mosquitoes were collected from two villages of Paradapattu and Veeranampattu. The per hour density of *Cx. quinquefasciatus* was 1.00 and 2.00 in Paradapattu and Veeranampattu villages, respectively. In the month of March 2013, overall 194 adult female *Cx. quinquefasciatus* mosquitoes were collected (20 man hours spent) from three villages of Thathanur, Paradapattu and Veeranampattu. The per hour density of *Cx. quinquefasciatus* ranged between 8.00 and 14.50 in Paradapattu and Thathanur villages, respectively.

16.10.2.2 JE Vectors

A total of 754 mosquitoes comprising 12 species under 5 genera were collected utilizing 31 man hours in G. Kudikadu. The per man hour density (PMHD) for *Cx. tritaeniorhynchus* was found to be dominant with 19.26, followed by *Cx. vishnui* (0.74), *Cx. gelidus* (0.71), *Cx. fuscocephala* (0.19) and *Cx. quinquefasciatus* (0.58). A total of 2617 mosquitoes comprising 14 species belonging to 6 genera were collected utilizing 31 man hours in Ko. Kollathankurichi. The PMHD for *Cx. tritaeniorhynchus* was found to be dominant as 66.74, followed by 1.39 of *Cx. vishnui*, and 6.74 of *Cx. gelidus*. PMHD of *Cx. tritaeniorhynchus* was increased twofold from 31.83 to 66.74 compared with pre- and post-net distribution period. A total of 2707 mosquitoes comprising of 16 species from 6 genera were collected by utilizing 31 man hours in three index villages in Nandhapadi. Among culicines, *Cx. tritaeniorhynchus* was found to be dominant with 87.32, followed by 1.10 of *Cx. vishnui* and 1.03 of *Cx. gelidus*. Among anophelines, *An. subpictus* was found to be most dominant with 4.42 and 4.29 of *An. peditaeniatus*. PMHD of *Cx. tritaeniorhynchus* was increased threefolds from 27.55 to 87.32 compared with pre- and post-net distribution period.

Every month, indoor mosquitoes were collected from three index villages by hand-catch method. After utilizing 6 man hours, a total of 117 mosquito specimens (15 males, 102 females) belonging to 2 different species were captured. *An. subpictus* was the dominant mosquito (female PMHD 12.50), followed by *Cx. quinquefasciatus* (4.50). In Ko. Kollathankurichi village, a total of 173 (42 male, 131 female) mosquito specimens belonging to 3 different species were captured (6 man hours). *An. subpictus* was dominant here too (female PMHD 13.67), followed by *Cx. quinquefasciatus* (8.00) and 0.16 of other species, respectively. After utilizing 6 man hours, a total of 101 (40 male, 61 female) mosquito specimens belonging to 4 different species were captured from Nandhapadi. *An subpictus* was the ever-persisting dominant mosquito (8.33), followed by *Cx. quinquefasciatus* (1.50).

16.10.3 Entomological Monitoring for LF and JE Vectors During Post-intervention Period

16.10.3.1 LF Vector

The entomological monitoring was carried out from April 2013 to November 2014. During the month of April and May 2013, only one adult female *Cx. quinquefasciatus* mosquito was collected, and per man hour density was 0.8 in Thathanur. During this period, a total of 257, 347 and 516 mosquitoes were collected investing 96, 92 and 95 man hours in the 3 study villages. In other study villages, no mosquitoes could be recorded. In the month of June 2013, a total of 17 adult females of *Cx. quinquefasciatus* mosquito were collected from all the 3 study villages. The overall per man hour density of female *Cx. quinquefasciatus* was 2.50 in Thathanur, 2.00 in Paradapattu and 4.00 in the remaining village Veeranampattu. The parity rate was 60.00 in Thathanur village alone and 33.33 in both Paradapattu and Veeranampattu villages. The infection rate, infectivity rate and transmission intensity index were nil in all three villages.

Similarly in July 2013, a total of 12 adult female *Cx. quinquefasciatus* mosquitoes were collected from LLIN (Thathanur) village and the non-LLIN (Paradapattu) village. In the LLIN village, PMHD was 2.50, whereas in non-LLIN village, it was 3.50. No mosquito was collected in Veeranampattu village. In the month of August 2013, a total of 51 mosquitoes were collected from all the 3 study villages in Tirukoilur block. In Thathanur (LLIN) village, PMHD and parity rate were recorded as 7.50 and 50.00, respectively. In Paradapattu (non-LLIN) village, the PMHD was 3.00, whereas in Veeranampattu (control) village, PMHD was 2.25. During the month of September 2013, one mosquito was collected, and PMHD was 1.00, whereas in the other village, no mosquito could be recorded. During the month of October 2013, PMHD was nil in all the three study villages. During November 2013, overall per man hour density of female *Cx. quinquefasciatus* ranged from 5.00 (Thathanur) to 14.00 (both Paradapattu and Veeranampattu). The parity rate was

33.33, 100 and 50.00 in the Thathanur, Paradapattu Veeranampattu, respectively. During the month of December 2013, 140 adult female *Cx. quinquefasciatus* mosquitoes were collected from all the 3 study villages. The overall PMHD of female *Cx. quinquefasciatus* was 8.33 in Thathanur, 18.00 in Paradapattu and 20.33 in Veeranampattu. The PMHD was observed lowest (1.38) during March 2014 and highest in January 2014 (5.25) in Thathanur village, whereas in Paradapattu (non-LLIN) village, the PMHD varied from 2.67 to 3.50. In the control village (Veeranampattu), the PMHD was 3.00 in March 2014 and 7.50 in January 2014. No infection was observed in any mosquito from the three index villages. The month-wise PMHD was gradually reduced in all villages (except during December 2013) in LLIN villages and reached 1.00. During April and May 2013, PMH density was 0.25 in LLIN village but nil in the non-LLIN and control village, respectively. The highest PMH density was observed in control villages during December 2013. Overall PMHD was 2.68, 3.77 and 5.43 in LLIN village, non-LLIN village and control, respectively. The reduction was 87.38% (21.21–2.68) in LLIN village, 15% (4.44–3.77) in non-LIIN village and 12.6% (6.21–5.43) in control village (Fig. 16.3). During the post-intervention, infectivity and TII were nil in all the three villages.

16.10.3.2 JE Vector

In dusk collection, in G. Kudikadu (treated) village, a total of 1183 mosquitoes comprising 12 species under 5 genera were collected utilizing 49 man hours. The PMHD for *Cx. tritaeniorhynchus* was found to be dominant (33.58). Mosquito collection was carried out in two parameters such as dusk collection and indoor resting collection in three index villages, viz. G. Kudikadu (treated), Ko. Kollathankurichi (placebo) and Nandhapadi (control). The PMHD for *Cx. tritaeniorhynchus* was compared before and after distribution of Defender Net™ (LLIN). The PMHD of the treated village of G. Kudikadu was 33.22 during this period (April 2013–Nov 2014). In Ko. Kollathankurichi (placebo) villages, a total of 5650 mosquitoes comprising 17 species belonging to 6 genera were collected utilizing 49 man hours. The PMHD of *Cx. tritaeniorhynchus* in pre-survey was 31.83, and in the post-survey, it was 92.24. Furthermore, in Nanthapadi (control) village, a total of 7864 mosquitoes comprising 17 species from 6 genera were collected by utilizing 55 man hours in 3 index villages in Nandhapadi. The PMHD of *Cx. tritaeniorhynchus* in pre-survey was 31.83, and in the post survey, it was 122.15. Throughout the month, a good reduction of mosquitoes was noted in treated net village comparable with other two villages, i.e. placebo and control.

In indoor resting collection of JE vectors, in G. Kudikadu (Treated) during the post-net distribution period (April to November 2014), a total of 458 (132 male 326 female) mosquito specimens belonging to 6 different species were captured. *Anopheles subpictus* was the dominant mosquito (female PMHD 6.83). The PMHD density of the *An. subpictus* decreased from 12.50 to 6.83 when compared from pre- and post-treatment period. Further, in Ko. Kollathankurichi (no net control), mosquito specimens belonging to five different species were captured. *Anopheles subpictus* was dominant here too (female PMHD 20.73), followed by *Cx. quinquefasciatus* 1.71. The PMHD of *An. subpictus* was calculated, and it was found to be escalating from 13.67 to 20.73. In Nandhapadi, six different species were captured. Here, too, *An. subpictus* was a dominant mosquito (33.45). The PMHD of the *An. subpictus* increased from 8.33 to 33.45 during pre- and post-net distribution period. On comparison with pre- and post-net distribution periods of January–March, a drastic reduction in the vector abundance was found in the treated village (Kudikadu) for both *Cx. tritaeniorhynchus* during dusk collection and *An. subpictus* during indoor resting collection. The JE vector mosquitoes collected during the survey were pooled and preserved in liquid nitrogen for JEV detection/ isolation. Those collected till September 2013 were tested for presence of JE virus antigen. All samples tested were found negative.

16.11 Bioefficacy Studies of Defender Net™ Against Mosquitoes

At least 50 mosquitoes on each net with 10 replicates were tested, and the results were pooled. Mosquitoes exposed to untreated nets were used as control. Bioassays were carried out at 25 ± 2 °C and 75 ± 10% RH. After washing, nets were dried at room temperature for 72 h for regeneration and stored at room temperature (30 ± 2 °C) (WHO 2005).

The percent knockdown of mosquitoes in cone bioassays was 92% in Vridhachalam samples and 84% in Cuddalore samples. The percent mortality in 24 h of post-exposure was respectively 96% and 88% in Vridhachalam and Cuddalore. In WHO cone bioassay, the field samples of Defender Net™ from Vridhachalam and Cuddalore caused ≥80% mortality or ≥95% KD against *Ae. albopictus*, and thus the samples met the criteria of efficacy as per WHO guidelines.

The LLIN Defender Net™ was highly effective in the control of *Cx. quinquefasciatus*, *Cx. tritaeniorhynchus* and *Aedes* and *Anopheles* spp. There was a substantial reduction in the vector density immediately after deployment of Defender Net™ in the experimental houses. The tested LLIN was wash-resistant, and there was no loss in the biological integrity even after 20 standard washes. No single blood-fed mosquito was found inside the net during the use throughout the study. The acceptance of the Defender Net™ was high, and the users reported significant reduction in nuisance due to mosquito bites. The net was safe to users as no adverse effects were observed and/or reported. Overall the Defender Net™ trials have indicated a high level of biological efficacy and user acceptance.

16.12 Field Evaluation in Army Locations (Anonymous 2015c)

Soon after the Defender Net™ was developed and tested advantageous by DRDO, the India Army MGO branch had moved to replace the old-time Mosquito Net Khaki with the indigenously developed LLIN, Defender Net™. To enact the replacement, an evaluation was carried out in three different geoclimatic and ecological zones: one at 323 Fd Regt, Western Command (Kathua), and other two units at 2 Dogra Regt (Karbi Anglong) and 71 Inf DOU under Eastern Command (Missamari). The primary objective was to assess comprehensive effectiveness of the Defender Net™ in different harsh conditions and areas.

1. Monitoring of mosquito density in the Defender Net™ and plain net barracks.
2. Anti-mosquito efficacy of Defender Net™ after different washings.
3. Acceptance of Defender Net™.

16.12.1 Study Design

Defender Net™ (LLIN) and control nets (plain nets) were distributed among the troops staying in suitably distant barracks (Fig. 16.1) in the units, and the mosquito density was observed using light traps. The LLINs after different washes were retracted time to time and evaluated in the laboratory for their efficacy against different vector mosquitoes using WHO cone bioassay. The soldiers were asked to give their feedback about any adverse effect after the use of net and overall acceptability.

16.12.2 Observation and Results

16.12.2.1 Anti-mosquito Activity

All the three selected areas were highly prone to mosquito-borne infections and had been reporting much malaria cases annually.

In Missamari, anopheline density (mean ± SD) in Defender Net™ barracks was found to have reduced to nil as compared to that in the plain net barracks where the average density was 5.5 ± 2.1 ($p < 0.05$). Similarly, the density (mean ± SD) of culicine and Mansonioides mosquitoes was also found to be reduced significantly ($p < 0.05$) in the Defender Net™ barracks than in the plain net barracks (Table 16.4). The traps installed in the plain net barracks recorded repeatedly high number of blood-fed mosquitoes (22.5 ± 4.2) as compared to that in the Defender Net™ barracks, and the difference was statistically significant (<0.0001; $t = 10.3$;

Table 16.4 Vector mosquito density in Defender Net™ and plain net barracks at Missamari (*Source*: This work)

	Defender Net™			Plain		
	Anophelines	Culicines	Mansonia	Anophelines	Culicines	Mansonioides
Mean	0	0.5	0.5	5.5	20.5	13.5
SD	0	0.71	0.71	2.1	3.5	2.1
COV	0	141.4	141.4	38.6	17.3	15.7
F (p)	0.8 (0.6)			1.8 (0.3)		
R square	0.38			0.59		

SD standard deviation, *COV* coefficient of variance

Table 16.5 Vector mosquito density in Defender Net™ and plain net barracks at Kathua (*Source*: This work)

	Defender Net™		Plain net	
	Anophelines	Culicines	Anophelines	Culicines
Mean	1	19	8.5	203.5
SD	1.4	14.1	2.1	84.2
COV	141.4%	74.4%	24.9%	41.4%
t (p)	1.8 (0.2)		3.3 (0.08)	

SD standard deviation, *COV* coefficient of variance

df = 6). Defender Net™ had provided 98.9% inhibition against the blood-feeding mosquitoes.

In Kathua (Jammu & Kashmir), the density (mean ± SD) of both anopheline and culicine mosquitoes was significantly reduced in the Defender Net™ barracks than in the plain net barracks (Table 16.5). Although the density (mean ± SD) of culicines was little higher during the survey, still the mosquitoes were not able to feed on the human blood as only one blood-fed mosquito specimen (0.25 ± 0.5) was sampled which was inconsiderable owing to the very high density of blood-fed mosquitoes (62.5 ± 20.3) in the plain net barracks (0.0009; $t = 6.14$; df = 6). The blood-feeding inhibition against the mosquitoes achieved was 99.6%.

Similarly in Karbi Anglong, the mosquito density (mean ± SD) was significantly reduced in the Defender Net™ barracks when compared to that in the plain net barracks. In the Defender Net™ barracks, the mean density of anopheline and culicine mosquitoes was found to be 1.5 ± 0.7 and 0.5 ± 0.7, respectively, while in the plain net barracks, the density (mean ± SD) of anopheline, culicine and Mansonioides mosquitoes was recorded as 5 ± 1.4, 7.5 ± 0.7 and 5.5 ± 2.1, respectively. The density of each of the mosquito genera was significantly reduced in the Defender Net™ barracks ($p < 0.05$) (Table 16.6). During the study, 100% blood-feeding inhibition against the mosquitoes was recorded.

Table 16.6 Vector mosquito density in Defender Net™ and plain net barracks at Karbi Anglong

	Defender Net™		Plain		
	Anophelines	Culicines	Anophelines	Culicines	Mansonia
Mean	1.5	0.5	5	7.5	5.5
SD	0.71	0.71	1.4	0.71	2.1
COV	47.1%	141.4%	28.3%	9.4%	38.6%
t/*F* (*p*)	1.4 (0.3)		1.5 (0.4)		

SD standard deviation, *COV* coefficient of variance

16.13 Wash Resistance Activity of Defender Net™

The corrected morality after 24 h and knockdown after 60 min (KD_{60min}) values were determined after 20 times of washes of the Defender Net™ retracted from the units after use. The corrected mortality obtained after 24 h was 95%, while the KD_{60min} was 97.5%. The results suggested that the Defender Net was equally effective even after 20 washes during the use in operational conditions.

16.14 Adverse Reaction, Reporting and Social Acceptability

During the study, volunteers, who were provided Defender Net™, were asked several questions about the ill effects, adverse skin irritation or reaction, satisfaction to the use of net and their experience about the use of net. Almost all the respondents replied in affirmative that they did not feel any ill effects after using the Defender Net™ and were completely satisfied with the use of the net. They shared their observations willingly about the mosquitoes when came in contact with the net even ephemeral and got knocked down to death after some time. The soldiers informed that the nets were comfortable to use and provided sound sleep after rigorous duty hours. Some soldiers in Karbi Anglong told that since the nets are light in weight, they were able to wrap and carry it into the deep jungle during the emergent operations. The overall reports from the users suggested that the nets were effective and provided high protection than the conventional khaki net which is conventionally supplied to them.

The test results clearly indicated that the Defender Net™ was found to be highly effective in reducing vector mosquito density. Defender Net™ was wash-resistant, and no significant loss in chemical and biological integrity was observed even after repeated washes in operational conditions. The study reported that the Defender Net™ was well accepted and appreciated by the users in the three trial units in different field areas. It also emphasizes that it is fit for the use in armed forces in varying geoclimatic and ecological situations and various operational conditions.

16.15 Discussions

Malaria is one of the major mosquito-borne diseases whose worst face is witnessed in Africa and Asia. Out of a long array of control mechanisms, processes and tools, LLIN have in recent decades provided much succour by controlling vector populations, on one hand, and impacting the disease dynamics, on the other. In India, although the LLINs are deployed through the government channel, still the timely and copious availability of the LLINs to the users inhabiting malaria-infested areas is almost entirely wanting especially in difficult mountainous and forested terrains in far eastern and northeastern states (Dhiman and Veer 2014; Dev et al. 2016). The LLINs are popular among rural masses in the endemic areas, and these are important segments of the country's malaria control programme.

Defender Net™ was developed in India by the Defence Research and Development Organisation (DRDO) specifically for the use of armed forces and spin-off application for the civilian use. These nets were evaluated in different geo-ecological regions of India following WHOPES criteria. The results have suggested that Defender Net™ were equally effective against malaria, lymphatic filariasis and Japanese encephalitis vectors under field conditions. In Assam, there was significant reduction in the overall density of mosquitoes during post-intervention period. Major malaria vectors were reduced to naught during this period which suggest that the LLINs are highly effective in preventing the entry of potential vectors into human dwellings (Dhiman et al. 2009, 2010a, 2012; Yadav et al. 2012; Nath et al. 2013). Previous studies reported that the LLINs were highly effective in northeastern region of India in deterring potential malaria vector *An. minimus*, although it was doubted if the serviceable life of such nets could last 3 years due to both the hot and moist climatic conditions and the various socio-ecological practices among the communities particularly with a view to meet the demands of annual festivities full of give-and-take pleasantries (Dev et al. 2010, 2011, 2016). Not only the Defender Net™ significantly reduced the vector mosquito density immediately but was also found to retain biological as well as chemical integrity even after 15 standard washes after use in the operational conditions.

Defender Net™ consists of a combination of deltamethrin and piperonyl butoxide (PBO), a synergistic compound to improve the bioefficacy against the synthetic pyrethroid-resistant mosquito vectors. The consistent optimal bioefficacy of the tested LLIN indicates that this combination of PBO in the LLIN represents a viable option for areas reporting pyrethroid resistance in the common mosquito vectors (Haji et al. 2013). Monthly collection of vectors shows that there was immediate reduction in the vector population corresponding to anopheline, culicines and Mansonioides mosquitoes. The vector mosquito population decreased significantly, perhaps below the threshold level required to trigger disease transmission, after 3 months of the regular use of LLIN. The LLIN tested here efficiently killed the potential dengue vector mosquito even after 15 washes and use in the field conditions, alluding towards an enhanced knockdown effect. Defender Net™ was well accepted by the users as none of them complained of any side effect. Use of the

Defender Net™ provided some collateral benefits to the households by controlling bed bugs and head lice.

In Madurai, the evaluation results displayed that Defender Net™ was highly effective in the control of culicines (*Culex quinquefasciatus* and *Cx. tritaeniorhynchus*), anophelines and *Aedes* vectors. Studies have shown that these vectors prevail in huge density in endemic settings (Dhiman et al. 2009; Sexena et al. 2014; Yadav et al. 2017). There was a substantial reduction in the vector density immediately after the beginning of the usage of Defender Net™ in the experimental houses. The tested LLIN was wash-resistant, and there was no loss in the biological integrity even after 20 standard washes. No single blood-fed mosquito was found inside the net during the use in the entire study. The acceptance of the Defender Net™ was high, and the users reported significant decrease in nuisance from mosquito bites. The net was safe to users as no adverse effects were observed and/or reported. Overall the Defender Net™ trials in the study areas indicated high levels of biological efficacy and user acceptance.

The feedback report from the soldiers of all the three units operating in different geo-climates undisputedly suggested that the Defender Net™ was safe and very well accepted (Dhiman et al. 2010a, 2011, 2015; Das et al. 2015). Therefore, such mobile populations as that of the armed forces require intervention measures which are effective, easy to use and user-friendly (Yadav et al. 2014). Defender Net™ was well accepted and appreciated by the users in three trial locations in different field areas, which emphasizes that it is fit for the use in the armed forces in different climates and various operational conditions.

The results of such a vast evaluation largely rely on the use pattern, compliance rate, sociodemographic limitations and religio-cultural beliefs of the user population. During the present trials, the compliance rate was good, and >95% of the users slept under Defender Net™ during the study period. The appropriate thickness and mesh size of Defender Net™ provide better ventilation, which was probably an important factor for better compliance during the study. The high compliance rates result in mass effect on deterring the mosquitoes and other vectors, thus providing herd protection against mosquito bites.

16.16 Conclusion

The indigenous LLIN Defender Net™ was developed while primarily considering the emerging needs of the Indian troops deployed in mosquito- and disease infested frontier areas and also, in general, the serious constraints of development of insecticide resistance in vector mosquitoes as well as other arthropod vectors. This LLIN was highly effective against a variety of mosquito vectors. The net not only deterred mosquitoes from entering houses for blood feeding but also offered several collateral benefits. The net has long-term service life, and there are obviously no safety concerns for using it. The LLIN, whose terminology is essential for a beginner to acquaint with (see block below), has been found effective against mosquito vectors regardless of the study locations, ethnic limitations and local vectors.

The Indian Armed Forces after carrying out their own field trials in army units have placed a supply order for 12,500 pieces, while paramilitary forces (BSF) placed a supply order for 135,000 pieces on limited tender basis of Defender Net™ from the DRDO technology holder industry.

Some Important Terms Associated with LLIN

Mesh: the Number of holes per square inch. For Example, mesh 156 has 12 × 13 holes per square inch. Hole shape may be rhomboidd or hexagonal and Raschel knitted.

Mesh size: the size of the openings in a net. It is determined by the number of holes per square inch (the mesh) and the thickness of the threads with which the netting is made. The mesh size recommended for the tropical countries is between 1.2 and 1.5 mm.

Denier: an indication of the weight (and therefore the strength) of the thread. It is defined as the weight in grams of 9000 m of a single thread. Commonly used mosquito net threads have a denier between 40 and 100 but denier 40 is easily torn and 70 or more is recommended.

Strength: an indication of the pulling strength of a thread, expressed in grams per denier if 1 m of 40 denier thread breaks with a load of 160 g, the strengths is 4 g per denier.

Monofilament/multifilament yarn: monofilament yarn is made of single fibre with high density polyethylene (HDPE) while mulitfilment yarn are made up of several fibres, usually 36 fibres of polyester.

Net border: nets are often provided with a strong border of cotton or synthetic fabric. This protects the net from wear due to daily tucking in of the net under the mattress if the border is wide enough (30 cm) the extra material will also reduces bites from insects that may make contact with the lower part of the net while the occupant is asleep,

Colour: white material is most commonly preferred but other colours are available. In a white net, it is easier to see any mosquitoes that have entered. A darker colour like blue may be preferable because nets are less likely to appear soiled.

Acknowledgements We are thankful to the villagers who volunteered for participation in the present study both in Assam and Tamil Nadu. Village headmen and selected village health workers are sincerely thanked for their keen interest and assistance in the study. Help of health staff of Missamari sub-centre and Gorubandha sub-centre as well as that of Centre for Research in Medical Entomology, Madurai, is hereby deeply thanked. Logistic support provided by the 71 infantry division and other army units is gratefully acknowledged.

References

Anonymous (2015a) Field evaluation (Phase-II Trial) of Defender Net™ a long lasting insecticidal net (LLIN) in a malaria endemic area of Assam. Technical Report No: DRLT/ME/Tech Report/Defender Net™ (LLIN)/1/2015

Anonymous (2015b) Field bio-efficacy evaluation of DRDO Defender Net against vectors of Japanese Encephalitis and Lymphatic Filariasis. Technical Report, Centre for Research in Medical Entomology (ICMR), Madurai (TN)

Anonymous (2015c) Field evaluation of Defender Net™ a long lasting insecticidal net (LLIN) in Army Units. Technical Report No: DRLT/ME/Tech Report/Defender Net (LLIN)/2/2015

Bhatt RM, Sharma SN, Uragayala S, Dash AP, Kamaraju R (2012) Effectiveness and durability of Interceptor long-lasting insecticidal nets in a malaria endemic area of Central India. Malar J 11:189

Das NG, Dhiman S, Talukdar PK, Goswami D, Rabha B, Baruah I, Veer V (2015) Role of asymptomatic carriers and weather variables in persistent transmission of malaria in an endemic district of Assam, India. Infect Ecol Epidemiol 5:25442. https://doi.org/10.3402/iee.v5.25442

Dev V, Raghavendra K, Barman K, Phookan S, Dash AP (2010) Wash-resistance and field efficacy of Olyset® net, a permethrin incorporated long-lasting insecticidal netting against *Anopheles minimus* transmitted malaria in Assam, Northeastern India. Vector Borne Zoonotic Dis 10:403–410

Dev V, Phookan S, Padhan K, Tewari GG, Khound K (2011) Laboratory wash-resistance and field evaluation of deltamethrin incorporated long-lasting polyethylene netting (Netprotect®) against malaria transmission in Assam, north-east India. Acta Trop 119:172–177

Dev V, Barman K, Khound K (2016) A cross-sectional study assessing the residual bio-efficacy and durability of field-distributed long-lasting insecticidal nets in malaria endemic ethnic communities of Assam, Northeast India. J Infect Public Health 9:298–307

Dhiman S (2019) Are malaria elimination efforts on right track? An analysis of gains achieved and challenges ahead. Infect Dis Poverty 8:14

Dhiman S, Veer V (2014) Culminating anti-malaria efforts at long lasting insecticidal net. J Infect Public Health 7:457–464

Dhiman S, Gopalakrishnan R, Goswami D, Das NG, Baruah I, Rabha B, Talukdar PK, Singh L (2009) Diversity, spatio-temporal distribution and biting activity of mosquitoes in Tripura State, India. Entomon 34(4):223–232

Dhiman S, Baruah I, Singh L (2010a) Military malaria in northeast region of India. Def Sci J 60 (2):213–218

Dhiman S, Goswami D, Rabha B, Gopalakrishnan R, Baruah I, Singh L (2010b) Malaria epidemiology along Indo-Bangladesh border in Tripura state, India. Southeast Asian J Trop Med Public Health 41(6):1279–1289

Dhiman S, Gopalakrishnan R, Goswami D, Rabha B, Baruah I, Singh L (2011) Malaria incidence among paramilitary personnel in an endemic area of Tripura. Indian J Med Res 133:665–669

Dhiman S, Bhola R, Goswami D, Rabha B, Kumar D, Baruah I, Singh L (2012) Polymerase chain reaction detection of human host preference and *Plasmodium* parasite infections in field collected potential malaria vectors. Pathogen Glob Health 106(3):177–180

Dhiman S, Goswami D, Rabha B, Yadav K, Chattopadhyay P, Veer V (2015) Absence of asymptomatic malaria in a cohort of 133 individuals in a malaria endemic area of Assam, India. BMC Public Health 15:919

Dhiman S, Veer V, Dev V (2018) Chapter-11. Declining transmission of malaria in India: accelerating towards elimination. In: Manguin S, Dev V (eds) Towards malaria elimination - a leap forward. Intech Open, London, UK, pp 257–280

Djènontin A, Moiroux N, Bouraïma A et al (2018) Field efficacy of a new deltamethrin long lasting insecticidal net (LifeNet©) against wild pyrethroid resistant *Anopheles gambiae* in Benin. BMC Public Health 18:947. https://doi.org/10.1186/s12889-018-5876-9

Erlanger TE, Enayati AA, Hemnigway J, Mshinda H, Tami A, Lengeler C (2004) Field issues related to effectiveness of insecticide-treated nets in Tanzania. Med Vet Entomol 18(2):153–160

Haji KA, Khatib BO, Smith S, Ali AS, Devine GJ, Coetzee M, Majambere S (2013) Challenges for malaria elimination in Zanzibar: pyrethroid resistance in malaria vectors and poor performance of long-lasting insecticide nets. Parasite Vectors 6:82

Jambulingam PK, Gunasekharan K, Sahoo SS, Hota PK, Tyagi BK, Kalyanasundaram M (1989) Effect of permethrin-impregnated bednets in reducing population of malaria vector, *Anopheles culicifacies* in a tribal village of Orissa State (India). Indian J Med Res 89:48–51

Lengler C (2004) Insecticide treated bednets and curtains for preventing malaria. Cochrane Database Syst Rev (2):CD00036. https://doi.org/10.1002/14651858.CD00036

Mboma ZM, Festo C, Lorenz LM, Massue Dennis J, Kisinza William N, John B, Moore Jason D, Renata M, Ikupa A, Jo L, Overgaard Hans J, Sarah Moore J (2021) The consequences of declining population access to insecticide-treated nets (ITNs) on net use patterns and physical degradation of nets after 22 months of ownership. Malar J 20:171. https://doi.org/10.1186/s12936-021-03686-2

Nath MJ, Bora AK, Yadav K, Talukdar PK, Dhiman S, Baruah I et al (2013) Prioritizing areas for malaria control using geographical information system in Sonitpur district, Assam. India Public Health 127(6):572–578

Pennetier C, Bouraima A, Chandre F, Piameu M, Etang J, Rossignol M et al (2013) Efficacy of Olyset® Plus, a new long-lasting insecticidal net incorporating permethrin and piperonyl-butoxide against multi-resistant malaria vectors. PLoS One 8(10):e75134. https://doi.org/10.1371/journal.pone.0075134

Rabha B, Goswami D, Dhiman S, Das NG, Talukdar PK, Nath MJ, Baruah I, Bhola RK, Singh L (2011) A cross sectional investigation of malaria epidemiology among seven tea estates in Assam, India. J Parasit Dis 36:1–6

Sexena R, Nagpal BN, Singh VP, Srivastava A, Dev V, Sharma MC et al (2014) Impact of deforestation on known malaria vectors in Sonitpur district of Assam, India. J Vector Borne Dis 51:211–215

Tami A, Mubyazi G, Talbert A, Mshinda H, Duchon S, Lengeler C (2004) Evaluation of Olyset™ insecticide-treated nets distributed seven years previously in Tanzania. Malar J 3:19

World Health Organization (2005) Guidelines for laboratory and field testing of long-lasting insecticidal mosquito nets. World Health Organization. https://apps.who.int/iris/handle/10665/69007. Accessed 24 Mar 2021

World Health Organization (2006) Guidelines for testing mosquito adulticides for indoor residual spraying and treatment of mosquito nets. World Health Organization. https://apps.who.int/iris/handle/10665/69296. Accessed 24 Mar 2021

World Health Organization & WHO Pesticide Evaluation Scheme (2013) Guidelines for laboratory and field-testing of long-lasting insecticidal nets. World Health Organization. https://apps.who.int/iris/handle/10665/80270. Accessed 24 Mar 2021

Yadav K, Nath MJ, Talukdar PK, Saikia PK, Baruah I, Singh L (2012) Malaria risk areas of Udalguri district of Assam, India: a GIS based study. Int J Geogr Inform Sci 26(1):123–131

Yadav K, Dhiman S, Rabha R, Saikia PK, Veer V (2014) Socio-economic determinants for malaria transmission risk in an endemic primary health centre in Assam, India. Infect Dis Poverty 3:19

Yadav K, Dhiman S, Rabha B, Goswami D, Saikia PK, Veer V (2017) Disappearance of *Anopheles minimus* and *Anopheles dirus* from certain malaria endemic areas of Assam, India. J Arthropod Borne Dis 11(1):27–35

Chapter 17
Nanotechnology and Polymer Science: A Novel Approach in Vector Control

Probal Basu and Sajal Bhattacharya

Abstract Vector-borne diseases have a great importance in the global health context. The biological dynamics of arthropod vector is sustained by the selection pressure from the environment. Vector control is thus a challenging approach to control the vectors through managing their dynamic physiological attributes within the changing environmental condition. Hence, the study and exploration of novel vector control methods are necessary in parallel to the sustained surveillance of mosquito vector behavior in the changing climate. In this context, nanobiotechnology can become an efficient approach for vector control, as it has the advantages of precise and specific utilization of technologies like nanoscience and polymer science that can affect specifically on vector physiology and facilitates the vector control. This work is an endeavor to explore different aspects of nano-biotechnological approaches for vector control using the nanoscience and polymer science.

Keywords Nanotechnology · Polymer science · Vector control

17.1 Introduction

Vector-borne diseases have a great importance in the global health. The significant invasion of disease-causing vectors in the past two decades has complicated the human health situation (Khan 2015). The arthropod vector-borne diseases account for around 17% of the estimated global burden of communicable diseases and lead to serious mortality in the tropical and sub-tropical areas (WHO 2017). Mosquito vectors belonging to genera like *Anopheles*, *Aedes*, and *Culex* disseminate severe vector-borne diseases like malaria, dengue, chikungunya, Zika, and Japanese

P. Basu (✉)
Centre of Polymer Systems, University Institute, Tomas Bata University in Zlin, Zlín, Czech Republic

S. Bhattacharya
Post Graduate Department of Zoology, Asutosh College, Kolkata, West Bengal, India

B. K. Tyagi (ed.), *Genetically Modified and other Innovative Vector Control Technologies*, https://doi.org/10.1007/978-981-16-2964-8_17

encephalitis. Different methods like application of indoor residual spraying (IRS) to kill vectors in the habit of resting indoors (endophagic and endophilic) and insecticiding breeding habitats of vector species, draining of water-logged areas, as well as varied biological tools have been already in vogue or decades but without desired outputs (Rozendaal et al. 1997; WHO 2006, 2019a). Even though insecticides are still useful particularly during episodes of epidemics, they are generally opposed due to their wide-ranging scenarios of ill effects on different animal groups, including man, and the environment (Carson 1962). Moreover, vector mosquitoes have already developed resistance against most of the insecticides, both larvicides and adulticides, leaving very few compounds such as temephos for use against the vectors preferring to breed in rather fresh water, e.g., *Anopheles stephensi*, *Aedes aegypti*, etc. To a lesser extent, though, certain growth hormone regulators like methoprene and diflubenzuron have been utilized as mosquito vector larval control agents (Cheng et al. 2004).

Entomopathogen bacteria like *Bacillus thuringiensis* var. *israelensis* and *B. sphaericus* have excelled in controlling vector mosquitoes in inaccessible breeding habitats such as marshes and bogs; however, in recent times, certain vectors have developed resistance against these biocides. *Bacillus thuringiensis israelensis* (*Bti*), a gram-positive bacterium, has got a significant attention in the field of microbial control of disease vectors. This bacterium produces insecticidal crystal protein toxins during its sporulation. *Bacillus sphaericus* (*Bs*) has an excellent larvicidal activity against larva of certain mosquito species of *Culex* (Bhattacharya and Basu 2018). An intracellular symbiotic bacterium, *Wolbachia*, has also been on the menu of current candidates of new and innovative technologies that influence host reproduction by cytoplasmic incompatibility both in the vector host body and the pathogen's (Bhattacharya and Basu 2018). Other microbes like baculovirus, densovirus, iridovirus, and cytoplasmic polyhedrosis virus (CPV) have been reported to significantly interfere with the mosquito physiology. Although slow in producing desired results, biological tools are preferred in certain vector-breeding environments because of their sustainability and economy-raising quality, besides no ill effects on the environment, mosquitofish (*Gambusia affinis*), amphibians (tadpoles of *Polypedates cruciger*), and aquatic copepods (*Cyclops vernalis*, *Megacyclops formosanus*, and *Mesocyclops aspericornis*) (Benelli et al. 2016). Different environmental management methods have been recommended for the destruction of the larval habitats and control vector mosquito (WHO 2017). Vector control methods for the arthropod vectors like *Ae. aegypti* require significant management. Environmental modification like "piped water supply" to the household areas and environmental manipulation through management and changing of water from the water-logged household vessels have found significant approaches towards vector control. Additionally, efficient designing of water storage containers has been suggested to be an effective approach to obstruct the mosquito oviposition (WHO 2017). Household vegetation management is also a notable method for controlling the human-vector interaction (AMCA 2019). However, all these conventional methods, whether insecticidal, biological, or environmental, also have notable limitation of application, storage, and application (WHO 2006). Thus there is an urgent need for evolving novel and sustainable strategies for the proper and effective control of vectors, and a multidisciplinary approach is solicited to overcome this difficulty (WHO 2019b).

In this context, nanobiotechnology can become an efficient approach for vector control, as it has the advantages of precise and specific utilization of technologies like nanoscience and polymer science that can affect vector physiology, in particular, and facilitate the vector control as a sustainable component of the integrated vector management (IVM) (Benelli 2018; Benelli et al. 2017a; Narkhede et al. 2016). Varieties of nanoparticle formulations such as, for example, polymer-coated nanoparticles have been developed as efficient vector control agents in recent times (Benelli 2018; Werdin González et al. 2017; Balaji et al. 2017). Since the entry of this new technology into the realms of vector control rather recent and much is still desired to be completely understood for its foolproof application, it is considered opportune to succinctly bring on record all the progress thus far made in exploring different aspects of nano-biotechnological approaches for vector control.

17.2 Nanotechnology and Nanoparticles in Vector Control

Nanoparticles are 10–1000 nm sized particles in dispersion or in solid state with significant properties like notable chemical reactivity, exceptional strength, etc. Nanotechnology involving nanoparticles, nanosphere, and nanocapsules has a variety of applications in environmental, agricultural, biomedical, and pharmaceutical industries (Priya and Santhi 2014). Nanobiotechnology applications have also been utilized in arthropod vector control like mosquitoes (Athanassiou et al. 2018; Benelli 2016, 2018; Benelli and Lukehart 2017; Banumathi et al. 2017) (Table 17.1).

Different nanoformulations involving metal-based nanoparticles like aluminum (Al), silica (Si), iron (Fe), copper (Cu), chromium (Cr), manganese (Mn), and zinc (Zn) have been developed (Merritt and Bewick 2017; Hua et al. 2015). Studies indicated that metal oxide nanoparticles like ZnO nanoparticles develop significant physiological anomalies in third instar larvae of primary dengue vector, *Ae. aegypti* (Benelli 2018). Additionally, exopolysaccharide-coated ZnO nanoparticles were also found effective on *Ae. aegypti* and fourth instar larvae of *An. stephensi* (Abinaya et al. 2018; Ishwarya et al. 2018). Nanorods of ZnO were found significantly efficient against *Ae. aegypti* (Benelli 2018).

Recently, nanoparticles developed from plant origin gained much attention due its significant advantages (Kumar et al. 2015; Goodsell 2004). Such nanoparticles synthesized from plant-borne compounds have significant mosquitocidal effects (Benelli et al. 2017b). Silver-(protein-lipid) (Ag-PL) nanoparticles developed from Indian almond tree (*Sterculia foetida*) are found to be notably effective against the larvae of mosquitoes like *An. stephensi* and *Ae. aegypti* (Benelli et al. 2017b). Other studies have reported that silver nanoparticles synthesized from green source are effective in even fresh water ecosystems where the mosquito *An. stephensi*

Table 17.1 Different nanoparticle formulations against arbovirus disease vectors (*Source*: Benelli 2018; Athanassiou et al. 2018)

Nanoparticle formulation	Vector	Disease
Ag, Au, and ZnO nanoparticles (green origin)	*Aedes aegypti*	Dengue
	Anopheles stephensi	Malaria
Ag nanoparticles prepared using *Cassia fistula* extract	*Aedes albopictus*	Dengue, yellow fever, chikungunya, Zika virus, etc.
	Culex pipiens pallens	Filariasis, encephalitis
Ag nanoparticles prepared using salicylic acid and 3,5-dinitrosalicylic acid	*Aedes albopictus*	Dengue
Au nanoparticles developed using latex of *Jatropha curcas*	*Aedes aegypti*	Dengue
Carbon-dot-Ag nanohybrid	*Anopheles stephensi*	Malaria
	Culex quinquefasciatus	Filariasis

invariably breeds (Soni and Prakash 2014). In addition, gold nanoparticles are also found to be effective against *Ae. aegypti* (Suganya et al. 2017).

The application of nanoparticles as mosquito ovicide, oviposition deterrents, and adulticides have also been explored by Benelli et al. (2016) and Benelli et al. (2017b). Their studies reported that after a single exposure of 30 ppm silver nanoparticles (synthesized from *Sargassum muticum*), the egg hatching ability of mosquitoes like *Ae. aegypti*, *An. stephensi*, and *Cx. quinquefasciatus* was lowered by about 100% (Madhiyazhagan et al. 2015). Additionally, nanoparticles like nanosilica have shown oviposition deterrent effects for these mosquitoes (Benelli et al. 2017b). Interestingly, their findings demonstrated that gold and silver nanoparticles enhance the predation efficiency of *G. affinis* (mosquito eating fish) and cyclopoid crustacean against mosquito larval populations (Benelli 2016; Benelli et al. 2017b).

17.3 Nanobiotechnology with Polymers in Vector Control

Polymer science or macromolecular science has a great input on nanobiotechnology, and different polymers have been used for developing nanoformulations from biomedical to industrial applications. Different polymers have also been used for polymer-based compositions for the mosquito vector control. Synthetic polymers like polyvinylpyrrolidone (PVP), poly(ethylene oxide) (PEO), poly(vinyl alcohol) (PVA), and cellulose derivative like ethyl cellulose (EC) have been utilized as vehicles for the successful delivery of nanoparticles (Akhtar 2014; Chattopadhyay et al. 2013). Polyhydroxyethylmethacrylate (pHEMA) was also significantly used

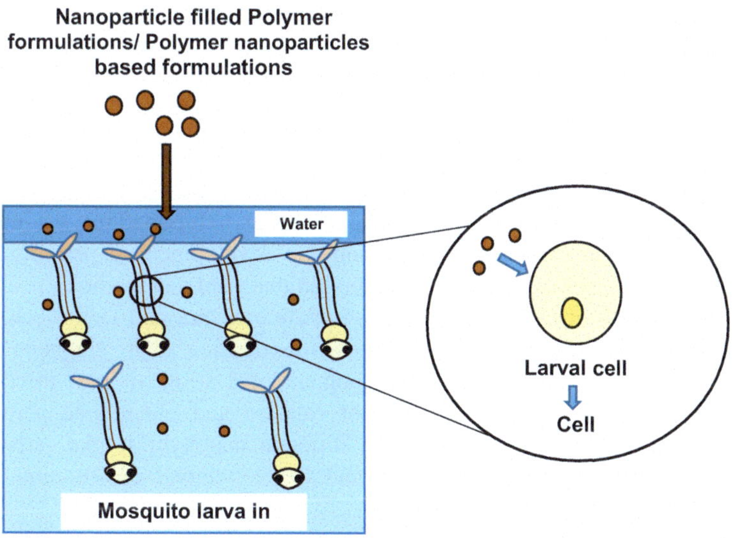

Fig. 17.1 Effect of nanoparticle-filled polymer formulation/polymer nanoparticle-based formulations on mosquito vector larva. (*Source*. Benelli 2018)

for mosquito repellent delivery formulation (Delong et al. 2016). On the other hand, natural polymer like chitosan has been notably used as a polymeric matrix to prepare nanoparticle-based formulation for controlling *Culex pipiens* larval populations (Werdin González et al. 2017). Mosquito repellent with polymer additive has been utilized to protect the potential human victims from the mosquito bites (Rozendaal et al. 1997). Polymer nanoparticles with essential oil have been shown to be a promising approach to mosquito control (Werdin González et al. 2017). However, other studies demonstrate the application of synthetic polymers in controlling the population of filarial vector (Curtis et al. 2002). However, further research showed that polyethylene glycol (PEG)-coated nanoparticle and essential oil containing pesticides were developed and shown nearly 80% efficiency in pest insect control (Hua et al. 2015), and thus synthetic polymers could also be useful for mosquito larval control (Fig. 17.1).

The PEG-based formulations have recently been used to control mosquito larval populations (Balaji et al. 2017). The polystyrene nanoparticles affect adversely on insect physiology by interacting with cytochrome P450 isozymes that are expressed in the form of baculosomes (i.e., microsomes developed from insect cells) (Fröhlich et al. 2010). Recently, researchers like Chuaycharoensuk et al. (2012) have developed monoterpene-incorporated polymer disks to act as an insecticide for vector mosquitoes. Monoterpene-like geraniol was incorporated into polymer disks to prepare polymer-based larvicide which has 92–100% efficiency against *Ae. albopictus*.

17.4 Effects and Mode of Action of Nanoparticles and Polymers: Nanobiotechnology in Vector Control

A variety of nanomaterials and nanoformulations could affect mosquito vector physiology through affecting on the molecular level (Fig. 17.2). The nanoparticles cause notable destruction of the epithelial layer and midgut and loss of head structure and caudal hairs (Suganya et al. 2017). Small et al. (2016) reported that the ingestion of gold nanoparticles damaged reproduction and development of pest insect like cockroaches, whereas Patil et al. (2016) reported that gold nanoparticles resulted in lower concentration of trypsin in the serum of *Ae. aegypti*, and Dziewięcka et al. (2016) published data on the interesting observation that graphene oxide nanoparticles were found to affect development of a string of pest insects like crickets through increasing the activity of catalase and glutathione peroxidase enzymes and HSP proteins. Interestingly, different nanohybrids like carbon-dot-silver nanohybrids also induce toxicity and develop larval deformation of the

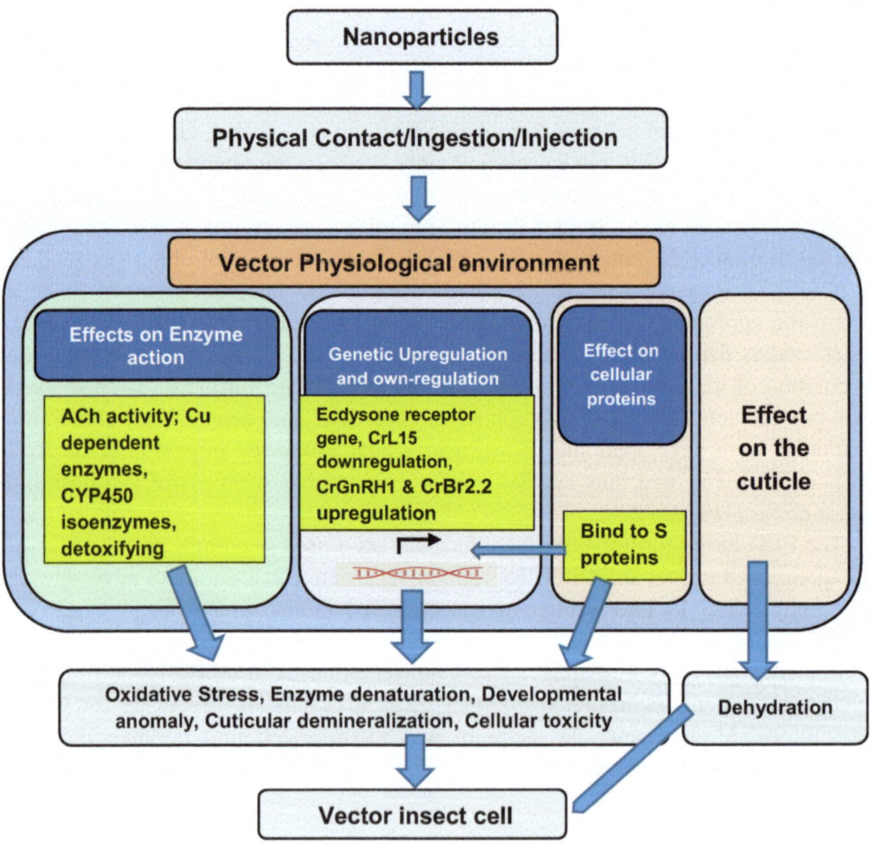

Fig. 17.2 Effect of nanoparticles on insect physiology. (Modified from Benelli 2018)

mosquito vectors like *An. stephensi* and *Cx. quinquefasciatus* (Sultana et al. 2018). Arumugam et al. (2016) found out that nanosilica was also found effective against insect control as they have been found to cause cell damage and cell death. In this context, it should be mentioned that the use of silica as nanopesticide have been considered safe as WHO (Athanassiou et al. 2018).

Different metal responses have a known effect on gene regulation (Merritt and Bewick 2017). After penetration, the nanoparticles either interact with sulfur of proteins or bind to the proteins or the phosphorous of DNA, which ultimately leads to altered genome response specific to enzyme denaturation or cellular organelle damage. In addition to this, impair membrane permeability and perturbation in proton motive force accelerate the eventual cellular dearth (Benelli 2018; Benelli et al. 2017b). Silver nanoparticles which generally induce toxicity through oxidative stress inside the arthropod tissues also induce aggregation and accumulation of the reactive oxygen species (ROS) in the fruit fly, *D. melanogaster*, which ultimately resulted in ROS-mediated apoptosis and autophagy. This knowledge prompted to study effect of silver nanoparticles on mosquito vectors where they result in a decrease in total protein level and reduce activity of acetylcholine esterase and α and β-carboxylesterase (Fouad et al. 2018). Nanoparticles activate Nrf2-dependent antioxidant pathway (Mao et al. 2018). 3,5-Dinitrosalicylic acid-synthesized Ag nanoparticles result in decreased level of mosquito vector total protein and phosphatase enzymes (Ga'al et al. 2018). Moreover, the cytotoxicity of the nanoparticles can be seen after their penetration through the arthropod exoskeleton (Rai et al. 2014). The metal-based nanoparticles generally displace essential metals and cause direct cellular damage or cause altered genetic expression through the redox state change specific to insect death (Merritt and Bewick 2017).

17.5 Biopolymer-Based Systems and Vector Control

Controlled release systems based on the biopolymers like cellulose, chitosan, agarose, alginate, and carrageenan have been considered quite useful in vector control (Kanis et al. 2018; Badawy and Rabea 2016). Nanoparticle delivery through biodegradable polymers like cellulose and chitosan has been considered a safe and environmentally hassle-free approach for vector control (Ramzan and Yousaf 2018). These release systems can be prepared through different formulations like micro- or nanoparticle based system loaded with insecticide or pesticide (Kanis et al. 2018; Chaudhary et al. 2017). Additionally, the polymer-based biodegradable formulations can become potential larvicides of vector mosquitoes as is evidenced by the cellulose-based formulations such as ethyl cellulose that have been demonstrated as efficient controlled release systems for pesticide (Mathew and Kalyanasundaram 2004; Pérez-Martínez et al. 2001). According to Kanis et al. (2018), Custódio et al. (2016), and Basu et al. (2019), polymeric hydrogel, a three-dimensional cross-linked polymeric structure, can be used in different biomedical and industrial applications. Studies demonstrated that natural biopolymer like cellulose has the biocompatibility and structural significance to be utilized in biomedical industry (Eo et al. 2016). Cellulose can be obtained from plant sources and can also be found in bacterial sources. Bacterial or microbial cellulose (BC) is generally produced by aerobic bacteria *Acetobacter*, and the advantage of using BC microfibrils is that these are free from contaminations like lignin and hemicellulose unlike plant cellulose. Thus, BC is a pure form of cellulose that can be obtained from bacterial source. The BC membranes, characterized with different macroporous structures and can be prepared in the form of hydrogels (Roy et al. 2012; Basu et al. 2019), could be used as an efficient controlled released system which is currently being explored for application in vector control method. Thus the bacterial cellulose-based hydrogel membranous structures have a great possibility for utilization in vector control particularly in the open water bodies or water storage devices where the membranous structure can be directly applied as a covering without significantly hampering the water quality.

17.6 Conclusion

In the absence of vaccines and specific antidotes for many vector-borne diseases, the overall management of the disease often rests on the plank of vector control. Alternative to conventional vector control methodology is so badly felt that there is a strong integration of interdisciplinary approach so as to control the vector population through either replacement or suppression of sizable populations of the vector whereafter it cannot potentially transmit a disease. Of the many novel technologies being explored, polymer-based nanobiotechnology approach is one potential innovative technology that can safely interplay with other components of the integrated vector management (IVM). Nanobiotechnology is an emerging area, which can efficiently facilitate vector mosquito control. Nanoparticles originated from "green source" have been found highly efficient. Additionally, silver and gold nanoparticles have been found effective against mosquito larval populations. On the other hand, polymer-based formulations have also been found effective. Nanoparticles also increase the predation efficiency of the natural predators of the mosquito larvae in the aquatic environment. Further studies are required to develop a suitable nano-biotechnological approach for vector control.

Acknowledgments The first author extends his gratitude to Prof. Ing. Petr Sáha, Associate Professor Nabanita Saha, and Prof. Ing. Vladimír Sedlařík (current director of the Centre of Polymer Systems, Tomas Bata University in Zlin, Czech Republic) for their kind support and encouragement. The authors also extend their gratitude to Prof. A.K. Hati (former Director, Calcutta School of Tropical Medicine, India) for his valuable suggestions.

References

Abinaya M, Vaseeharan B, Divya M, Sharmili A, Govindarajan M, Alharbi NS, Kadaikunnan S, Khaled JM, Benelli G (2018) Bacterial exopolysaccharide (EPS)-coated ZnO nanoparticles showed high antibiofilm activity and larvicidal toxicity against malaria and Zika virus vectors. J Trace Elem Med Biol 45:93–103. https://doi.org/10.1016/j.jtemb.2017.10.002

Akhtar MU (2014) Towards controlled release of a natural mosquito repellent from polymer matrices. Master of Science Dissertation, Faculty of Engineering, University of Pretoria. https://repository.up.ac.za/bitstream/handle/2263/46236/Akhtar_Towards_2015.pdf;jsessionid=DD92F2E25BD267AEDC5D1BF7CD74911D?sequence=1. Accessed 12 Apr 2019

AMCA factsheet (2019). https://www.mosquito.org/page/control. Accessed 08 Apr 2019

Arumugam G, Velayutham V, Shanmugavel S, Sundaram J (2016) Efficacy of nanostructured silica as a stored pulse protector against the infestation of bruchid beetle, Callosobruchus maculatus (Coleoptera: Bruchidae). Appl Nanosci 6:445–450. https://doi.org/10.1007/s13204-015-0446-2

Athanassiou CG, Kavallieratos NG, Benelli G, Losic D, Usha Rani P, Desneux N (2018) Nanoparticles for pest control: current status and future perspectives. J Pest Sci 91:1–15. https://doi.org/10.1007/s10340-017-0898-0

Badawy MEI, Rabea EI (2016) Chapter 7—Chitosan and its derivatives as active ingredients against plant pests and diseases. In: Bautista-Baños S, Romanazzi G, Jiménez-Aparicio A (eds) Chitosan in the preservation of agricultural commodities. Academic Press (Elsevier), USA, pp 179–219. https://doi.org/10.1016/B978-0-12 802735-6.00007-0

Balaji APB, Ashu A, Manigandan S, Sastry TP, Mukherjee A, Chandrasekaran N (2017) Polymeric nanoencapsulation of insect repellent: evaluation of its bioefficacy on *Culex quinquefasciatus* mosquito population and effective impregnation onto cotton fabrics for insect repellent clothing. J King Saud Univ Sci 29:517–527. https://doi.org/10.1016/j.jksus.2016.12.005

Banumathi B, Vaseeharan B, Periyannan R, Prabhu NM, Ramasamy P, Murugan K, Canale A, Benelli G (2017) Exploitation of chemical, herbal and nanoformulated acaricides to control the cattle tick, Rhipicephalus (Boophilus) microplus—a review. Vet Parasitol 244:102–110. https://doi.org/10.1016/j.vetpar.2017.07.021

Basu P, Saha N, Saha P (2019) Inorganic calcium filled bacterial cellulose based hydrogel scaffold: novel biomaterial for bone tissue regeneration. Int J Polym Mater Poly Biomater 68 (1–3):134–144. https://doi.org/10.1080/00914037.2018.1525733

Benelli G (2016) Plant-mediated biosynthesis of nanoparticles as an emerging tool against mosquitoes of medical and veterinary importance: a review. Parasitol Res 115:23–34. https://doi.org/10.1007/s00436-015-4800-9

Benelli G (2018) Mode of action of nanoparticles against insects. Environ Sci Pollut Res 25:12329–12341. https://doi.org/10.1007/s11356-018-1850-4

Benelli G, Lukehart CM (2017) Special issue: Applications of green synthesized nanoparticles in pharmacology, parasitology and entomology. J Clust Sci 28:1–2. https://doi.org/10.1007/s10876-017-1165-5

Benelli G, Jeffries CL, Walker T (2016) Biological control of mosquito vectors: past, present, and future. Insects 7(52):2–18. https://doi.org/10.3390/insects7040052

Benelli G, Maggi F, Romano D, Stefanini C, Vaseeharan B, Kumar S, Higuchi A, Alarfaj AA, Mehlhorn H, Canale A (2017a) Nanoparticles as effective acaricides against ticks—a review. Ticks Tick-borne Dis 8:821–826. https://doi.org/10.1016/j.ttbdis.2017.08.004

Benelli G, Caselli A, Canale A (2017b) Nanoparticles for mosquito control: challenges and constraints. J King Saud Univ Sci 29:424–435. https://doi.org/10.1016/j.jksus.2016.08.006

Bhattacharya S, Basu P (2018) Microorganisms in the management of mosquito vectors: a sustainable eco friendly approach. In: Tyagi BK, Dhanasekaran D (eds) Microbial control of vector-borne diseases. CRC Press (Francis & Taylor), Boca Raton, FL. NLM ID: 101711461

Carson R (1962) Silent Spring, Boston: Houghton, pp. 323. https://scholar.google.com/scholar_lookup?title=Silent+Spring&author=Carson,+Rachel&publication_year=1962)

Chattopadhyay P, Dhiman S, Devi KA, Banerjee S, Rabha B, Chaurasia A, Veer V (2013) Ultra low concentration deltamethrin loaded patch development and evaluation of its repellency against dengue vector *Aedes* (S) *albopictus*. Parasit Vectors 6(284):1–10. https://doi.org/10.1186/1756-3305-6-284

Chaudhary S, Kanwar RK, Sehgal A, Cahil DM, Baroow CJ, Sehgal R, Kanwar JR (2017) Progress on *Azadirachta indica* based biopesticides in replacing synthetic toxic pesticides. Front Plant Sci 8:610. https://doi.org/10.3389/fpls.2017.00610

Cheng S-S, Liu J-Y, Tsai K-H, Chen W-J, Chang S-T (2004) Chemical composition and mosquito larvicidal activity of essential oils from leaves of different Cinnamomum osmophloeum provenances. J Agric Food Chem 52:4395–4400. https://doi.org/10.1021/jf0497152

Chuaycharoensuk T, Manguin S, Duvallet G, Chareonviriyaphap T (2012) Assessment of geraniol-incorporated polymers to control Aedes albopictus (Diptera: Culicidae). Parasite 19:427–432. https://doi.org/10.1051/parasite/2012194427

Curtis CF, Malecela-Lazaro M, Reuben R, Maxwell CA (2002) Use of floating layers of polystyrene beads to control populations of the filaria vector *Culex quinquefasciatus*. Ann Trop Med Parasitol 96(Suppl 2):S97–S104. https://doi.org/10.1179/000349802125002446

Custódio KM, Oliveira JG, de Moterle D, Zepon KM, Prophiro JS, Kanis LA (2016) A biodegradable device for the controlled release of *Piper nigrum* (Piperaceae) standardized extract to control *Aedes aegypti* (Diptera, Culicidae) larvae. Rev Soc Bras Med Trop 49:687–692. https://doi.org/10.1590/0037-8682-0340-2016

Delong W, Weibin M, Mingchen J, Zhonglin Y, Juntao F, Xing Z (2016) pHEMA hydrogels with pendant triazinyl-β-cyclodextrin as an efficient and recyclable reservoir for loading and release of plant-based mosquito repellents: a new aqueous mosquito repellent formulation. RSC Adv 6: 27301–27312. https://doi.org/10.1039/C5RA27942A

Dziewięcka M, Karpeta-Kaczmarek J, Augustyniak M, Majchrzycki Ł, Augustyniak-Jabłokow MA (2016) Evaluation of in vivo graphene oxide toxicity for *Acheta domesticus* in relation to nanomaterial purity and time passed from the exposure. J Hazard Mater 305:30–40. https://doi.org/10.1016/j.jhazmat.2015.11.021

Eo MY, Fan H, Cho YJ, Kim SM, Lee SK (2016) Cellulose membrane as a biomaterial: from hydrolysis to depolymerization with electron beam. Biomater Res 20:16. https://doi.org/10.1186/s40824-016-0065-3

Fouad H, Hongjie L, Hosni D, Wei J, Abbas G, Ga'al H, Jianchu M (2018) Controlling *Aedes albopictus* and *Culex pipiens pallens* using silver nanoparticles synthesized from aqueous extract of Cassia fistula fruit pulp and its mode of action. Artif Cells Nanomed Biotechnol 46: 558–567. https://doi.org/10.1080/21691401.2017.1329739

Fröhlich E, Kueznik T, Samberger C, Roblegg E, Wrighton C, Pieber TR (2010) Size-dependent effects of nanoparticles on the activity of cytochrome P450 isoenzymes. Toxicol Appl Pharmacol 242:326–332. https://doi.org/10.1016/j.taap.2009

Ga'al H, Fouad H, Tian J, Hu Y, Abbas G, Mo J (2018) Synthesis, characterization and efficacy of silver nanoparticles against *Aedes albopictus* larvae and pupae. Pestic Biochem Physiol 144:49–56. https://doi.org/10.1016/j.pestbp.2017.11.004

Goodsell DS (2004) Bionanotechnology: lessons from nature. Wiley, Hoboken, NJ, p 352. https://doi.org/10.1002/0471469572

Hua K-H, Wang H-C, Chung R-S, Hsu J-C (2015) Calcium carbonate nanoparticles can enhance plant nutrition and insect pest tolerance. J Pestic Sci 40(4):208–213. https://doi.org/10.1584/jpestics.D15-025

Ishwarya R, Vaseeharan B, Kalyani S, Banumathi B, Govindarajan M, Alharbi NS, Kadaikunnan S, Al-anbr MN, Khaled JM, Benelli G (2018) Facile green synthesis of zinc oxide nanoparticles using Ulva lactuca seaweed extract and evaluation of their photocatalytic, antibiofilm and insecticidal activity. J Photochem Photobiol B Biol 178:249–258. https://doi.org/10.1016/j.jphotobiol.2017

Kanis LA, Rabelo BD, Moterle D, Custódio KM, de Oliveira JG, de Lemos AB, da Silva OS, Zepon KM, Magnago RF, Prophiro JS (2018) Piper ovatum (Piperaceae) extract/starch-cellulose films to control Aedes aegypti (Diptera: Culicidae) larvae. Ind Crop Prod 122:148–155. https://doi.org/10.1016/j.indcrop.2018.05.055

Khan M (2015) Important vector-borne diseases with their zoonotic potential: present situation and future perspective. Bangl J Vet Med 13(2):1–14. https://doi.org/10.3329/bjvm.v13i2.26614

Kumar S, Viney L, Deepti P (2015) Green synthesis of therapeutic nanoparticles: an expanding horizon. Nanomedicine 10:2451–2471. https://doi.org/10.2217/nnm.15.112

Madhiyazhagan P, Murugan K, Kumar AN, Nataraj T, Dinesh D, Panneerselvam C, Subramaniam J, Mahesh Kumar P, Suresh U, Roni M, Nicoletti M, Alarfaj AA, Higuchi A, Munusamy MA, Benelli G (2015) Sargassum muticum-synthesized silver nanoparticles: an effective control tool against mosquito vectors and bacterial pathogens. Parasitol Res 114:4305–4317. https://doi.org/10.1007/s00436-015-4671-0

Mao BH, Chen ZY, Wang YJ, Yan SJ (2018) Silver nanoparticles have lethal and sublethal adverse effects on development and longevity by inducing ROS-mediated stress responses. Sci Rep 8(1):2445. https://doi.org/10.1038/s41598-018-20728-z

Mathew N, Kalyanasundaram M (2004) Development of biodegradable aluminium carboxymethylcellulose matrices for mosquito larvicides. Pest Manag Sci 60:685–690. https://doi.org/10.1002/ps.855

Merritt TJS, Bewick AJ (2017) Genetic diversity in insect metal tolerance. Front Genet 8:172. https://doi.org/10.3389/fgene.2017.00172

Narkhede CP, Suryawanshi RK, Patil CD, Borase HP, Patil SV (2016) Use of protease inhibitory gold nanoparticles as a compatibility enhancer for Bt and deltamethrin: a novel approach for pest control. Biocatal Agric Biotechnol 8:8–12. https://doi.org/10.1016/j.bcab.2016.07.004

Patil CD, Borase HP, Suryawanshi RK, Patil SV (2016) Trypsin inactivation by latex fabricated gold nanoparticles: a new strategy towards insect control. Enzym Microb Technol 92:18–25. https://doi.org/10.1016/j.enzmictec.2016.06.005

Pérez-Martínez JI, Morillo E, Maqueda C, Ginés JM (2001) Ethyl cellulose polymer microspheres for controlled release of norfluazon. Pest Manag Sci 57:688–694. https://doi.org/10.1002/ps.339

Priya S, Santhi S (2014) A review on nanoparticles in mosquito control – a green revolution in future. Int J Res Appl Sci Eng Technol 2(12):378–387

Rai M, Kon K, Ingle A, Duran N, Galdiero S, Galdiero M (2014) Broadspectrum bioactivities of silver nanoparticles: the emerging trends and future prospects. Appl Microbiol Biotechnol 98:1951–1961. https://doi.org/10.1007/s00253-013-5473-x

Ramzan H, Yousaf Z (2018) Chapter 4. Green fabrication of metallic nanoparticles. In: Grumezescu AM, Andrew W (eds) Inorganic frameworks as smart nanomedicines. Elsevier, Amsterdam, p 173

Roy N, Saha N, Kitano T, Saha P (2012) Biodegradation of PVP-CMC hydrogel film: a useful food packaging material. Carbohydr Polym 89(2):346–353. https://doi.org/10.1016/j.carbpol.2012.03.008

Rozendaal JA, World Health Organization (1997) Vector control: methods for use by individuals and communities/prepared by Jan A. Rozendaal. World Health Organization. http://www.who.int/iris/handle/10665/41968. Accessed 20 Apr 2019

Small T, Ochoa-Zapater MA, Gallello G, Ribera A, Romero FM, Torreblanca A, Garcerá MD (2016) Gold-nanoparticles ingestion disrupts reproduction and development in the German cockroach. Sci Total Environ 565:882–888. https://doi.org/10.1016/j.scitotenv.2016.02.032

Soni N, Prakash S (2014) Green nanoparticles for mosquito control. Sci World J 2014:496362., 6 pages. https://doi.org/10.1155/2014/496362

Suganya P, Vaseeharan B, Vijayakumar S, Banumathi B, Govindarajan M, Alharbi NS, Kadaikunnan S, Khaled JM, Benelli G (2017) Biopolymer zein-coated gold nanoparticles: synthesis, antibacterial potential, toxicity and histopathological effects against the Zika virus vector *Aedes aegypti*. J Photochem Photobiol B 173:404–411. https://doi.org/10.1016/j.jphotobiol.2017.06.004

Sultana N, Raul PK, Goswami D, Das B, Gogoi HK, Raju PS (2018) Nanoweapon: control of mosquito breeding using carbon-dot-silver nanohybrid as a biolarvicide. Environ Chem Lett 16(3):1017–1023. https://doi.org/10.1007/s10311-018-0712-0

Werdin González JO, Jesser EN, Yeguerman CA, Ferrero AA, Band BF (2017) Polymer nanoparticles containing essential oils: new options for mosquito control. Environ Sci Pollut Res 24:17006–17015. https://doi.org/10.1007/s11356-017-9327-4

WHO (2006) Report on indoor residual spraying (global malaria program). https://apps.who.int/iris/bitstream/handle/10665/69386/WHO_HTM_MAL_2006.1112_eng.pdf;jsessionid=2060F1EE8A3984FD5DB1A3E0577E9994?sequence=1. Accessed 05 Feb 2019

WHO (2017). https://www.who.int/en/news-room/fact-sheets/detail/vector-borne-diseases. Accessed 10 Apr 2019

WHO (2019a) Malaria vector control. Insecticides for indoor residual spraying. https://www.who.int/whopes/resources/who_cds_whopes_2001.3/en/. Accessed 5 Feb 2019

WHO (2019b). https://www.who.int/neglected_diseases/vector_ecology/mosquito-borne-diseases/en/. Accessed 10 Apr 2019

Chapter 18
Nanosized Emulsion System: A Comprehensive Tool Towards Controlling Vector Mosquito Populations

Prabhakar Mishra, Natarajan Chandrasekaran, Amitava Mukherjee, and Brij Kishore Tyagi

Abstract In the current scenario, the world is facing a massive struggle with mosquito-borne diseases. These dipteran insects are the major cause of vector-borne diseases prevailing in society in turn creating havoc through morbidity and mortality. Mosquito poses a significant threat to human society, and therefore this scenario urges the proper combating strategy towards these deadly vectors. The current advancement in the field of nanotechnology provides the solution towards this issue in form of nanoemulsion-based pesticides. These nanometric emulsion systems possess a significant impact on portraying its ovicidal, larvicidal, pupicidal, and adulticidal activities against the deadly species of *Anopheles*, *Culex*, and *Aedes*. The nanoemulsion is a colloidal system comprising of the organic phase and aqueous phase, wherein the active ingredient present in the organic phase provides the potential efficacy to the system. Various research studies conducted in terms of screening of mosquitocidal efficacy of nanoemulsion prove the potent efficacy of the nanoemulsion.

Keywords Nanotechnology · Mosquito · Vector · Control

P. Mishra (✉)
Department of Biotechnology, School of Applied Sciences, REVA University, Bengaluru, Karnataka, India
e-mail: prabhakar.mishra@reva.edu.in

N. Chandrasekaran · A. Mukherjee
Centre for Nanobiotechnology, Vellore Institute of Technology, Vellore, Tamil Nadu, India
e-mail: nchandrasekaran@vit.ac.in

B. K. Tyagi
SpoRIC, Vellore Institute of Technology, Vellore, Tamil Nadu, India

B. K. Tyagi (ed.), *Genetically Modified and other Innovative Vector Control Technologies*, https://doi.org/10.1007/978-981-16-2964-8_18

18.1 Introduction

Mosquitoes are among the lethal creatures existing in this nature since they not only petrify human beings but also transmit many mortifying infections to them. For instance, malaria alone causes about one million deaths worldwide every year, 90% of which are infants in Africa. The vector- or mosquito-borne diseases are deadliest, although few of them have been remarkably brought under control in many countries. However, several other nations do not only face greater prevalence and distribution, but also new diseases emerge de novo (Fig. 18.1). Most arboviral and other arthropod-borne diseases have neither a vaccine to protect from infection nor any specific drug to treat the patient against the infection. In the present condition, the vector control stratagem is critical and obligatory (Benelli 2015).

To overcome this eternal issue of mosquito control with precise and limited resources, nanotechnology involving diverse nanomaterial applications proves to be a novel platform. One of the recent advancements in the field of nanopesticides is nanoemulsion. The varied formulated nanopesticides have a greater specificity towards the target species and exhibit negligible environmental hazards (Anjali et al. 2010; Kumar et al. 2013). We describe in detail the nanoemulsions and their potential to overcome the increased vector population.

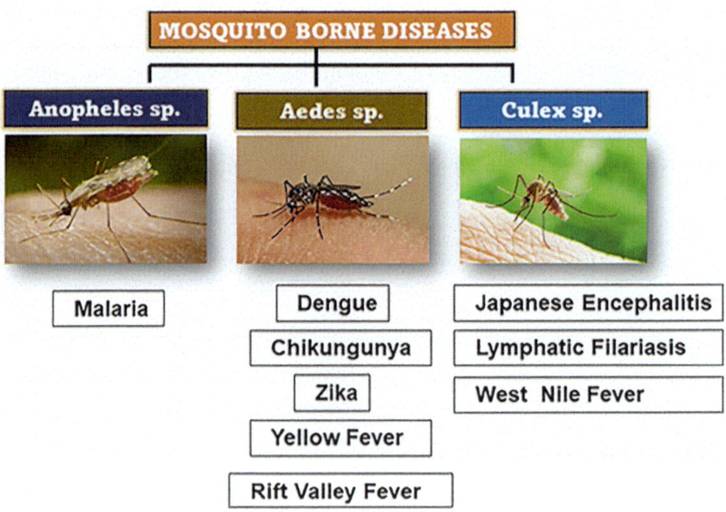

Fig. 18.1 Different mosquito vectors and their transmitted mosquito-borne diseases. (*Source*: This work)

18.2 Nanotechnology and Nanoemulsions

The development in nanotechnology delivers the potential to transfigure a diverse range of applications from drug delivery to pest control methods (Ymeti et al. 2007; Kim et al. 2007; Rai et al. 2009; Amerasan et al. 2016). These nano derivative pesticides tend to exhibit enormous competence when compared with conventional pesticides. The green synthesis of nanopesticides is beneficial over other varied forms of pest/vector control strategies. The surge in the number of plant-mediated compounds has been expected for an efficient and target-specific formulation of nanopesticides (Shankar et al. 2004; Priyadarshini et al. 2012; Ponarulselvam et al. 2012). Efficient and benign pesticidal properties displayed by these nanosized insecticides make these compounds a promising approach for future vector control strategies (Santhoshkumar et al. 2011; Marimuthu et al. 2011; Panneerselvam et al. 2012, 2013; Dinesh et al. 2015; Suresh et al. 2015; Murugan et al. 2015; Benelli 2015). This plant-based nanometric insecticide has essential oils originated from various plants, which has the advantage of controlling vector population and providing biosafety environment for non-target species.

The emulsion that consists of droplet size in the nanosized scale (20–200 nm) is often termed as nanoemulsion (El-Aasser and Sudol 2004; Nakajima 1997; Sonneville-Aubrun et al. 2004). As the nanosized emulsions attain a suitable size, they seem to be transparent or translucent and, therefore, attain stability in field conditions. It includes properties that become applicable to various fields like pharmaceutical, cosmetics, and pest management (Tadros et al. 2004). These nanosized emulsions include organic and aqueous phases. However, these systems being the non-equilibrium system cannot be formed spontaneously (Rang and Miller 1999), so they necessitate external energy for its formation (Binks 1998). The energy is provided to the system either by the chemical potential or through the mechanical devices. The nanoemulsion formulation is so-called a dispersion or high-energy emulsification system. This is achieved using the high-shear stirring and ultrasound generators or through the high-pressure homogenizers. Figure 18.2 describes the various formulation strategies behind the nano-emulsification process. The smallest size droplets in nanoemulsion can be formed in the shortest time by providing different energy inputs (Walstra 1996). The two different approaches in formulating nanoemulsion are low-energy and high-energy methods. Spontaneous emulsification, phase inversion temperature, and emulsion inversion point are few examples of low-energy methods. The high-energy emulsification techniques include ultrasonication and high-pressure homogenization.

Nano-emulsification (Landfester et al. 2004) can be achieved through another form of high-energy emulsion formation technique, i.e., high pressure homogenization, in which mechanical energy is applied to shear down macro-droplets of emulsion system, forming the nanometric emulsion (Floury et al. 2003). The nanometric emulsion system depicts its higher degree of optical transparency with kinetic stability. Nanoemulsion in comparison with conventional emulsions accomplishes a higher degree of stability which assists its varied and distinctive application

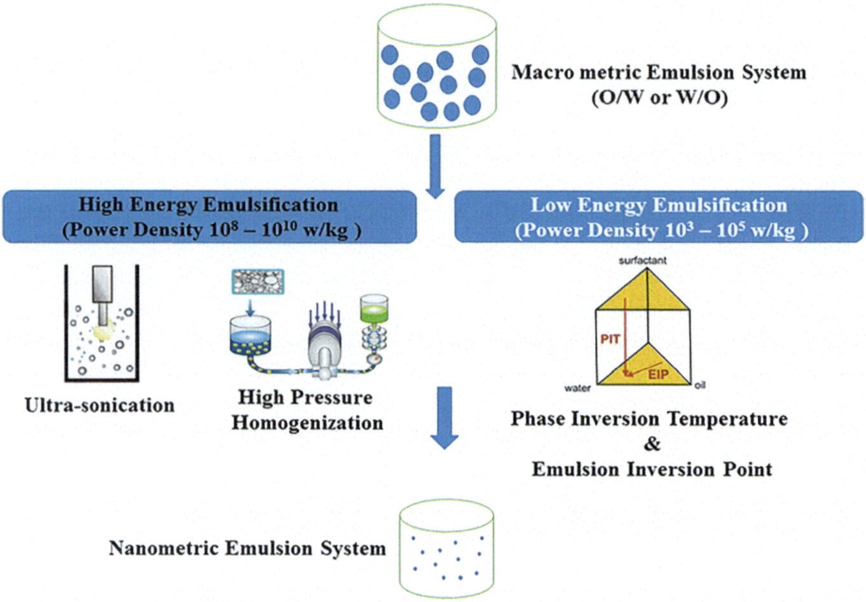

Fig. 18.2 Various mechanistic approaches towards nanoemulsion formulation. (*Source*: This work)

(Shinoda and Saito 1968). Current researches focus on potential applications of nanoemulsion as a conventional tool for controlling vector population that spreads dreadful diseases. Essential oil nanoemulsions, for example, basil oil nanoemulsion, eucalyptus oil nanoemulsion, and neem oil nanoemulsion, provide a significant contribution as ovicidal, larvicidal, and pupicidal agent against mosquito vectors.

The presence of the biochemical components in the essential oils (e.g., azadirachtin in neem oil, eugenol in clove oil, and eucalyptol in eucalyptus oil) enhances the larvicidal potency of the nanometric emulsions against *Aedes aegypti* and *Culex quinquefasciatus* (Veerakumar et al. 2014). Nanoparticles are also used for controlling the vector mosquito population. These nanoparticles are formulated using the polymerization technique (Pavel 2004; Gasco et al. 2009), direct solvent evaporation (Margulis-Goshen and Magdassi 2012), and precipitation in an aqueous droplet (Destree and Nagy 2006).

18.3 Mechanism of Nanoemulsion for Vector Control

Eucalyptus oil nanoemulsion, neem oil nanoemulsion, and basil oil nanoemulsion are few of the examples of nanometric emulsion systems which possess larvicidal, ovicidal, and pupicidal properties against the mosquito population. The phytochemicals present in these essential oils tend to exhibit their potential as an effective larvicide, ovicide, and pupicide, especially when integrated with certain more residual insecticides. Figure 18.2 presents the different strategies followed towards making different nanoemulsions from essential oils and their mechanistic approach towards mosquitocidal properties. Compounds like eugenol in basil oil, azadirachtin in neem oil, and eucalyptol in eucalyptus oil have been applied as nanoemulsions against *Culex quinquefasciatus* and *Aedes aegypti* (Veerakumar et al. 2014).

Neem oil possesses anti-ecdysone activity, which makes it a desirable natural product to study its effect mosquitocidal property. Besides, the nanoemulsified form of neem oil is one of the promising strategies to enhance the potential of neem oil in controlling the vector population. In previous studies, the neem nanoemulsion was prepared by ultrasonication of the mixture of neem oil, nonionic surfactant Tween 20, and Milli-Q water for 20 min. The obtained final nanosized emulsion system showed a mean hydrodynamic size of 30.12 ± 1.3 nm and the polydispersity index of 0.262 (Anjali et al. 2012; Mishra et al. 2018). The formulated nanoemulsion exhibited the lethal indices (LC50) of 11.75 mg/L against the dreadful filarial vector *Cx. quinquefasciatus* in 24 h study period. Efficient larvicidal activity of neem oil was attained in its nanoemulsion form oil. The smallest droplet diameter of neem oil nanoemulsion than its bulk counterpart, i.e., neem oil, is the reason for an increased surface area of contact that led to elevated absorption causing improved larvicidal activity. This property of the nanosized emulsion system can become an efficient alternative for the vector and vector-borne diseases' control.

Similarly, another essential oil that is reported to exhibit mosquitocidal activity against the *Cx. quinquefasciatus* is eucalyptus oil. Moreover, eucalyptus oil in nanoemusified form showed improved larvicidal property against the larvae of *Cx. quinquefasciatus*. It was formulated using eucalyptus oil: Tween 80 in 1:3 ratio to obtain a translucent nanoemulsion system. The mean hydrodynamic size of the nanoemulsion was estimated to be 9.4 nm. The nanosized emulsion treatment on larvae of *Cx. quinquefasciatus* showed an efficient larvicidal property at a concentration of 250 mg/L resulting in 98% mortality within 4 h. The components of essential oil such as azadirachtin, eucalyptol, etc. in the nanoemulsion attribute to its larvicidal activity against the filariasis vector. The chemical constituents in these components include terpene hydrocarbons, such as monoterpenes and sesquiterpenes, and oxygenated compounds, such as phenols, alcohols, aldehydes, ketones, esters, lactones, ethers, and oxides. This distinct variety of kinetically stabilized formulations comprises the two different, immiscible phases. In nanoemulsions, the smaller droplet diameter of 20 200 nm gives it a translucent appearance with a bluish tint (Forgiarini et al. 2000; Solans et al. 2005). These nanosized emulsions of essential oils tend to attain ameliorated insect control properties (Wang et al. 2007).

Neem urea nanoemulsion (NUNE) is another example of nanoemulsion being applied as a mosquitocidal agent. The mean droplet size of the nanoemulsion was 19.3 ± 1.34 nm, which was attained by blending neem oil, Tween 20, and urea by micro-fluidization method. The ratio of oil/surfactant/urea was optimized using the response surface modelling method. The formulated NUNE showed stability for a period of 4 days in the field conditions. This feature aids in its potent ovicidal and larvicidal activity against *A. aegypti* and *C. tritaeniorhynchus* vectors. As evidenced

Table 18.1 Mosquitocidal application of nanometric emulsion system in recent years (*Source:* This work)

Essential oil	Active ingredient	Mode of action	Host vector	References
Crithmum maritimum	Dillapiole	Larvicidal	*Aedes aegypti*	Suresh et al. (2020)
Ocimum basilicum	Eugenol	Larvicidal	*Culex quinquefasciatus*	Sundararajan et al. (2018)
Pterodon emarginatus	β-Caryophyllene	Adulticidal	*Aedes aegypti*	Oliveira et al. (2016)
Mentha spicata	Cineole	Larvicidal	*Culex pipiens*	Mohafrash et al. (2020)
Ayapana triplinervis	Thymohydroquinone dimethyl ether	Larvicidal	*Aedes aegypti*	Rodrigues et al. (2020)
Juniperus virginiana	Terpinen-4-ol	Pupicidal	*Anopheles stephensi*	Tahghighi et al. (2019)
Kaempferia galanga	Ethyl *p*-methoxycinnamate	Larvicidal	*Aedes vittatus*	AlSalhi et al. (2020)

by the histopathological analysis of the NUNE-treated larvae and biochemical profile of the target host, the NUNE ameliorated the mosquitocidal property of its bulk counterpart (Mishra et al. 2018). The chemical components in the NUNE exert beneficial properties; for example, azadirachtin present in neem oil possesses the growth inhibitory property, and urea provides fertility to the soil. Azadirachtin was reported to affect ecdysone hormone, the key hormone for the growth of insects. Thus, NUNE inhibits the molting of insects causing mortality. The enhanced specificity and target delivery were attributed to the nanometric size of the formulated nanoemulsion, which was found to be more effective than bulk pesticides.

Besides mosquitocidal properties of nanoemulsions, nanoemulsions should be bio-friendly, i.e., it has to be safe against non-target species. Considering this prerequisite, several studies were conducted to affirm the biosafety property of formulated nanoemulsion. The biosafety study of the eucalyptus oil nanoemulsion against the *Rhizobium leguminosarum* showed no zone of inhibition. It depicts the eco-safe property of the nanoemulsion (Sugumar et al. 2014). Similarly, the neem oil nanoemulsion exhibited the bio-safe property against the fingerlings of *Labeo rohita* (Mishra et al. 2014). Therefore, these studies validate the eco-safe property of formulated nanopesticide, which is an efficient tool to control the vector population, *Cx. quinquefasciatus*. The biosafety of formulated NUNE towards the non-target species like plant beneficial bacterium (*E. ludwigii*) and the paddy plant (*O. sativa*) was also reported. Thus, nanometric emulsion at the concentration used for the mosquitocidal application was found to be potentially safe towards the environment (Mishra et al. 2018). Therefore, the nanometric neem-laced urea emulsion tends to be an efficient mosquito control agent with an environmentally benign property. The biosafety property of these essential oil nanoemulsions makes them a noteworthy insect control agent. Table 18.1 describes the potential application of various forms of nanometric emulsion systems in recent years towards controlling the mosquito vector population.

18.4 Conclusion

The increasing threat of mosquito-borne diseases is becoming a major challenge to both humans and animals. The various nano-derived products are emerging as promising materials for effective vector control. One of such nano-derived products is nanoemulsion. These nanoemulsions are the colloidal solution which is prepared by blending organic phase, i.e., essential oil, and aqueous phase, i.e., surfactant and deionized water. The nanoemulsion of essential oils possesses a potent ovicidal, larvicidal, and pupicidal activity due to the active phytochemical ingredients present in essential oils. These nanoemulsions have shown their effective mosquitocidal property against *Ae. aegypti*, *Cx. tritaeniorhynchus*, and *Cx. quinquefasciatus*. Furthermore, the green synthesis of nanoemulsions makes it an eco-friendly and innovative technology for long-term control of vector mosquito populations.

Acknowledgments We acknowledge REVA University, Bengaluru, and VIT University, Vellore, for all the support.

References

AlSalhi M, Elumalai K, Devanesan S, Govindarajan M, Krishnappa K, Maggi F (2020) The aromatic ginger *Kaempferia galanga* L. (Zingiberaceae) essential oil and its main compounds are effective larvicidal agents against *Aedes vittatus* and *Anopheles maculatus* without toxicity on the non-target aquatic fauna. Ind Crop Prod 158:113012

Amerasan D, Nataraj T, Murugan K, Panneerselvam C, Madhiyazhagan P, Nicoletti M, Benelli G (2016) Myco-synthesis of silver nanoparticles using Metarhizium anisopliae against the rural malaria vector Anopheles culicifacies Giles (Diptera: Culicidae). J Pest Sci 89(1):249–256

Anjali CH, Khan SS, Margulis-Goshen K, Magdassi S, Mukherjee A, Chandrasekaran N (2010) Formulation of water-dispersible nanopermethrin for larvicidal applications. Ecotoxicol Environ Saf 73(8):1932–1936

Anjali CH, Sharma Y, Mukherjee A, Chandrasekaran N (2012) Neem oil (Azadirachta indica) nanoemulsion—a potent larvicidal agent against Culex quinquefasciatus. Pest Manag Sci 68 (2):158–163

Benelli G (2015) Research in mosquito control: current challenges for a brighter future. Parasitol Res 114(8):2801–2805

Binks BP (1998) Modern aspects of emulsion science. Royal Society of Chemistry, Cambridge

Destree C, Nagy JB (2006) Mechanism of formation of inorganic and organic nanoparticles from microemulsions. Adv Colloid Interf Sci 123:353–367

Dinesh D, Murugan K, Madhiyazhagan P, Panneerselvam C, Kumar PM, Nicoletti M, Jiang W, Benelli G, Chandramohan B, Suresh U (2015) Mosquitocidal and antibacterial activity of green-synthesized silver nanoparticles from *Aloe vera* extracts: towards an effective tool against the malaria vector *Anopheles stephensi*? Parasitol Res 114(4):1519–1529

El-Aasser MS, Sudol ED (2004) Miniemulsions: overview of research and applications. JCT Res 1 (1):21–32

Floury J, Desrumaux A, Axelos MA, Legrand J (2003) Effect of high pressure homogenisation on methylcellulose as food emulsifier. J Food Eng 58(3):227–238

Forgiarini A, Esquena J, González C, Solans C (2000) Studies of the relation between phase behavior and emulsification methods with nanoemulsion formation. In: Trends in colloid and interface science XIV. Springer, Berlin, pp 36–39

Gasco MR, Priano L, Zara GP (2009) Solid lipid nanoparticles and microemulsions for drug delivery: the CNS. In: Progress in brain research, vol 180. Elsevier, Amsterdam, pp 181–192

Kim JS, Kuk E, Yu KN, Kim JH, Park SJ, Lee HJ, Cho MH (2007) Antimicrobial effects of silver nanoparticles. Nanomedicine 3(1):95–101

Kumar RS, Shiny PJ, Anjali CH, Jerobin J, Goshen KM, Magdassi S, Chandrasekaran N (2013) Distinctive effects of nano-sized permethrin in the environment. Environ Sci Pollut Res 20 (4):2593–2602

Landfester K, Eisenblätter J, Rothe R (2004) Preparation of polymerizable miniemulsions by ultrasonication. JCT Res 1(1):65–68

Margulis-Goshen K, Magdassi S (2012) Organic nanoparticles from microemulsions: formation and applications. Curr Opin Colloid Interface Sci 17(5):290–296

Marimuthu S, Rahuman AA, Rajakumar G, Santhoshkumar T, Kirthi AV, Jayaseelan C, Kamaraj C (2011) Evaluation of green synthesized silver nanoparticles against parasites. Parasitol Res 108 (6):1541–1549

Mishra P, Suresh Kumar RS, Jerobin J, Thomas J, Mukherjee A, Chandrasekaran N (2014) Study on antimicrobial potential of neem oil nanoemulsion against *Pseudomonas aeruginosa* infection in *Labeo rohita*. Biotechnol Appl Biochem 61(5):611–619

Mishra P, Samuel MK, Reddy R, Tyagi BK, Mukherjee A, Chandrasekaran N (2018) Environmentally benign nanometric neem-laced urea emulsion for controlling mosquito population in environment. Environ Sci Pollut Res 25(3):2211–2230

Mohafrash SM, Fallatah SA, Farag SM, Mossa ATH (2020) Mentha spicata essential oil nanoformulation and its larvicidal application against Culex pipiens and Musca domestica. Ind Crop Prod 157:112944

Murugan K, Benelli G, Ayyappan S, Dinesh D, Panneerselvam C, Nicoletti M, Jiang-Shiou H, Palanisamy MK, Jayapal S, Suresh U (2015) Toxicity of seaweed-synthesized silver nanoparticles against the filariasis vector *Culex quinquefasciatus* and its impact on predation efficiency of the cyclopoid crustacean *Mesocyclops longisetus*. Parasitol Res 114(6):2243–2253

Nakajima H (1997) Microemulsions in cosmetics. In: Solans C, Kunieda H (eds) Industrial applications of microemulsions. Marcel Dekker, New York, pp 175–197

Oliveira AEMFM, Bezerra DC, Duarte JL, Cruz RAS, Souto RNP, Ferreira RMA, Nogueira J, Conceição EC, Leitão S, Bizzo HR et al (2016) Essential oil from *Pterodon emarginatus* as a promising natural raw material for larvicidal nanoemulsions against a tropical disease vector. Sustain Chem Pharm 6:1–9

Panneerselvam C, Murugan K, Kovendan K, Kumar PM (2012) Mosquito larvicidal, pupicidal, adulticidal, and repellent activity of *Artemisia nilagirica* (Family: Compositae) against Anopheles stephensi and Aedes aegypti. Parasitol Res 111(6):2241–2251

Panneerselvam C, Murugan K, Kovendan K, Kumar PM, Subramaniam J (2013) Mosquito larvicidal and pupicidal activity of Euphorbia hirta Linn. (Family: Euphorbiaceae) and Bacillus sphaericus against Anopheles stephensi Liston. (Diptera: Culicidae). Asian Pac J Trop Med 6 (2):102–109

Pavel FM (2004) Microemulsion polymerization. J Dispers Sci Technol 25(1):1–16

Ponarulselvam S, Panneerselvam C, Murugan K, Aarthi N, Kalimuthu K, Thangamani S (2012) Synthesis of silver nanoparticles using leaves of Catharanthus roseus Linn. G. Don and their antiplasmodial activities. Asian Pac J Trop Biomed 2(7):574

Priyadarshini KA, Murugan K, Panneerselvam C, Ponarulselvam S, Hwang JS, Nicoletti M (2012) Biolarvicidal and pupicidal potential of silver nanoparticles synthesized using Euphorbia hirta against Anopheles stephensi Liston (Diptera: Culicidae). Parasitol Res 111(3):997–1006

Rai M, Yadav A, Gade A (2009) Silver nanoparticles as a new generation of antimicrobials. Biotechnol Adv 27(1):76–83

Rang MJ, Miller CA (1999) Spontaneous emulsification of oils containing hydrocarbon, nonionic surfactant, and oleyl alcohol. J Colloid Interface Sci 209(1):179–192

Rodrigues ABL, Lopes RM, Rabelo ÉM, Tomazi R, Santos LL, Brandão LB. Search articles by 'Lethícia Barreto Brandão' Brandão LB, Faustino CG, Farias ALF, dos Santos CBR, de Castro Cantuária P, Galardo AKR, da Silva de Almeida SSM (2020) Development of nano-emulsions based on Ayapana triplinervis for the control of Aedes aegypti larvae. bioRxiv. https://doi.org/10.1101/2020.07.09.194985. PPR: PPR185951

Santhoshkumar T, Rahuman AA, Rajakumar G, Marimuthu S, Bagavan A, Jayaseelan C, Kamaraj C (2011) Synthesis of silver nanoparticles using Nelumbo nucifera leaf extract and its larvicidal activity against malaria and filariasis vectors. Parasitol Res 108(3):693–702

Shankar SS, Rai A, Ahmad A, Sastry M (2004) Rapid synthesis of Au, Ag, and bimetallic Au core–Ag shell nanoparticles using neem (Azadirachta indica) leaf broth. J Colloid Interface Sci 275 (2):496–502

Shinoda K, Saito H (1968) The effect of temperature on the phase equilibria and the types of dispersions of the ternary system composed of water, cyclohexane, and nonionic surfactant. J Colloid Interface Sci 26(1):70–74

Solans C, Izquierdo P, Nolla J, Azemar N, Garcia-Celma MJ (2005) Nano-emulsions. Curr Opin Colloid Interface Sci 10(3–4):102–110

Sonneville-Aubrun O, Simonnet JT, L'alloret F (2004) Nanoemulsions: a new vehicle for skincare products. Adv Colloid Interf Sci 108:145–149

Sugumar S, Clarke SK, Nirmala MJ, Tyagi BK, Mukherjee A, Chandrasekaran N (2014) Nanoemulsion of eucalyptus oil and its larvicidal activity against *Culex quinquefasciatus*. Bull Entomol Res 104(3):393–402

Sundararajan B, Moola AK, Vivek K, Kumari BR (2018) Formulation of nanoemulsion from leaves essential oil of Ocimum basilicum L. and its antibacterial, antioxidant and larvicidal activities (Culex quinquefasciatus). Microb Pathog 125:475–485

Suresh U, Murugan K, Benelli G, Nicoletti M, Barnard DR, Panneerselvam C, Kumar PM, Jayapal S, Devakumar D, Chandramohan B, Chandramohan B (2015) Tackling the growing threat of dengue: Phyllanthus niruri-mediated synthesis of silver nanoparticles and their mosquitocidal properties against the dengue vector Aedes aegypti (Diptera: Culicidae). Parasitol Res 114(4):1551–1562

Suresh U, Murugan K, Panneerselvam C, Cianfaglione K, Wang L, Maggi F (2020) Encapsulation of sea fennel (Crithmum maritimum) essential oil in nanoemulsion and SiO_2 nanoparticles for treatment of the crop pest Spodoptera litura and the dengue vector Aedes aegypti. Ind Crop Prod 158:113033

Tadros T, Izquierdo P, Esquena J, Solans C (2004) Formation and stability of nano-emulsions. Adv Colloid Interf Sci 108:303–318

Tahghighi A, Maleki-Ravasan N, Djadid ND, Alipour H, Ahmadvand R, Karimian F, Yousefinejad S (2019) GC–MS analysis and anti-mosquito activities of Juniperus virginiana essential oil against Anopheles stephensi (Diptera: Culicidae). Asian Pac J Trop Biomed 9(4):168

Veerakumar K, Govindarajan M, Rajeswary M, Muthukumaran U (2014) Low-cost and eco-friendly green synthesis of silver nanoparticles using Feronia elephantum (Rutaceae) against Culex quinquefasciatus, Anopheles stephensi, and Aedes aegypti (Diptera: Culicidae). Parasitol Res 113(5):1775–1785

Walstra P (1996) Emulsion stability. In: Becher P (ed) Encyclopaedia of emulsion technology. Marcel Dekker, New York, pp 1–62

Wang L, Li X, Zhang G, Dong J, Eastoe J (2007) Oil-in-water nanoemulsions for pesticide formulations. J Colloid Interface Sci 314(1):230–235

Ymeti A, Greve J, Lambeck PV, Wink T, van Hövell SW, Beumer TA, Kanger JS (2007) Fast, ultrasensitive virus detection using a young interferometer sensor. Nano Lett 7(2):394–397

Chapter 19
DEPA (*N*, *N*-Diethyl Phenylacetamide): Commercial Journey of India's First Indigenous Mosquito Repellent with Novel Properties

C. Kandasamy

Abstract DEPA (*N*,*N*-diethyl phenylacetamide) is less explored in India for its usage as an innovative mosquito repellent product in India. This article glances through the properties of DEPA, various kinds of formulations, the process of manufacturing, ingredients and equipment required, etc. to make a commercially viable formulation. In addition, the regulatory requirements and the current status of registration of DEPA in India, especially as a pharma compound, are briefly dealt with. More importantly, how to access the Technology from DRDO with special reference to TOT and how successful the commercialization of DEPA in India for the supply to Indian Armed forces and also to common public. The USPs of DEPA are also briefly touched upon.

Keywords DEPA · Mosquito · Novel · Repellent

19.1 Introduction

DEPA (*N*,*N*-diethyl phenylacetamide) is known for several decades as a safe and effective repellent for personal protection against disease-causing vectors, especially mosquitoes. At a time when several chemical repellents were ruling the health segment of personal protection against mosquito species, DEPA, which is based on the principles of pharma product, came up as a brainchild of an extensive research at the ICMR-Vector Control Research Centre (Ministry of Health, Government of India), Pondicherry, about four decades ago (Kalyanasundaram 1982). Subsequently, the research activities were intensified at the Defense Research and Development Organization (DRDO—Ministry of Defense, Gwalior) to explore wide-spectrum activities against a variety of hematophagous organisms, such as mosquitoes, sand flies, ticks, leeches, etc. (Vijayaraghavan et al. 1991). While the VCRC

C. Kandasamy (✉)
Agrochemicals, Chennai, India

B. K. Tyagi (ed.), *Genetically Modified and other Innovative Vector Control Technologies*, https://doi.org/10.1007/978-981-16-2964-8_19

established the properties of repellency of DEPA (Kalyanasundaram and Nisha 2006), the DRDO took it forward for data generation on toxicity and safety parameters with an objective to further distribute among the defense personnel (Rao et al. 1988, 1989a, b). Also, the DRDO achieved their objective of "Lab to Land" by initiating transfer of DEPA technology for commercialization, which was a great success during later years. This author has almost two decades of arduous journey of coordination with the DRDO and pesticide industries in the country, with an opportunity to make DEPA reach out to fulfill needs of not only the defense personnel, especially in the difficult terrains of frontier areas infested with blood-sucking creatures, but also the common man in India. It is, therefore, considered most opportune to reminisce here all the pathbreaking details innovative enough to profusely protect humans from the vexing bites of multitudes of hematophagous arthropods, particularly pestilent and vector mosquitoes, on one hand, and save the national expenditure otherwise meant for competitive repellents abroad.

19.2 Repellent Properties

Though DEPA was broadly screened initially for its repellent properties against blood-sucking organisms like mosquitoes, black flies, sand flies, bed bugs, leeches, etc., the focus was restricted to testing against adult mosquitoes which act as vectors for various deadly and highly incapacitating diseases like malaria, dengue, filariasis, chikungunya, etc. There are several publications from both the ICMR-VCRC and the DRDO establishing repellent properties of DEPA against adult mosquitoes. The DRDO extensively evaluated DEPA for repellency on defense personnel under field conditions in North-East Frontier area of India. Toxicological studies were done by DRDO to confirm that DEPA formulation is safe. Further, DEPA 20% cream formulation manufactured in India was tested against mosquito vector *Aedes aegypti* at the Institute of Medical Research, Kuala Lumpur, Malaysia, which is one of the WHO-approved Labs, and found that the repellency lasted for 6–8 h.

Studies on the repellent properties of DEPA against adult mosquitoes showed that a single aerosol spray on the human skin, preferably on the forearm, will last for about 4–6 h from biting. The same is the case when applied on the linen screens so that mosquito resting on cloths is prevented. Almost similar protection is reported with cream formulation also, when applied on the human skin. Results from DEPA are comparable to that of the world's commercially most successful mosquito repellent, DEET (*N*,*N*-diethyl-*m*-toluamide). Both DEPA and DEET were found to be equally effective with a protection time from 4.37 ± 0.08 to 4.45 ± 0.15 h at 0.1 mg/cm^2. A mélange of a pharmacologically safe polymer-based liquid and a liposphere lotion were developed into a vanishing cream formulation of DEPA that, in comparison with DEET, was found to enhance protection time from 4.4 to 6.5 and 7.13 h, respectively, but also with an alcohol solution applied to rabbits exposed to *Aedes aegypti* at 0.5 mg/cm^2 augmented the protection time from 4.0 to 6.0 h and 4.0 to 5.0 h, respectively. Against DMP (dimethyl phthalate), another significant mosquito repellent, DEPA, was more effective against all test organisms (Kalyanasundaram and Nisha 2006).

19.3 DEPA Chemistry

DEPA ($C_{12}H_{17}NO$; Mol. wt. 191.27 g/mol) belongs to amine group, well-known for a large number of effective insect repellents (e.g., DEET, DMP, etc.). Its process chemistry and standard operating procedures (SOPs) are succinctly offered below:

1. **Process Chemistry**

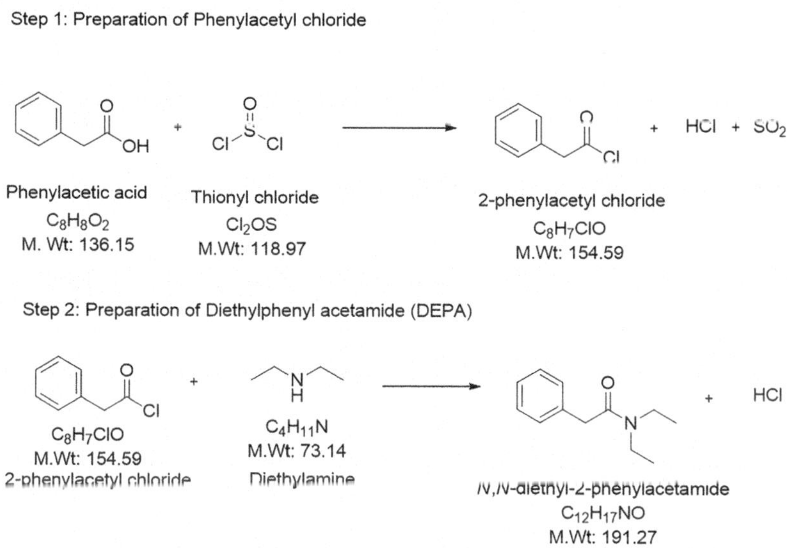

(a) **SOP**

Step 1: Preparation of phenylacetyl chloride

- 136 g of phenylacetic acid is added to 600 ml of solvent EDC under stirring. At 30–35 °C, 142.8 g of thionyl chloride is added slowly.
- The mass was maintained at 30–35 °C for 2 h and checked for reaction completion under GC.
- If the unreacted phenylacetic acid is less than 0.2%, then the phenylacetyl chloride-EDC solution is taken for the next stage reaction.

Step 2: Preparation of diethyl phenyl acetamide (DEPA)

- To the phenylacetyl chloride-EDC solution, 161 g of diethylamine is added at 40–45 °C for 2 h and maintained at the same temperature for further 2 h.
- Check a sample for reaction completion (phenylacetyl chloride unreacted will be less than 0.2%).
- Distill out EDC under vacuum at 60–65 °C and then continue distillation of DEPA at 155–170 °C at 750 mmHg vacuum. Yield of pure DEPA (99% purity) is 180 g (94%).

19.4 Formulations

Having determined repellent properties of DEPA *hook*, *line*, and *sinker*, the next task was to develop a safe and effective formulation which should be commercially viable as well. The DRDO again came forward with all its human resource and infrastructure for developing various formulations like aerosol spray, cream, lotion, and impregnated mosquito nets and fabrics (Rao et al. 1991). The efficacy of DEPA-impregnated fabrics was evaluated in the laboratory and field against the dengue vector, *Aedes aegypti* (L.), and filariasis vector, *Culex quinquefasciatus*. Say, the technology provided repellency for 30 and 36 days and had a half-life of 11 and 5 days, respectively. DEPA-treated fabric patches were found to be a practical and novel approach and were effective for 90 days in the laboratory against *Ae. aegypti* and 77 days in the field against *Cx. quinquefasciatus*. In case of bed nets, after testing various concentrations, a 20% DEPA active ingredient was found to be effective and economical for use as aerosol spray and cream. However, impregnation in mosquito nets was a complex process of rather industrial skill, and only concerned technocrats were au fait with the proper technology. Compared to this, spray technology was simpler and, thus, commercialized earlier. The toxicity of DEPA formulation was a prime concern for introducing the product into the market, and, therefore, aerosol formulation was studied in detail, especially in comparison with another known product, DEET (*N*,*N*-diethyl-meta-toluamide), which is mostly used in pharma sector. DRDO has conducted several toxicity studies with DEPA formulations and found it safe for man's own use.

In the case of aerosol spray formulations, isopropyl alcohol (IPA) is used as solvent for dilution of DEPA, whereas the cream formulation is developed using vanishing cream base. It is emphasized here that the quantity of IPA for the spray and cream used was maintained about 70% which is recommended as a sanitizer for microbial decontamination including coronavirus. Also, another pharmacologically safe liposphere lotion was formulated, but the efficacy of this could not be thoroughly established. The formulation recipe includes citronella or other similar essential oils for pleasant smell. In the case of cream, aloe vera gel was also incorporated for better skin protection.

Table 19.1 Recipe for DEPA 20% solution V/V

S. NO	Ingredients	Recipe (g)
1	DEPA technical (99%)	20.20
2	IPA (isopropyl alcohol)	61.80
	Total	82.00

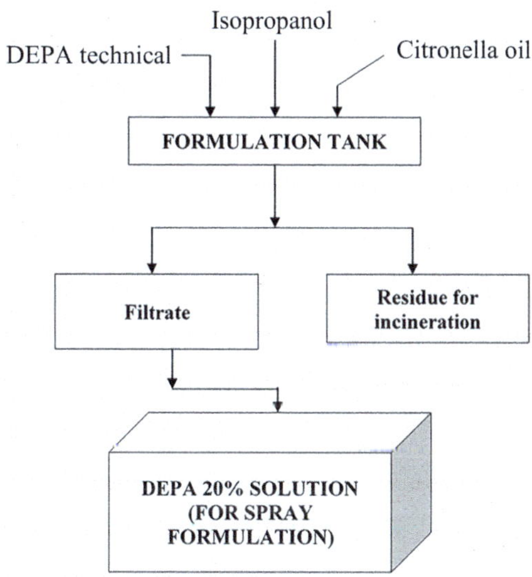

Fig. 19.1 Graphical representation of manufacture process of DEPA 20% liquid formulation

So far all the formulations were developed at a small and laboratory scale only and the challenges laid ahead to produce the formulation in a commercial industrial unit which is altogether a different cup of tea. Once again, the DRDO, with all its resources, came to rescue and extended all the technical support, and finally DEPA 20% liquid solution was granted to be brought out. The recipe, flow chart, and the manufacturing process for DEPA spray formulation standardized after several trial-and-error experiments in pilot plants are given below in Table 19.1 and Fig. 19.1.

19.5 DEPA 20% Manufacture

A step-by-step manufacturing process of DEPA 20% liquid formulation is described below.

To manufacture DEPA 20% v/v solution, we have to take DEPA technical of 99% purity, which is to be dissolved in 618 g of isopropyl alcohol solvent to get the yield of 826.6 g (specific gravity of formulation is 0.826 g). The process is as follows:

- Charge required quantity of diethyl phenyl acetamide (DEPA) to the reactor and start mild agitation.

- Charge calculated quantity of isopropanol under mild agitation and stir the mass at ambient temperature for 30–45 min for total homogeneity.
- Charge calculated quantity of citronella oil and stir the mass at ambient temperature for 30–45 min.
- Filter the mass for any undissolved particles and send the residue for incineration.
- Subject the clear transparent liquid for total analysis as per standard specifications.
- Pack the analyzed DEPA 20% solution in required/suitable packs size for spray filling.
- As a practice, about 70% of IPA is used which is a recommended sanitizer for microbial contamination.

19.6 Regulatory (Registration) Requirements

After successful completion of the above steps, registration of the product was the next mandatory regulatory requirement for use of DEPA either by the army or by the public.

DEPA is a well-known pharma product, and its active ingredients are used largely in pharma industry. To meet the regulatory requirement for use as a public health product, DRDO applied and obtained the license from the Drugs Controller General of India as insect repellent. Also, DEPA molecule got included in the Indian Pharmacopoeia, and an Indian patent was also granted. Further, with the continuing efforts of the DRDO, DEPA 20% spray solution has been well accepted and recommended by the Director General Armed Forces Medical Services (DGAFMS) for use in armed forces who started using the product instantaneously (https://www.drdo.gov.in/diethyl-phenyl-acetamide-depa-20-spray).

Incidentally, DEPA is also listed in the "Schedule to the Insecticides Act" as an insecticide (Gazette Notification—The Schedule—List of insecticides, updated on 01/01/2021, Sl No. 386 DEPA). But, till date, the same is neither registered with Central Insecticides Board (CIB) nor delisted from the Schedule (3). This means that DEPA appears as both pharma product and insecticide, which is a rare situation because the regulatory guidelines drastically vary between pharma product and an insecticide. DRDO has generated voluminous data on toxicity (acute inhalation and bio-efficacy) and other related matters to meet the pharma regulatory requirements, which are available in DRDO websites.

19.7 Transfer of Technology

When it comes to distribution of formulations to either Indian army or the general public, the major requirement is commercial production of active ingredients and manufacturing of formulations in large scale. For this purpose, DRDO, in association with Defense Ministry, formed guidelines for transfer of technology (TOT) under which the Indian manufacturing industries are encouraged to obtain TOT and produce DEPA technical as well as formulations for supply to army personnel and also for sale to the public. This TOT is also known as "Certificate of Registration" issued to the company, as a vendor for supply of DEPA.

19.8 Commercialization

Under the above guidelines, eligible manufacturers were given TOTs by the DRDO, which are renewable periodically. DEPA aerosol spray is a preferred formulation for distribution and use than the formulation of cream. Therefore, the manufacturers focused on installing a facility with adequate capacity to produce DEPA technical and spray formulation as per the specifications prescribed in the TOT by the DRDO.

The commercial packings include 200, 500, and 1000 ml in metal container with pressurized nozzle, for easy application as aerosol spray. At least three companies were given registration to manufacture DEPA technical, and five to six companies were registered for manufacturing 20% aerosol spray formulation. The cost depends on the raw material and packing material used. The approximate cost, for example, was maintained around INR 200 for DEPA 20% aerosol solution packed in 200 ml spray can. These TOT holders are only eligible to supply the formulation to the Indian Defense Ministry. However, subsequently permission was accorded to the manufacturing companies to sell this formulation in public domain also.

The procurement of DEPA by the Indian Defense Ministry is strictly as per prevailing government rules applicable to tender notification process, and manufacturing and supply of the product are made under stringent quality control inspection and auditing by a team of experts from the offices of Director General of Quality Assurance (DGQA).

19.9 USPs of DEPA

USPs are vital points for commercialization of any product. Following are the highlights for DEPA (Kandasamy 2010a, b, 2012, 2013):

- DEPA contains no pesticides—based on pharma compound.
- Repels adult mosquito vectors causing diseases like malaria, dengue, chikungunya, etc.
- Protection up to 6 h.
- Non-poisonous and safe to use.
- Environmentally safe.
- Also effective against blood-sucking organisms like bed bugs and leeches.
- Non-irritant to the eyes, skin, and nose.

- Listed in Indian Pharmacopoeia and registered with the Drugs Controller General of India.
- Added citronella oil for pleasant smell.
- Indian army uses this in big volumes and found highly effective against mosquitoes.

19.10 Lab to Land

The feedback on the efficacy of DEPA documents excellent reporting by the Indian army personnel. Thus, DEPA is a typical example of how a mosquito repellent technology developed by an Indian organization can be commercialized successfully in terms of "Lab to Land" program. It is relevant here to mention that the Indian Defense Ministry has given the prestigious "Defense Technology Absorption Award" to Sree Ramcides Chemicals Pvt Ltd (now known as Ramcides CropScience Pvt Ltd), Chennai, during the year 2010 for their significant contribution in adoption of DEPA technology.

DRDO has also permitted the sale of DEPA spray formulation to the public. The application of spray is recommended on the skin so that the adult mosquitoes will be prevented from biting. However, other formulations are yet to reach the market. There is tremendous scope of further research in developing commercially successful cream and lotion formulations, in addition to impregnation of DEPA in mosquito bed net. DEPA is truly a formidable arsenal in our armory to combat mosquito menace, on one hand, and prevent transmission of vector-borne diseases, on the other. DEPA is a safe product for a healthy life free from the fear of mosquito bites.

Acknowledgments I gratefully acknowledge Dr R. Vijayaraghavan, Director (retired.), DRDO-DRDE, Gwalior, Dr M. Kalyanasundaram, Scientist (retired), ICMR-VCRC, Puducherry, and Mr. R Padmanabhan, Managing Director, Ramcides CropScience Pvt. Ltd., Chennai, for their support and inputs given for this article on DEPA.

References

Kalyanasundaram M (1982) A preliminary report on the synthesis and testing of mosquito repellents. Indian J Med Res 76:190–194

Kalyanasundaram M, Nisha M (2006) N,N-Diethyl Phenylacetamide (DEPA): a safe and effective repellent for personal protection against hematophagous arthropods. J Med Entomol 43 (3):518525

Kandasamy C (2010a) The success of DEPA, an excellent mosquito repellent from DRDO. In International symposium on recent advances in ecology and management of vectors and vector borne diseases, 1–3 Dec 2010

Kandasamy C (2010b) DEPA (Diethyl Phenyl Acetamide) – an innovative mosquito repellent. In 5th conference of Medical Arthropodology, 18–19 Oct 2011

Kandasamy C (2012) DEPA cream – a new mosquito repellent from DRDO. In 6th conference of Medical Anthropodology, 18–19 Oct 2012

Kandasamy C (2013) Diethyl Phenyl Acetamide (DEPA) repellency against Aedes aegypti – brain storming conference on dengue scenario in India, 25–26 Jul 2013

Rao SS, Rao KM, Ramachandran PK (1988) Chemistry, efficacy and toxicology of insect repellents. J Sci Ind Res 47:722–735

Rao SS, Agarwal MK, Rao KM, Swamy RV (1989a) Study on dermal toxicity and urinary metabolites of the new insect repellent N,N-diethylphenylacetamide in rabbits. Toxicol Lett 45:67–72

Rao SS, Swamy RV, Ramachandran PK (1989b) Toxicity and metabolism of a new insect repellent N,N-diethylphenylacetamide in mice, rats and guinea pigs on cutaneous application. Toxicology 58:81–89

Rao KM, Prakash S, Kumar S, Suryanarayana MVS, Bhagwat MM, Gharia MM, Bhavsar RB (1991) N,N-diethylphenylacetainide in treated fabrics as a repellent against Aedes aegypti and Culex quinquefasciatus (Diptera: Culicidae). J Med Entomol 28(1):142–146. https://doi.org/10.1093/jmedent/28.1.142

Vijayaraghavan R, Rao SS, Suryanarayana MVS, Swamy RV (1991) Acute and sub acute inhalation toxicity studies of a new broad spectrum insect repellent N,N-diethyl phenyl acetamide. Toxicology 67:85–96

Chapter 20
Identification and Mapping of Mosquito Breeding Habitats in Tiruchirappalli City Using Remote Sensing and GIS Technologies

M. Govindaraju, J. Fowmitha Banu, K. Sridhar, P. Vignesh, and K. Rangarajan

Abstract Mosquitoes of all important genera, such as, for example, *Anopheles*, *Culex*, and *Aedes*, prefer to breed in stagnant water of varied nature, often very large to be conveniently covered for sampling on ground. Remote sensing and GIS, a combined advanced technology useful to identify and map vector habitats, are considered potential alternative tools for disease vector surveillance. In this study, we have identified vector habitats with their location in 11 locations in the Tiruchirappalli City Municipal Corporation limits using satellite Sentinel-2 imageries for the identification of mosquitoes breeding sites. Using Global Positioning System (GPS), the location was geo-tagged with the recorded value. The GPS reading was imported in GIS platform for mapping of vector habitats. The study reveals that major breeding habitats are located nearby the residential areas in the form of numerous cesspools continually recharged from domestic wastewater and are also major sources for the breeding of human lymphatic filariasis vector, *Culex quinquefasciatus*. To control their spread with disease vectors, health information system (HIS) can function efficiently. Application of GIS is more useful in managing health data for betterment of public health, and it becomes a great supporting system for health officials in the decision-making process.

Keywords Vector · Mosquitoes · Breeding sites · GPS · Remote sensing · Stagnant water

M. Govindaraju (✉) · J. F. Banu · K. Sridhar · P. Vignesh · K. Rangarajan
Centre for Climate Change Research, Department of Environmental Biotechnology, School of Environmental Sciences, Bharathidasan University, Tiruchirappalli, Tamil Nadu, India
e-mail: mgrasu@bdu.ac.in

B. K. Tyagi (ed.), *Genetically Modified and other Innovative Vector Control Technologies*, https://doi.org/10.1007/978-981-16-2964-8_20

20.1 Introduction

Since all mosquito vectors need water in some form and quality to breed, Hopp and Foley (2001) have found that egg production and multiplication of larval population increase copiously during the rainy season. Mosquitoes, particularly *Culex quinquefasciatus* (the vector for lymphatic filariasis), *Aedes aegypti* (the vector for dengue), and *Anopheles stephensi* (the vector for malaria in urban set-ups), prefer to lay their eggs in stagnant and water bodies generally clean and/or polluted with vegetative richness, all over the place near roadside, in drains, storm canals, water spilled near wells and conduit taps, domestic water storing containers like metal drums, automobile tires, water coolers, and swampy and marshy areas where the human settlements and improper disposal of waste materials are likely to be over compiled on their periphery. The mapping of vector habitats using remote sensing and GIS is both practical and handy for the stakeholders such as health departments and municipal corporations while planning for comprehensive mosquito vector reduction so as to impact also the human morbidity and mortality due to various vector-borne diseases (Purse and Golding 2015). Spatial and temporal data of remote sensing are useful in the identification of distribution of both the vector mosquito breeding and diseases they transmit (Singh et al. 2015). GIS data are useful in mapping, modeling, and forecasting outbreaks of malaria, dengue, etc. (Parselia et al. 2019). Remote sensing and GIS analysis are helpful in the identification and spread of diseases in a particular region, the vulnerable populations at risk, patterns of disease outbreaks, available healthcare facilities, and planning and assessment in a disease outbreak (Cleckner et al. 2011). For the management of vector-borne disease outbreaks, health information system is essential. With the help of GIS, the health data can be managed meaningfully as it will help plan in advance the anti-mosquito operation more accurately to reduce the mortality rate. The aim of this study was to identify and map varied vector habitats in the Tiruchirappalli City Municipal Corporation area using remote sensing and GIS.

20.2 Role of Remote Sensing and GIS in Mapping Vector Breeding and Controlling Vector-Borne Diseases

Health information system (HIS) is one of the best model-based preventive methods to help control the vector-borne diseases (Bhunia et al. 2010; Govindraju et al. 2011; Kumar and Agrawal 2020). This is a mechanism to archive the health data in GIS platform. It may be analyzed and manipulated, and forecasting could be made using the advanced software. It consists of many characteristics such as the disease type, dissemination, outbreaks, and control tools (WHO 2008). Remote sensing and GIS are the tools that can successfully be utilized to identify the vector breeding habitats, which can be employed in designing optimal mosquito control strategies based on precise spatial/temporal information databases (Agarwal et al. 2012). The remote sensing data of Landsat TM, IRS-LISS III, IRS-CARTOSAT, SPOT, IKONOS, NOAA-AVHRR, etc. are useful to map vector breeding habitats and analyze vector abundance as major parameters to correlate with local climate vicissitudes (Palaniyandi et al. 2014). For example, Advanced Very High Resolution Radiometer (AVHRR) satellite sensor, used as a background map to overlay with another map, is used to delineate breeding sites and mosquitoes ecology as well as transmission index (Nizamuddin et al. 2013). The application of satellite images (Landsat and National Oceanic and Atmospheric Administration, NOAA, satellite sensor) is principally applied to study epidemiology of endemic diseases in the field and is an added advantage (Zhang et al. 2013). Also, remote sensing information about the moisture content of the soil rich with vegetation cover can provide an efficient utility value in the application of modeling to generate a selective vector disease transmission risk map, using specific environmental variables (Suganthi et al. 2018). Use of spatial data to monitor the dynamics of diseases and vectors can assist in disease prediction and identification of hotspots where de novo pathogens are likely to emerge (Morse et al. 2012). Risk-based approaches on Geographic Information System can improve the efficiency of disease surveillance that signals for early warnings in areas with increased risk for a disease transmission (Christaki 2015). It emphasizes the need to implement decision support capabilities in disease mapping and management using GIS in developing countries like India.

In the present study, Sentinel-2A satellite, launched in June 2015 by the European Space Agency (ESA), imagery was used. It is a freely accessible with multispectral data in 13 bands comprising visible, near-infrared and shortwave-infrared spectrums. Sentinel-2 satellites are the second constellation of the ESA Sentinel missions and carry onboard multispectral scanners. The primary objective of the Sentinel-2 mission is to provide high-resolution satellite data for land cover/land use monitoring especially during urban surveillance, water quality monitoring, water demand modeling, groundwater management, and flood mapping, in addition to complement other satellite missions such as the Landsat. Since the launch of Sentinel-2 multispectral instruments in 2015, it has been used for several mosquito-borne disease mapping programs (Phiri et al. 2020).

20.3 Materials and Methods

20.3.1 Study Area

Tiruchirappalli City, the fourth largest Municipal Corporation of Tamil Nadu State having coordinates 10° 48′ 18″ N–78° 41′ 08″ E, is situated in the Cauvery river basin. As per the 2021 census, the population of the city is 11.82 lakhs. It is classified as a medium-sized city. The average annual rainfall of the city is 808 mm. Tiruchirappalli City falls in one of the most agriculturally significant regions in the Tamil Nadu State.

20.3.2 *Data and Software*

20.3.2.1 Satellite Imagery and Software Used

High resolution satellite imagery was downloaded from NASA, USGS, and Earth Explorer. The satellite imageries such as Landsat-8, Sentinel-2 image sets red band-2, Green band-3, and Blue band-4 image resolution 10 × 10 m were prepared with the use of ArcGIS10.7.1.

20.3.2.2 Field Data Collection

The high-resolution imageries were randomly prepared for the mosquito breeding sites in and around Tiruchirappalli City. Areas with waterlogging, poor sanitation, and improper disposal of the waste were selected for mosquito density and preponderance determination. The sampling sites are listed in Table 20.1. Mainly there are four zones in the study area such as Srirangam, Ariyamangalam, K. Abishekapuram, and Ponmalai. Totally 11 sites were selected within these zones, where the human settlement and the household waste discharge is more that helps to inveigle stupendous mosquito breeding. Using Global Positioning System (GPS), geo-tagged photographs were taken in the field. In Fig. 20.1, the satellite image composite band mosaic shows the breeding sites corresponding to breeding sites as documented by field photographs.

20.3.2.3 Identification of the Water Bodies/Stagnant Water

Water-stagnated areas in the study area were delineated by satellite imageries with Sentinel-2A imageries which, aided with geo-tagged photographs, were used to identify, map, and evaluate the mosquito breeding potential of the sites, imported in the software. Image pre-processing was used first for removal of clouds or errors.

Table 20.1 Geo-reference data for sampling sites in the study area

S. No	Location	Longitude	Latitude
1	Thuvakudi	78° 23′ 12″	10° 44′ 39″
2	Thiruverumbur	78° 42′ 22″	10° 46′ 25″
3	Ariyamangalam	78° 43′ 2″	10° 46′ 34″
4	Srirangam	78° 40′ 1″	10° 49′ 2″
5	Ponmalaipatti	78° 43′ 36″	10° 48′ 17″
6	Kumaran Nagar	78° 41′ 5″	10° 47′ 47″
7	Karumandapam	78° 40′ 265″	10° 46′ 2″
8	Railway junction	78° 40′ 1″	10° 49′ 3′
9	Sempattu	78° 49′ 21″	10° 44′ 46″
10	ThiruvalarSolai	78° 41′ 32″	10° 46′ 32
11	Mellur	78°40′ 13″	10° 52′ 31″

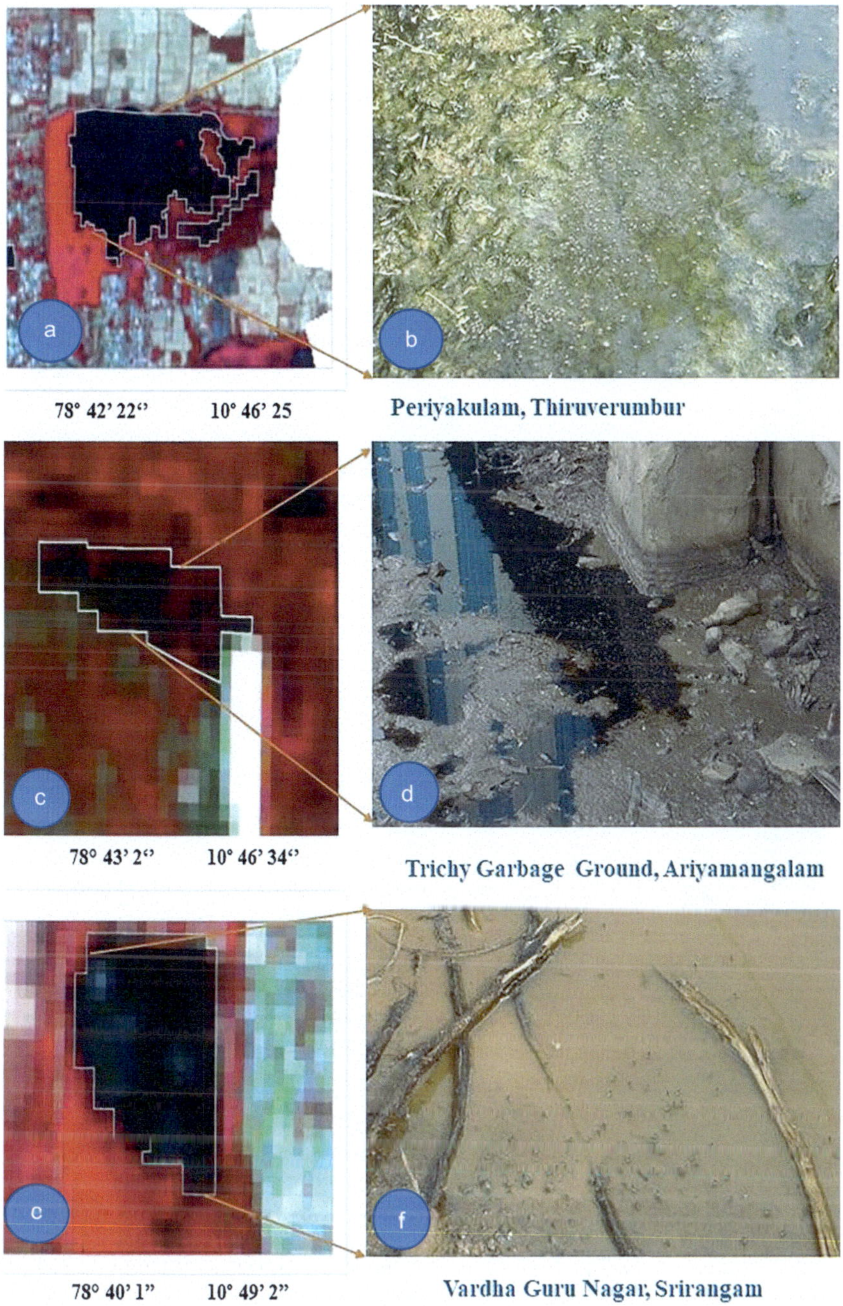

Fig. 20.1 (**a–l**) Satellite images (**a**, **c**, **e**, **g**, **i**, **k**) vis-à-vis field photographs (**b**, **d**, **f**, **h**, **j**, **l**) for different localities with mosquitogenic scenarios (January 2021)

Radiometric and geometric corrections were carried out for further analysis. Identification of the water quality was done by supervised classification of the image—an

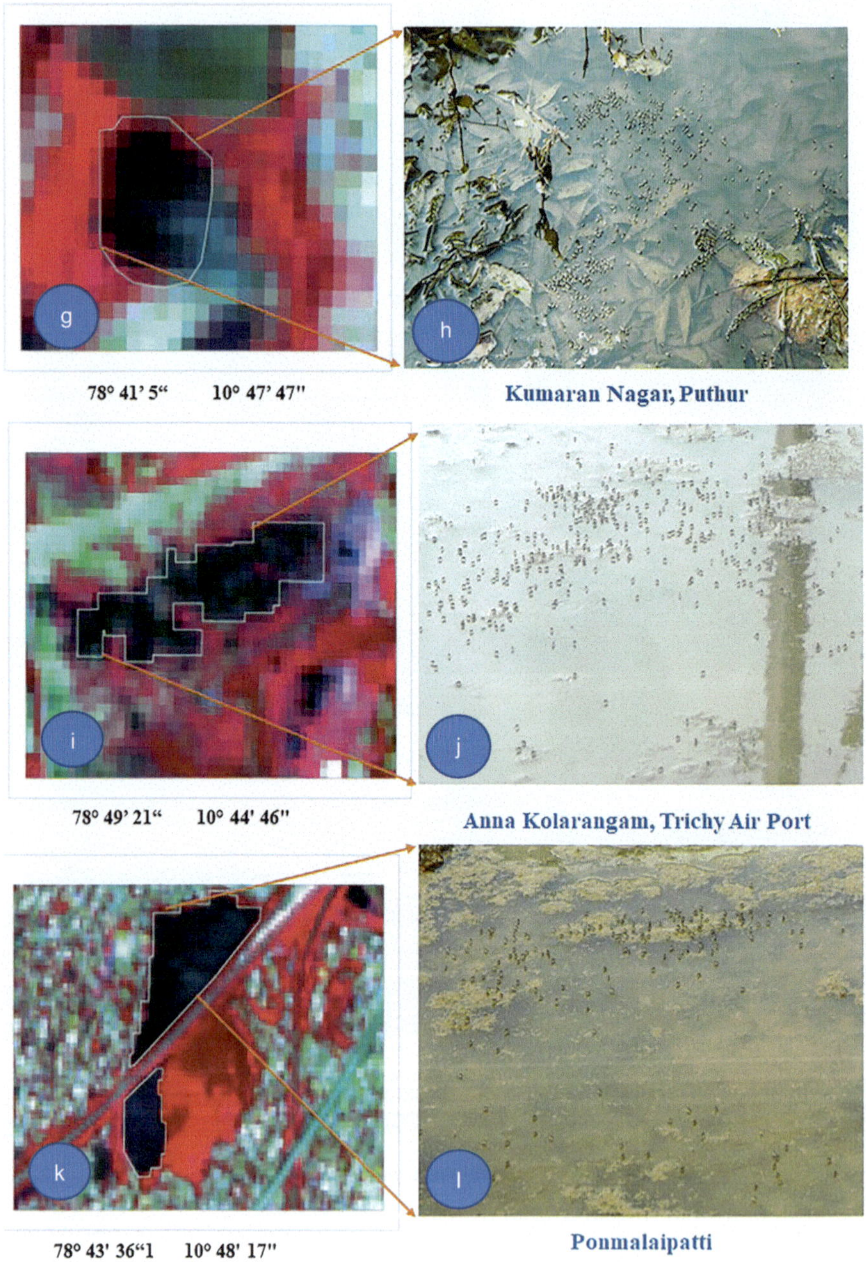

Fig. 20.1 (continued)

exercise performed by using the Gaussian maximum likelihood classifier. The dark blue color or black color in the imagery was clearly the area with stagnated water. This alludes about areas that are the probable breeding sites of mosquitoes (Fig. 20.1a–l).

20.4 Observations, Results, and Discussion

The recent satellite imagery Sentinel-2A, 21 January 2021, was processed using the Arc GIS 10.7.1 software. The false color composite satellite image was used as mosaic image to identify the vector breeding habitats. Stored and stagnated water areas appeared in dark blue or black in the imagery which was easily delineated. The recent heavy rain in the study area has led to the creation of small ponds and open ditches in the sampling sites. Due to discharge of household wastewater into the water bodies, they were converted to suitable vector breeding habitats for *Culex quinquefasciatus* mosquitoes, in particular.

Characteristically zones such as Ariyamangalam, Srirangam, and Ponmalai were relatively densely populated when compared to that in K. Abishekapuram. Ariyamangalam zone (E 78° 43′ 2″, N 10° 46′ 34″) consists of Thiruverumbur (E 78° 42′ 22″, N 10° 46′ 25″) and Thuvakudi (E 78° 23′ 12″, N 10° 44′ 39″) which fall under industrial or commercial areas. These places are therefore much congested and enduring with large dumping ground for solid waste from the town.

Large dumping grounds with solid waste, etc. attract huge pest and vector problems including vexatious and disease-transmitting mosquitoes. It was observed that improper disposal of plastic containers and tin cans increased water storage in large open areas and resulted in increased density of mosquitoes. The sudden rainfall in recent days increased the waterlogging all around that also helped in rise of humongous adult mosquitoes. The dark blue and black color in the given Fig. 20.1 shows the water is polluted which is a preferred site for breeding of the filariasis vector, *Cx. quinquefasciatus*.

Ponmalai (E 78° 43′ 36″, N 10° 48′ 17″) zone is mainly a residential area, though smitten with plenty of commercial near railway junction (E 78° 40′ 1″, N 10° 49′ 3′). It consists of vacant lands with or without houses and continuously receive domestic sewage from nearby residential areas that make it more mosquitogenic; the dengue vector *Aedes aegypti* breeds profusely in such situations. Commercial places such as those at the central bus stand and railway junction have virtually converted into waste dumps. The other most important mosquitogenic zone of Tiruchirappalli City is Srirangam (E78° 40′ 1″, N10° 49′ 2″) which is a densely populated area with more apartments. The improper drainage facilities and daily release of unfathomable domestic waste are the main reasons for vector-borne diseases in this area. The K. Abishekapuram zone is completely residential and consists of Kumaran Nagar (E 78° 41′ 5″, N10° 47′ 47″) and Karumandapam (E 78° 40′265″, N10° 46′ 2″) which are of less populated with better provision of amenities, hence also less prone to mosquito nuisance. So, the three zones, viz., Ariyamangalam, Srirangam, and

Ponmalai, are the areas delineated for more mosquitogenic situations and vector-borne diseases.

According to Ikpeama et al. (2017) and Suganthi et al. (2018), the principal vector of filariasis *Culex quinquefasciatus* and the malaria vector *Anopheles culicifacies* breed in turbid waters like those collected in ditches, etc. *Culex quinquefasciatus* breeds efficiently in all kinds of polluted waters, while the dengue vector *Aedes aegypti* is heavily adapted to the rainwater collections in any place. Therefore, most of the areas in Tiruchirappalli City are rampant with *Culex* and *Aedes* mosquito species. In such situations, technologies like the remote sensing and GIS can provide accurate data about vector habitats so that a proper action can be taken to control vectors and vector-borne diseases.

20.5 Conclusion

Unplanned urbanization and poor land use practices enhance the mosquitogenic habitats. The present study has revealed that accumulation of polluted water through domestically generated wastewater is the major source of vector breeding. It is possible that remote sensing and GIS can provide early warnings about mosquito breeding sites that can be used to our benefit of planning and controlling the vector populations effectively. It has been observed that role played by the community volunteers, school children, and other stakeholders is of immense value and their active engagement in prevention and control of vector-borne diseases cannot be overemphasized.

Acknowledgments The authors are thankful to the MoEF&CC, UGC-SAP, DST-FIST, and RUSA2.0 for their financial and other supports to carry out the research work.

References

Agarwal SA, Sikarwar SS, Sukumaran D (2012) Application of RS & GIS in risk area assessment for mosquito borne diseases - a case study in a part of Gwalior City (MP). Int J Adv Technol Eng Res 2(1):1–4

Bhunia GS, Kesari S, Jeyaram A, Kumar V, Das P (2010) Influence of topography on the endemicity of Kala-azar: a study based on remote sensing and geographical information system. Geospat Health 4(2):155–165

Christaki E (2015) New technologies in predicting, preventing and controlling emerging infectious diseases. Virulence 6:558–565

Cleckner HL, Allen TR, Bellows AS (2011) Remote sensing and modeling of mosquito abundance and habitats in coastal Virginia, USA. Remote Sens 3(12):2663–2681

Govindraju M, Ganeshkumar RS, Selvaraj M, Macklin Rani L, Kumar P, Suganthi P, Sunis IP, Thenmozhi V, Tyagi BK (2011) Development of new methodologies using Remote Sensing and GIS technique to control vector borne diseases for Sitheri Hills, Dharmapuri district, Tamil Nadu. In: 12th Esri India User Conference, pp 1–8

Hopp MJ, Foley JA (2001) Global-scale relationships between climate and the dengue fever vector, *Aedes aegypti*. Clim Chang 48(2–3):441–463

Ikpeama CA, Obiajuru IO, Nwoke BE, Ezike MN (2017) A survey of the relative abundance of mosquitoes in their preferred breeding micro-habitats in refuse dumps in Owerri, Imo State, Nigeria. Univ J Agric Res 5(6):323–328

Kumar S, Agrawal S (2020) Prevention of vector-borne disease by the identification and risk assessment of mosquito vector habitats using GIS and remote sensing: a case study of Gorakhpur, India. Nanotechnol Environ Eng 5(2):1–15

Morse SS, Mazet JAK, Woolhouse M, Parrish CR, Carroll D, Karesh WB (2012) Prediction and prevention of the next pandemic zoonosis. Lancet 380:1956–1965

Nizamuddin M, Kogan F, Dhiman R, Guo W, Roytman L (2013) Modeling and forecasting malaria in Tripura, India using NOAA/AVHRR-based vegetation health indices. Int J Remote Sens Appl 3(3):108–116

Palaniyandi M, Anand PH, Maniyosai R (2014) Spatial cognition: a geospatial analysis of vector borne disease transmission and the environment, using remote sensing and GIS. Int J Mosquito Res 1(3):39–54

Parselia E, Kontoes C, Tsouni A, Hadjichristodoulou C, Kioutsioukis I, Magiorkinis G, Stilianakis NI (2019) Satellite earth observation data in epidemiological modeling of malaria, dengue and West Nile virus: a scoping review. Remote Sens 11(16):1862

Phiri D, Simwanda M, Salekin S, Nyirenda VR, Murayama Y, Ranagalage M (2020) Sentinel-2 data for land cover/use mapping: a review. Remote Sens 12(14):2291

Purse BV, Golding N (2015) Tracking the distribution and impacts of diseases with biological records and distribution modelling. Biol J Linn Soc 115(3):664–677

Singh R, Ranjan K, Verma H (2015) Satellite imaging and surveillance of infectious diseases. J Trop Dis Public Health:1–6

Suganthi P, Govindaraju M, Kangabam RD, Suganthi K, Thenmozhi V, Tyagi BK (2018) Identification and mapping of breeding habitats of the filariasis vector, Culex quinquefasciatus say, 1823, using remote sensing and GIS technologies: a case study from the endemic Tamil Nadu state, India. In: Tyagi BK (ed) Lymphatic filariasis. Springer-Nature, Singapore, pp 259–269

WHO (2008) Information systems, 2nd edn. World Health Organization, Geneva

Zhang Z, Ward M, Gao J, Wang Z, Yao B, Zhang T, Jiang Q (2013) Remote sensing and disease control in China: past, present and future. Parasit Vectors 6(1):1–10

Chapter 21
Social and Ecological Factors Responsible for Changing Habitat of Kala-Azar Vector (*Phlebotomus argentipes*) and Its Control Through Locally Innovated Environment-Friendly Technology

Diwakar Singh Dinesh, Roshan Kamal Topno, and Krishna Pandey

Abstract Kala-azar (KA) is one of the neglected tropical diseases of the impoverished communities across the world that is caused by a protozoan parasite *Leishmania donovani* (Kinetoplastida: Trypanosomatidae) and transmitted by the sandfly vector, *Phlebotomus argentipes* (Diptera: Psychodidae). A sandfly is a tiny insect which characteristically prefers dark and humid places. Presently in India Kala-azar is in elimination mode by getting the target <1 case/10,000 population at PHC level by 2020. The vector control measures like indoor residual spray of the synthetic pyrethroid, alphacypermethrin, has been continuing since 2016, after replacing DDT following development of resistance in the vector against the pesticide. The shift in insecticide use too failed appreciably to bring the down the sandfly density significantly, albeit the vector showing no trace of resistance development against this insecticide. The factors behind this debacle were a mélange of social, economic, educational, and environmental conditions. The level of information, education, and communication (IEC) was found to be poor among the target population suffering with the disease. Environmental management for destroying the breeding sites of sandfly was found suitable for significant reduction of sandfly density by skirting the indoor wall at base level 9″h × 9″b. In the highly endemic cohort of different grades (high, meso, and low) of endemicity of Kala-azar, the economic status was found up to one lakh per annum in 98% of family. However the literacy rate was 20–79% in all three grades of endemicity. The integrated vector management (IVM), though old-time and multidisciplinary approach, is nevertheless innovative enough with accurate high level of IEC among the Kala-azar-affected communities.

Keywords Kala-azar · Vector · Sandfly · Control

D. S. Dinesh (✉) · R. K. Topno · K. Pandey
ICMR-Rajendra Memorial Research Institute of Medical Sciences, Patna, Bihar, India

B. K. Tyagi (ed.), *Genetically Modified and other Innovative Vector Control Technologies*, https://doi.org/10.1007/978-981-16-2964-8_21

21.1 Introduction

Visceral leishmaniasis (VL)/Kala-azar (KA) is one of the neglected tropical diseases that is caused by the protozoan parasite *Leishmania donovani* (Kinetoplastida: Trypanosomatidae) and transmitted by the sandfly vector, *Phlebotomus argentipes* (Diptera: Psychodidae) (Swaminath et al. 1942) (Fig. 21.1). There are 88 countries affected from this disease, with an estimate of two million new cases each year, of which Indian subcontinents (India, Bangladesh, Bhutan, Nepal, Sri Lanka, and Thailand) share 0.3 million cases (Alvar et al. 2012). In India, four states, viz., Bihar, Uttar Pradesh, West Bengal, and Jharkhand, are endemic for Kala-azar. The VL control program in the country was subsumed within the National Vector Borne Disease Control Programme (NVBDCP) in 2003 with a well-laid National Road Map for Kala-azar Elimination, 2014. The disease is targeted to be eliminated by focusing on the target less than 1 case/10,000 population at sub-district/block level by the year 2010 (National Health Policy 2002), which was revised to 2015

Fig. 21.1 (**a**) An adult sandfly (*P. argentipes*) and (**b**) the larva

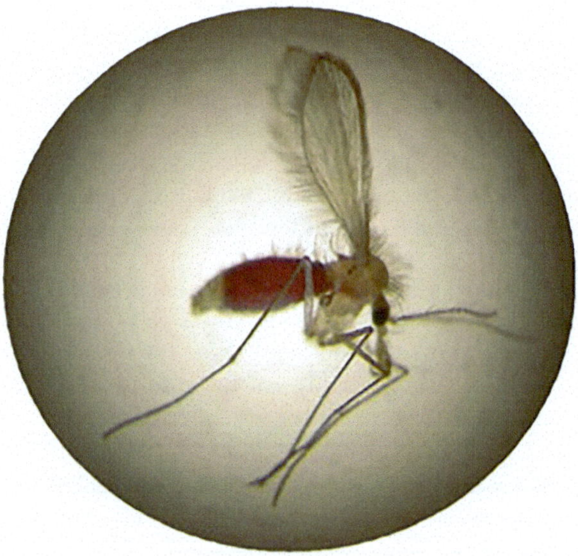

(a)

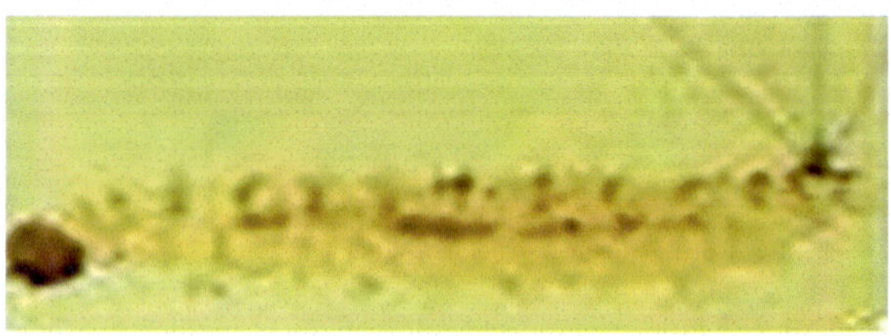

(b)

(National Health Policy 2015) and still further revised up to 2017 (WHO 2014) and, with objective failing completely to be achieved, was even extended to 2020. The key interventions presently include prompt diagnosis using an antigen detection rapid test, treatment, and, above all, vector control by indoor residual spray (IRS) of 5% alphacypermethrin. In 1958, DDT was launched to control malaria vectors under National Malaria Eradication Program (NMEP) for indoor residual spray (IRS) which also deemed to bring about control of Kala-azar as a collateral effect (Sanyal 1979). As a result, the incidence of VL was reduced to a very low level by the early 1960s. However, due to the emergence of resistance in the malaria vectors with DDT, IRS was discontinued in some areas, including those affected by Kala-azar, from 1964 onwards. This caused in to resurgence of sandflies, and Kala-azar reappeared in an epidemic form with about 100,000 cases in the late 1970s (Sen Gupta 1975; Dutta and Ghosh 1983).

The National Kala-azar Control Programme was initiated by the government of India was transformed into National Kala-azar Elimination Programme under the targeted diseases of NMEP, in 1990–1991, and sodium antimony gluconate was the main drug for the treatment, while DDT-IRS was the mainstay for the vector sandfly control.

Sandflies are tiny insects, with fragile wing venation. They prefer dark corners for resting indoors, with typical hopping and short flight movements within the dwellings, and do not fly long distances in nature. *Phlebotomus argentipes* is by nature endophilic and therefore rests indoors at cattle sheds and human and mixed dwellings. Outdoor resting was found in vegetation which is higher than indoors (Poche et al. 2012).

The main strategy for sandfly control was two rounds (February to March and May to June) DDT-IRS (1 g active ingredient/m^2) annually up to 1.8 m height from ground till 2015. Due to development of resistance in the sandfly against DDT, application of 5% alphacypermethrin at 25 mg/m^2 was started instead (Dinesh et al. 2010; Coleman et al. 2015; Dhiman and Yadav 2016). It was implemented in all four VL endemic states of Bihar, Jharkhand, Uttar Pradesh, and West Bengal from 2016 onwards. Use of hand compression sprayer replaced the conventional stirrup pumps fitted with a control flow valve and 8002E flat-fan nozzle tip for uniform spraying of the insecticide. Cases of Kala-azar have been appreciably reduced since then which is facilitating to achieve the target even when the disease transmission was still continuing and there was not much significant reduction in the density of sandflies (Poche et al. 2017). The study was thus conducted to find out the social and ecological factors responsible for changing habits and habitats of the vector and its control through innovative IEC-based environmental management.

21.2 Materials and Methods

21.2.1 Selection of Study Area

There are 38 districts in Bihar, of which 33 are affected from Kala-azar. Based on the Kala-azar elimination criteria, i.e., less than 1 case/10,000 population at block level, the districts were arranged in decreasing order. Top 11 districts were considered as high, next 11 were meso, and last 11 were low endemic, as follows:

1. *High endemic district*: Muzaffarpur, block: Motipur
2. *Meso endemic district*: Sheikhpura, block: Sheikhpura
3. *Low endemic district*: Nalanda, block: Hilsa

21.2.2 Socioeconomic, Ecological, and Entomological Status

The study villages were selected at block level from endemic zone of Kala-azar from different districts having hot spot villages (>5 cases). The study villages preferred showing appearance of Kala-azar cases since the last 3 years. To find out the social, demographic, educational, and economic factors responsible for existence of sandfly population in the area and causing constant transmission of the disease, the study was conducted in all three grades of endemicity, i.e., high, meso, and low categories. Sandflies behavior is reportedly transformed from pedophilic (indoors blood feeding) to exophilic (outdoors blood feeding). The conventional sandfly control method in vogues indoor residual spray of 5% alphacypermethrin which is effective although the duration of effectiveness is limited up to 3 months after spray. However, in contrast to this, the environmental management approaches were found durable and effective by destroying the indoor breeding sites through skirting the wall using brick chimney fly ash mixed with lime (Dinesh et al. 2017).

21.3 Results

21.3.1 Socioecological and Demographic Status

The demographic data was collected using standard developed questionnaire with the help of ASHA Worker of the locality. In high, meso, and low endemic areas, 84, 98, and 80 houses, respectively, were covered showing the population density 4.3, 6.3, and 7 per house with appearance of two to three cases per year. The male and female ratio was found similar, i.e., 1.1:1, and the age group distribution as 0–14, 15–24, 25–64, and >65 was found similar. Illiterate families ranged between 20% and 79%. The annual income of the family was >1 lakh/annum. Association with animal in outdoor and indoor situations was found to be 10–50% of the houses.

Table 21.1 Demographical, socioeconomic, socioecological, socio-educational, and entomological status of different grades of endemicity of Kala-azar in study districts

Indicators	High	Meso	Low
Density/HH	4.3	6.3	7
Male:female	1.1:1	1.1:1	1.1:1
KA cases (last 5 years)	8 (2011–2019)	36 (2015–2019)	5 (2017–2019)
Animal association			
Animal indoor (HH%)	0	10	53
Animal outdoor (HH%)	50	18	4
Income Rs. <1 lakh (HH%)	99.20	93.40	98.10
Education			
Illiterate (HH%)	23.30	78.60	20
Use of mosquito net (HH%)	26.70	85.20	0
Sandfly status			
(PTPN) indoors-CDC-LT	8.4	7.2	12.4
(PTPN) outdoors-CDC-LT	1.1	0.8	0.6

Abbreviations: *HH* household, *PTPN* per trap per night, *CDC* Centers for Disease Control and Prevention, *LT* light trap

Indoor-resting sandflies were collected using flashlight and an aspirator during pre-dawn and later in the overnight CDC light traps installed in domestic and peridomestic areas. Peridomestic population was found less than that of the indoors. In the indoor density, sandflies were in high density (12, per trap per night) in low endemic in comparison to 7 and 8 in meso and high endemic areas. It shows that sandfly density is not responsible for different grades of endemicity. Gravid females ($n = 30$) were processed for detection of natural infection with the aid of microscopy and molecular tools. Dissected sandflies were inoculated in NNN culture medium after microscopy. The samples were found negative for *Leishmania* parasite. Four samples were found positive for *Leishmania* parasite through PCR on specimens collected from high endemic zone of Kala-azar. Detailed studies are underway of progress (Table 21.1).

21.3.2 Entomological Investigations

The sandfly-genic houses require environmental management to seal the potential indoor breeding sites below the base of wall and ground surface of animal shelter. Indoor residual spray of alphacypermethrin was done in low and meso endemic areas 1 month back, and even sandflies were captured. This might be due to untreated microhabitat, i.e., in-depth crevices in wall. This can be removed by adopting environmental management along with IRS. The temperature was found in the range of 17–45 °C and the relative humidity 55–100% in the locality of investigation. The poor hygienic condition and nature of house were considered responsible for endemicity of the disease.

21.3.3 Environmental Management

The environmental management was conducted with the training to the house owner to use brick chimney fly ash with lime, which was found cost-effective and easily available, and people were using them in their houses with the established techniques and found this new addition, a local innovation quite effective in controlling sandfly density (Fig. 21.2a–d) (Dinesh et al. 2017).

21.4 Discussion

Public awareness in disease control program is an essential component for the successful implementation of control strategies. Kala-azar is a disease of the poor communities living in heavy moist and unlighted habitation. Their hygienic conditions, nature of houses, micro-ecological conditions, educational status, etc. all reflect the cause of transmission of the disease under optimal environmental conditions with the presence of vector and the reservoir. *Phlebotomus argentipes* is the only vector for visceral leishmaniasis, and human is the only reservoir in the life cycle of the parasite *Leishmania donovani*. In this regard the social and ecological factors responsible for changing habitat of Kala-azar vector study were found most relevant in the present scenario where all conventional technologies have remained partially successful.

In a pilot study, the result is comparable with the previous study in another highly endemic villages of Kurhani block of Muzaffarpur district in Bihar; a majority of the respondents (70%) lived in thatched/mud-plastered houses, and 60% of the inhabitants were illiterate (Siddiqui et al. 2010). In the present study, too, the male and female ratio was found almost equal in three endemic villages of different districts where the illiterate family were found up to 79%. Regarding the knowledge about transmitting vector of Kala-azar rampant in an endemic area of Kala-azar at Muzaffarpur, ironically 62.8% responded about mosquito bite being responsible for the malady, 32.9% had no idea at all, and only 4.3% associated the disease with the bite of the sandfly (Siddiqui et al. 2010). In the present study, they were not aware with sandfly habitat and causative agent of Kala-azar. The ill perception was at large that by killing mosquitoes, and not sandflies, Kala-azar will be controlled. The conventional method of chemical control through indoor residual spray of insecticide, firstly DDT and now the synthetic pyrethroid, alphacypermethrin, is but partially effective, and yet both the vector and the disease continue to thrive in the endemic districts. There was evidence of resistance against DDT in *P. argentipes* in sandflies population of endemic areas of Bihar, West Bengal, and Jharkhand (Dinesh

Fig. 21.2 People are managing environmental engineering to control sandflies by skirting indoor wall and by (**a**) identifying brick chimney fly ash, (**b**) collecting fly ash, (**c**) carrying fly ash, (**d**) mixing fly ash with lime, (**e**) plastering/skirting the wall to prevent sandfly to reside in the crevices and larval developmental stages

et al. 2010). The bioassay test of sandflies in the bordering countries, Bangladesh and Nepal, however, showed that sandflies were susceptible to malathion, pirimiphos-methyl, and bendiocarb with different degree of knockdown and 100% corrected mortality was observed with deltamethrin and alphacypermethrin (Chowdhury et al. 2018). *Phlebotomus argentipes* was found to develop resistance against malathion 5% in Bihar. Development of resistance in phlebotomine sandflies (*P. neglectus*, *P. tobbi*), vectors of *Leishmania infantum*, was found against deltamethrin and permethrin which had long been applied at Mughal Province. In the insecticide-free area of Aydin Province at Turkey, insecticide susceptible vector strains were found. The development of resistance due to continuing insecticide load cannot be ignored in future. The alternative insecticide needs to be tested. The doses of insecticide in the impregnated papers meant for mosquitoes were used for sandflies as there was no specific dose insecticide ever determined for sandflies. The continuing two rounds spray of alphacypermethrin is not able to reduce the density of sandflies at significant level in spite of approximately 80% coverage (Poche et al. 2018). IRS is being conducted in both human dwellings and cattle sheds which were isolated or closely associated in all places. Human dwelling was shared with cattle in almost all houses in Jharkhand to prevent from animal theft. Sandflies were often found out in great number in cattle sheds than those in human dwellings. However, *P. argentipes* was found susceptible to alphacypermethrin in this study. Exophilic behavior of *P. argentipes* might be one of the possible reasons for the abundance of the vector (Dinesh et al. 2008; Poche et al. 2012, 2017). Although in India alphacypermethrin IRS was not found effective in controlling sandfly population significantly, in Morocco the reduction in abundance of gravid sandfly was found to a remarkable degree (Faraj et al. 2013). *P. argentipes* was found susceptible to most of the insecticides listed by WHO except DDT.

The environmental management according to the bio-ecology of sandfly vector is more effective and eco-friendly. Poverty is the main root cause of prolonged endemicity of Kala-azar due to suitable sandfly-genic conditions, like imperfect construction of dwellings, association with animal, lack of literary rate, and very low economic status. Most of the family was having the annual income less than Rs. 1 lakh. As they can manage their daily need not in a fixed manner or definite source, they cannot think over the health program and the continuing control program of Kala-azar. The Kala-azar control program need to be supported by IEC activity in equal status of chemical control method and should also adopt environmental management to destroy the breeding sites of sandflies. In a study aimed to destroy breeding, four types of intervention materials were used in four different arms by skirting the wall ($9'' \times 9''$: base, high), viz., (1) EM-I, brick + cement plastering; (2) EM-II, rabish (fly ash from brick chimney) + lime; (3) EM-III, wire mesh; and (4) EM-IV, glazed tiles in ten households each with control having similar ecotype. The pre and post density of sandflies was evaluated. The percentage reduction in the density of sandflies was found in the order of EM-I > EM-II > EM-IV > EM-III: 46.2%, 29.6%, 8.5%, and 0% respectively. Rabish + lime was found to be the most cost-effective and user-friendly in comparison to others. In the second phase of the study for validation at village level along with DDT spray,

Sirsa Ram Rai village for Rabish + lime and Sahtha village for cement plastering were considered in Vaishali district. The resting density dropped up to basal level. However, the reduction in density was found 60.2% (mean ± S.D. = 2.48 ± 2.78, 95% CI = 1.93–3.02, $P = 2.27$) with cement plastering and 48.2% (mean ± S. D. = 1.98 ± 2.20, 95% CI = 1.55–2.41, $P = 3.18$) with Rabish + lime skirting for moving sandflies collected from CDC light trap. Significant reduction in density of *P. argentipes* was found as compared to control in both the interventions ($P < 0.05$) for 198 degree of freedom. However, there was no significant reduction between these two interventions ($P > 0.05$) indicating equal efficacy. The sealing of sandfly breeding sites by skirting the indoor walls with rabish was found user friendly in reducing the sandfly density in the community. This can be implemented in the programme following this technique of (Dinesh et al. 2017). The peridomestic breeding sites of *P. argentipes* are needed to be investigated further. In Rajasthan by skirting the wall with cement at the ground level 9″ vertical and 9″ horizontal on the ground surface, 70% *P. papatasi* was successfully controlled (Dhiman 1995). Significant reduction in sandfly density was found by plastering the wall inside and painting with 30% lime water up to 1.22 m height from the ground in ten houses in Bihar (Kumar et al. 1995). In India Bangladesh, and Nepal, environmental management with mud plaster and painting with lime was found effective in reducing the density of *P. argentipes*, but mud plaster alone could not affect in controlling sandflies in Bangladesh (Joshi et al. 2009). Both interventions, i.e., cement and ash skirting, were found effective to control sandflies throughout the year. In a study of India, Nepal, and Bangladesh, the effect of integrated vector management tools like DDT-IRS and LLINs and environmental management with lime plastering was found to be effective in controlling sandflies density by 72.4%, 43.7%, and 42%, respectively (Joshi et al. 2009). So far limited studies are available in genetic studies of sandfly particularly to modify and genetically to inhibit parasite growth inside or to control the sandflies by genetic engineering. Paratransgenic bacterial manipulation of *P. argentipes* gut biota appears feasible to render adult sandflies refractory to *L. donovani* infection (Hurwitz et al. 2011). Two nonpathogenic organisms, *Bacillus megaterium* and *Brevibacterium linens*, were also isolated from gut biota of *P. argentipes*, and it was found that *B. megaterium* and *B. linens* are possible candidates for use in a model of paratransgenesis to prevent transmission of *Leishmania*. The easy method to be employed by common people in controlling sandfly vector of Kala-azar is the integrated vector management, of which environmental management and social factors are important for implementation in the program.

References

Alvar J, Vélez ID, Bern C, Herrero M, Desjeux P, Cano J, Jannin J, den Boer M, the WHO Leishmaniasis Control Team (2012) Leishmaniasis worldwide and global estimates of its incidence. PLoS One 7:e35671. https://doi.org/10.1371/journal.pone.0035671

Chowdhury R, Das ML, Chowdhury V, Roy L, Faria S, Priyanka J, Akter S, Maheswary NP, Khan RK, Argaw D, Kroeger A (2018) Susceptibility of field-collected Phlebotomus argentipes (Diptera: Psychodidae) sand flies from Bangladesh and Nepal to different insecticides. Parasit Vectors 11:336. https://doi.org/10.1186/s13071-018-2913-6

Coleman M, Foster GM, Deb R, Singh RP, Ismail HM, Shivam P, Ghosh AK, Dunkley S, Kumar V, Coleman M, Hemingway J, Paine MJI, Pradeep D (2015) DDT-based indoor residual spraying suboptimal for visceral Leishmaniasis elimination in India. Proc Natl Acad Sci U S A 112(28):8573–8578. https://doi.org/10.1073/pnas.1507782112

Dhiman RC (1995) Effect of minor engineering intervention in the control of breeding of *Phlebotomus papatasi* (Scopoli) sandflies. Southeast Asian J Trop Med Public Health 26:368–370

Dhiman RC, Yadav RS (2016) Insecticide resistance in phlebotomine sandflies in Southeast Asia with emphasis on the Indian subcontinent. Infect Dis Poverty 5:106. https://doi.org/10.1186/s40249-016-0200-3

Dinesh DS, Bhattacharya SK, Das P (2008) Peridomestic breeding and resting sites of sandflies (Diptera: Psychodidae) in Bihar, India. Entomological News 120:496–501. https://doi.org/10.3157/021.120.0505

Dinesh DS, Das ML, Picado A, Roy L, Rijal S, Singh SP, Das P, Boelaert M, Coosemans M (2010) Insecticide susceptibility of Phlebotomus argentipes in visceral Leishmaniasis endemic districts in India and Nepal. PLoS Negl Trop Dis 4:e859. https://doi.org/10.1371/journal.pntd.0000859

Dinesh DS, Kumari S, Hassan F, Kumar V, Singh VP, Das P (2017) Efficacy and evaluation of environmental management system to control sandfly vector of Kala-azar. J Environ Manag 201:366–368. https://doi.org/10.1016/j.jenvman.2017.06.066

Dutta M, Ghosh TK (1983) Reviewed of current status of Leishmaniasis epidemiology. In: Mahajan RC (ed) Proceeding of the Indo-UK, workshop on Leishmaniasis. Indian Council of Medical Research, New Delhi, pp 97–102

Faraj C, Adlaoui EB, Quahabi S, Elkohli M, Elrhazi M, Laqraa L, Ameur B (2013) Evaluation of alphacypermethrin in indoor residual spraying for Leishmaniasis control in an endemic area, northern Morocco. Parasit Vectors 6:354

Hurwitz I, Hillesland H, Fieck A et al (2011) The paratransgenic sand fly: a platform for control of *Leishmania* transmission. Parasit Vectors 4:82. https://doi.org/10.1186/1756-3305-4-82

Joshi AB, Das ML, Akhter S et al (2009) Chemical and environmental vector control as a contribution to the elimination of visceral Leishmaniasis on the Indian subcontinent: cluster randomized controlled trials in Bangladesh, India and Nepal. BMC Med 7:54. https://doi.org/10.1186/1741-7015-7-54

Kumar V, Kesari SK, Sinha NK et al (1995) Field trial of an ecological approach for the control of *Phlebotomus argentipes* using mud and lime plaster. Indian J Med Res 101:154–156

National Health Policy (2002) Available via DIALOG. http://www.mohfw.nic.in/npcb2002.htm

National Health Policy (2015) Draft. In Ministry of Health and Family Welfare, December 2014. Available via DIALOG. http://www.nhp.gov.in/sites/default/files/pdf/draft_national_health_policy_2015.pdf

National Road Map for Kala-azar Elimination Delhi (2014) In Directorate of National Vector Borne Disease Control Programme (NVBDCP), Directorate General of Health Services, Minister of Health & Family Welfare. Available via DIALOG. http://www.nvbdcp.gov.in/Doc/Road-map-ka_2014.pdf

Poche RM, Garalpati R, Dia-Eldin A et al (2012) The role of Palmyra palm tree (Borassus flabellifer) and sandfly distribution in northeastern India. J Vector Ecol 37(1):148–153. https://doi.org/10.1111/j.1948-7134.2012.00211.x

Poche DM, Poche RM, Mukherjee S et al (2017) Phlebotomine sandfly ecology on the Indian subcontinent: does village vegetation play a role in sandfly distribution in Bihar, India? Med Vet Entomol 31:207–213. https://doi.org/10.1111/mve.12224

Poche DM, Garalpati RB, Mukherjee S et al (2018) Bionomics of *Phlebotomus argentipes* in villages in Bihar, India with insights into efficacy of IRS-based control measures. PloS Negl Trop Dis 1:e0006168. https://doi.org/10.1371/journal.pntd.0006168

Sanyal RK (1979) Some observations on epidemiology of current outbreak of Kala azar in Bihar. J Commun Dis 4:170–182

Sen Gupta PC (1975) Return of Kala-azar. J Indian Med Assoc 65:89–90

Siddiqui NA, Kumar N, Ranjan A et al (2010) Awareness about Kala-azar disease and related preventive attitudes and practices in a highly endemic rural areas of India. Southeast Asian J Trop Med Public Health 41(1):1–12

Swaminath CS, Short HE, Anderson LAP et al (1942) Transmission of Indian Kala-azar to man by the bite of P. argentipes Ann and Brun. Indain J Med Res 30:473–477

WHO (2014) Health ministers commit to eliminating Kala-azar. Available via DIALOG. http://www.searo.who.int/mediacentre/releases/2014/pr1581/en/

Chapter 22
Safety Assessment of Novel Genetic Technologies for Vector Control: National and International Perspectives

Vibha Ahuja

Abstract Novel genetic technologies provide an alternative approach for control of vectors particularly those carrying deadly pathogens. Genetic control technologies aim to either suppress target populations or modify the vector by introducing a heritable factor that reduces or blocks their ability to transmit the diseases. These technologies are thus referred to as population suppression or population modification approaches. Both these approaches offer significant advantages for vector control; however, there are associated biosafety concerns related to possible ecosystem interactions. Therefore, extensive testing on a case-by-case basis is required before these can be used as a public health intervention. This paper provides details of the international initiatives towards development of guidelines and status of regulations in India.

Keywords Safety · Assessment · Genetic technologies · Vector control · National and international

22.1 Introduction

With about 700,000 deaths annually in more than 129 countries, and accounting for over 17% of all infectious diseases, vector-borne diseases particularly malaria (219 million cases; 400,000 deaths) and dengue (96 million symptomatic cases; 40,000 deaths) have put half of the world's population on risk of contracting one or another infection (WHO 2019). Mosquito-borne infections such as dengue, chikungunya, and Zika have in recent decades caused several outbreaks globally. These vector-borne diseases are a serious public health problem in developing nations like India (Tyagi 1994). Since most of these vector-borne diseases have neither a vaccine to prevent their onslaught nor specific drugs to cure the patients, the

V. Ahuja (✉)
Biotech Consortium India Limited, New Delhi, India
e-mail: vibhaahuja@biotech.co.in

B. K. Tyagi (ed.), *Genetically Modified and other Innovative Vector Control Technologies*, https://doi.org/10.1007/978-981-16-2964-8_22

importance of vector control can never be overemphasized. To further strategize disease control with vector abatement, the World Health Assembly in 2017 approved the "Global Vector Control Response (GVCR) 2017–2030" which undertakes to offer strategic guidance to fortify vector control with a twin goal, viz., (1) to prevent disease and (2) respond to outbreaks and to significantly contribute to the achievement of the Sustainable Development Goals and Universal Health Coverage. Over the last two decades, 11 countries, e.g., United Arab Emirates (2007), Morocco (2010), Turkmenistan (2010), Armenia (2011), Sri Lanka (2016), Kyrgyzstan (2016), Paraguay (2018), Uzbekistan (2018), Algeria (2019), Argentina (2019), and El Salvador (2021), have successfully eliminated the deadly malaria, a bane in the humanity for centuries. On the other hand, arboviral infections like dengue have spread rapidly to newer areas across the world, with outbreaks occurring more frequently. It is noteworthy that the main vector for dengue, chikungunya, and Zika Virus is the same mosquito, *Aedes aegypti*, which is highly anthropophilic, that is, preferring to live in the human living environment.

The conventional vector control methodologies, despite useful results during epidemics, are generally constrained with operational and other shortcomings. Most vector control strategies in vogue are based on suppressing or eliminating the insect populations making use of insecticides and pesticides. With biotechnological advances in recent past, novel strategies targeting vectors have been developed in multiple ways including the use of genetic modification and *Wolbachia* endosymbiont-triggered cytoplasmic incompatibility. Although most of these advanced technologies have produced significant results in varied testing, some of these technologies, particularly those involving genetic transformation, require safety assessments and regulatory approvals prior to their use. Initiatives have been undertaken at various global fora, most notably the World Health Organization (WHO), Cartagena Protocol on Biosafety (CPB), and Organisation for Economic Co-operation and Development (OECD), to primarily support safety assessment of mosquitoes and testing for their safe use. The use of genetic control technologies is also governed by national regulatory authorities under the biosafety frameworks applicable to genetically modified organisms (GMOs)/living modified organisms (LMOs).

22.2 Genetic Technologies for Vector Control

Advancements in biotechnology have led to the development of genetic technologies such as genetic engineering, gene editing, gene drives, etc. Some of these technologies have attracted interest for applications in vector control as supplements or alternatives to traditional methods within the gamut of integrated vector management (IVM) strategy. Quite a few approaches have been extensively investigated and many have already reached the field.

22.3 Key Features

Genetic control technologies require introduction of a heritable element into the vectors so that these organisms become a biocontrol agent against unmodified species (Gilna et al. 2013). These technologies are based on mating of modified insects with wild population leading to vertical transmission. Hence, genetic control technologies are extremely species-specific with minimal off-target effects (Alphey 2014). However, multiple control tools may be required where different vector species are present.

As compared to conventional vector control technologies, genetic control strategies have certain advantages. They are area-wide methods accessible to everyone regardless of their socioeconomic level. This is in contrast to vaccines, drugs, and insecticide-treated long-lasting bed nets (LLIN), which are directed at individual human. Whereas conventional control methods such as use of insecticides require direct physical contact with the vector to be effective, the genetic technologies are based on the mosquito's natural mate-seeking behavior. Therefore genetic control methods are able to reach all mosquito populations and their larval breeding sites which were otherwise difficult to reach using conventional vector control methods.

22.4 Classification

Genetic control strategies are classified based on the intended outcome, i.e., population suppression or population replacement, as indicated below:

22.4.1 *Population Suppression Strategies*

Population suppression strategies aim to reduce the population of competent vectors, in the same manner as use of insecticides. For example, genetically modified sterile males are released into the wild population, and when they mate with wild females, no offspring are produced, leading to reduction in population size. Release of insects carrying a dominant lethal gene (RIDL) methodology falls in this category (Alphey 2014). These technologies require repeated releases of GM mosquitoes in relatively large numbers. GM mosquitoes based on release of insects carrying a dominant lethal (RIDL) methodology of the Oxitec whereby the vector population is targeted to be reduced in size belong to this group.

22.4.2 *Population Replacement Strategies*

The objective of population replacement strategies is to replace species or populations of mosquitoes capable of transmitting pathogens by those that cannot transmit such pathogens. This is achieved by introducing new genes or altering endogenous genes so as to inhibit pathogen multiplication within the target vector population. Population replacement approaches involve the release of both male and female vectors which will spread the genetic modification by mating with wild population. Gene drive mechanism using CRISPR/Cas9 system has been developed for population replacement.

Benefits and risks for both approaches are different for individuals, communities, and the environment (Alphey 2014).

22.5 Safety Assessment Methodologies

The development of genetic control technologies has also led to discussion regarding their safety particularly possible ecosystem effects. Potential issues include establishment of new vectors as the disease vectors are removed/suppressed; development of resistance over time, on part of either the insect or the pathogen; etc. However, several of these issues are similar to other control methods such as insecticides and drugs. Accordingly, safety assessment forms an integral part of the overall risk analysis process to make informed decisions regarding the release and use of genetically engineered vectors. In addition to specific testing of each technology/species, safety assessment also includes using published data, study reports, and other data generated through evaluations in contained facilities. Information from environmental releases that have been previously approved through regulatory frameworks are also extremely useful as part of safety assessment. Safety assessment methodologies for genetic control technologies have been put in place under various initiatives at the international level through consultative process including experts from multiple disciplines.

22.6 Guidance Framework for Testing of Genetically Modified Mosquitoes

A “guidance framework for testing of genetically modified mosquitoes (GMMs)” has been prepared under the aegis of WHO (Special Programme for Research and Training in Tropical Diseases referred as WHO-TDR) and the Foundation for the National Institutes of Health (FNIH). The objective of guidance is to ensure quality and consistency for testing and regulations of new genetic technologies as per standards of efficacy and safety testing in the same way as used for testing of

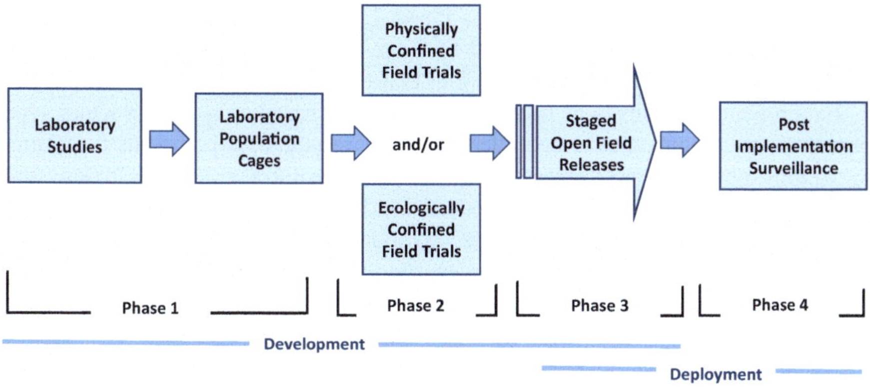

Fig. 22.1 Schematic diagram illustrating the unidirectional pathway activities pertaining to development of basic data and deployment of a GMO in the field. (*Source*: WHO 2014)

other new public health products. A phased testing pathway progressing from laboratory-based experiments to semi-field or confined testing and then open-field releases has been recommended. This is similar to approach used for assessment of other public health tools requiring safety and efficacy at each step.

The phased testing pathway requires systematic assessment of safety and efficacy at each step. These include the following:

Phase 1—Laboratory/cage studies
Phase 2—Confined field trials
Phase 3—Stage open field releases
Phase 4—Post-implementation surveillance

The following illustration describes the above as a unidirectional pathway. However, in practice, repetitions of some segments of the pathway may be required in order to improve the technology and refine the procedure until the requirements for moving to the next phase are met (WHO 2014) (Fig. 22.1).

The guidance indicates that all new genetic control technologies are first studied in laboratory settings referred to as Phase 1. This is followed by testing under confined conditions to limit release into the environment, at the same time providing a more natural setting referred to as Phase 2. The confinement may involve physical confinement, such as testing in large cages/insectaries equipped to stimulate required disease-endemic settings or ecological confinement, that refers to geographic, spatial, or climatic isolation. As physical confinement and ecological confinement may have varied environment exposure, the safety assessment and regulatory requirements will also be different. Testing under confined conditions is followed by a series of staged open-field releases, referred to as Phase 3. These releases are undertaken to test efficacy in field conditions and designed on a case-by-case basis, in line with the study objectives. Testing in open environment provides detailed information to take decision for moving to next phase, i.e., deployment as a public health intervention referred as Phase 4. The initial deployment may be at pilot scale on a limited area followed by large-scale release. Monitoring plan for safety and efficacy is a key requirement in Phase 4.

Experimental designs to be used at each phase are based on specific vectors, type of technology, and studies to be undertaken. Progression of experiments for testing in various phases, from laboratory to the field, requires extensive review and planning at each stage. Site characteristics and appropriate comparators are key considerations and depend on objectives and experimental design. Efficacy is measured using entomological and epidemiological outcomes. These measurements depend on the vector species, disease, and ecological and epidemiological circumstances. Efficacy measurements also differ in various testing phases.

Expression analysis, stability of phenotype, reduction in population, spread and frequency of transgenes, etc. are some of important parameters that serve as meaningful indicators of efficacy of a specific genetic technology being studied. In addition, entomological outcomes are monitored throughout the testing phases. Increased attention to epidemiologic outcomes needs to be given, be when testing is undertaken in settings in which humans may be present. Mathematical modeling can also be used to strengthen safety assessment of new genetic control technologies on a case-by-case basis. The objective of mathematical modeling in the context of safety assessment is to predict behavior of genetically modified vectors and assessing the likelihood of events based on their properties and various assumptions.

This technology will be tested under confined conditions that provide a more natural setting but still limit release into the environment referred to as Phase 2. Studies in Phase 2 may involve testing under physical confinement, as in a large cage equipped to simulate a disease-endemic setting, or under ecological confinement, as under geographic, spatial, or climatic isolation. Regulatory requirements for physical and ecological confinement will differ because of the different levels of environmental exposure. Risk assessment and prior experience with the technology will be the basis to decide on the type of testing. Following confined testing, GMMs will be subjected to a series of staged open-release trials in Phase 3. These trials are designed to measure performance under different conditions and to assess the ability of GMMs to reduce infection and/or disease in human populations. Based on results from Phase 3, a decision may be made to deploy GMMs as a public health intervention (Phase 4). Phase 4 would be accompanied by a plan for long-term monitoring of safety and efficacy. The specific experimental designs to be used may vary widely based on the type of technology. Testing and progression of experiments from the laboratory to the field will require reconsideration at each stage including selection of study sites (WHO 2014).

In addition to the scientific evaluation, the guidelines lay emphasis on the community engagement throughout the process of testing and evaluation.

22.7 Cartagena Protocol on Biosafety

The Cartagena Protocol on Biosafety is an international agreement under the aegis of Convention on Biological Diversity that applies to transboundary movement, transit, handling, and use of all living modified organisms (LMOs) that may have adverse effects on the conservation and sustainable use of biological diversity, taking also into account risks to human health. The Protocol includes the Nagoya-Kuala Lumpur Supplementary Protocol on Liability and Redress addressing response measures in the event of damage caused by the transboundary movement of LMOs. Articles 15 and 16 and Annex III of the CPB outline requirements that relate to risk assessment and risk management of LMOs. After an elaborate process spanning more than 10 years, in December 2016, the parties to the CPB have adopted a voluntary Guidance on Risk Assessment of Living Modified Organisms and Monitoring in the Context of Risk Assessment. This guidance also includes a section highlighting specific issues that may need special consideration for the environmental release of LM mosquitoes (CPB 2016).

22.8 Biology Documents by OECD

The biosafety assessment of GMOs is usually based on the information collected on the characteristics of the host organism, the introduced traits, and the environment into which the organism will be released. The OECD has prepared a series of biosafety consensus documents that are intended to be a "snapshot" of current information on a specific host organism or trait, for use during regulatory assessments of organisms considered for their release in the environment. Though such documents have been made by OECD on various plants, in 2018, biosafety consensus document has been prepared on *Aedes aegypti*, addressing for the first time the biology and ecology of an insect species (OECD 2018). *Aedes aegypti* has been a subject of biotechnological research and applications (including genetic engineering), aiming to contribute to the control of its population, reduce its capacity to spread diseases, and thus limit its drastic impact on human health. This publication contains information relating to the mosquito taxonomy, morphology, reproductive biology, genetics, ecology, and other aspects. Similar initiative is underway for *Anopheles gambiae*, the deadliest vector of malaria.

22.9 Regulatory Requirements in India

Various countries regulate the development and use of genetic technologies according to their national laws and governance mechanisms.

In India, there is a systematic and structured regulatory framework for biosafety evaluation of genetically modified organisms (GMOs) and products thereof under the Environment (Protection) Act, 1986. The regulations for GMOs are "Rules for manufacture, use/import/export and storage of hazardous microorganisms/genetically engineered organisms or cells"1989 (MOEFCC 1989).

The Ministry of Environment, Forest and Climate Change (MoEFCC) is the nodal ministry for implementation of Rules, 1989. MoEFCC implements the Rules jointly with the Department of Biotechnology (DBT), Ministry of Science and Technology, and respective state governments. The scope of Rules, 1989 is broad and covers all activities involving GMOs and derived products. These rules also have provision to regulate new gene technologies, in addition to genetic engineering. Various competent authorities and their composition and functions have been provided for in Rules, 1989. These include the following:

1. Recombinant DNA Advisory Committee (RDAC)
2. Institutional Biosafety Committee (IBSC)
3. Review Committee on Genetic Manipulation (RCGM)
4. Genetic Engineering Appraisal Committee (GEAC)
5. State Biotechnology Coordination Committee (SBCC)
6. District Level Committee (DLC)

The administrative agencies and role of each of these committees are summarized in Table 22.1.

Rules, 1989 also provide for reporting mechanisms of various committees. IBSCs are required to submit their recommendations and reports to RCGM. RCGM gives its recommendation to GEAC. DLCs are required to submit their reports to SBCC, who in term report to GEAC. Rules, 1989 are supported by various guidelines issued by MoEFCC and DBT. These guidelines are issued keeping in view advancements in development of and use of GMOs in the country. The guidelines issued so far cover guidance for biosafety compliance and data generation at various stages of development process of GMOs. Guidelines and handbook have also been notified for effective functioning of IBSCs.

Regulations and Guidelines on Biosafety of Recombinant DNA Research and Biocontainment 2017 provide for guidance on setting up Insect Biosafety Level Facilities (DBT 2017). The purpose of establishment of Insect Biosafety Level (IBSL) Facilities is to prevent escape and establishment of the experimental arthropods including genetically engineered arthropods into the natural environment and ensure the safety of the laboratory personnel in the facility. Facility design, safety instruments, personal protection equipment, and procedures have been elaborated for four levels of IBSL, i.e., IBSL1–4.

The guidelines for field testing and release are under development by the regulatory authorities in India.

Effective use of these novel genetic technologies for vector control requires development of rationale guidance to ensure decision-making for their large-scale implementation for public health purposes.

Table 22.1 Six competent authorities, with their composition and function as per Rules, 1989 (MoEFCC 1989)

Competent authority	Role as per Rules, 1989	Administrative agency
rDNA Advisory Committee (RDAC)	Takes note of development in biotechnology at national and international levels and provides advice on biosafety regulatory requirements	Department of Biotechnology, Ministry of Science and Technology
Institutional Biosafety Committee (IBSC)	Nodal point for interaction within the institutions engaged in recombinant DNA research; responsible for implementation of biosafety guidelines	Set up in all organizations engaged in recombinant DNA research; reports to RCGM in DBT
Review Committee on Genetic Manipulation (RCGM)	Review ongoing GM research activities including high-risk category organisms and controlled field experiments. Also responsible for bringing out manuals of guidelines specifying various processes	Department of Biotechnology, Ministry of Science and Technology
Genetic Engineering Appraisal Committee (GEAC)	Approval of activities involving large-scale use of hazardous microorganisms and recombinant products in research and industrial production	Ministry of Environment, Forest and Climate Change
State Biotechnology Coordination Committee (SBCC)	Inspection and take punitive action in case of violations	State Governments
District Level Committee (DLC)	Local supervision to ensure biosafety compliance	

22.10 Conclusion

Vector-borne diseases exact a heavy toll worldwide, and new infections have been regularly emerging to pose great challenges for their control. Control of vectors is considered an effective way to reduce the burden of these diseases. Conventional methodologies have so far failed in weeding out the significant vector populations, and there has been a frantic search for alternative approaches. Novel genetic technologies provide an alternative approach for control of vectors transmitting deadly and/or debilitating pathogens. These genetic control strategies aim to either suppress or replace the target populations so as to finally impact the disease dynamics. However, despite their significant advantages, there are biosafety concerns related to possible ecosystem interactions. Thus, case-specific testing is required to comprehend the benefits and ill effects of these novel approaches, keeping in mind both potential benefits and risks. International laws exist to care the biosafety measures associated with these biotechnological advancements, and many countries including India have their own robust assessment mechanisms to regulate biosafety steps in each case of specific technologies.

References

Alphey L (2014) Genetic control of mosquitoes. Annu Rev Entomol 59:205–224. https://doi.org/10.1146/annurev-ento-011613-162002

Cartagena Protocol on Biosafety (2016) Guidance on risk assessment of living modified organisms and monitoring in the context of risk assessment. https://www.cbd.int/doc/meetings/bs/mop-08/official/bs-mop-08-08-add1-en.pdf

DBT (2017) Regulations and guidelines for recombinant DNA research and biocontainment. Ministry of Science and Technology, Government of India, New Delhi. http://dbtindia.gov.in/sites/default/files/uploadfiles/Regulations_%26_Guidelines_for_Reocminant_DNA_Research_and_Biocontainment%2C2017.pdf

Gilna B, Kuzma J, Otts SS (2013) Governance of genetic biocontrol technologies for invasive fish. Biol Invas. https://doi.org/10.1007/s10530-012-0367-x

MoEFCC (1989) Rules for the manufacture, use, import, export and storage of hazardous microorganisms, genetically engineered organisms or cells. https://geacindia.gov.in

OECD (2018) Safety assessment of transgenic organisms in the environment, OECD consensus document of the biology of mosquito Aedes aegypti, harmonisation of regulatory oversight in biotechnology, vol 8. OECD Publishing, Paris. https://doi.org/10.1787/9789264302235-en

Tyagi BK (1994) Distribution of arthropod vector-borne communicable diseases and control of their vectors in India. Indian Rev Life Sci 14:223–243

World Health Organization (2014) Guidance framework for testing of genetically modified mosquitoes. FNIH, WHO, TDR, Geneva, Switzerland. https://www.who.int/tdr/publications/year/2014/Guidance_framework_mosquitoes.pdf

World Health Organization (2019) World malaria report 2019. World Health Organization, Geneva, Switzerland

Chapter 23
Measuring Public Attitudes to Releases of Transgenic Mosquitoes for Disease Control, with Special Reference to Dengue and Malaria

Lilian A. De Las Llagas and Mary Sylvette T. Gunigundo

Abstract Since the advent of DDT in public health and agriculture, science leaped forward with revolutionary technology such as gene drive or editing, thus making it possible to develop alternative approaches to address vector-borne diseases. However, their utilization and sustenance in public life are dependent on public attitude, i.e., societal awareness and social acceptance. In the face of strong skepticism against genetically modified organisms in both developed and developing countries, public acceptance is therefore a requirement (Boete and Beisel 2013, and Bohannon 2002, as cited in De Souza et al. Understanding the requirements and factors necessary for the acceptance of genetically modified mosquitoes as a potential malaria control tool in Ghana: a questionnaire survey, AsPac J Biol Biotechnol 21 (3):76–88, 2013).

This present article throws light on the inevitable necessity to measure public attitudes before undertaking release of the transgenic mosquitoes for disease control, with special reference to dengue and malaria.

Keywords Public attitude · Transgenic · Mosquitoes · GMM · Dengue · Malaria

L. A. De Las Llagas (✉)
Commission on Higher Education, Quezon City, Philippines

University of the Philippines, Manila, Philippines

M. S. T. Gunigundo
Commission on Higher Education, Quezon City, Philippines

University of the Philippines, Diliman, Philippines

B. K. Tyagi (ed.), *Genetically Modified and other Innovative Vector Control Technologies*, https://doi.org/10.1007/978-981-16-2964-8_23

23.1 Introduction: The Basic Imperatives

Since the advent of dichlorodiphenyltrichloroethane (DDT) in public health and agriculture, science has advanced a great leap forward with revolutionary technology such as gene drive or editing, particularly the clustered regularly interspaced short palindromic repeats (CRISPR)-associated protein 9 (Cas9) format, but this had to bear the brunt of public attitude, i.e., societal awareness and social acceptance, towards its utilization and sustenance in public life. As goes the famous quote, "Public acceptance is required—in the face of strong skepticism against genetically modified organisms in both developed and developing nations" (Boete and Beisel 2013, and Bohannon 2002, as cited in De Souza et al. 2013, p. 77).

According to Touré et al. (2003), "regardless of the quality of the science, the public confidence and acceptance will be the key factors to drive the tool in use (p. 221)." This echoes the cautious advice by the world authorities on genetic manipulation. This quote is an important driving point for investigators who are doing researches involving human participation or innovation for disease control. This is met with either acceptance or rejection of the community. The families residing in the selected study sites who could be exposed to a potential health hazard are in the best position to be consulted and share their beliefs, doubts, concerns, and fears as they themselves are nearest to the problem—the problem of acquiring the disease. Review of some informed consent forms showed the insufficiency of information regarding the risks and benefits, with more benefits described compared to specific risks. Obtaining the signature or conformity to the forms is both a process and an outcome. The consultation should always be documented and processed to protect both the investigator and the population participating in the study. Engagement is crucial as the health of the human population is at stake. Improving health is not only an outcome of development but also a prerequisite for development.

The researchers have equal obligations to society. The purpose of public health is the protection of life, the preservation of human existence, and the protection and improvement of the environment. Government consent, the agency responsible for the protection of the lives of the people, must also be sought. The core concept embedded in these obligations is a basic question that one is expected to answer. Is the research ethical? Will the researcher be able to put safeguards in place to protect the human rights and the use of law (regulations) to claim and assert it? The basic obligation, therefore, for an ethical researcher is to uphold and enhance public perception of the fundamental principle in health and practice of medicine, "to do no harm" (Hippocrates 400 BCE).

To contextualize this within the purview of mosquito control specifically, the field release of genetically modified mosquitoes (GMM), genetically sterile mosquitoes (GSM), or transgenic mosquitoes (TM), we need a framework within which these products of modern biotechnology can be safely assessed.

23.2 Vector-Borne Diseases (VBD): The Target Diseases

Among all the infectious diseases globally, VBD account for more than 17% of them and cause more than 700,000 deaths every year. One of those classified as a vector-borne disease is malaria—a protozoan parasite infection transmitted by anopheline mosquitoes. On the other hand, the *Aedes* mosquitoes serve as a vector for most epidemiologically significant infections including dengue fever, chikungunya, yellow fever (YF), Zika virus, and Japanese encephalitis (JE). Among these, dengue fever is the most important public health problem in the Philippines and elsewhere in the world for mainly a complete lack of any vaccine to protect human population and a specific drug to treat patients, albeit many promising candidate products are making rounds for several years now. These diseases are all preventable with the utilization of protective measures and effective community participation.

23.3 Current Mosquito Control Practices

Mosquito preventive and control measures such as container management, adulticiding with chemicals, mosquito larvicidal trapping (MLT), biological control, and self-protection measures against dengue are being used with varying effectiveness. Insecticide-treated nets and/or long-lasting insecticidal nets (LLINs) and the indoor residual spraying (IRS) are common vector control practices for malaria-endemic areas. Limited resources and difficulties in some countries delayed the elimination of transmission of the disease and have not made the strategies very successful.

23.4 The Need for New Method: Application of Transgenic Mosquito

As mosquitoes become more resistant to available insecticides with attached environmental health concerns, and their escalating costs prove to be inhibitive, many countries' societies have expressed the need for an alternative approach that can be integrated with other existing control methods, commonly called the integrated vector control (IVC) approach.

Recently, there has been an increased interest towards the development and application of sterile insect technique (SIT) to control *Aedes* mosquitoes either by using ionizing radiation resulting in the males' inability to reproduce or with the release of insects carrying a dominant lethal (RIDL) system engendering transgenic mosquito whose mating with the wild female counterparts produces eggs lacking viable hatching to a great degree or the early instars die out without having had the opportunity to transform into next aquatic stage of development.

A company named Oxitec Ltd., founded by the University of Oxford, has developed a genetically sterile male *Aedes aegypti* mosquito (strain OX513A) which mates with wild females to produce offspring that will die as larvae or pupae. Field trials of its release were done in the Cayman Islands and Brazil and, more recently, in the United States, among others. At the same time, the OX513A transgenic *Aedes aegypti* had been evaluated in the laboratory and caged simulated field conditions for its biological traits as compared to native wild mosquito in both India and Malaysia (OXITEC 2016).

23.5 The Framework

Trust in the researchers or key players has a distinct influence on attitudes and apprehensions of any target participant being recruited in one research project. Success in convincing any volunteer-participant is built on perceived benefits and privileges obtained during the recruitment process. Experiences brought us to recommend, to anybody exploring a community study, to ensure that participants are needed to be well informed on the study and such engagement would also require that they be given token "bother fee" most especially when the community is poor and wanting of resources and the most vulnerable to be unhealthy and needed to be disturbed because of the intensive procedures essential to be done for the project.

Product safety must be assessed and must conform with the standards of the international instruments for genetic modification of organisms (GMOs) relative to the Cartagena Protocol on Biosafety, Codex Alimentarius Commission, and local regulation of the government. Regulatory ethical, social, and cultural dimensions of the study should be factored in when desired sites for release of transgenic mosquitoes (TM) are to be finalized. These considerations are as crucial as the technical soundness of the project.

Additional lessons and insights have similarly taught us that community engagement through empowerment of families and their neighbors who were sick due to malaria enabled them to make decision for themselves, and not to be scared of malaria, but rather to use their decision-making options on how to fight malaria. This research provided empirical evidence that Family Health Empowerment Model (FHEM) worked reasonably well (De Las Llagas and Portus 2016).

23.6 Methodology

23.6.1 Qualitative Approach

Methodologically, the qualitative approaches aid researchers to bring out the various sociocultural factors that were not captured in a survey questionnaire (i.e., quantitative approach). Opposite to quantitative approach, Portus et al. (2018) have described the qualitative research in very exquisite terms which landmark to the whole success of GMM release: "Qualitative research is a method of data or information generation that, instead of asking for the 'What' and the 'How,' seeks out the 'Why' of a particular phenomenon. Researchers use qualitative research methods to gain insights into peoples' attitudes, behaviour, value systems, concerns, motivations, aspirations, contexts and lifestyles. The data that are produced take the form of interview, or meeting-transcripts, readings or other documentation outputs, emails, notes, feedback forms, photos, videos and other visual texts." This gives emphasis to the suitability of the Focus Group Discussion (FGD), Key Informant Interviews (KIIs), and Participant Observation (PO) when trying to understand the factors for the success and failure of any program or intervention being and have been tested. More studies have shown that qualitative methods fit in the study of communities and the general public.

1. **Focus Group Discussion**
   This is a qualitative method of gathering data wherein a relatively single number of knowledgeable informants (from 6 to 12) under the guidance of a skilled moderator or facilitator talk about special subjects, topics, or themes, within a specified period from 1 to 1.5 h.
2. **In-Depth Interview, Focus Interview, and Key Informant Interview (KII)**
   This is to seek out an informant's deep insights, together with his/her detailed and extensive knowledge and experiences.
3. **Participant Observation (PO)**
   Participant Observation as a qualitative research method involves the direct personal observation of a social phenomenon in the natural setting of the population studied. It prolongs visits or living in the community of the subject observation. This is a hallmark of ethnographic research. It is increasingly used as a research method outside anthropology.

23.6.2 Survey Research (Quantitative)

This is a system procedure of planning, collecting, organizing, and analyzing data generated from a sample survey. This is a viable methodology of inquiry from representative units of the population. Different types of data (factual, perception, knowledge, attitude, behavioral, etc.) can be generated from the sample and used in making generalizations about the population.

A quantitative researcher uses carefully crafted and structured questionnaire or a validated scale. A Likert scale questionnaire is the most widely used instrument to measure attitudes. The Likert scale is a method of ascribing quantitative value to qualitative data. The mean value for all the responses is computed at the end of the survey. The choices range from strongly disagree to strongly agree, typically in a 5-point scale:

1 = Strongly disagree
2 = Disagree
3 = Neither disagree or agree
4 = Agree
5 = Strongly agree

The researcher can get a holistic view of peoples' opinions and their level of agreement. This is an ordinal level of measurement which is evaluative in nature. One value is greater or better than the other, but the exact difference between the ranks is unknown.

23.7 Marketing Research

The importance of giving utmost value to community participation in any health development program is the same intent as when one engages in social marketing, seeking to develop and integrate marketing concepts and other approaches for societal change (Regacho 2020).

The term marketing research refers to the systematic gathering and analysis of quantitative and qualitative data about issues relating to the marketing of a product. It links the producers with end users through information used to identify and define marketing opportunities and problems, generate and evaluate marketing actions, and monitor sales performance.

Quantitative market research is the collection of numerical data often resulting in statistical analysis to understand trends in the data. It allows for comparisons and trends in the data to be easily found and understood. It is a more structured market research process that involves large number of respondents to answer surveys or questionnaires. Malhotra and Briks (2007), as cited in Zaborek (2015), listed six stages in a marketing research process. For this article, a modified version by Regacho (2020) is presented. It is also composed of six steps in a quantitative market research after defining the marketing problem and research design (Fig. 23.1).

A marketing survey is an example of a qualitative-quantitative survey that aims to determine customers' satisfaction to a certain product (Table 23.1). The use of marketing survey will not only aid a businessman in determining the attractiveness of a product, but also it can confirm the social and cultural acceptability of a new biotechnology—in this particular case, the acceptability of transgenic mosquitoes (TM). A marketing survey can also determine when a certain product will be purchased by an end user, or, in the case of the project on transgenic mosquitoes

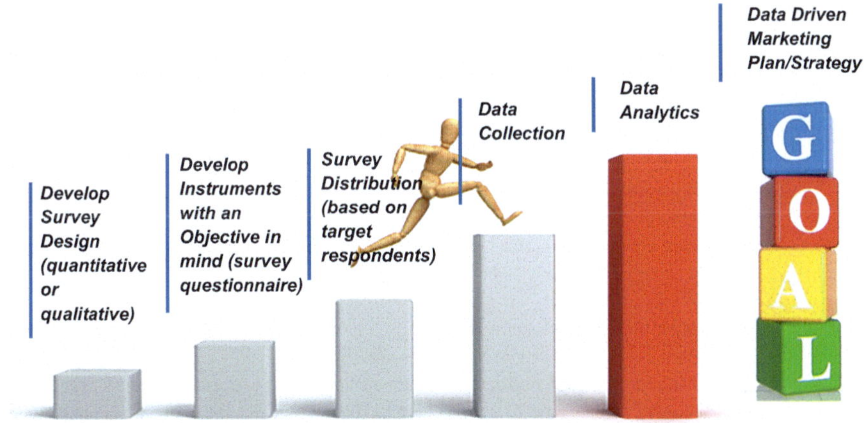

Fig. 23.1 Steps in quantitative market research (Regacho 2020)

(TM), it will determine the action of a community in the interruption and eventually decrease the incidence of mosquito-borne diseases such as dengue and malaria.

Undertaking research across socioeconomics and cultures is not an easy task. Thus, by using marketing research, the question of acceptability of a new biotechnology can be culled from the marketing research on the attractiveness of existing anti-mosquito products. In marketing, the bottom-up approach is used to measure the attitude of the respondents towards the use/purchase of a product. The bottom-up approach is done by summing the evaluative rating of each factor multiplied by the level of importance of that factor. The results can be subjected to statistical treatment to show the differences between respondents (Table 23.2).

23.8 Measurements of Attitudes

Perhaps, an introduction of any product or technology in a given site assumes the preparedness of the community to accept the intervention knowing fully the benefits that will accrue in the neighborhood. Determination of every household's opinion and attitude will lead to cooperation of the community.

De Las Llagas and Portus (2016) carried out an in-depth research on entomological and epidemiology of dengue in the Philippines, with a strong input on sociological and cultural components. The entomological component determined the effectiveness of an intervention of an anti-mosquito device that could be installed inside homes. They utilized a triangular approach in order to generate various perspectives on the sociocultural dimension of the project. FGDs, KIIs, and surveys were used to collect the data for appropriate statistical treatment. To assess the readiness to adopt the device of the target household participants during the study, the following were determined:

Table 23.1 Sample marketing survey form (Source: Regacho 2020)

1. Are you scared of mosquito bites? Yes ☐ No ☐

2. If yes, what do you do to prevent from being bitten? (check ALL that apply)
   - ☐ I use mosquito coil.
   - ☐ I use electric mosquito mat.
   - ☐ I use electric mosquito liquid.
   - ☐ I apply mosquito repellant lotion / spray on my skin.
   - ☐ I use an anti-mosquito patch.
   - ☐ I use an anti-mosquito wrist band / strap.
   - ☐ I use a mosquito trap.
   - ☐ I use a UV/electric mosquito killer.
   - ☐ I search and destroy containers with mosquito larvae.

3. How often would you use your preferred anti-mosquito repellant strategy or product? (check only one)
   - ☐ Everyday
   - ☐ Once every two days
   - ☐ Once a week
   - ☐ Irregular

4. What makes you use/purchase your preferred anti-mosquito strategy or product? (check from the scale)

Characteristics	Very Important	Important	Not Important
Tested Brand			
Effectivity / Protection			
Price			
Safety Package			
Colorless/Stain on Skin			
Odourless/Fragrance			
Active Ingredient			
Packaged Promotion			
User-friendly			
Popularity (advertising/endorser)			

5. Do you have any fear or reservation on your choices? (check only one)
   - ☐ Yes, I have fear on its harmful effects on people/environment.
   - ☐ No, I trust the effectivity of my choice/s.

- Commitment (e.g., attitudes and values towards maintaining healthy household and "barangays" (village)) towards adopting anti-dengue methods and technologies.
- Competence (i.e., knowledge, understanding, predisposition, and confidence) in adopting new technologies.
- Clout (i.e., practices and behaviors, assessment of responsibility in undertaking dengue reduction methods and practices).
- Resistance (i.e., stated objections, alternatives considered) against adopting new technology to prevent dengue.

Table 23.2 Form for measuring attitude using the bottom-up approach (*Source*: Regacho 2020)

Factors	Very important	Important	Not important
Tested brand			
Effectivity/protection			
Price (affordable/reasonable)			
Safety package (label/warning)			
Colorless/stain on skin			
Odorless/fragrance			
Active ingredient (organic/synthetic)			
Packaged promotion (e.g., buy one, take one)			
User-friendly			
Popularity (advertising/endorser)			
TOTAL SCORE			

- Sociodemographic data of the households in the intervention and control groups.
- Socioeconomic condition of the households and barangays as benchmark data.

Some significant studies about public attitude pertaining to release of transgenic mosquitoes for disease control are presented below for better visualization on how to measure such attitudes:

23.8.1 *Positive Attitude*

Arham et al. (2020) examined the attitudes of Malaysians towards *Wolbachia*-infected *Aedes* mosquitoes to control dengue. This was a quantitative study wherein they used a 7-point Likert scale questionnaire with the following constructs:

1. *General Factors*—These included trust in key actors (scientists, government, industries), attitude towards nature vs. material, attitude towards technology, and religiosity.
2. *Specific Factors*—These pertained to perceived benefits and perceived risks.
3. *Main Construct*—This measured attitude towards *Wolbachia*-infected *Aedes* mosquitoes.

Results in this Malaysian study showed that overall, the respondents had positive attitude towards *Wolbachia*-infected *Aedes* mosquitoes as a dengue control technique. Mean scores suggested that there was a high level of religiosity, trust in the key actors, perceived benefits, and positive attitude towards the *Wolbachia*-infected *Aedes* mosquitoes.

23.8.2 Negative Attitude

Kimura and Macer (2008) studied the Japanese attitude towards genetically modified mosquitoes to control malaria and Japanese encephalitis. They conducted two qualitative surveys in 2003 and 2004. In the 2003 survey, they randomly distributed to households across all prefectures of Japan an open-ended questionnaire. In the 2004 survey, respondents were delimited to selected areas. The answers to the survey were categorized into three groups: (1) positive feeling, (2) mixed feeling, and (3) negative feeling. Majority of the respondents had negative feelings towards GMM, and the top three answers of those who felt negatively towards this technique were (a) possibility of disaster and risks that would happen, (b) fear of the unknown in terms of side effects and new emerging diseases, and (c) ecological concerns, i.e., effects on the ecosystem and the environment.

In another landmark qualitative study, Bloss et al. (2017) conducted content and thematic analyses of 2624 comments submitted to the US Food and Drug Administration (FDA) from March 11, 2016 to May 16, 2016 on a draft environmental assessment for a proposed field trial in Key Haven, Florida, of a genetically engineered mosquito designed to suppress wild-type *Aedes aegypti* mosquitoes that can transmit Zika, dengue, and chikungunya. The comments were categorized into four themes mainly ecological safety, human health implications, genetically modified organisms (GMO), and mistrust of government and industry. Majority of those who submitted comments were not in favor of the proposed field trial. For those who opposed the trial, the number one concern was the GMOs in general. For those who supported or remained neutral, most of their comments focused on human health implications.

Yet in another significant investigation conducted on a community's opinion survey on the use of genetically modified mosquitoes (GMMs) to reduce mosquito-transmitted diseases in Florida (Adalja et al. 2016), the major objective was to increase the understanding of community knowledge, attitudes, and beliefs regarding mosquito control and GMMs. An 18-question self-administered survey was mailed and fielded to all households in identified Key West, Florida, where a GMM trial had been proposed from July 20, 2015 to November 1, 2015. Results showed that majority of the respondents were against or did not support the use of GMMs as a mosquito control method because of fears about possible harmful impacts of this intervention, impacts on human and animal health, and potential negative effects on the ecosystem or environment.

23.8.3 Mixed Attitude

Okorie et al. (2014) determined how receptive Nigerian scientists were to the possible release of genetically modified mosquitoes (GMMs) for the control of malaria. Altogether, 164 scientists representing 3 disciplines (science, medicine, and agriculture) working in the academic and research institutes, or tertiary health facilities, were given a semi-structured and self-administered questionnaire. Some scientists were also interviewed for clarification purposes only thus making this study more quantitative. The questionnaire was composed of ten items on a 5-point scale. The findings suggested that majority of the scientist-respondents were "skeptical" about genetically modified mosquitoes in general; nonetheless, most of them encouraged the use of GMMs on the condition that evidence of contingency measures is in place, such as the removal of the GMMs should a hazard becomes evident during the course of the release.

The above results were corroborated by a similar earlier survey by de Souza et al. (2013) who conducted a pilot questionnaire study to understand the needs, requirements, and factors necessary for the acceptance of GMMs as a potential malaria control tool in Ghana. Their results showed that some of the respondents were open to GMMs despite the perceived risks; however, majority were against the release of GMMs as malaria control measure in Ghana.

23.9 Discussion and Conclusion

Vector-borne diseases such as malaria and dengue are transmitted to people through blood-sucking arthropods like mosquitoes. Rampant in most tropical countries, these diseases affect millions of people and remain a serious public health concern. These diseases can be prevented through mosquito control practices. However, conventional methods of controlling mosquito population, which involve insecticide fogging, aerosol space spraying, larviciding, and indoor residual insecticide spray, were found to be costly and environmentally hazardous. These have also become ineffective in reducing vector density because mosquitoes seemed to have developed resistance to insecticides.

With the advancement of science and technology, alternative approaches have been developed, one of which is genetically modifying mosquitoes. In spite of this innovative and promising development, based on the review of some studies measuring public attitude towards GMMs, results showed that majority of the public were skeptical in supporting the releasing of transgenic mosquitoes to control mosquito population. And as mentioned at the start of this article, when there is strong skepticism, public acceptance of this GMM alternative approach is now an imperative.

This article gave an overview of various ways to measure public attitude. Researchers may use either qualitative or quantitative approaches. Based on the previous studies measuring public perception, the most ideal research design is the combination of the qualitative and quantitative techniques in order to get a better holistic picture or understanding of public attitude. Marshall et al. (2010) suggested that any quantitative study should be preceded by qualitative interviews, careful design, and extensive pilot work. They also pointed out that prior to distributing survey questionnaires, informational materials should also be distributed because public attitude may be influenced by public awareness.

We do agree with these suggestions especially in the careful designing of the research methodology which needs to be done meticulously and ethically and with cultural sensitivity and in the distribution of GMM informational materials to give respondents more knowledge on GMM prior to answering any question. However, for the sequencing of the research methods, whether qualitative research should precede or succeed a quantitative approach, such sequence really depends on the research questions to be asked and profile of the target sample population, among others.

The use of transgenic mosquitoes is a novel and innovative tool to address vector-borne diseases. However, despite rapid advances, there are still no widely accepted biosafety guidelines that provide guidance to consumers on all aspects including risk analysis. On the other hand, even if such guidelines are available, there would still be requirements from different countries to develop national policies, as well as to build local capacity to safely assess the risk of the environmental use of transgenic mosquitoes. Further, the risks in communication and public perception along with the acceptability of such risks, balanced against the potential benefits of sustainable vector control, will ultimately decide the acceptance of transgenic mosquitoes.

References

Adalja A, Sell TK, McGinty M, Boddie C (2016) Genetically modified (GM) mosquito use to reduce mosquito-transmitted disease in the US: a community opinion survey. PLOS Curr Outbreaks. https://doi.org/10.1371/currents.outbreaks.1c39ec05a743d41ee39391ed0f2ed8d3. https://currents.plos.org/outbreaks/index.html%3Fp=67624.html. Accessed 26 May 2020

Arham AF, Amin L, Mahadi Z, Yaacob M, Razman MR (2020) Predicting factors and attitude towards *Wolbachia*-infected *Aedes* mosquitoes in Malaysia. Acad Strat Manage J 19(2). https://www.abacademies.org/articles/predicting-factors-and-attitude-towards-wolbachiainfected-aedes-mosquitoes-in-malaysia-9095.html. Accessed 26 May 2020

Bloss CS, Stoler J, Brouwer KC, Bietz M, Cheung C (2017) Public response to a proposed field trial of genetically engineered mosquitoes in the United States. JAMA 318(7):662–664. https://pubmed.ncbi.nlm.nih.gov/28810013/. Accessed 26 May 2020

De Las Llagas LA, Portus LM (2016) Promoting health communication and empowerment: intervention towards malaria mosquito control. Int J Health Wellness Soc 6(3):39–53

De Souza DK, Brown CA, Ahorlu CK, Suzuki T (2013) Understanding the requirements and factors necessary for the acceptance of genetically modified mosquitoes as a potential malaria control tool in Ghana: a questionnaire survey. AsPac J Biol Biotechnol 21(3):76–88. https://www.researchgate.net/publication/264001400. Accessed 26 May 2020

Hippocrates (400 B.C.E.) Of the epidemics, as translated by Adams F, Library of Alexandria. http://classics.mit.edu/Hippocrates/epidemics.1.i.html. Accessed 26 May 2020

Kimura N, Macer D (2008) Japanese attitudes towards genetically modified mosquitoes. In: Calderbank D, Macer DRJ (eds) Asia Pacific perspectives on biotechnology and bioethics, pp 48–57. https://eubios.info/APPBB.pdf. Accessed 26 May 2020

Marshall JM, Toure MB, Traore MM, Taylor CE (2010) Towards a quantitative assessment of public attitudes to transgenic mosquitoes: question based on a qualitative survey in Mali. AsPac J Biol Biotechnol 18(2):249–270

Okorie PN, Marshall JM, Akpa OM, Ademowo OG (2014) Perceptions and recommendations by scientists for a potential release of genetically modified mosquitoes in Nigeria. Malar J 13(154). https://malariajournal.biomedcentral.com/articles/10.1186/1475-2875-13-154. Accessed 26 May 2020

OXITEC (2016) Aedes aegypti OX513-A: draft environmental assessment for investigational use of Aedes aegypti OX513A. https://www.fda.gov/files/animal%20&%20veterinary/published/Oxitec-Mosquito%2D%2D-Draft-Environmental-Assessment.pdf. Accessed 14 Jan 2021

Portus LM, Barrios EB, Conaco MCG, Go SP (2018) Doing social science research: a guidebook. Philippine Social Science Council, Manila

Regacho BC (2020) Personal communication

Touré YT, Oduola AMJ, Sommerfeld J, Morel JCM (2003) Biosafety and risk assessment in the use of genetically modified mosquitoes for disease control. In: Takken W, Scott TW (eds) Ecological aspects for application of genetically modified mosquitoes, vol 2, pp 217–222. https://library.wur.nl/ojs/index.php/frontis/article/view/851. Accessed 26 May 2020

Zaborek P (2015) Elements of marketing research. Warsaw School of Economics. https://www.academia.edu/11512479/Elements_of_Marketing_Research. Accessed 26 May 2020

Chapter 24
Safe Application of Genetically Modified Mosquito (GMM) to Combat Dengue and Chikungunya Depends on Socioeconomic Status and Social Acceptance in the Developing Countries: A Comprehensive Analysis

Mohammad Nazrul Islam

Abstract The emerging and re-emerging vector-borne diseases are a serious public health problem throughout the world. It has been observed that more than 100 countries and approximately half of the world's population are at risk on vector-borne diseases (VBDs). The global burden of the vector-borne diseases is unacceptably high. It alludes toward their functional inappropriateness, untimeliness, and irrelevance in controlling vectors and vector-borne diseases. Modern technologies, coupled with other appropriate ones within the precincts of integrated vector management (IVM), can tide over this situation posed by conventional, mostly insecticide-based, methodologies. A lot of challenges, obstacles, and interruptive factors have warranted urgent deployment of new approaches for the control of VBDs keeping in mind the inbuilt ethical, social, and regulatory issues. Genetically modified mosquito (GMM) technology is a complex and highly sophisticated biotechnological intervention for suppression of vector populations. *Wolbachia*-associated sterile insect technique (SIT) has been proved highly significant and effective for replacement of mosquito populations. Adopting a highly sophisticated GMM technology to suppress or replace the mosquito populations' density is a big question in developing countries because their priority is directed to foremost fulfill the basic human rights to sustain. Yet, notwithstanding foreseeable bottlenecks, of paramount importance is the need to deploy GMM technology with due consideration to socioeconomic factors and availability of advanced biotechnological facilities during the application of GMM in the developing countries.

Keywords GMM · Dengue · Chikungunya · Society · Developing countries

M. N. Islam (✉)
Hamdard University Bangladesh, Dhaka, Bangladesh

B. K. Tyagi (ed.), *Genetically Modified and other Innovative Vector Control Technologies*, https://doi.org/10.1007/978-981-16-2964-8_24

24.1 Introduction

Application of genetically modified mosquitoes (GMMs) for controlling vector-borne diseases (VBDs) depends on social, economic, and ethical issues. The World Health Organization (WHO) has been planning to apply the genetically modified mosquitoes to combat VBDs in endemic regions since 2010. Their main targets are to identify issues, challenges, and opportunities to develop internationally acceptable guidelines for GMM testing (who.int/tdr/publications/documents/gmm-report). Globally, though, safety and efficacy of GMM for human health are considered a paramount subject for human's attention, along with the regulatory issues. The GMM can suppress the vector mosquito populations and reduce the prevalence of diseases by sterile insect technology (SIT) and population gene replacement therapy. The GMMs by sterile insect technique (SIT) and *Wolbachia*-associated biocontrol tool for *Aedes aegypti* are already applied in the field trials in many countries. During field trials, however, they faced different challenges and obstacles irrespective of products, process, and ethical-social-cultural and political issues. The issue related to acceptability of GMM in field trials was of paramount debate. GMM technology was developed on the basis of two approaches: (1) the first approach is called "gene drive," a process by which a gene or group of genes spreads from one generation to another generation, boosting the frequency of the trait to the specific population, and (2) the second approach is called "population suppression," in which strategy mosquito populations are reduced immensely so that there are fewer mosquitoes which can probably pass the pathogen. *Wolbachia* is a well-known intracellular symbiont bacteria subjected to gene drive technology that has been deployed in many countries, including Australia and China, to replace vector mosquito populations and to eventually control the disease like dengue. It interrupts the reproduction of insect to spread itself into the population. When *Wolbachia* is transferred into a previously uninfected mosquito, it often makes the mosquito more resistant to infection with pathogen. It has been proven that *Wolbachia* species are more potent modulators of pathogenic infection than other organisms (journals.plos.org/plospathogens/article?id=10.1371/journal.ppat.1002043).

24.2 Ecological Aspect of GMM

Mosquito-borne diseases such as dengue, chikungunya, malaria, and filariasis pose a global burden which sap off economy out of the developing nations. Unfortunately, there is no single panacea to intervene these scourges (Scott et al. 2002). Different factors such as effective drugs, vaccine, and insecticides are interrupting to reduce the morbidity and mortality rate, but they are gradually proving insufficient and inappropriate (Takken and Knols 2009). For this very reason, mosquito population replacement technique has come into limelight for release in nature to allow treated males to compete and control wild vector mosquito populations in different ecological settings (James 2000). Diversified ecology has also been playing a vital role on GMMs' efficacy. To introduce transgenes into natural habitat for effective suppression of mosquito population is a humongous challenge and is still being studied. To make it work optimally, issues such as the gene flow, mosquito mating patterns, reproductive behavior, genetic exchange between neighboring populations, fitness and phenotypic effects of colonization, and mass rearing will have to be investigated thoroughly. In population replacement technique, refractory transgenes are introduced on the basis of mosquito reproduction. The transgene drive mechanism is more rapid than normal Mendelian tenet, and this gives the technology a great edge over others. Its success depends on spread of transgene and size of target mosquito populations. The most plausible natural circumstance is assortative mating, i.e., the tendency for certain phenotypes to mate with one another originating from distant stocks. It is, therefore, very important to estimate the population size before GMM release, which is full of challenges due to seasonal fluctuations (Taylor et al. 1993).

It is well established that various populations' size may have beneficial or detrimental effect on spread and stability of transgenes with varying degree of impacts. In wild where natural mélange of various different populations prevail, if mosquitoes from different lineages do not mingle and mate assortative, the resultant populations are structured into reproductively separate subdivisions, implying that different unpredicted result may be found. The genetically modified mosquitoes thus carry several unresolved questions. Before releasing GMM, the adequate numbers of genetically manipulated mosquitoes must be amplified and acclimatized to natural conditions since GM mosquitoes produced in a laboratory setting can affect their reproductive fitness. Aside from the isolated mosquito populations in nature, factors such as the geographic and ecological variations can also impact their success rate to adapt to new habitats or ecosystems (Lorimer et al. 1976). It is observed that transgene drive from GMM to wild mosquito populations depends on their strength and fidelity of the drive mechanism and their fitness (Boëte and Koella 2002). Fitness of GMM may also affect the spread of parasite refractory transgenes into wild mosquito populations. The transgenes which reduce host fitness can be inextricably linked to the drive mechanism in order to avoid elimination by selection. There are endless questions alluding to the fitness of the GM mosquitoes when readying for release in the nature (Boëte et al. 2002), but the following ones need urgent attention and addressing on priority:

1. What are the evolutionary costs of genetic modification to mosquitoes?
2. How will these costs shape plans for interfering with pathogen transmission?
3. What effects will natural environmental conditions have on the expression of refractoriness of GMM?
4. Will GMM have an enhanced capacity to transmit pathogens other than the one that they are intended to block?
5. What extent parasites will evolve resistance to GMM?
6. Can we predict the virulence characteristics of resistance phenotypes?

7. Will increase in a mosquito's immune response result in an increase in parasite-induced immunosuppression?
8. Will changes in parasite populations in response to GMM affect the efficacy of vaccines or anti-parasitic drugs?

All the above questions are important from the public health outcomes and require to be addressed with alacrity for reasons of posterity and transparency of technology details. The difficulty, however, is that the relationship between vector density and human infection may vary with time and the particular geographic location. In southern Tanzania, the risk of human infection increased with entomological inoculation rate (EIR), while human infection rate was low. On the other hand, when infection rate was high, the transmission became saturated. It was also observed that increase EIR did not raise parasitemia in infants, implying that when transmission is low (the predicted situation after GMM release), the "rebound effect" must be avoided. It is desired to analyze the impact of GMM on human health after the release. Further investigations are required to evaluate the relationships among the different measures as these will further strengthen the prediction of the success of GMM strategy.

The risk factor assessment tools for dengue transmission are still being developed. A successful application of GMM for dengue control program depends on a well-informed release and post-release monitoring of vector population size and structure, besides human herd immunity to the pathogen and the low transmission threshold level. A goal for future quantitative analysis should be prepared. Simulation model is used to predict entomological thresholds for dengue transmission (Focks et al. 2000). Mathematical model was developed to identify parameters required to predict the dynamics of transgene drive mechanisms in vector populations (Kiszewski and Spielman 1998). It is needed to develop uniform tools for dealing with the ethical, legal, and social issues on GMM technology (Alphey et al. 2002).

The adverse effects of GMM were observed due to biological diversity and consequences of gene flow. Hayes et al. (2002) illustrated a hierarchical holographic model for analysis of adverse effects. According to this model, fault tree analysis is considered at first. For this purpose, it is needed to list all possible harms whether they are caused by the stressor or not. After that it is followed by event tree analysis. It is well accepted that ecological risks of GM mosquitoes can be analyzed by event tree approach. To address the second possibility by using transparent, expert-driven qualitative prioritization processes are used to limit the numbers.

24.3 Socioeconomic Aspects of GMM

The global economic burden becomes unpredictably high due to emerging and re-emerging vector-borne diseases. It interrupts the fundamental economic development. It has been observed that the average days lost for ambulatory patient due to dengue infection is 14.8 in eight countries; the average cost invested is US$ 514. It is also observed that average lost days for non-fatal hospitalized dengue patient is 18.9 days; the average cost here is US$ 1491 (Suaya et al. 2009; Guha-Sapir and Schimmer 2005; Meltzer et al. 1998; WHO 2012). In 2012, dengue was the top-ranked economic burden due to mosquito-borne viral disease with epidemic potential worldwide. It has increased by approximately 30-folds in the last 50 years. At present, most of the vector species are resistant to existing insecticides. It is mandatory to increase fund and the political commitment to scale up access of existing vector control tools. It is estimated that more than 2.5 billion of the world's population are at risk of dengue, while the World Health Organization has estimated that every year, more than 100 million dengue infections are observed worldwide. An estimated 500,000 people with severe dengue infections were hospitalized each year, among them a large proportion being children. Approximate case fatality rate was 2.5%. Before 1970, only nine countries had experienced severe dengue epidemics. The disease is now endemic in more than 100 countries in Africa, the Americas, the Eastern Mediterranean, Southeast Asia, and the Western Pacific. In Asia and the Indian Ocean region, the main vectors of chikungunya are *Aedes albopictus* and *Ae. aegypti*. In Africa, arboviruses are abundantly transmitted by *Aedes* species, in addition to *Culex annulirostris*, *Mansonia uniformis*, and *Anopheles* species.

Application of GMM depends on socioeconomic condition of the community. Any intervention into the community depends on knowledge and cultural acceptability, community engagement, ecological or political assumptions, and ethical issues. It is needed to strengthen capacity building (individual and community) and transfer of biotechnology to endemic countries. For better addressing, ethical, legal, and social issues are needed to be considered beforehand. Community participation and community involvement are mandatory to introduce GMM. Information, communication, education, and social mobilization are required to implement new method to the community. Before introducing new method to the community, some factors should be considered, such as the following:

1. Different regions in the world have their specific ethical dogmas to new approaches, and all such extant ethical paradoxes and dilemmas need be recognized beforehand.
2. Even if several paradoxes and dilemmas within a given society are settled amicably, it is still indispensable to identify contentious points among the parties so that they too could be addressed and settled with good understanding.
3. Since different GM approaches have specific modus operandi, it becomes inevitable to examine the risks and advantages of each option under reference.
4. All over the world, it is clearly understood that religious heads or leaders certainly have a great degree of influence on the people and, therefore, it is considered opportune to examine how traditional religions might influence ethical reasoning for a successful release of the GM mosquitoes.
5. Transparency, credibility, and timeliness of the parties involved in the GM release play an important role to roll out the new technology. Therefore, every

information on the new approach chosen must be duly brought to the attention of the concerned parties for effecting the most suitable solution.

Consequently, the GM mosquito approach can be followed by conforming to the laws of biosecurity set by the Convention of Cartagena.

24.4 Conclusion

Application of GM mosquitoes to combat vector-borne diseases like malaria, dengue, chikungunya, and yellow fever is a controversial issue till now in many countries, albeit some nations have already green-signaled for their application, and one such nation is the United States where most recently, in the month of May 2021, GM mosquitoes developed through RIDL technology have been released in Florida Keys. Most developing nations are awaiting permission by their respective regulators, as also the release depends on social, economic, ethical, and cultural acceptance. It is observed that the poorest segments of society and least developed countries are most affected by the VBDs. It is proved that urban, peri-urban, and rural communities are more prone for infection due to lack of access to adequate housing, safe drinking water, and sanitation. Malnourished with weakened immunity peoples are now more vulnerable to the VBDs than earlier. These diseases also exacerbate poverty. At present, very few developed countries have agreed to adopt GMM technology. So wide application of GMM to combat VBDs like dengue and chikungunya will depend on local needs, socioeconomic conditions, and social acceptance of the technology by the community.

References

Alphey L, Beard CB, Billingsley P, Coetzee M, Andrea C, Chris C et al (2002) Malaria control with genetically manipulated insect vectors. Science 298(5591):119–121

Boëte C, Koella JC (2002) A theoretical approach to predicting the success of genetic manipulation of malaria mosquitoes in malaria control. Malar J 1:3. https://doi.org/10.1186/1475-2875-1-3

Boëte C, Paul REL, Koella JC (2002) Reduced efficacy of the immune melanization response in mosquitoes infected by malaria parasites. Parasitology 125:93–98. https://doi.org/10.1017/s0031182002001944

Focks DA, Brenner RJ, Hayes J, Daniels E (2000) Transmission thresholds for dengue in terms of Aedes aegypti pupae per person with discussion of their utility in source reduction efforts. Am J Trop Med Hyg 62:11–18

Guha-Sapir D, Schimmer B (2005) Dengue fever: new paradigms for a changing epidemiology. Emerg Themes Epidemiol 2(1):1

Hayes KR (2002) Identifying Hazards in Complex Ecological Systems. Part 1: Fault-tree Analysis for Biological Invasions. Biological Invasions 4, 235–249. https://doi.org/10.1023/A:1020979914453

James AA (2000) Control of disease transmission through genetic modification of mosquitoes. Chapter 18. In: Handler AM, James AA (eds) Insect transgenesis - methods and applications. CRC Press, Boca Raton, FL, pp 319–333

Kiszewski AE, Spielman A (1998) Spatially explicit model of transposon-based genetic drive mechanisms for displacing fluctuating populations of anopheline vector mosquitoes. J Med Entomol 35:584–590

Lorimer N, Lounibos LP, Petersen JL (1976) Field trials with a translocation homozygote in Aedes aegypti for population replacement. J Econ Entomol 69:405. https://doi.org/10.1093/jee/69.3.405

Meltzer MI, Riqau-Péres JG, Clark GG, Reiter P, Gubler DJ (1998) Using disability-adjusted life years to assess the economic impact of dengue in Puerto Rico: 1984–1994. Am J Trop Med Hyg 59(2):265–271

Scott TW, Takken W, Knols BGJ, Boëte C (2002) Ecology of genetically modified mosquitoes. Science 298(5591):117–119. https://doi.org/10.1126/science.298.5591.117

Suaya JA, Shepard DS, Siqueira JB, Martelli CT, Lum LC, Tan LH et al (2009) Costs of dengue cases in eight countries in the Americas and Asia: a prospective study. Am J Trop Med Hyg 80(5):846–855

Takken W, Knols BGJ (2009) Malaria vector control: current and future strategies. Trends Parasitol 25(3):101–104

Taylor CE, Touré Y, Coluzzi TM, Petrarca V (1993) Effective population size and persistence of Anopheles arabiensis during the dry season in West Africa. Med Vet Entomol 7(4):351–357. https://doi.org/10.1111/j.1365-2915.1993.tb00704.x

WHO (2012) Global strategy for dengue prevention and control, 2012–2020, vol 35. World Health Organization, Geneva, pp 3–7

Chapter 25
Experiences and Outcomes from a Worldwide Training Programme on Genetically Modified Vectors (GMVs) Related Biosafety for Human Health and the Environment

B. K. Tyagi

Abstract Partial to virtual lack of any impact on control of vectors of human diseases, especially mosquitoes, warranted urgent search for new alternate technologies which will be safe, economical and environment-friendly, on one hand, and integrate with other tools and methodologies of the integrated vector management (IVM), on the other. Past few decades have witnessed surge of many effective and sustainable genetically and biotechnologically developed de novo technologies which tend to control mosquito vectors by working either to suppress (transgenesis) or replace (paratransgenesis), besides an array of other physiological interventions, on the vector populations. Several technologies such as, for example, Release of insect carrying Dominant Lethal (RIDL) gene system, *Wolbachia* (an endocellular symbiotic bacterium naturally present in many arthropods) induced cytoplasmic incompatibility (CI) resulting in unviable egg production and transforming dengue vectors (*Aedes* spp.) and malaria vectors (e.g., *Anopheles stephensi*) into resistant to respective pathogens, i.e., viruses and Plasmodium, have offered promise in controlling vector-borne diseases. Notwithstanding unchallengeable significance, these technologies have also raised many questions from both societies and governments of many countries. To alleviate their scepticism and other queries, many international organizations conducted meetings to generate consensus for guidelines, but even this helped marginally to pacify global interrogations. It was, therefore, considered opportune by the Tropical Disease Research (TDR)/WHO to set up a series of multi-regional training workshops in Africa (Bamako, Mali), Asia (Madurai, India) and Latin America (Medellin, Colombia) between 2008 and 2011 (WHO 2015). About 150 trainees were drawn from as diverse disciplines/walks of life as science, health departments, academics, social, legal, non-governmental organization. The outcome, inculcated from the experiences expressed by the trainees themselves post-workshops, has been very encouraging as they all found the training

---

B. K. Tyagi (✉)
SpoRIC, VIT University, Vellore, Tamil Nadu, India

B. K. Tyagi (ed.), *Genetically Modified and other Innovative Vector Control Technologies*, https://doi.org/10.1007/978-981-16-2964-8_25

courses highly beneficial to comprehend genetically modified vectors/mosquitoes (GMV/GMM) related biosafety to the human and the environment and thus become a potential ambassador in their areas or countries to strongly communicate and advocate about the lasting benefits of the various genetically evolved technologies in the control of mosquitoes responsible for transmission of dengue and malaria, in particular.

Keywords GMM · Biosafety · Human · Environment · Africa · Asia and Latin America

25.1 Introduction

Mosquitoes rank number one on the list of most dreadful hematophagous arthropods in transmitting some of the world's most devastating vector-borne diseases to human and have thus justifiably earned the sobriquet "man's deadliest foe on the planet" (Spielman and D'Antonio 2001; Tyagi 2004)! These vector-borne infections severely not only affect the health of people but also emaciate their economy and impoverish a growing nation's intelligentsia (Covell 1927; Sinton 1939). Ironically, despite enormous advancement in science, there is no vaccine or specific antidote yet available commercially for most of these infections, and therefore the disease management in such cases is largely dependent on the vector control. During recent past the world has witnessed emergence and re-emergence of several vector-borne diseases, with or without zoonosis to further complicate their biology and pose fresh challenges to public health globally (Sharma 1999). It is common understanding that factors like increased global travel, increased human populations (7.8 billion), increased urbanization at the cost of proper planning and urban management (51% of the world's population is urban), encroachment of human activity into new ecosystems especially agriculture fields and forests (India's forest cover is now reduced to <10% against mandatory 33%), dissemination of vectors to areas hitherto considered terra incognita for these largely due to inclement climatic changes and, above all, collapse of vector control and public health programmes have contributed to a pathetic and uncontrollable growth of most common vector-borne diseases recently.

Because the conventional methods of controlling mosquitoes such as insecticide space fogging, Aerosol space spray, indoor resting spraying, larviciding of breeding habitats, and manual elimination of breeding sites proved ineffective to either check the onslaught of the epidemics or cost-inhibitiveness during the long run of application, alternative tools, processes and technologies were urgently warranted which could be both effective and cost-effective, besides environment-friendly. Insecticides such as DDT, malathion and various pyrethroids—the main plank for vector control during early years of their discoveries—were gradually opposed both in health and agriculture applications due to many hazardous activities in human, animals as well as the environment, and one major setback was the evolution of resistance in most mosquito vectors against the chemicals. Although every national mosquito-borne disease control programme includes abatement of vectors, these remain for most part of the programme only an auxiliary exercise to tackle outbreaks. Thus, the worsening scenario of the mosquito-borne diseases world over has warranted scientific fraternity in earnest to not only innovate to strengthen the conventional methods but also unearth new and novel approaches as an inseparable component of the multidisciplinary and multidimensional integrated vector management (IVM) to check the disease transmission among the communities. In spite of the fact that all these novel technologies (transgenesis, paratransgenesis, gene drives, etc.) are rigorously experimented, tested and broached both in field and simulated field conditions, yet these are not without imperfections, and therefore they encounter severe opposition by the people or societies for the unresolved fears alluding unseen, unfounded and calamitous future impacts which pose a great challenge before their implementation (WHO 2014). Genetic manipulation as a vector control tool has undergone a long peregrination since its origin. Although the Russian scientist Serebrovsky (1940) is credited with the initiation of the thought process for the genetic engineering/manipulation of arthropods and English-speaking duo Edward Knipling and Raymond Bushland (Knipling 1955) had practically developed the SIT to eliminate screwworms preying on warm-blooded animals, especially cattle, yet the credit for logical explanation of genetic manipulation using the sterile insect technique (SIT) goes to Curtis (1968) who first suggested possibility of using translocations to fix desirable genes in insect pest populations (Dyck et al. 2005) and was one of the consultant scientists at the Genetic Control of Mosquitoes Research Unit (GCMRU), financed by the United States for execution through the World Health Organization with the cooperation of Indian Council of Medical Research, during the mid-1970s, for controlling the deadly yellow fever mosquito *Aedes aegypti* in villages just outside Delhi. Several research publications were brought out on investigations of biology and ecology of *Ae. aegypti*, and the project initially progressed very satisfactorily. However, suddenly the project was aborted just before it was all geared up to meet its scientific objective (Curtis and Reuben 2007). The SIT had earlier already revolutionized control of certain agriculture pests in the western world, e.g. the screwworm fly (*Cochliomyia hominivorax*) was successfully controlled in North and Central America. Similarly, many successes were achieved for control of fruit fly pests, most particularly the Mediterranean fruit fly (*Ceratitis capitata*) and the Mexican fruit fly (*Anastrepha ludens*), in the 1960s onwards. The premise on which the project was founded was that sterilized males would mate with wild females while competing with their natural wild male population, knowingly resulting in unviable egg production, so that there would be no mosquitoes born at all, and thereby controlling the disease. The sterile insect technique (SIT) requires mosquito's mass rearing, production, sex separation, sterilization and subsequent release of large amount of sterile male mosquito into targeted population where the sterile male mosquitoes mate with female mosquito and produce non-viable offspring. The massive responsibilities posed serious challenges associated with massive production, sex separation, sterilization and distribution to the managers of the laboratories, and, though this method was feasible by demonstrating limited impact on control of vector mosquitoes, it nevertheless turned out to be cost-inhibitive largely because the sterilized males produced after irradiation were far too week as compared to their counterparts in the wild and failed to compete successfully to attract females. The SIT is considered to be originator for all modern genetic engineering tools and techniques be it transgenesis, paratransgenesis or gene drive technology.

Quite some sustainable genetic engineering-based technologies are currently available to do their bit within the gamut of the IVM and result either in suppression or replacement of the vector, e.g. release of insects carrying a dominant lethal (RIDL) gene system; *Wolbachia* (an endocellular symbiotic bacterium)-induced cytoplasmic incompatibility (CI) dramatically reducing the longevity of adult female mosquitoes; RNA interference (RNAi) aiming at improving the natural defence system of the mosquito to improve the virus resistance and suppress virus replication; site-specific DNA lesion (transcription activator-like effect or nucleases (TALENs), still in infancy, and, of course, the most dynamic and promising gene drives (particularly CRISPR/Cas9), etc. However of paramount importance are the kinds of uncertainty associated with regulatory and societal acceptance incessantly hovering like the sword of Damocles on the positive conclusion of scientists' rigorously developed new technologies. This serious constraint is, however, conquerable by instilling proper training to the scientists who should foresee the public attitude and response to their discoveries, inventions and innovations based on genetic manipulations, besides, least to mention the mandatory cautions to be taken care of relating the regulatory and biosafety aspects, capacity building and development of best practice guidance (Beech Camilla et al. 2009). Here under, therefore, a sincere effort is made to recount unique experiences and outcomes from possibly the world's foremost training on the "Biosafety for human health and the environment in the context of the potential use of genetically modified mosquitoes (GMMs), 2008–2011", trans-continentally conducted in Africa, Asia and Latin America.

25.2 Scope

During the end of the twentieth century, the genetic transformation of vector mosquitoes opened a new era for control of viral and parasitic diseases such as dengue and malaria by significantly reducing the ability of their vectors to transmit pathogens. The twin strategies which were subject to large-scale investigations and discussions included (1) release of insects carrying a dominant lethal (RIDL) gene for control of dengue vector *Aedes aegypti* through population reduction and (2) *Wolbachia*-induced cytoplasmic incompatibility (CI) in both the dengue-transmitting *Aedes* species and the malaria vector *Anopheles stephensi* through population replacement as well as making existing populations refractory to transmit dengue viruses and *Plasmodium*, respectively. The simple-looking process is however highly challenging to implement, and a lot of exercises need to be carried out to win over ultimate stumbling blocks in the form of regulatory approval and societal acceptance before entering into release in nature. Therefore, this worldwide three-site continental training was designed to impart through an array of workshops the knowledge about biosafety needs to human being and the environment in connection with the GM mosquitoes through multimedia channels including classroom theoretical and practical lectures on handling, feasibility, efficiency and release of such genetically modified organisms in specific ecological settings, laboratory-based practical lessons, visit to biosafety laboratories, mock biosafety rehearsals, etc., with special reference to legal frameworks and regulation, public engagement and capacity building.

25.3 Objectives

With the pivotal mandate of creating a pool of regional scientists well trained in the assessment and management of biosafety related to the implementation of genetically modified disease vectors (e.g. mosquitoes) for the control of vector-borne diseases, the training workshop carried a threefold objective:

1. To increase the awareness of researchers and decision-makers to issues and challenges such as ethical, legal and social implications related to the development and implementation of this technology.
2. To ensure the feasibility and safety of genetically modified disease vectors in countries.
3. To build capacity in each region (Africa, Asia and the Americas) for the safe development and implementation of this technology.

25.4 Target Trainees

Researchers, vector biologists, local leaders and decision-makers in vector control were the key targeted trainees for getting hands-on training in the process of GMM-related assessment and management of biosafety for human health and the environment. Nearly 150 participants were trained in a series of workshops in all the 3 continents over a period of 3 years, 1 training course every year was mandated for each region (Figs. 25.1, 25.2, 25.3 and 25.4).

The training programme on GM mosquitoes and related biosafety regulations was an enormous success as we continue to see over next several years the trained personnel taking leading role in conducting research on GM mosquitoes in their respective countries since they have been handsomely aware of the science behind developing the GM mosquitoes, along with their assessment and management of biosafety as well as being *au fait* in the setup and management of regulatory principles and bodies (Fig. 25.5).

Fig. 25.1 Dr V.M. Katoch, Director General, Indian Council of Medical Research, inaugurating the TDR/WHO-funded Asian Biosafety Training Course on genetically modified vectors (15–26 June 2009; Madurai, TN, India), as Dr B.K. Tyagi, Course Director, Dr S.S. Vasan, Oxitec, UK, and Dr Kittayapong Pattamaporn, Director, Centre of Excellence for Vectors and Vector-Borne Diseases, Mahidol University at Salaya, Thailand, look on. (*Source*: The Hindu, 17 June, 2009; Dr Tyagi's pers. archive)

Fig. 25.2 Dr B.K. Tyagi addressing during the first WHO/TDR Biosafety Training Course on genetically modified vectors (15–26 June 2009; Madurai, TN, India). (*Source*: Dr Tyagi's pers. archive)

Fig. 25.3 (**a–c**) Trainees at the Asia Biosafety Course on genetically modified vectors held in 2009, 2010 and 2010. (*Source*: Dr Tyagi's pers. archive)

Fig. 25.4 (**a**) Trainees and faculty at the African Biosafety Training Course (Bamako, Mali) and (**b**) at the Latin America Biosafety Training Course (Medellin, Colombia). (*Source*: Dr Tyagi's pers. archive)

In order to harmonize training material among all the three African, Asian and the Latin American Regional Centres, the course protocol had also included organizing at least one GM Coordination Meeting at each of its centres. The Second Coordination Meeting, along with the second TDR/WHO Biosafety/Biosecurity Training Course, was organized within the framework of Asian Biosafety Training Courses on 21–27 November 2009 (Fig. 25.6).

Fig. 25.5 (**a**) Classroom lessons by Ms Camilla Beech, UK; (**b**) visit by trainees to the ICMR-CRME laboratories and interactive discussion with Dr B.K. Tyagi, Asian Training Coordinator, and Prof. Madama Bouare (African Training Coordinator); and (**c, d**) group discussions among trainees and visit to the BSL-II at the Microbiology Department, Madurai Medical College, Madurai. (Source: (**a**) WHO/Tyagi 2015; Dr Tyagi's own contribution; (**b–d**) Dr Tyagi's personal archive)

Fig. 25.6 The Second Coordination Meeting, along with the second TDR/WHO Biosafety/ Biosecurity Training Course, was organized within the framework of Asian Biosafety Training Courses on 21–27 November 2009; seen are (from left) Dr Madama Bouare, Dr B.K. Tyagi, Dr Pattamaporn Kittayapong, Dr Yeya Touré and Dr Camilla J. Beech. (*Source*: Dr B.K. Tyagi's pers. archive)

25.5 Training Course Outline

A given course was guided by a director who was also the coordinator for organizing lectures, etc. and inviting applications from the prospective trainees in the respective region. Each training course involved by and large a strong cohort of about 15 trainees. A detailed advertisement was put up in the TDR/WHO's newsletter, "TDR News" as well as in other local channels of communication. In as far as possible, representative trainees were drawn from as many countries in a given region as possible, and, depending on application strength, due representation was given to the female trainees. The duration allotted to each regional course was 2 weeks. The following major topics were covered in the courses:

1. *Genetically modified organisms/living modified organism (GMO/LMO)*: International treaties or consensus (Cartagena Protocol, OECD, etc.)
2. *Genetic manipulation of vectors*: History, basic principles and potential impact of genetic manipulation on humans and environment.
3. *Laboratory and simulated field studies*: Appropriate designing of Arthropod Containment Level II (ACL II) and simulated field studies to demonstrate comparable reproductive behaviour of GM mosquito.
4. *Field application of GM mosquitoes*: Ethical, legal and social implications.
5. *Potential hazards*: Risk assessment, risk management and risk/benefit analysis.
6. *Laboratory-based biosafety and biosecurity*: Principles and practices for the assessment and management.
7. *Management committees*: Guiding principles for function of institutional biosafety and ethics review committees and national biosafety review boards.
8. *Preparation of application to regulatory body*: Guidance on biosafety regulation and legal principles at national levels for securing the development and use of GM-based vector control method.

Table 25.1 Faculty members involved in training the students-trainees at the African, Asian and Latin American centres

I. Africa
Dr Abdourahamane Sangare, Dr Amidou Dembele, Dr John Marshall, Dr Ken Vernick, Madama Bouare, Dr Abdoulaye M. Toure, Dr Samba Diop, Dr Willy K. Tonui, Dr B.K. Tyagi, Camilla Beech, Dr S.S. Vasan and Yeya Tiemoko Toure
II. Asia
Dr B.K. Tyagi, Dr S.S. Vasan, Dr Madama Bouare, Dr Camilla Beech, Dr Vijay Veer, Dr Selva Kumar, G. Kumaresan, Dr Jhansi Charles, Dr Kittayapong Pattamaporn, Dr L. Kriangkrai, Dr P. Paul Kumaran, Dr T. Jeyalakshmi, Dr S. Visalakshi, Dr Lee Han Lim, Dr T.S. Saraswathy, Dr Sarala Subbarao, Dr Bharat Char, Dr Ritesh Mishr and Dr Worachart Sirawaraporn
III. Latin America
Dr Manuel Lluberas, Dr Manuel Lluberas-Pilar Corena, Dr Ann Kramer, Dr Marcelo Jacobs-Lorena, Dr B.K. Tyagi, Dr Rene Gato, Dr Camilla Beech, Dr Hervé Bossin, Dr Ivan D. Velez, Dr Elizabeth Hodson, Dr Anita Villacis and Andre da Silva and Anita Villacis

Table 25.2 A Complete list of trainees during the three training courses delivered by the Asian Biosafety Training Centre

S. No.	Name of the trainee	City	Country of origin
1	Ali Reza Chavshin	Tehran	Iran
2	Dr Lazuardi Mochmad	Surabaya	Indonesia
3	Dr Jitendra Prasad Mathuria	Pokhara	Nepal
4	Ms Manorenjitha Sivanathan Malar	Penang	Malaysia
5	Dr Chaiyaphruk Pilakasiri	Bangkok	Thailand
6	Dr Sinnathamby Noble Surendran	Jaffna	Sri Lanka
7	Dr Nazni Wasi Ahmad	Kuala Lumpur	Malaysia
8	Prof. Lillian A. De Las Llagas	Manila	The Philippines
9	Dr Bannoth Reddya Naik	Hyderabad	India
10	Dr Murlidhar Mendki	Gwalior	India
11	Dr Rakesh Singh	Varanasi	India
12	Dr Partap S. Kataria	Bikaner	India
13	Dr P. Thiyagarajan	Coimbatore	India
14	Dr Sujata Mohanty	Noida	India
15	Dr Buyankhishig Burneebatar	Ulaanbaatar	Mongolia
16	Ms Xu Libo	Beijing	China
17	Dr Kusumawathie, P.H.D.	Kandy	Sri Lanka
18	Dr Kanutcharee Thanispong	Colombo	Sri Lanka
19	Dr Tran Vu Phong	Hanoi	Vietnam
20	Dr I.P. Sunish	Madurai	India
21	Dharam Singh	Madurai	India
22	Dr Siraj Ahmed Khan	Dibrugarh	India
23	Dr Renaud Lacroix	Kuala Lumpur	Malaysia
24	Dr Moe Myint Aung	Yangon	Myanmar
25	Mohammad Shafiul Alam	Dhaka	Bangladesh
26	Rajib Chowdhury	New Delhi	India
27	Dr Prakash Ghimire	Kathmandu	Nepal
27	Dr Prabhakargouda B. Patil	Jalna	India
28	Dr Ratna Joseph Thalathoti	Hyderabad	India
29	Saber Gholizadeh	Tehran	Iran
30	Dr Nguyen Van De	Hanoi	Vietnam
31	Saafi LaOde	Kendari	Indonesia
32	Dr Mohammad Nazrul Islam	Dhaka	Bangladesh
33	Dr Narankhajid Myadaguren	Ulaanbaatar	Mongolia
34	Ms Jeevanie Harischandra	Colombo	Sri Lanka
35	Ms Aikaterini Mandaltsi	Kuala Lumpur	Malaysia
36	Chandru Angamuthu	Kuala Lumpur	Malaysia
37	Dr Mohamed Mohd. Salleh	Kuala Lumpur	Malaysia
38	Ashok Kumar Visvanathan	Tirukoilur	India
39	Ms ChellamuthuVasugi	Coimbatore	India
40	Ms Simarjit Kaur	Patiala	India
41	Dr Ritesh R. Mishra	Jalna	India
42	Gowrishankar	Tiruchirappalli	India

25.6 Faculty Contents of Courses and the Region

A very special GM research oriented and experienced faculty was drawn from various countries just as the trainees were also enrolled from different underdeveloped, developing and developed countries across the globe (Tables 25.1 and 25.2). The medium of training lectures was English, although local language was often resorted to when such a moment arose for convenience of facilitation to the trainees, implying that sometimes lecturers took help from French in Mali (Africa) and those in Medellin (Colombia) spoke in Spanish and/or Portuguese. Keeping in mind a rather newly evolving science of genetic engineering and biosafety, it was decided to keep informal the teacher-taught relationship so that more thorough discussions could emerge and extensive discourses could take place for a better clarity of the subject. The topics selected were both timely and futuristic. A total of 110 lectures were delivered with the help of power point slides, a flip chart and a black board. A list of lectures presented in the three regional training centres is concisely presented in Table 25.3.

25.7 In-Training/Course Publications and Documentations

The training provided ample opportunities to broach on the subject of genetically modified vector (GMV)-related biosafety for human health and the environment and bring out several important publications and/or reports. As a result, three summaries, besides two research publications one each by a group of the various faculty members and the other by the trainees, were brought out during the training courses held for Asia region. However, the most significant documentary contribution came out in the form of a WHO-published monograph that was, as a specific project, drafted as the first book manuscript, assembled from various chapters contributed by authors from all over world and edited by B.K. Tyagi. All these documents/publications are listed below chronologically.

1. [Tyagi, B.K., 2009] 2009. Executive Report: 1st Asian Biosafety Training Course on Asian Centre for Training in Biosafety Assessment for Human Health and Environment using Genetically Modified Vectors, June 15–26, 2009. Centre for Research in Medical Entomology, Madurai, 72 pp.
2. [Tyagi, B.K., 2009] 2010. Executive Report: 2nd Asian Biosafety Training Course on Asian Centre for Training in Biosafety Assessment for Human Health and Environment using Genetically Modified Vectors, February 22–March 5, 2010. Centre for Research in Medical Entomology, Madurai, 89 pp.
3. [Tyagi, B.K., 2009] 2010. Executive Report: 3rd Asian Biosafety Training Course on Asian Centre for Training in Biosafety Assessment for Human Health and Environment using Genetically Modified Vectors, October 18–29, 2010. Centre for Research in Medical Entomology, Madurai, 76 pp.

Table 25.3 Topics of lectures delivered at the three regional centres

I. African region
1. Lessons learnt from the use of Genetically Modified Plants in Agriculture: Risk and advantages of GMP
2. Transgenes: How are GMP made up?
3. La convention des nations unies sur la diversite biologique (CDB)
4. Protocole de Cartagena sur la prévention des risques biotechnologiques
5. Cadre juridique national de Biosécurité (CNB)
6. Regime de Responsabilite dans l'utilisdation des OGM
7. Introduction a la propriete intellectuelle
8. La brevetabilite du vivant: (a) Bref rappel de l'état de la question, (b) Position du droit malien, and (c) Analyse
9. Accès aux ressources génétiques et les droits de propriété intellectuelle: Le biopiratage
10. La Procédure malienne de prise de décision relative à la libération dans l'environnement d'un OGM
11. Présentation sommaire du droit de la proproriété intellectuelle en relation avec les OVM
12. L'accès au ressources génétiques et l'interface avec le système actuel du droit de la propriété intellectuelle
13. Besoin d'amélioration du système de régulation de la biosécurité
14. Responsabilité en cas de dommage causé par un OGM
15. Présentation sommaire de la réglementation (internationale/régionale) en matière de biosécurité
16. Identification de structures et de cadres juridiques relatifs aux insectes génétiquement modifiés
17. Mise en place et suivi d'une structure nationale de biosécurité
18. Missions of the Control Structures
19. Pouvoirs des structures de contrôle
20. Containment issues during planned field cage trials
21. Overview of disease vector control: issues and challenges
22. Gene drive systems for spreading refractory genes
23. Public perspectives to genetically modified organisms in Western nations and Africa
24. The Cartagena Protocol and GM Mosquitoes
25. Ethical issues related to GM mosquitoes
26. Gene drive systems and containment
27. GM mosquitoes for malaria control: Perspectives of people in Mali, West Africa to a transgenic release
28. Engineering mosquitoes refractory to malaria and dengue fever in the laboratory
29. Identification of hazards and risks, and
30. Overview of genetic control methods for insect vectors
31. Introduction to biosafety for humans and the environment
32. Creation et gestion d'un Comité national d'ethique: structure et role
33. Transparence, participation et communication avec le public
34. Implications éthiques et sociales dans l'utilisation des OGM
35. Risk management
36. Introduction to Biosafety & Biosecurity in Laboratories, and Overview on Biosafety

(continued)

Table 25.3 (continued)

37. Overview of the Asian Biosafety Training Course format, objectives and General logistics
38. Medical Arthropodology: Biosafety Risk Assessment Overview
39. Risk Assessment for Arthropod Vectors: GMVs and Biosafety issues
40. Medical Arthropodology: Biosafety Risk Assessment Overview
41. Transgenic Insects: From Laboratory to Field
42. Innovative control using modified insect vectors
43. Identification of legal frameworks and guidance documents in relation to GM vectors
44. Foundations of Risk Assessment and Risk Management
45. Risk Management and Development of Emergency Response Plan
II. Asian region
1. Monitoring and Environmental Impact Assessment (Session 20A)
2. From Lab to Field and use stepwise
3. Identification of legal frameworks and guidance documents in relation to GM vectors
4. Selecting a field site
5. Cartagena Biosafety Protocol under the Convention on Biological Diversity
6. Communications
7. Sterile Insect GM Strategies Status of RIDL
8. Identification of legal frameworks and guidance documents in relation to GM vectors
9. Risk Management and Development of Emergency Response Plan
10. Genetically Modified Vectors (GMV): Biodefence & Bioterrorism
11. Basic safety measures in biological laboratories
12. Molecular Biology of Transgenesis and Heterologous Gene expression
13. Molecular Biology of Transposon Mediated Transgenesis Strategies
14. Biosafety Issues in Genetically Modified Organisms
15. Importance of Biosafety and Medical Microbiology—In a Practitioner's perspective
16. Packaging and Transport of Sputum Specimens from the Districts to the reference laboratory
17. Best practice guidance for deployment of genetic control methods against mosquito vectors in disease endemic countries
18. Importance of Biosafety: Ethical Issues
19. Biosafety, Regulatory and Laboratory Experience of IIBAT
20. Overview of the Cartagena Protocol: PART-1: Regional Initiatives Under Cartagena Protocol
21. First field release of transgenic *Aedes aegypti*: What needs to be done?
22. Biosafety review process
23. Communication Plan
24. Regulation and coordination required for a first transgenic release?
25. Ethical, Socio-economic, cultural issues in relation to use of GM vectors
26. Review of sterile male techniques in India during the 1960s & 1970s
27. Biotech crops in India: From lab to reality
28. Effective use of modern biotechnology: GM crops
29. Best practice guidance for deployment of genetic control methods against mosquito vectors in disease endemic countries
III. Latin American region

(continued)

Table 25.3 (continued)

1. A review of the current vector control methods and strategies from the scientific and practical point of view
2. A critical assessment of transgenic vector risks and impact on health and the Environment based on previous experiences with conventional vector control programmes
3. Hypothetical release exercise 1: from the laboratory to the field
4. Intro to GM vectors and relevance to human health
5. Why is it important to address biosafety in the context of GM insects of medical importance?
6. Intro to biosafety and biosecurity and their relevance to humans and the environment
7. Principles and practices of Biosafety and biosecurity under different conditions
8. Intro to genetic control methods in the laboratory (population suppression, population replacement and para-genesis) and risk assessment considerations
9. Challenges for development and implementation of laboratory control methods using GMVs. Infrastructure, equipment and materials
10. Introduction to strategies for transgene containment and site selection
11. Genetic drive mechanism
12. Hazards and risks associated with handling of GMVs: regulatory and ethical issues
13. Transgenesis, paratransgenesis and other modifications of insects
14. Physical and biological characterization of the release site
15. Containment management systems including packaging and transport of GMVs
16. Ethical, social and legal implications of transgenic release and implications for cross-border movement of GMVs
17. Lessons learnt from Tsunami experience: what needs to be done prior to firs transgenic release
18. Rules, regulations, responsibilities, and training research and field personnel
19. Containment levels: facility design and practices
20. Conventional insect sterile technique (SIT) versus RIDL-SIT, with examples from public health and agriculture: Containment facility design and work practices
21. Mosqguide
22. Overview of the Cartagena protocol
23. Systematic risk assessment for GM vectors
24. Environmental risk management
25. Regulatory and legislative aspects of GMVs
26. Site selection/plan criteria
27. Principle of biosafety applied to genetically modified vectors and disease transmission
28. Moving GMVs from the lab to the field. Who is responsible for what in the event of an unintended accident
29. Vector behaviour and infection risk in the context with vector control in Latin America
30. Insectary design using biosafety principles
31. Risk assessment, management and communication
32. Ethical, socio-economic, cultural (ESC) and other implications of use of GMVs
33. Accidents in handling GMVs
34. Sterilization and disinfection in the laboratory

4. Beech Camilla J, SS Vasan, M Megan Quinlan, Margareth Lara Capurro, Luke Alphey, Vicente Bayard, Madama Bouaré, Maria Corena McLeod, Pattamaporn Kittayapong, James V. Lavery, Lee Han Lim, Mauro Toledo Marrelli, J. Nagaraju, Kenneth Ombongi, Rofina Yasmin Othman, Vilasini Pillai, Janine Ramsey, Rachel Reuben, Robert I. Rose, Brij Kishore Tyagi, and John Mumford (2009) Deployment of Innovative Genetic Vector Control Strategies: Progress on Regulatory and Biosafety Aspects, Capacity Building and Development of Best-Practice Guidance. AsPac J. Mol. Biol. Biotechnol 17(3): 75–85.
5. Patil Prabhakargouda B, Alam MS, Ghimire P, Lacroix R, Kusumawathie PHD, Chowdhury R, Sunish IP, Libo Xu, Vu PT, Myint AM, Burneebaatar B, Thanispong K, Dharamsingh DV, Khan SA (2010) Discussion on the proposed hypothetical risks in relation to open field release of a self-limiting transgenic Aedes aegypti mosquito strains to combat dengue. AsPac J. Mol. Biol. Biotechnol. 18(2): 241–246.
6. [Tyagi B.K.] (2015) Biosafety for human health and the environment in the context of the potential use of genetically modified mosquitoes. Training Manual. A Tool for biosafety training (CD and internet version) based on courses in Africa, Asia and Latin America, 2008–2011. World Health Organization. 240 pp.

25.8 Conclusion

A post-course survey brought out the fact that the wisdom gained in the three regional biosafety courses by representative trainees of all strata of public life on future application of genetically modified vectors in controlling vector-borne diseases had advanced immensely and formed a solid foundation in putting into use such a novel tool without any scepticism (WHO 2015). The success of any disease control programme is after all based on the knowledge and extent of preparation made in advance! In the last two decades, the development of application of molecular biology and genetic engineering in vector control is moving towards advancement stage with the support of funding from the international agencies and other bodies. In parallel, the consultation meet with scientists and different stakeholders has been undertaken mainly to assess the benefits and risks associated with various strategies in order to prepare the standardized risk assessment method and universal guidelines/act for regulating all the issues related to GM vectors. Although rapid development takes place towards the use GM insect as effective control tool on one side, the controversy/adversary arises against the GM insects, on the other hand. The fruits of the application of innovative control methods in reducing the disease burden can be realized if the scientific community and national or international regulatory authority would come forward as a joint forum to address the challenges/issues which arise around the usage of GM insects and to strengthen the community participation for the ultimate success and sustainment of the programme. The development of preparation of vaccination and drugs against mosquito-borne disease is also in progress. The reduction of disease burden and letting the people free from

VBDs can be reached effectively only if we could improve the power of usage of control strategies by combining the development of genetically engineered strategies and vaccination/drug along with the modified existing insecticide control methods.

References

Beech Camilla J, Vasan SS, Megan Quinlan M, Capurro ML, Alphey L, Bayard V, Bouaré M, McLeod MC, Kittayapong P, Lavery JV, Lim LH, Marrelli MT, Nagaraju J, Ombongi K, Othman RY, Pillai V, Ramsey J, Reuben R, Rose RI, Tyagi BK, Mumford J (2009) Deployment of innovative genetic vector control strategies: progress on regulatory and biosafety aspects, capacity building and development of best-practice guidance. AsPac J Mol Biol Biotechnol 17 (3):75–85

Covell G (1927) A critical review of the data recorded regarding the transmission of malaria by the different species of Anopheles, with notes on distribution, habits and breeding places. Indian Med Res Mem 7:87–99

Curtis CF (1968) Possible use of translocations to fix desirable genes in insect pest populations. Nature 218:368–369

Curtis CF, Reuben R (2007) Destruction in the 1970s of a research unit in India on genetic control of mosquitoes and a warning for the future management of transgenic research. Antenna 31:214–216

Dyck VA, Hendrichs J, Robinsonet AS (2005) Sterile insect technique: principles and practice in area-wide integrated pest management. Springer, New York. 787 pp

Knipling EF (1955) Possibilities of insect population control through the use of sexually sterile males. J Econ Entomol 48:459–462

Serebrovsky AS (1940) On the possibility of a new method for the control of insect pests. Zool Zh 19:618–630

Sharma VP (1999) Current scenario of malaria in India. Parassitologia 41(1–3):349–353

Sinton JA (1939) What malaria costs India. Government of India Press, Delhi, pp 4–22

Spielman A, D'Antonio M (2001) Mosquito: the story of man's deadliest foe. Faber & Faber, London. 247 pp

Tyagi BK (2004) The invincible deadly mosquitoes: India's health and economy enemy No. 1. Scientific Publishers, India. 266 pp

WHO (2014) Guidance framework for testing of genetically modified mosquitoes. World Health Organization, Geneva. 159 pp

World Health Organization [First Draft and Compilation by B.K. Tyagi] (2015) Biosafety for human health and the environment in the context of the potential use of genetically modified mosquitoes (GMMs) - A tool for biosafety training based on courses in Africa, Asia and Latin America, 2008–2011. World Health Organization, Geneva. 242 pp

The manufacturer's authorised representative in the EU is Springer Nature Customer Service Centre GmbH, Europaplatz 3, 69115 Heidelberg, Germany. If you have any concerns regarding our products, please contact ProductSafety@springernature.com

Printed and bound by CPI Group (UK) Ltd, Croydon, CR0 4YY
23/09/2026
02221533-0003